Nanotechnology and the Challenges of Equity, Equality and Development

Yearbook of Nanotechnology in Society

Volume 2

Series Editor

David H. Guston, *Arizona State University*

For further volumes:
http://www.springer.com/series/7583

Nanotechnology and the Challenges of Equity, Equality and Development

Susan E. Cozzens
Jameson M. Wetmore
Editors

# Springer

Editors
Susan E. Cozzens
Georgia Institute of Technology
School of Public Policy
685 Cherry Street
Atlanta
Georgia
3033-0345
USA
scozzens@iac.gatech.edu

Jameson M. Wetmore
Arizona State University
Center for Nanotechnology in Society
Tempe Arizona 85287
USA
Jameson.Wetmore@asu.edu

ISBN 978-90-481-9614-2   e-ISBN 978-90-481-9615-9
DOI 10.1007/978-90-481-9615-9
Springer Dordrecht Heidelberg London New York

Library of Congress Control Number: 2010937022

© Springer Science+Business Media B.V. 2011
No part of this work may be reproduced, stored in a retrieval system, or transmitted in any form or by any means, electronic, mechanical, photocopying, microfilming, recording or otherwise, without written permission from the Publisher, with the exception of any material supplied specifically for the purpose of being entered and executed on a computer system, for exclusive use by the purchaser of the work.

Printed on acid-free paper

Springer is part of Springer Science+Business Media (www.springer.com)

Preface

A yearbook isn't a yearbook until there are two of them. I am thus incredibly pleased to present to you the second *Yearbook of Nanotechnology in Society*, which by virtue of its existence alone validates the first volume as well as the series itself.

The Center for Nanotechnology in Society at Arizona State University (CNS-ASU), which is responsible for these yearbooks, has come a long way since the series was conceived and the first volume published. That volume, *Presenting Futures* (Fisher et al. 2008), was based in part on a series of seminars hosted by CNS-ASU in its very first year of work. As I write, we have secured the renewal of our center from the U.S. National Science Foundation and await notice of our allocated budget for 2010–2015. In the center's renewal application, we accounted for 55 peer-reviewed journal articles in print, forthcoming or under review, more than two dozen book chapters, nearly three dozen undergraduate and graduate theses, more than two dozen reports and working papers, and more than 200 presentations to academic, policy, industry, and public audiences. We have even filed nine invention disclosures!

The futures seminars we held that first year were predicated on the notion that because there were so few actual nanotechnologies around to scrutinize as objects, we first had to discuss how we conceived of—and how we possibly could conceive of—future nanotechnologies. In the time since I wrote the preface for Volume One, the number of nanotechnological products on the Woodrow Wilson International Center's Nanotechnology Consumer Products Inventory has roughly doubled, from more than 500 products to 1,015 in August 2009. While many of these are simply consumer products with a tad of nano sprinkled on or engineered in, as perhaps exemplified in the advertisement reprinted in this volume for a golf club made with nano-materials, there are tantalizing hints that nano is moving into a new stage more closely attuned to some of its revolutionary promise. A bibliometric analysis by Subramanian et al. (forthcoming) suggests that the transition from passive nanostructures to active nanostructures—as described in Mihail Roco's popular roadmap for nano—is afoot.

Even as nano is growing and changing, social science and humanities scholarship about nano is too: from 231 hits in the Web of Science for social science and humanities about nanotech* and nanoparticle* from 1990 to the time of the first preface, to 479 hits with the same search now.[1] This spectacular growth has been fueled by increasingly empirical treatments of nanotechnologies and nano-scale science and engineering researchers, perhaps most significantly in areas related to environmental health and safety.

These changes, however, are not yet registering much with the public. Newspaper coverage of nano had been accelerating, but that growth has slowed and has even dropped off; the share of articles devoted to issues like environmental health and safety—which have garnered increasing scholarly attention—has not kept pace with scholarship (Dudo et al. 2009). Moreover, despite the coverage nano has received in popular media; the formal educational programs instituted to teach and train people about nano; and outreach and informal education conducted by researchers, science museums and centers, non-governmental organizations, and others, the overall levels of public awareness and understanding of nano have not budged (Scheufele et al. 2009). But this lack of good news on public understanding masks what may be outright bad news: There is evidence that a gap is opening in public understanding of nano between more and less educated adults, in which those with a college degree are learning more about nano—often from online sources—and those with less than a high school diploma are learning less (Corley and Scheufele, forthcoming).

Thus, a volume on *The Challenges of Equity, Equality, and Development* is both timely and important.

Issues of equity, equality, and development have been on the social science and ethics agenda of nanotechnology from the very beginning of such scholarship. Many of the early articles that sketched out prospective research agendas identified equity issues as among the most important five or six issues that could be addressed. The issues have also been on the agendas of non-governmental organizations and, in distinct contrast to other large-scale research programs, of the nano-scale science and engineering enterprise itself. But "on the agenda" and receiving sufficient attention are not quite the same thing. Neither does being on the agenda mean that a robust and progressive research agenda exists for the topic.

The work of volume editors Susan Cozzens and Jameson Wetmore aims not only to ensure sufficient attention, but also to ground, extend, and energize that research agenda. This volume is therefore a natural extension of earlier lines of work by the editors, importantly including a major international collaboration on distributional consequences of emerging technology by Cozzens (Cozzens et al. 2008) and the editing of a special issue of *Science and Public Policy* on "Science, Policy and Social Inequity" by Wetmore (2007). The sophistication of their approach in this volume is a product, in part, of that experience. It is also partially a product of the role of CNS-ASU, which commits one of its several organized research programs, led by Cozzens and Wetmore, to Equity, Equality and Responsibility.

Through this research program, CNS-ASU sponsored a workshop in November 2008 that many of the authors in this volume used to develop their contributions. And through Cozzens and Wetmore's encouragement and vision, several other CNS-ASU researchers have drawn on their center-related work to explore themes of equity, equality, and development—thus contributing to the cross-fertilization of the center's programs as well as to the breadth of this volume. The Equity, Equality, and Responsibility program also provides critical, normative guidance to the center's strategic vision of anticipatory governance (Barben et al. 2008). That is, Susan and Jamey remind us frequently that our work should have important consequences for more than scholarship, and that we should always consider the ways those consequences are distributed as well as their absolute magnitude.

It is my fond hope that this volume will, like its predecessor, mark a milestone in scholarship on the role of nanotechnologies in society. In so doing, the volume should also carry forward the values that it espouses. I am therefore pleased to invite anyone—anyone—who cannot afford to purchase this book or to download its contents from the web to contact CNS-ASU at cns@asu.edu and we will make every effort to get you the book in a reasonable fashion.

Tempe, Arizona
David H. Guston

Note

1. Thanks again to Walter Valdivia for repeating this search for me.

References

Barben, Daniel, Erik Fisher, Cynthia Selin, and David Guston. 2008. Anticipatory governance of nanotechnology: Foresight, engagement and integration. In *The handbook of science and technology studies,* ed. Ed Hackett, Olga Amsterdamska, Michael Lynch, and Judy Wajcman, 979–1000. Cambridge: MIT.

Corley, Elizabeth A., and Dietram A. Scheufele. 2010. Outreach going wrong? When we talk nano to the public, we are leaving behind key audiences. *The Scientist* 24(1): 22.

Dudo, Anthony, Sharon Dunwoody, and Dietram Scheufele 2009. The emergence of nano news: Tracking thematic trends and changes in media coverage of nanotechnology. Paper presented at the annual convention of the Association for Education in Journalism & Mass Communication, August, Boston.

Fisher, Erik, Cynthia Selin, and Jameson M. Wetmore, eds. 2008. *Presenting futures.* The yearbook of nanotechnology in society. series ed. D.H. Guston. New York, NY: Springer, vol. 1.

Scheufele, Dietram A., Elizabeth A. Corley, Tsung-jen Shih, Kajsa E. Dalrymple, and Shirley S. Ho. 2009. Religious beliefs and public attitudes toward nanotechnology in Europe and the United States. *Nature Nanotechnology* 4: 91–94.

Subramanian, Vrishali, Jan Youtie, Alan L. Porter, and Philip Shapira. Forthcoming. Is there a shift to 'active nanostructures'? *Journal of Nanoparticle Research.* Doi: 10.1007/s11051-009-9729-4.

Wetmore, Jameson M. ed. 2007. Special issue on science, policy and social inequities. *Science and Public Policy* 34 (2) March.

Introduction

Susan E. Cozzens and Jameson M. Wetmore

For at least the past decade, nanotechnology has been touted as the Next Big Thing in technology, the successor to information and biotechnologies (Interagency Working Group on Nanoscience, Engineering, and Technology 1999). Scientists and policymakers both contend that investments in nanoscale science and engineering will create revolutions in areas as diverse as materials, drug delivery, cancer treatment, and space travel. The hope is that many of the problems of today can be addressed using nanotechnology enabled products.

While new technologies often do provide solutions to pressing issues, the outcomes are not always an unmitigated good. They can also exacerbate other social problems or create entirely new issues. Automobiles, for instance, made an entirely new form of individual transportation possible, but also brought with them new forms of pollution and congestion and made the streets unsafe for children and others. The Internet puts a seemingly limitless amount of information at our fingertips, but that information has been used in ways that both benefits society and changes the social order.

One issue that scholars of science and technology policy are increasingly interested in is the ways in which new technologies change the relationships between the "haves" and the "have nots" (Wetmore 2007; Cozzens et al. 2008). There is much hope that technologies can help us to build a more equitable world. And yet in most cases, new technologies do the opposite. Sometimes this is simply the result of the privileged having first access to the newest advances. But studies have also shown that even when technologies are specifically designed for the disadvantaged they can still hinder their development (Lemos and Dilling 2007). In short, new technologies can have a significant impact on a variety of equity issues.

Although the analysis of the connections among emerging technologies, equities, and inequalities is in the early stages, we can draw upon our experience with previous revolutionary technologies to begin to imagine the ways nanotechnologies will change the fabric of society. Nanotechnologies could alter distributional dynamics and raise important issues about fairness. They could greatly aid the economies and public health of poor countries, or they could increase the gulf between poor and rich ones. This volume begins to develop a better understanding of how those changes might play out and what we should be aware of as new nanotechnologies and industries are created.

What is Nanotechnology?

To begin this process it is useful to know a bit more about the technological focus of the book. "Nanotechnology" does not describe any specific technology or research area. Instead it is used as a catch-all term for techniques designed to manipulate matter on the nanoscale—roughly 1 billionth to 100 billionth of a meter. There are two major benefits to working at this scale. The first is that it may give inventors and designers control to build things from the most basic building blocks—atoms and molecules. This control can enable the creation of macro-scale objects that have been engineered with unprecedented precision as well as nano- and micro-sized materials (and potentially devices) that were never before possible. Second, many materials have very different characteristics at the nanoscale than they do at the macro scale. For instance, at the nanoscale, silver is a powerful antibiotic. Nanotechnologies allow engineers to tap into these characteristics and deploy old materials in new ways.

Thus far there are a relatively small number of products being marketed as nanotechnologies or "nano-enhanced." As of the end of 2009, the Woodrow Wilson Center's Project on Emerging Nanotechnologies has catalogued over 1,000 nano consumer products. Most of these products are slightly revised versions of somewhat traditional products like sporting goods, cosmetics, sunscreens, and active wear (Project on Emerging Nanotechnologies 2009). Few of these products will have a major impact on society—although they do demonstrate that the consumer applications of nanotechnology are currently aimed at wealthy Western consumers. Nanomaterials are also being incorporated into materials used in industry, such as paints and coatings. Applications in electronics are receiving a great deal of attention because of their potential to reduce size and energy costs below the current properties of silicon. And applications in medicine are on the foreseeable horizon.

But scientists promise that this is just the start; a nanotechnology "revolution" is coming. We know from experience that revolutions tend to be messy. They often create radical changes in areas well beyond the scope of the revolutionaries. Government officials, regulators, corporate executives, scientists, engineers, non-governmental organizations (NGOs), and scholars have already begun to study and reflect on the possible ramifications of an impending nanotechnology-based revolution. Most recently the loudest discussions in the United States about the potential downsides of nanotech research and development (R&D) have focused on the possible toxic properties of nanoparticles (U.S. House Committee on Science and Technology 2007, 2008). But these are far from the only concerns being presented. Increasingly both civil society and academic scholars are exploring the effects that nanotechnologies have, and may in the future have, on society, including issues of equity and equality.

Some of the interest in nanotechnology and equity has been focused on whether or not the new technologies will increase the gaps between global "haves" and the "have nots." If those with the access, power, and money to direct the development of nanotechnologies do so solely for their own ends, the gulf between them and the rest of the world could increase dramatically. Some of these effects may be somewhat trivial. For instance, those who are able to afford the latest lightweight tennis rackets made with carbon nanotubes might win more matches. But the issues could be significantly more serious if, for example, the low-paid workers that run the tennis racket assembly lines inhale nanoparticles and develop health problems.

Research on nanotechnology and equity not only aims to stem the negative effects of nanotechnologies on equity, but also to find ways to use nanotechnologies to redress existing inequities. The promise of new technologies, such as drugs and efficient water filtration devices, has generated optimism that the new ability to control matter at the nanoscale could decrease some of the inequalities and inequities in the modern world. Fabio Salamanca-Buentello and his colleagues, for example, have argued that developing countries can benefit enormously from the technologies and help them achieve many (if not all) of the UN Millennium Development Goals (Salamanca-Buentello et al. 2005). Others, including Noela Invernizzi and Guillermo Foladori (2005), have fired back, however, that rarely are new technologies designed for or available to the disadvantaged. More frequently they are designed for those with advantages already, whose power the new technologies frequently reinforce. Non-governmental organizations such as the Meridian Institute, the ETC Group, and Friends of the Earth Australia have also expressed concern that the benefits and risks of nanotechnologies will not be equally distributed. Similarly, a number of the participants in the Arizona State University's Center for Nanotechnology in Society's recent National Citizens Technology Forum expressed concern about the equity implications of nanotechnologies in their final reports (Bal and Cozzens 2008; Chapter 14).

The fact that nanotechnologies have not yet entirely arrived on the scene gives us an opportunity to anticipate, prepare for, and shape its effects. This volume is inspired by and done in conjunction with Real Time Technology Assessment (RTTA) (Guston and Sarewitz 2002). Through RTTA, it is possible to analyze the research currently being done in science and engineering labs, learn from scientists and engineers the types of technologies they hope to create, reflect on the ways the implementation of these technologies might change social structures and inhibit or promote certain values, and evaluate which changes are desirable and which are not. Based on this framework, the more we can understand the possible effects of nanotechnologies in the present, the more options we have for bringing about change in the lab, in policy, and throughout civil society.

The goal of this volume is not to ensure that nanotechnologies benefit everyone equally. Such a project is probably impossible and may not even be advisable. But it is based on the idea that striving for greater equity and helping to address the needs of developing areas are laudable goals that should play a much more important role in decision-making than they currently do. The kinds of analysis represented in this volume can help us work for systems that distribute the benefits of nanotechnologies more widely and the costs more evenly.

Multiple Dimensions of Equity

The question of equity arises at multiple levels, including the personal, national, and global. At the personal level, questions of fairness arise: Will nanotechnologies open equal career and business opportunities for women and for men, for advantaged and disadvantaged ethnic groups? Will nanotechnologies create capabilities that exacerbate existing personal advantages or create new disadvantages? For whom will nano products be designed? Will they affect how culturally defined groups are reinforced or broken down, privileged or disadvantaged?

At the national level, it is relevant to ask how nanotechnologies will affect the shape of national economies and their social and political relations. What institutions and processes will control the development of these technologies, and with what participants? Will the benefits of nanotechnologies be distributed widely, or will they be concentrated in particular places or groups? How will the risks generated in production and disposal of new technologies be distributed? Who will receive the benefits including jobs and economic growth?

At the global level, some of the same kinds of questions arise, but with different possible answers. How will the global commercialization of nanotechnologies intersect with processes of human development within countries? What roles will national governments, global institutions, and multinational firms play? Will global competitiveness dominate the R&D agenda, or will local needs be given significant attention? Will the processes engage and benefit the world's poor or leave them out? How will relations of power and privilege change, if at all, as various political actors take up the application of nanotechnologies? Will developing countries see proportional benefits from nanotechnologies? Will strategies like "pro-poor" technology actually make a difference?

Despite the fact that this is only a partial list of questions that could be asked about the impacts of nanotechnologies on equity and equality, it is clear that the implications will not be isolated in any one place. In fact if the predictions of nanotechnology proponents are even close to accurate, it is likely that nanotechnology will touch nearly every person and nation across the globe. It would be impossible for a single volume to address, let alone answer, all of these questions. Thus this volume does not provide a systematic explanation of how nanotechnologies will affect equity. Instead it provides a number of case studies at different levels of analysis. We hope that these analyses will demonstrate the depth and pervasiveness of the issue and thereby motivate additional research and broader discussion about the implications of nanotechnologies. The earlier we integrate equity concerns into our decision-making the more likely we are to achieve equitable solutions.

Inequalities and Inequities

The number and variety of the questions listed in the previous section can be a bit overwhelming. Our past work has shown that it is easy to get lost talking about these issues unless we have a mental map of this complicated space and place ourselves somewhere within it. To this end we have found it useful to distinguish between *inequality* and *inequity*. We use *inequality* as a term to describe a given empirical distribution. Thus, for instance, an inequality exists between two people if one has ten rupees and the other has a hundred rupees. We use *inequity* to denote a normative judgment, perhaps rooted in a philosophical theory of justice, about a given inequality (Cozzens 2007). Thus if someone deems the distribution of rupees in the above example to be a wrong that should be righted, they are claiming that it is an *inequitable* or unjust distribution.

We further break down inequalities into two subgroups: vertical and horizontal. Vertical inequality is the rich-poor dimension. Vertical inequalities abound in twenty-first century society. For instance, differences in income and wealth between countries are huge, and some countries show as much income inequality within their borders as the globe encompasses.[1] The distribution of other valued items often line up in the same unequal patterns as income and wealth. For example, families with little money are more likely to suffer poor health and be affected by polluted environmental conditions.

The second dimension of inequality is horizontal, a term that refers to fault lines in general distributional patterns that follow culturally-defined differences, such as gender, ethnicity, or religion. An example of a horizontal inequality is the difference in life expectancies between black and white Americans, which is not entirely accounted for by income difference (Institute of Medicine 2002). Another horizontal inequality is the fact that in the U.S. it is much more likely for black and brown families than white families to live near toxic waste dumps—with ethnicity again a stronger predictor of that likelihood than income (Commission for Racial Justice 1987).

One reason why it is useful to break down inequalities in this way is because horizontal inequalities are also manifestations of inequities. No theory of distributive justice justifies unequal distribution based on such culturally-defined characteristics (Cozzens 2007). Most theories, however, provide justification for some level of vertical inequality. In that dimension, reducing inequality is more likely to be the moral goal than achieving equality (the horizontal objective). Politically, of course, both dimensions are important.

The utility of these terms can be seen when we begin to apply them to an analysis of nanotechnologies. For instance in the development of nanotechnologies there would probably be general agreement that risks should not be distributed unequally, either vertically or horizontally. Exposure to toxic nanoparticles should not be greater for people who are poor, female, Hispanic, or Lutheran. But adherents of different theories of distributive justice would take different positions on appropriate vertical distributions of the wealth that will be generated from new nanotechnology businesses. Libertarians might stress the importance of maintaining property rights in nanotechnology development—thus ensuring that wealth will accumulate to those who control intellectual property. Utilitarians might stress the overall contribution of nanotechnologies to economic growth, leaving the question of who benefits from the growth to others. A contractarian (or Rawlsian) approach would strive to insure that the least advantaged members of society benefit, regardless of whether others benefit more. A communitarian approach might seek to use nanotechnologies to reduce overall inequality, including in the vertical dimension.

In short, in a world of uneven development and resources, achieving equity and equality is not a simple goal. The different dimensions of inequality have different but often interacting dynamics. Finding the place of nanotechnologies in that complicated flow is an intellectual as well as a practical challenge.

Yearbook Focus and Goals

As mentioned above, the relative newness of nanotechnology offers an important opportunity to engage with a technology in its early stages. But it also means that there have been few people who have had the time, inclination, and ability to probe the equity implications of nanotechnologies. To begin to address this situation the Center for Nanotechnology in Society at Arizona State University held a workshop entitled "Nanotechnology, Equity, and Equality" in Tempe, Arizona from November 20–22, 2008. We were able to find a handful of scholars and practitioners who already had addressed specific issues of nanotechnologies and equity, but we wanted to expand beyond this group. To spread the word and get more people engaged we also invited a number of academics with a deep knowledge of equity issues in other technologies and engineers and scientists who work at the nanoscale. We then charged the group to address two central questions:

> First, given what we currently know about nanotechnologies, what are the likely consequences for equality and fairness of the technologies on the near-term horizon? What features of the nanoscience and nanotechnology enterprise are contributing to those consequences?

> Second, given what we know about equality and fairness, what is the variety of possible consequences of nanotechnologies over the longer term? What can be done now to affect the probabilities of the various possibilities?

These questions produced fruitful discussions and a number of excellent papers. This Yearbook is largely based on work that was presented at that workshop, with the additional benefit of a round of peer review. Sixteen of this book's chapters were originally presented at the workshop and are so noted on their opening pages.

We were not able, however, to include the variety of perspectives and viewpoints that we would have ideally liked at that initial workshop. As can be seen from even this brief introduction, the issues raised by nanotechnology for equity are broad and pervasive. Any successful effort to begin to address them will require an equally broad and pervasive set of participants and perspectives. To remedy this we sought out additional materials.

Because of the recent interest in the topic there has been some academic work that directly addresses nanotechnology and equity. We have republished three of these excellent academic articles. In other cases scholars working in the area who did not attend the workshop developed some of their latest work into chapters that appear in the book. And finally because the volume is a yearbook we wanted to

Introduction

include a few pieces that can give future readers some idea of how organizations today are dealing with or affecting equity and nanotechnologies. Thus we have also included a resolution issued by an NGO, an advertisement, and a newspaper article.

Drawing from these multiple sources, this yearbook brings together social scientists, engineers, natural scientists, policymakers, NGOs, and corporate perspectives from six continents and presents a wide variety of approaches to and methods by which to address nanotechnology, equity, equality, and development. Of course not every perspective could be captured, and not every group could be given a voice. But we have tried to provide as broad a scope as possible.

A Volume of Options

Dimensions of Nano Fairness

This Yearbook is divided into five sections. It begins in the horizontal dimension, examining nanotechnologies from the viewpoints of various disadvantaged groups: women, the disabled, and African and Latino/a Americans. The setting for the analysis is mostly affluent countries, but the cultural dynamics described could have even more prominent effects in lower income countries. Across the various culturally-defined categories, the analysts represented in these chapters see the same future pattern: nanotechnology on its current trajectory will disproportionately benefit currently advantaged groups, probably even more than was the case with biotechnology, the last Next Big Thing. But we know more now than we did in the early years of biotechnology about how to address horizontal inequities, and the authors use that knowledge to offer suggestions for changing nanotechnology's trajectory to create a more equal future.

Laurel Smith-Doerr, for example, points out that nanotechnology as currently constituted is likely to reflect the particularly male composition of engineering and the physical sciences. Eventually, as its life science applications develop, it is likely to attract more women. Smith-Doerr, however, does not recommend waiting for that solution to male dominance in the field. Instead, she urges nanotechnology organizations to "avoid machismo and to keep a gender neutral, open-frontier, open-to-comers culture intact along with organizational flexibility." In other words, she asks for a commitment to female-friendliness. This commitment would also help address the emerging pattern Yu Meng and Philip Shapira introduce: the patenting gap between men and women in nanotechnology.

Likewise, Sonia Gatchair projects gaps in hiring and wages for African Americans and Latino/a Americans, based on patterns identified in high technology industries in the past, which hold even after controlling for education. These gaps, too, need to be addressed with organizational commitment and change in culture in the workplace. Catherine Slade points to complementary action at the policy level. Slade identifies another gap, this time between rhetoric and program plan in the lofty goals government has set for the contributions of nanomedicine to reducing health disparities in the United States. Fill the gap with specific, implementable, and plausibly effective program actions, she urges, or stop claiming that the benefits of nanotechnologies are being spread to members of ethnic minority groups.

Gregor Wolbring projects contributions of nanotechnologies to widening the gaps between groups with different sets of abilities. Ableism now gives cultural dominance to those with one set of abilities while labeling others "disabled." If nanotechnologies open the door to human enhancement—as promised by its advocates—then the new levels of ability may become the new standard, and even those currently in the "abled" category will be disempowered. Wolbring recommends analysis and discussion of the ethical issues involved. Another path for addressing this issue appears in a later section, in Dean Nieusma's chapter on design. Participatory design is one route for different ability groups to get human variability literally built in to the new nano-enabled environment.

The first section ends with an advertisement for a nano-enhanced golf club built by the Yonex Corporation. This advertisement is included as an example of the contemporary use of nanomaterials. It is one of the vast majority of nano-enhanced consumer products now available that have been developed largely for the wealthy Western world. It is also a reminder of the innate desire to capitalize on inequalities—in this case the lure of buying the latest sporting equipment to give an edge over one's competitors.

Unequal Structures

The second section of the book shifts the focus from horizontal to vertical inequalities, particularly as they appear in the structure of the economy. Large-scale dynamics such as long waves of technological change, the diffusion of general purpose technologies, or regional agglomerations of new industries, may appear to be out of the control of those who shape nanotechnologies currently. Furthermore, those in charge may not want to change the patterns. This is the view of Georgia Miller and Gyorgy Scrinis, whose highly critical, pessimistic essay opens the section. They note that nanotechnology is arising from actions that align it with powerful economic and political interests in the global North. Despite paying lip service to studying the "ethical, legal, and social implications" of nanotechnology, those who are driving the rapid expansion of nanotechnology have not shown any genuine commitment to reorienting the enterprise to human needs or a more equal society. Given the power disparities between nano advocates and critics, Miller and Scrinis find it improbable that there will be any fundamental realignment.

Yet many of the chapters in this section explicitly or implicitly suggest actions that could make a difference, including awareness among unions and engagement by non-governmental organizations. Mark Knell calls attention to the need for new distributional institutions after technological revolutions, and Walter Valdivia points out that investing in skills can have different effects for inequality at different points in the introduction of a new technology. Jan Youtie and Philip Shapira track the concentrations of nanotechnology activities in metropolitan areas, where they are following some different patterns from previous emerging technologies; many communities are investing in their own nanocapabilities to shape that pattern.

Guillermo Foladori and Edgar Zayago Lau urge inclusion of workers in technology assessment processes to avoid historical patterns of neglect of their safety issues by the corporations who are almost always at those tables. The resolution from the European Trade Union Council that concludes this section demonstrates the appearance and forcefulness of the voice of unions already in the public debate around nanotechnologies.

Equalizing Processes

The third section examines additional pathways by which technology assessment processes could be carried out to change the trajectory of nanotechnology towards equity and equality. Again, the chapters do not provide an exhaustive set of strategies, but develop illustrative ideas that could be implemented in a broad range of areas. Dean Nieusma reviews the extensive experience in the participatory design tradition with incorporating views of women, ethnic and racial minorities, and people with different physical and mental abilities into the product development process. Ravtosh Bal describes another participatory process where marginalized voices might affect nanotechnology development trajectories: citizens consensus conferences. Matthew Harsh describes the actors involved in discussions about biotechnology regulation in Kenya as a process analogous to what could take place around nanotechnologies in the context of a developing country. He clearly advocates bringing more actors into the process, including those who will be most affected by the decision. A reprinted newspaper article from Uganda illustrates the differences of understanding that could come to the table under such a broadening—bringing the diversity of views that must be heard for real development to proceed.

Nanotechnology in the World System

In the fourth section, the volume turns explicitly to issues in developing countries. Nanoscience research and nanotechnology development are both largely phenomena of the global North. The incorporation of nanotechnologies into innovation systems and everyday lives in the global South thus reflects a set of global relationships. The chapters in this section explore those relationships from a variety of viewpoints in the South.

The opening chapter, an extract from a report by the Meridian Institute on a set of dialogues they facilitated called "Nanotechnology and the Poor: Opportunities and Risks for Developing Countries," introduces basic questions about nanotechnologies and equitable development. Where will development and production take place? What are the environmental and health risks? Will they increase inequalities?

Who will hold the intellectual property, and what are the implications of ownership? Will the public be engaged in governance and decision making? Is there sufficient local capacity to make the systems sustainable?

The next two chapters take up these questions in the context of Brazil, one of the so-called "emerging economies"—that is, countries that are still at low or middle per capita income levels but are experiencing rapid growth. Noela Invernizzi provides perspective on the Brazilian nanotechnology initiative, which has both strong and weak points from the viewpoint of increasing equality. Luciano Kay and Philip Shapira take a more detailed look at its consequences. Using publication and patent data, they examine Brazil's performance on four criteria of equitable economic development—agenda setting, R&D investment, R&D outcomes, and risk awareness and allocation—and conclude that the potential for nanotechnologies to contribute to equity in Brazil is not yet being fulfilled. Both papers point to a number of ways that Brazil could link its nanotechnology initiative more closely to its social inclusion goals.

Some observers have claimed that tight intellectual property protection will keep developing countries from receiving the benefits of emerging technologies. Dhanaraj Thakur explores the possibilities for developing countries that might come from open access nanotechnologies. Donald MacLurcan's report on interviews with Thai and Australian scientists on the "nano-divide" also points to the potential for Southern innovation. Participating in applications is not beyond the capacity of developing countries, and they may be particularly active at the smaller scales that are important in their economies.

To illustrate the process of matching local needs in a developing country with nanotechnology capabilities, we reprint a chapter by David Grimshaw, Lawrence Guzda, and Jack Stilgoe, reporting on their efforts to find such a match in one community in Zimbabwe. The section closes with an essay by Raghubir Sharan, Yashowanta Mohapatra, and Jameson Wetmore on making the study of values an explicit part of engineering education. Each chapter, then, points to a different kind of action that the nanoscience and engineering community can take to spread the benefits of its products more widely.

Lessons for Action

The last section of the book offers lessons for action from yet another viewpoint, that of policymakers and funders. Rinie van Est, a staff member at the Dutch parliamentary technology assessment organization, the Rathenau Institute, proposes that the nanotechnology enterprise needs investigation and interaction as well as intervention. In van Est's view, scholarship on this topic will have to be engaged with political action in order to have an impact. Drawing on his experience at the Woodrow Wilson Center Project on Emerging Nanotechnologies, as well as his current position in strategic planning for the Rockefeller Foundation, Evan Michelson also urges engagement. Now is the time to get pro-poor issues on the nanotechnology agenda, he suggests, while relationships and institutions are still emerging. Finally, Susan Cozzens summarizes lessons from the existing research on distributional consequences of emerging technologies in the form of steps an innovation policymaker could take to orient a nanotechnology initiative towards equity and equality outcomes. Several kinds of program designs can be incorporated in the initiative to help it produce benefits for everyone.

In Summary

Across the contributions, then, the authors see a significant risk that nanotechnology, in all of its various application areas, will increase inequalities in both vertical and horizontal directions. While some see this result as inevitable given the way the technology is embedded in current economic and political systems, many can envision ways to change the outcome. New institutions may need to be invented, new capabilities developed, and new relationships formed, but nanotechnology is nothing if not new. The authors in the yearbook call for informed action. Now is the time to seek, invent, and implement the possibilities.

Note

1. Inequality among the world's households is estimated at a Gini index of about .65. Brazil and South Africa show similar levels of domestic inequality.

References

Bal, Ravtosh, and Susan Cozzens. 2008. Public perceptions of NBIC technologies. Paper presented at the PRIME-Latin America Conference, September 24–26, Mexico City.
Commission on Racial Justice. 1987. *Toxic wastes and race in the United States: A national report on the racial and socioeconomic characteristics of communities with hazardous waste sites.* New York, NY: United Church of Christ.
Cozzens, Susan. 2007. Distributive justice in science and technology policy. *Science and Public Policy* 34 (2) March: 85–94.
Cozzens, S., I. Bortagaray, S. Gatchair, and D. Thakur. 2008. Emerging technologies and social cohesion: Policy options from a comparative study. Paper presented at PRIME-Latin America conference, September 24–26, Mexico City. http://prime_mexico2008.xoc.uam.mx/papers/Susan_Cozzens_Emerging_Technologies_a_social_Cohesion.pdf. (accessed August 23, 2010).
Lemos, Maria Carmen, and Lisa Dilling. 2007. Equity in forecasting climate: Can science save the world's poor? *Science and Public Policy* 34 (2) March: 109–116.
Guston, David H., and Daniel Sarewitz. 2002. Real-time technology assessment. *Technology in Society* 24:93–109.
Institute of Medicine. 2002. *Unequal treatment: Confronting racial and ethnic disparities in healthcare.* Washington, DC: Author.
Interagency Working Group on Nanoscience, Engineering, and Technology. 1999. Nanotechnology: Shaping the world atom by atom. September. Washington, DC.

Invernizzi, Noela, and Guillermo Foladori. 2005. Nanotechnology and the developing world: Will nanotechnology overcome poverty or widen disparities? *Nanotechnology Law and Business* 2 (3) September/October: 294–303.

Project on Emerging Nanotechnologies. 2009. Consumer product inventory. Woodrow Wilson International Center for Scholars. http://www.nanotechproject.org/inventories/consumer. (accessed August 4, 2010).

Salamanca-Buentello, Fabio, Deepa L. Persad, Erin B. Court, Douglas K. Martin, Abdallah S. Daar, and Peter A. Singer. 2005. Nanotechnology and the developing world. *PLoS Medicine* 2 (5) May: 0383–0386.

U.S. House Committee on Science and Technology. 2008. *The national nanotechnology initiative amendments act of 2008*, Hearing, April 16.

U.S. House Committee on Science and Technology. 2007. *Research on environmental and safety impacts of nanotechnology: Current status of planning and implementation under the national nanotechnology initiative*, Hearing, October 31.

Wetmore, Jameson M. ed. 2007. Special issue on science, policy and social inequities. *Science and Public Policy* 34 (2) March.

Contents

Part I   Dimensions of Nano Fairness

1   **Contexts of Equity: Thinking About Organizational and Technoscience Contexts for Gender Equity in Biotechnology and Nanotechnology**   3
    Laurel Smith-Doerr

2   **Women and Patenting in Nanotechnology: Scale, Scope and Equity**   23
    Yu Meng and Philip Shapira

3   **Potential Implications for Equity in the Nanotechnology Workforce in the U.S.**   47
    Sonia Gatchair

4   **Exploring Societal Impact of Nanomedicine Using Public Value Mapping**   69
    Catherine P. Slade

5   **Ableism and Favoritism for Abilities Governance, Ethics and Studies: New Tools for Nanoscale and Nanoscale-enabled Science and Technology Governance**   89
    Gregor Wolbring

6   **i Will Go Further**   105
    Yonex Corporation USA

Part II   Uneven Structures

7   **Nanotechnology and the Extension and Transformation of Inequity**   109
    Georgia Miller and Gyorgy Scrinis

8   **Nanotechnology and the Sixth Technological Revolution**   127
    Mark Knell

9   **Innovation, Growth, and Inequality: Plausible Scenarios of Wage Disparities in a World with Nanotechnologies**   145
    Walter D. Valdivia

| 10 | Metropolitan Development of Nanotechnology: Concentration or Dispersion? | 165 |

Jan Youtie and Philip Shapira

| 11 | The Role of Organized Workers in the Regulation of Nanotechnologies | 181 |

Guillermo Foladori and Edgar Zayago Lau

| 12 | ETUC Resolution on Nanotechnologies and Nanomaterials | 199 |

European Trade Union Confederation

Part III Equalizing Processes

| 13 | Materializing Nano Equity: Lessons from Design | 209 |

Dean Nieusma

| 14 | Public Perceptions of Fairness in NBIC Technologies | 231 |

Ravtosh Bal

| 15 | Equity and Participation in Decisions: What Can Nanotechnology Learn from Biotechnology in Kenya? | 251 |

Matthew Harsh

| 16 | Nanotechnology: How Prepared Is Uganda? | 271 |

Kikonyogo Ngatya

Part IV Nanotechnology and the World System

| 17 | Nanotechnology and the Poor: Opportunities and Risks for Developing Countries | 277 |

Todd F. Barker, Leili Fatehi, Michael T. Lesnick, Timothy J. Mealey, and Rex R. Raimond

| 18 | Science Policy and Social Inclusion: Advances and Limits of Brazilian Nanotechnology Policy | 291 |

Noela Invernizzi

| 19 | The Potential of Nanotechnology for Equitable Economic Development: The Case of Brazil | 309 |

Luciano Kay and Philip Shapira

| 20 | Open Access Nanotechnology for Developing Countries: Lessons from Open Source Software | 331 |

Dhanaraj Thakur

| 21 | Southern Roles in Global Nanotechnology Innovation: Perspectives from Thailand and Australia | 349 |

Donald C. Maclurcan

| 22 | How Can Nanotechnologies Fulfill the Needs of Developing Countries? | 379 |

David J. Grimshaw, Lawrence D. Gudza, and Jack Stilgoe

| 23 | Technical Education and Indian Society: The Role of Values . . . . | 393 |

Raghubir Sharan, Yashowanta N. Mohapatra,
and Jameson M. Wetmore

Part V Lessons for Action

| 24 | Keeping the Dream Alive: What ELSI-Research Might Learn from Parliamentary Technology Assessment . . . . | 409 |

Rinie van Est

| 25 | Nanotech Ethics and the Policymaking Process: Lessons Learned for Advancing Equity and Equality in Emerging Nanotechnologies . . . . | 423 |

Evan S. Michelson

| 26 | Building Equity and Equality into Nanotechnology . . . . | 433 |

Susan E. Cozzens

Index 447

About the Authors

Ravtosh Bal is a doctoral student in Public Policy in the Joint PhD program of Georgia Institute of Technology and Georgia State University. She is majoring in Science and Technology Policy and is a research assistant in the Technology Policy and Assessment Center at Georgia Tech. Her research interests are science, technology and innovation policy, public participation, and public budgeting.

Todd F. Barker is a Partner at Meridian Institute with more than 15 years designing and managing collaborative problem solving processes. Mr. Barker has worked extensively on issues related to science and technology, including the implications of emerging technologies such as nanotechnology and biotechnology for developing countries. Mr. Barker has worked on a broad array of other issues including agriculture and food security, energy and climate change, water quality issues, natural resource management, and cleanup of hazardous waste sites.

Susan E. Cozzens is Associate Dean for Research in the Ivan Allen College of humanities and social sciences at the Georgia Institute of Technology, and Professor of Public Policy in the School of Public Policy there. Her research is on innovation and inequality, emphasizing developing countries. She is a former Director of the Office of Policy Support at the National Science Foundation and consults widely on science and technology policies and research evaluation.

Leili Fatehi is a graduate student at the University of Minnesota School of Law and Editor-in-Chief of the *Minnesota Journal of Law, Science & Technology*. She was a Research Assistant and the Editor of *Nanotechnology and Development News* at Meridian Institute from 2005 to 2008. She has contributed to several publications on the linkages between nanotechnology and international development.

Guillermo Foladori is an anthropologist with a Ph.D. in Economics who has published over fifteen books and one hundred articles. He is a professor and researcher at the Autonomous University of Zacatecas, Mexico, where he specializes in environmental, health, and nanotechnology studies. Foladori co-coordinates the Latin American Nanotechnology and Society Network.

Sonia Gatchair is an independent researcher and affiliate of the Technology Policy and Assessment Center at Georgia Institute of Technology. Her research interests are on the effects of science, technology, and innovation policies and programs on social and economic outcomes.

David J. Grimshaw is head of International Programme: New Technologies at Practical Action, an NGO based in the United Kingdom that has used simple technology to fight poverty for the past 40 years, and Senior Research Fellow for the United Kingdom's Department for International Development, a decade old government department also focused on world poverty. He is also currently on the Steering Group of the Responsible Nano Forum.

Lawrence D. Gudza is Head of International Programme: New Technologies in Southern Africa, Practical Action. Prior to joining Practical Action he worked for many years as an ICT consultant. During this time he advised Government on ICT regulations and policy environments.

David H. Guston is Director of the Center for Nanotechnology in Society at Arizona State University. He also serves as Associate Director of the Consortium for Science, Policy & Outcomes and Professor of Political Science at Arizona State. Among other books, Guston authored *Between Politics and Science: Assuring the Integrity and Productivity of Research* (Cambridge University Press 2000) and co-authored *Informed Legislatures: Coping with Science in a Democracy* (with Megan Jones and Lewis M. Branscomb, University Press of America 1996).

Matthew Harsh is a Postdoctoral Associate at the Consortium for Science, Policy and Outcomes and the Center for Nanotechnology in Society at Arizona State University. Much of his research focuses on how decisions about emerging technologies are made in Africa. A Marshall Scholar, he holds a BSc in materials science and engineering from Northwestern University and an MSc and PhD in science and technology studies from the University of Edinburgh.

Noela Invernizzi is an anthropologist with a Ph.D. in Science and Technology Policy. She works at the Education Faculty of the Federal University of Parana, Brazil. She researches the impacts of industrial innovation on workforce skills and employment conditions and the development and potential positive and adverse implications of nanotechnologies for Latin American countries. Invernizzi co-coordinates the Latin American Nanotechnology and Society Network.

Luciano Kay is a Ph.D. candidate in Public Policy with a concentration in Economic Development at the Georgia Institute of Technology's School of Public Policy. He is a research assistant with the Georgia Tech Program in Science, Technology and Innovation Policy. He primarily investigates scientific and corporate activities in nanotechnology in the U.S., Latin America, and globally, to understand the impacts of emerging technology on industry and society. Kay has published in the *Journal of Nanoparticle Research* and the *International Journal of Innovation and Regional Development*.

Mark Knell is a research professor at the Norwegian Institute for Studies in Innovation, Research and Education (NIFU STEP) in Oslo. An economist by training, he has published widely on various aspects of emerging technologies, innovation, and economic growth, as well as in the history of economic thought and the economic transformation of Eastern Europe.

Michael T. Lesnick is a founder and Senior Partner of the Meridian Institute. He has over 25 years of experience designing and facilitating collaborative processes and strategy assessment and planning activities. Dr. Lesnick has worked domestically and internationally across a range of issues including: environmental quality, national and homeland security, international development, science and technology policy, agriculture, public health, natural resource management, and sustainable development. His work with decision makers and stakeholders from government, corporations, civil society, international organizations, and scientific bodies has focused on bringing practical solutions to some of society's most controversial and complex problems.

Donald C. Maclurcan is a PhD candidate at the University of Technology, Sydney where he is working on a dissertation that assesses nanotechnology's global consequences. He is currently working on two books—*Nanotechnology as if the World Mattered* and *Nanotechnology and Global Equality*—both to be published in 2010.

Timothy J. Mealey is a founder and Senior Partner of Meridian Institute. He serves as a convener, facilitator, and mediator of multi-party policy dialogues, negotiations, and collaborative problem solving processes on a wide variety of national and international environmental and sustainable development issues—including issues related to nanotechnology research, development and utilization.

Yu Meng is a doctoral student in the School of Public Policy at Georgia Institute of Technology, and also a visiting researcher in the Department of Policy and Regions at Fraunhofer ISI, Germany. Her research focuses on science, technology, and innovation (STI) policies, gender issues in STI, and interdisciplinary sciences.

Evan S. Michelson is a doctoral candidate at the Robert F. Wagner Graduate School of Public Service at New York University. Previously, he served as a research associate for the Project on Emerging Nanotechnologies at the Woodrow Wilson International Center for Scholars, where he co-developed the first publicly available inventory of nanotechnology consumer products. He has published on a variety of nanotechnology policy issues in *Nanotechnology and Society, Nanotechnology Applications for Clean Water, Journal of Industrial Ecology, Ecotoxicology*, and *Converging Technologies for Human Progress*.

Georgia Miller has been the national coordinator of the Friends of the Earth Australia Nanotechnology Project since 2005. Miller has a strong interest in working towards technology development which prioritizes societal and environmental needs. She has worked with environment and social justice organizations since 1994. Georgia has an Honors degree in Environmental Science.

Yashowanta N. Mohapatra is a Professor of Physics at the Indian Institute of Technology Kanpur, India. He heads the Department of Physics at IIT Kanpur and is a member of the faculty of the Materials Science Programme and Samtel

Centre for Display Technologies. Professor Mohapatra's major research interests are in physics of problems connected with development, characterization and applications of electronic and optoelectronic materials, specifically in inorganic and organic semiconductors and their nanotechnological applications.

Kikonyogo Ngatya is a veteran freelance reporter based in Uganda who writes for a number of Ugandan newspapers including the *Daily Monitor* and the *New Vision*. He frequently writes about science and technology issues that have important impacts on the Ugandan people including genetically modified crops, HIV/AIDS, and nanotechnology.

Dean Nieusma is an assistant professor in Science and Technology Studies at Rensselaer Polytechnic Institute. His research and teaching focus on interdisciplinary design collaboration and the expertise that enables it. This work centers especially on strategies that align technoscientific production with democratic process and social justice goals.

Rex R. Raimond is a Senior Mediator at Meridian Institute where he designs and manages collaborative problem solving processes aimed at helping people solve complex and controversial societal problems. He has worked on local, national, and international projects regarding a range of issues related to: agriculture and food security; innovation, science and technology; international and sustainable development; forestry; water and watershed management; climate and energy; and security. He is currently working on innovative processes that engage scientists, businesses, and African entrepreneurs in improving the efficiency of agricultural value chains for smallholder farmers in sub-Saharan Africa.

Gyorgy Scrinis is an honorary fellow in the School of Philosophy, Anthropology and Social Inquiry at the University of Melbourne, Australia. His research focuses on the ways the technosciences shape structural, cultural and ecological relations, particularly across the food system. His publications have addressed the issues surrounding the introduction of genetically modified foods and nano-foods, and a critique of the ideology of nutritionism, or nutritional reductionism, within nutrition science. He is currently working on a book on nutritionism.

Philip Shapira is Professor of Innovation, Management and Policy at the Manchester Institute of Innovation Research, Manchester Business School, University of Manchester, UK; and Professor of Public Policy at Georgia Institute of Technology, Atlanta, USA. His research interests include science, technology, and innovation management and policy; industry analysis; regional innovation; R&D and knowledge measurement; and policy evaluation. Shapira has been associated with the Center for Nanotechnology in Society (CNS-ASU), Tempe, USA, since 2005, contributing to the real-time technology assessment of nanotechnology research and innovation systems. He is a co-editor of *The Theory and Practice of Innovation Policy—An International Handbook* (Edward Elgar, 2010).

Raghubir Sharan has been a distinguished professor at LNM Institute of Information Technology in Jaipur, India since 2004. From 1969 to 2004 he was a member of faculty of Electrical Engineering at IIT Kanpur, India. His interests are in display engineering, engineering education, and meaningful uses of technology. He obtained a Ph.D. in Electrical Engineering from the University of Waterloo, Canada in 1968.

Catherine P. Slade is a Postdoctoral Associate at the Consortium for Science, Policy and Outcomes at Arizona State University. She is also a Research Scientist at the Department of Public Administration and Policy at the University of Georgia. Much of her research focuses on science policy and equity, with an emphasis on health disparities. She has a PhD in public policy from a joint program of the Georgia Institute of Technology and Georgia State University.

Laurel Smith-Doerr is Associate Professor of Sociology at Boston University, and current co-Chair of the BU Women In Science and Engineering (WISE). She studies tensions in the institutionalization of science including gender equity in different organizational forms and life scientists' responses to ethics education requirements. She is author of *Women's Work: Gender Equality vs. Hierarchy in the Life Sciences* (Lynne Rienner Publishers 2004).

Jack Stilgoe is a Senior Researcher at Demos, a think-tank focused on power and politics based in London. He works on science and technology projects and specializes in issues of science, expertise, and public engagement. He is co-author of *The Received Wisdom* (Demos 2006) and *The Public Value of Science* (Demos 2005).

Dhanaraj Thakur is a Ph.D. Candidate in the School of Public Policy, Georgia Institute of Technology. His research interests include online deliberation and the application of information and communications technologies for development.

Walter D. Valdivia is a doctoral student at the School of Public Affairs and graduate research associate at the Center for Nanotechnology in Society at Arizona State University. His research interests encompass policy evaluation theory and governance, history of science policy in relation to development, and technology transfer policy.

Rinie van Est is research coordinator and "trendcatcher" with the Rathenau Institute's Technology Assessment division. He studied applied physics at Eindhoven University of Technology and political science at the University of Amsterdam. His PhD-thesis "Winds of Change" (1999) examined the interaction between politics, technology, and economics in the field of wind energy in California and Denmark. Van Est joined the Rathenau Institute in 1997 and is primarily concerned with emerging technologies such as nanotechnology, cognitive sciences, persuasive technology, robotics, and synthetic biology. He has many years of hands-on experience with designing and applying methods to involve expert, stakeholders, and citizens in debates on science and technology in society. In addition to his work for the Rathenau Institute, he lectures Technology Assessment and Foresight at the School of Innovation Sciences of the Eindhoven University of Technology.

Jameson M. Wetmore is an Assistant Professor at the Consortium for Science, Policy & Outcomes and the School of Human Evolution & Social Change at Arizona State University. He studies the history, sociology, politics, and ethics of technology to better understand how to reflect on, shape, and direct social and technological change. He co-edited *Technology & Society: Building our Sociotechnical Future* (MIT Press 2008) with Deborah Johnson and the first *Yearbook of Nanotechnology in Society: Presenting Futures* (Springer 2008) with Erik Fisher and Cynthia Selin.

Gregor Wolbring is an Assistant Professor at the University of Calgary in the Faculty of Medicine with the Department of Community Health Sciences' Program in Community Rehabilitation and Disability Studies; part time professor in the Faculty of Law at the University of Ottawa, Canada; Senior Investigator at the Center for Nanotechnology in Society at Arizona State University, USA; and Adjunct Faculty in Critical Disability Studies at York University in Canada. He is the President elect of the Canadian Disability Studies Association and Chair of the Bioethics Taskforce of Disabled People's International. He has published extensively on issues of nanoscale science and technology governance and inequity and inequality issues.

Jan Youtie is Manager of Policy Services at the Enterprise Innovation Institute, adjunct to the School of Public Policy, and co-founder of the program in Science, Technology, and Innovation Policy at Georgia Institute of Technology. Her research focuses on manufacturing competitiveness, technology-based economic development, emerging technology assessment, and innovation and knowledge measurement.

Edgar Zayago Lau has a Ph.D. in Development Studies from the Universidad Autonoma de Zacatecas. He is currently a researcher for the Latin American Nanotechnology & Society Network (ReLANS). Lau was an invited scholar at the Forum of Nanotechnology of Latin America and Europe (NANOFORUM_EULA) at the Mesa Plus Lab, University of Twente (Enschede, the Netherlands).

Part I
Dimensions of Nano Fairness

Chapter 1
Contexts of Equity: Thinking About Organizational and Technoscience Contexts for Gender Equity in Biotechnology and Nanotechnology

Laurel Smith-Doerr

One of the most visible inequities in scientific research and development is the small number of women involved. But while this has traditionally been the case, there are groups working to remedy this imbalance in the future. In this chapter, Laurel Smith-Doerr considers the probable place of women in nanotechnology research and production, viewing these developments through the lenses of feminist theories and past experience with biotechnology in the United States. She notes that very little is known so far about the participation of women in nanotechnology research or production, but its association with physical sciences and engineering suggests that participation rates will be lower, since women are better represented in the life sciences and biotechnology. In addition, the non-hierarchical organizational environments that characterized startup firms in biotechnology—environments where women thrived—appear less frequently in nanotechnology. Smith-Doerr identifies a number of questions that are ripe for research during nanotechnology's formative stages, questions about how nanotechnology work serves the broader society, interdisciplinarity, patenting, and authority relationships. Because so much is already known about organizational settings that foster female participation, studies of these topics in nanotechnology could lead to early female-friendly interventions by organizations and funding bodies.—eds.

L. Smith-Doerr (✉)
Department of Sociology, Boston University, Boston, MA, USA
e-mail: ldoerr@bu.edu

This chapter was peer-reviewed. It was originally presented at the Workshop on Nanotechnology, Equity, and Equality at Arizona State University on November 21, 2008.

1.1 Introduction

> Women scientists, long underrepresented, underpromoted and underpaid in their fields relative to men, have not achieved parity despite the forces of affirmative action.... American women have never been strongly represented in the physical sciences, although they have made up a significant portion of the bioscience workforce, even at the highest levels of education, and are well represented in the social and behavioral sciences.

The above epigraph quote may seem to give a fairly accurate picture of the current status of women in Science, Technology, Engineering, and Math (STEM) fields in the United States. The quote, however, is from an article written by Betty Vetter—a pioneering researcher of trends in science careers—in 1976. The "affirmative action" language may have been the one clue to the date of publication; a point that demonstrates that over a long three decades, policies have been applied to try to broaden participation in science. These decades have seen myriad recommendations from scholarly books, the Massachusetts Institute of Technology (1999) and other university reports, National Research Council (NRC) and National Academy (2007) reports, among others.

In spite of this long period of programmatic efforts to broaden the participation of women and underrepresented minorities in STEM, the rates of participation in many fields and in many work locations have barely budged. In some cases, they have declined. A report released in 2008 by the National Action Council for Minorities in Engineering (NACME) and the Commission on Professionals in Science & Technology noted that the already small number of African Americans in the engineering workforce in 2000—when African Americans made up 5.7% of engineers—had declined by over 10% in the intervening years. And while the percentage of college students completing engineering degrees has decreased in the last decade among all racial and ethnic groups, the decline is larger for Latinos and Latinas; engineering degrees accounted for only 4.2% of all bachelor's degrees to Latinos and Latinas by 2005. The National Science Board's Science and Engineering Indicators (2008) show that the percent of computer science degrees awarded to women was approaching 40% in 1985, yet has declined nearly by half to just a little over 20% by 2006.

The setbacks to equity and equality in STEM may not be merely a failure of policies, but of understanding. In the life sciences, the War on Cancer was supposed to have been won by the Bicentennial year of 1976. Biological scientists take this "failure" to mean that we still simply lack fundamental knowledge of the causes of cancer. In the same way, our lack of progress for equity in STEM may signal that a key problem is a lack of fundamental scientific knowledge on the barriers to broader, equitable participation.

This chapter employs feminist theories to provide a useful lens with which to understand issues of gender, and equity for women, in nanotechnology fields. It focuses on social processes at the organizational level, but shows how an integration of feminist theories can provide a better platform for building the fundamental knowledge needed to understand inequity. Nanotechnology, because it is still early in its development, seems a particularly ripe context for developing more tools to study inequalities in STEM.

The chapter's substantive focus is to compare nanotechnology to biotechnology. Thinking about the potential parallels between these two emergent fields (in contrast to more established STEM settings) offers a way to demonstrate the application of various theoretical lenses. Nanotechnology and biotechnology are both eclectic fields that contain a variety of scientific and technical knowledge areas, organizational actors, and products. Nanotechnology is generally defined by the small scale of the work, and biotechnology by the purposive shaping of biological processes and genetic materials. Biotechnology developed in the late 1970s with the discovery of recombinant DNA, while nanotechnology emerged more prominently in the late 1990s with further development of technologies for manipulating materials at the atomic scale. The U.S. National Nanotechnology Initiative estimates that about 20,000 workers are employed in nanotech worldwide (http://www.nano.gov/html/facts/faqs.html). The Biotechnology Industry Organization reports that 180,000 workers were employed in biotech just in the United States in 2006 (http://bio.org/speeches/pubs/er/statistics.aspa). Thus, the issues for nanotechnology are perhaps more early stage than the relatively more established biotechnology field, but both are high-tech arenas that require the work of highly educated scientists and engineers to develop new products and processes.

This chapter begins with what we know about women in other STEM fields, particularly biotechnology, in order to develop an informed research agenda on equity in nanotechnology settings. The sections below draw on sociological and Science, Technology & Society (STS) theory and research.

1.2 Theoretical Perspectives on Gender Equity

Nanotechnology presents interesting contexts for research—promising to develop industries where emerging, innovative technoscience will be located; can gender equity reside there as well? In order to understand where and how nanotechnology contexts might present patterns of gender equity that are different than (or similar to) other STEM areas, a brief summary of what past research has found is necessary background. This chapter follows a definition of inequity similar to Cozzens' (2007, 86): "a normative term denoting an unjust or unfair distribution," noting that addressing gender *inequality* (a "descriptive term denoting any uneven distribution, right or wrong") in STEM will have the effect of mitigating *inequities*. To understand why there has been such longstanding inequity, social science has considered processes spanning micro and macro levels of analysis, including meso, or organizational, levels. The theories considered below represent a range of perspectives that, like all feminisms, are dynamic and often defy categorization (Tong 2008). Still, because of the utility in thinking about various assumptions, agendas, and methods in the multiple approaches to feminist theory, two broadly aggregated types of feminist theories are employed below. Liberal feminist theories try to understand individual and institutional barriers to equity for underrepresented groups in science and engineering, and often employ standard social science methods. More critical theories—sometimes called "radical feminist" theories—come from neo-Marxist, multicultural, and science studies feminist perspectives; take critical views on the macro, patriarchal contexts of inequality; and may employ more experimental or action based/participatory methods of analysis.

1.2.1 Liberal Feminist Theories

1.2.1.1 An Organizational Perspective

Research focused on the organizational level finds that the contexts of scientific work matter for gender equity. Fox and Mohapatra (2007), for example, find women are more productive in academic departments with climates perceived to be creative and exciting, although this climate effect does not replace positive effects of working on gender-integrated research teams, and working on multiple projects at one time. In examining organizational variation across academic and industrial sectors in biotechnology, I have found that the difference is less by sector than by form: hierarchical organizations are less conducive to equity than the network form (Smith-Doerr 2004).

Powell (1990) contrasts network and hierarchical organizations on several dimensions. Where hierarchies' main mode of communication is routines, relationships provide communication in the network form. Conflict is resolved in hierarchy by authority ranking, but in network firms by reputation. Hierarchy has a formal tone where networks have an informal one. Networks are more open-ended and provide collective benefits rather than individualized promotions based on benchmarks.

Following other research, I define biotechnology firms as research-intensive, for-profit firms focused on human therapies rather than agricultural or other applications (Powell et al. 1996; Zucker et al. 1998). These biotechnology firms represent network organizations, in contrast to large pharmaceutical corporations and universities which are more hierarchical structures. Not only do women scientists find more opportunities for authority in networked biotech firms (Smith-Doerr 2004), but women in biotech firms also have equitable outcomes in patenting productivity, in marked contrast to large gender gaps in patenting in academic and large pharma contexts (Whittington and Smith-Doerr 2008).

In a study of over 2000 life scientists in the United States, women PhDs were nearly eight times more likely to hold leadership roles in biotechnology firms than in more hierarchical settings in academia and large drug companies (Smith-Doerr 2004). Interview data in the same study (Smith-Doerr 2004) with a smaller sample suggest that hierarchical settings are less conducive to equity in promotion than network firms because hierarchies lack collective rewards, lack collaborative choice, and have less transparency. The more formal authority structures seem to be places where inequities can be perhaps more easily hidden than are settings more reliant on collaborative research and development (R&D) networks between organizations. Although hierarchies can be reasonably conducive to gender equity when there is sufficient transparency (e.g., Fox et al. 2007; Reskin and McBrier 2000), hierarchies lack collaborative choice. Women do well in organizations where scientists have flexibility both to collaborate over time with good colleagues working in a variety of science-based organizations and to avoid working on subsequent projects with colleagues who turn out not to be so enlightened (Smith-Doerr 2004). Foreign-born life scientists also seem to find more opportunities for entrepreneurship in the biotech industry (McQuaid et al. 2009).

Theories of individual choice that do not take the organizational context into account can lead to flawed perceptions about the issues for women in science. Consider the "pipeline" idea. The basic idea of the pipeline model is that the science career track is like a long pipe. Along the pipe are different stages of an academic science career—perhaps beginning with an undergraduate major in a STEM field, then on to graduate school and a PhD in a STEM field, postdoctoral fellowship, assistant professor on tenure track, tenured associate professor, and finally full professor in a STEM discipline. The idea is that women "leak out" at each stage of the career so that very few women make it to full Professor. This model is said to explain why the percentage of full professors who are women remains very small.

The "pipeline" may be useful as a metaphor for raising awareness of the gap in representation of women and underrepresented minorities at all levels in STEM; it has been used to justify a variety of policies that intend to increase equality. But we know from social science research that the pipeline is not a useful theoretical model when considering all of the data on women in science (Schiebinger 1999). First, it cannot account for those who leave and then return to academia, or who work as adjunct faculty before gaining tenure (Wolfinger et al. 2009). In the life sciences, the supply of women PhDs, postdoctoral researchers, and even assistant professors has been robust for many years; but women are still underrepresented as tenured, senior faculty in the biological sciences. In 1983, women made up 35% of the postdocs in the life sciences, but women's share of senior faculty positions is still not up to that level twenty-five years later (National Science Board 2008). Having a critical mass of women occupying the middle of the pipeline to take on leadership roles is apparently not the problem.

A body of research on discrimination against women scientists shows that the image of passive "leaks" from a pipe is not an accurate picture of why women do not advance in STEM careers. For example, in a 2005 RAND study of decisions at the National Institutes of Health on research funding, researchers found that women investigators received about sixty-three cents for every dollar that male investigators did (Hosek et al. 2005). This study controlled for career level factors such as age, education, institution, and type of grant. Thus in a crucial point for STEM careers, there is evidence that women scientists are afforded fewer opportunity to fully pursue their research than are comparable male colleagues. Another study in Sweden shows that women scientists' achievements were not counted the same as men's at their Medical Research Council also. For a prestigious fellowship, women PhDs had to publish 2.5 times the amount of male colleagues (the equivalent of ten more peer reviewed articles) to receive the same subjective ranking as male colleagues (Wenneras and Wold 1997). These studies and others show that systematically, women scientists face discrimination in receiving the resources needed to do their work. The pipeline idea gives us, at best, little to work with in explaining these results, and at worst is misleading about the processes of inequity in STEM fields.

The pipeline metaphor likewise does not explain equity differences by organizational type; one of the problems with the "pipeline" idea is that it assumes a monolithic individual choice model outside of social context. We need further research to develop theories and methods that will help us better understand the individual, organizational, and cultural dynamics that present consistent barriers to equity in STEM.

1.2.1.2 Social Psychological Perspectives

At the individual, social psychological level, studies of cognitive bias based on psychological experiments (mostly in laboratory settings) have shown that across gender, age, education level, racial and other demographic groups, we all share a male-centric bias. One study (Steinpreis et al. 1999), for example, randomized groups from a sample of male and female faculty members. The groups were shown the exact same curriculum vitae (CV) with one difference: masculine names on CVs in one group were feminine in the other group, and vice versa. An identical CV with a male name was on average ranked as a more desirable candidate than when it was assigned a female name. A body of social psychological research (see Valian 1998 for review) shows consistent biases in evaluations of faculty candidates in such studies: Weaker men candidates are rated as having "potential," while stronger women candidates are described as "looking good on paper." Men candidates with many co-authored pieces are positively viewed as "collaborative," while similar women candidates "need to demonstrate that she can write articles on her own."

The implicit biases that both men and women scientists share toward masculinity result in systematic discrimination against women scientists. A study in high-energy physics provides one example among many of this kind of biased outcome (Towers 2008). In a Fermilab experiment group, the Run II Dzero, Towers (2008) found that women postdoctoral researchers in the group were on average more productive than their men postdoc colleagues, but were awarded one-third as many conference paper presentations on average. Because everyone in these huge physics collaborative groups are listed on the publications, the conference presentations help distinguish the postdocs who are seen as good assistant professor material. In this case, productivity in the experiments and publications were counted less for women scientists than for men, who were rewarded more for doing less.

1.2.1.3 Demographic Perspectives

Some research features demographic analysis of family and life course contexts. There is a long line of sociological research on men's and women's productivity controlling for family status. Zuckerman and Cole (1975, 1984), for example, noted that while women have fewer publications than men, married women publish more than single women, and having children does not decrease publication rates. More recently, Xie and Shauman (2003) have argued that when controlling for field, kind of institution, and faculty promotion level, productivity differences by gender disappear. Of course, STEM fields, like the rest of the U.S. workforce, tend to be quite gender segregated. Women academics are still more likely to be biological or social scientists rather than physical scientists or engineers, as Vetter noted in 1976.

Women scientists are more likely to work in teaching institutions and in non-tenure track positions than are their male colleagues. Gender gaps in productivity outside the United States vary by country, but women researchers' share of publications (unlike patents) across fourteen countries have not increased over the last decade (Frietsch et al. 2009).

And while gendered expectations about family roles are changing, the actual distribution of household labor for men and women has changed more slowly. In one study of faculty members (Suitor et al. 2001), male faculty report working the same number of hours on household tasks as the U.S. male average (ten hours). Women faculty "only" work 50% more hours in the household than men faculty, in comparison to the average gender gap in U.S. household labor in which women work 100% more hours than men. This type of finding appears to be related to Ginther and Kahn's (2006) research that found that single women scientists do better at each stage of their careers than single men scientists—single women have more time to spend on career related work if they are not doing so much household labor. Current demographic analyses show a shift from the 1970s and 80 s findings by Harriet Zuckerman and colleagues (e.g., Zuckerman and Cole 1984) concluding that motherhood did not present a penalty for women scientists. Having young children now seems to be a barrier to mother scientists' promotion, yet is related to increased promotions for father scientists (Ginther and Kahn 2006).

Both men and women PhD students in the United States may now be more likely to leave science and engineering because of the intense time demands of a research career. In a survey across the University of California system researching how graduate students' career goals changed from the start of their PhD programs, the percent of women who planned to be research professors declined from 39 to 27%, and the percent of men who planned to be research professors declined from 45 to 36% (Mason et al. 2009). But note that the percent of men planning on research careers after exposure to their grad school mentors' lifestyles in this study (36%) is close to the percentage of women grad students planning on research careers in academe before being discouraged by overworked professor role-models (39%). Women respondents were even more likely to want to leave the academic "fast track" than their men cohorts and cited the too-steep sacrifices to family life as a key reason.

1.2.2 Critical Feminist Perspectives

The above perspectives tend to fall under the general rubric of "liberal feminist" theoretical perspective; those studies try to understand individual and institutional barriers to equity for underrepresented groups in science and engineering. More critical theories—sometimes called "radical feminist" theories—come from neo-Marxist, multicultural, and science studies feminist perspectives and take critical views on the macro, patriarchal contexts of inequality (see Rosser (2006) for a review of feminist perspectives on women and technology). Often these critical perspectives focus on convergences between gender inequity and inequities for people of color, persons with disabilities, gay and lesbian workers, and members of other minority groups.

1.2.2.1 Socialist and Multicultural Feminist Perspectives

Tong's review of feminist theory (2008) distinguishes between Marxist, radical, socialist, multicultural and global/postcolonial feminisms. In her categorization, Marxist feminism focuses on capitalism as the source of inequality for women, radical feminism focuses on patriarchy as the source of inequality, while socialist feminists sees a two headed hydra of patriarchal capitalism that must be taken down for gender equity to occur. Hartmann's earlier work (1976: 138) illustrates this socialist/Marxist/radical approach: "before capitalism, a patriarchal system was established in which men controlled the labor of women and children in the family, and ... in so doing men learned the techniques of hierarchical organization and control."

More recent developments in feminist theory within this critical perspective have faulted earlier socialist feminist perspectives as not being sensitive enough to multicultural or global issues. Tong (2008) notes that while both multicultural and global/postcolonial feminisms focus on differences among women, she separates multiculturalism as focusing on differences among women in one region or nation (such as for women of color in the United States), and globalism on differences among rich and poor countries due to the legacy of colonialism. Mohanty (2003), for example, argues that the resilient "military/prison/cyber/corporate complex" works to recolonize marginalized people in complex ways, depending on how femininity/masculinity is racialized in their corner of the world. The feminisms that focus on differences among women share with the earlier socialist feminism a deep criticism of patriarchal-capitalism.

1.2.2.2 Feminist Science/Technology Studies

STS perspectives seek to open the "black box" of the objects of science/technology being constructed, and theorize how these are co-constructed with the identities of the designers. Feminist STS perspectives thus bring in the close critical study of the technoscience process and products—asking not only who is doing the work but whether the work process and its products are gendered.

STS perspectives are also reflexive. For example, Judy Wajcman's "technofeminism" perspective (2004) points out that the methods of STS may be a primary cause for the lack of attention to gender. Consider Latour's (1996) *Aramis*: although the title bespeaks "love," no discussion of the construction of masculinity as part of the love of technologies is discussed. Masculinity would seem central to loving machines, perhaps especially trains in the failed light rail system that Latour explores in that book. Agency centered methods like actor network theory may miss gendering by focusing on who is present in technoscience networks rather than who is missing (i.e., women, people of color, persons with disabilities). Donna Haraway, in discussing her work, notes about a feminist technoscience argument like hers:

It is neither technophobic, nor technophilic, but about trying to inquire critically into the worldliness of technoscience. It is about exploring where real people are in the material-semiotic systems of technoscience and what kinds of accountability, responsibility, pleasure, work, play, are engaged, and should be engaged. (Haraway 2004, 326)

These kinds of critical theoretical questions can be usefully applied to the study of equity in nanotechnology.

1.3 Convergences Between Theoretical Perspectives

All of the above discussed perspectives on gender equity in science and technology are compatible with a central assumption of this chapter: that gender is socially constructed. Gender is not an essential trait and not all men and women are the same. Still, this chapter focuses on gender because it is important to understand trends of inequity in culturally defined groups; social constructions have real consequences for people and for technology. Other culturally defined groups are just as important to study as gender—including race/ethnicity, nationality, dis/ability, and social class; fortunately other chapters in this volume ably cover issues for these groups in thinking about nanotechnology, equity and equality.

A heuristic for thinking about these two basic theoretical approaches to gender inequity in science might be to call them more institutional versus constructivist or more liberal feminist versus critical feminist or more structural versus more cultural. Certainly the approaches have been developed by fairly separate research communities (Thompson 2008). This chapter argues that it would be both useful and possible for these two seemingly incompatible approaches to be brought together. In considering the two broad feminist perspectives sketched above, rather than setting up a "theory contest" between them, this chapter argues that the more important intellectual agenda is thinking about how and whether liberal feminist perspectives on reducing barriers to equality could be integrated with critical STS feminism on the gendering of technoscience. In order to analyze gendered technoscience in the nanotechnology arena, we look for the signals of gendering in similar emerging technology areas like biotechnology. To approach the question of gendering in nanotechnology, we need to consider how to measure inequity in nanoscience productivity, leadership in nanotechnology enterprises, and in participation by the user-designers of nanotechnologies. Along these lines, Table 1.1 considers four dimensions of emerging technology areas—including nanotech and biotech—and the way gender issues relate to these four dimensions: for-profit industry employment settings, interdisciplinarity of the field, patenting practices, and authority in the technoscientific work.

Although it simplifies two complex theoretical perspectives, Table 1.1 shows how we can usefully compare liberal and critical feminist theories on some basic aspects of emerging technology areas. Illustrative citations are provided for liberal and critical views on the four dimensions of the table, but do not intend to provide a comprehensive list of research. In thinking about how these theories would apply

Table 1.1 Perspectives and research on gender equity in emerging technology areas like nanotech and biotech

Dimensions of Emerging Technology Areas	Liberal feminist/Organization/structure focus	Critical feminist/Technoscience/culture focus	Potential for converging liberal and critical feminist perspectives	Application to nanotech equity research
Industry	Women's opportunities in industrial settings v academic. Some applications seen as having larger social benefit (e.g., why women do biotech—social justice). E.g., Rayman 2001; Smith-Doerr and Croissant, 2009. A feminist approach to university-industry relations: Integrating theories of gender, knowledge and capital, unpublished.	Capitalist system inherently unequal/patriarchal—closer connection to markets creates alienation and devaluation of women's labor. E.g., Hartmann 1976; Metcalfe and Slaughter 2008.	Here perhaps fundamental conflict between views—opportunities within capitalist system v socialist critique.	Although convergence limited, emergent areas like nanotech more likely to find different varieties of for-profit organization—possible to compare nano orgs with different tech/goals for equity.
Interdisciplinary field	Women more likely to do interdisciplinary (ID) work—does this provide advantage or disadvantage; and what kind—specialists on ID teams v individuals doing ID. E.g., Rhoten and Pfirman 2007; Jacobs and Frickel 2009.	Women's standpoint as outsiders/multitaskers/boundary spanners affinities for ID work; hierarchies of ID-fields like women's studies marginalized, while ID research centers well resourced. E.g., Keller 1992; Hammonds and Subramanian 2003.	Overlap—interdisciplinarity and gender issues intertwined on many dimensions.	Nanotech equity research could compare teams of specialists v individuals who do ID nanotech, and other varieties/career stages of interdisciplinarity for which most equitable and innovative.

Table 1.1 (continued)

Dimensions of Emerging Technology Areas	Liberal feminist/Organization/structure focus	Critical feminist/Technoscience/culture focus	Potential for converging liberal and critical feminist perspectives	Application to nanotech equity research
Patenting	Gender gaps in commercialization of science. E.g., Ding, Murray and Stuart 2006; Whittington and Smith-Doerr 2005; Meng and Shapira, Chapter 2	Academic capitalism (AC) changing nature of knowledge funded/sought/who benefits from. E.g., Mohanty 2003; Slaughter and Rhoades 2004; Croissant and Restivo 2001.	Here some tension again, but resistance to AC or emphasis on quality of inventions may account for gender gap.	Convergence would go beyond nano patent/citation counts by gender to also include more qualitative measures of invention content and women's resistance to commercialization.
Authority	Authority gaps in academe, and large hierarchies; network biotech context may provide more equity. E.g., Smith-Doerr 2004; Eaton 1999; Whittington and Smith-Doerr 2008.	Gender inequalities created in the technoscientific objects, assumptions in the technological politics. E.g., Wajcman 2004, Oldenziel 1999.	Overlap—lack of resources and control over technoscience by women detrimental in a variety of ways.	Nanoequity research could simultaneously investigate kinds of organizations, kinds of technology outcomes, for greater gender equity.

to emerging technologies like nanotechnology, we can also consider whether the feminist theoretical perspectives could be brought together on the topic or not. The last column in Table 1.1 provides some concrete examples of nanoequity research questions that could usefully combine the liberal and critical feminist viewpoints.

In Row One of Table 1.1, the situation for women working in industry contexts in an emerging technology like nanotechnology is considered from the liberal/organizational and critical/technoscience studies perspectives. From the former perspective, women's opportunities in industrial settings may provide opportunities to work on projects that could benefit larger constituencies; this at least is a discourse that has been found in the biotechnology field. Women express a desire for working in biotech firms rather than academic labs because of the possibility of working on human therapies that can fight widespread diseases like AIDS and cancer (Rayman 2001). A critical feminist perspective that comes out of a neo-Marxist viewpoint (e.g., Firestone 1970) would likely see the closer connection to capitalist markets in industrial science as being inherently inequitable and patriarchal, and a way of creating more alienation and devaluation of the work. The two feminist perspectives, on this point of considering work in for-profit scientific settings, might not be reconcilable or find a place of convergence. Although convergence may be limited on this dimension, in emergent areas like nanotechnology, researchers are more likely to find different varieties of for-profit organization. Thus, feminist researchers could possibly investigate dimensions of equity in nanotech organizations with alternative technologies, and more socially communal goals. Both liberal and critical perspectives would be helpful on such a project.

In Row Two of Table 1.1, the interdisciplinary nature of emerging technologies like nanotech is examined for its gendered implications. From a liberal feminist view, the empirical question is whether women are more likely to do interdisciplinary scientific work and if this provides advantage or disadvantage (Rhoten and Pfirman 2007). Since nanotechnology is an interdisciplinary area, and becoming more so over time, this condition could be conducive to women's participation there. The kind of interdisciplinary scientific work is also a question: is being a specialist on an interdisciplinary team or an individual scientist doing work that crosses disciplines more or less advantageous for men and women scientists? Interdisciplinary work is difficult to evaluate via peer review, and the scholarship of women and underrepresented minorities often faces a double standard that adds to the hurdle of favorable evaluation if their work is interdisciplinary (Lamont 2009). From a more critical feminist view, women's standpoint as the outsiders on the margins of science means that individual women often have affinities for interdisciplinary work, as in Keller's (1983) biographical analysis of Barbara McClintock's career.

At the field level, feminist science studies (Hammonds and Subramanian 2003) and work on interdisciplinarity (Jacobs and Frickel 2009) has observed that interdisciplinary, feminized fields like Women's Studies are often marginalized and under-funded while interdisciplinary research centers headed by leading (male) scientists are well-resourced. These lines of analysis overlap between the feminist perspectives. Interdisciplinarity and gendering are intertwined on many dimensions and should be studied from a variety of perspectives when looking at emerging technology areas. For example, nanotech equity research from a liberal feminist stance could compare teams of specialists to individuals who do interdisciplinary nanotech. More critical feminist research could look for other varieties of interdisciplinarity and the reception for women researchers with more marginal positions (and perhaps a more critical edge) outside mainstream nanotech circles.

Row Three of Table 1.1 takes a look at patenting in emerging technology areas. In new fields like nanotechnology, research is often in early stages of development so patents may be the most visible (and countable) products. Organizationally focused studies in biotechnology have found significant gender gaps in patenting in academe (Ding et al. 2006; Whittington and Smith-Doerr 2005; Murray and Graham 2007). In an initial study of the nanotechnology arena reported in this volume, Meng and Shapira find the ratio of women to men inventors in nanotechnology patents database appears to be 1:9 (excluding those for whom sex could not be identified). From 2002 to 2006, about 17% of patents in nano had at least one female inventor.

More critical feminist analyses of patenting and the commercialization of science in academic as well as industry settings note that this new academic capitalism may be changing the nature of knowledge pursuit, as well as who benefits from academic research with health implications (Metcalfe and Slaughter 2008). These views have some tension between them (a focus on equity in patenting versus a neo-Marxist critique of commercial focus in science). But findings that women scientists seem more likely to patent for quality rather than quantity (Whittington and Smith-Doerr 2005; McMillan 2009), could signal that the gender gap in patenting may be due to women scientists resisting commercialization in academic settings. This finding could suggest some potential overlap between the critique of academic capitalism and women's lower patenting rates in universities.

Consider Row Four of Table 1.1, the authority issue. In the biotechnology realm, the laterally organized network organizations of independent firms focused on human health therapeutics have been found to be more conducive to gender equity than more hierarchical settings in industry and academic sectors (Smith-Doerr 2004; Whittington and Smith-Doerr 2008). What about nanotechnology? The discussion of organizational forms in nanotechnology seems to have focused more on discussion of questions for future research and what could be than on what exists. For example, Macnaughten et al. (2005, 16) ask about the potential governance of and global inequities latent in emerging nanotechnologies: "What new institutional and organizational forms may be appropriate to articulate these inchoate, globally-distributed concerns, conflicts, and democratic aspirations?" Some initial studies of nanotechnology industry contexts found that rather than forging networks in order to grow strategically, startup nano firms are more the product of interorganizational and informal networks between existing large firms, universities, professional associations, and government funding agencies (Meyer et al. 2005; Libaers et al. 2006). If these university-industry networks tend to follow the biotech model, then there may indeed be a divergence between more hierarchical modes of organizing nano in existing, fundamentally hierarchical scientific organizations and new industrial firms organized along network lines. If this were the case, there would likely be some of the same outcomes for gender equity as in biotech—network organizations providing more flexibility of movement around discriminatory barriers for women scientists. A more critical feminist perspective would focus more closely on the technoscientific process and products in the different settings. In this research, less hierarchical approaches and more inclusion of women scientists could result in different kinds of projects and different products. Both of these gender sensitive approaches (the more structurally focused and the more science culture focused) would expect that hierarchical contexts that excluded and/or shifted resources away from women and underrepresented minorities in science would be detrimental to the emerging field of nanotechnology.

As of early 2009, the literature on women's participation in nanotechnology settings is very small or nonexistent. There do appear to be a few works in progress listed in conference proceedings, and there are a few published articles discussing what the issues might be for women in nanotechnology in the future. For example, Bainbridge (2002) found in a survey of public attitudes toward nanotechnology that about 62% of the men surveyed had favorable attitudes toward nanotechnology, while a significantly smaller proportion of the women surveyed had a favorable attitude—about 48%. He surmised that women might be less interested in an engineering based area like nanotech than men are, following past gender patterns in STEM.

In a special issue of *Development* published in December 2006, several scholars discussed issues related to gender and nanotechnology, mostly from the future user's standpoint, and focused on ethical issues related to reproduction and disabilities. In that issue, for example, Hans (2006) questions the ethical use of new technologies like nanotech for women with disabilities in developing countries of the global South, where their health and human rights are routinely disregarded. Harcourt (2008, 42) notes that nano-biotech (the merging of living and nonliving at the nanoscale) concerns feminist advocates in that nano-biotechnologies could change "the way women experience the world, the choices they make and the work they do" particularly with regard to women's heath and reproduction, agriculture/food and environmental issues.

I was, however, unable to find peer reviewed literature providing data on women's work in nanotechnology. Meng and Shapira's chapter in this volume is one empirical study of patenting in nanotech that begins this effort. Generally, this is an area of research that appears to be wide open and is in need of development. While this chapter lays out a possible empirical agenda on four key aspects of an emerging technology area from different feminist perspectives, the work of designing research and collecting data is largely yet to be done on issues of gender equity in nanotechnology.

1.4 Discussion and Conclusion

Nanotechnology seems like a particularly ripe context for developing more tools to study inequalities in STEM. Nanotech is still early in its development, and social scientists have good early access to observe this development and perhaps even to shape it in productive ways, as Cozzens and colleagues (2008) note.

This chapter has focused on the complementarities and comparisons between biotech and nanotech, as emergent technology fields in the late twentieth century located primarily in the United States. Other approaches to equity may also want to think about the differences between nano and bio. For one, nanotech and biotech fields have very different origins. While both fields are interdisciplinary, biotechnology resides within the domain of the life sciences and nanotechnology's roots are more strongly planted in the physical sciences.

The competitive tensions between nanotechnology and biotechnology that stem from their different origins could be a fertile area for future research. Nanotechnology is mostly an invention of physicists (perhaps most famously attributed to Richard Feynman's (1960) lecture) and material engineers. The framing of nanotechnology's emergence (i.e., "nano-hype") seems to have been staged to take back some of the limelight that had focused on biotechnology in the late twentieth century, particularly in the United States, at a time when the life sciences were being heralded as the key scientific priorities for discoveries related to human health and the National Institutes of Health's (NIH) budget saw major increases. In contrast, physics was viewed as old news based on early twentieth century successes in nuclear physics, while events like the failed Superconducting Supercollider project heralded a shift in priorities for scientific funding from the 1980s-early 2000s. The public face of science changed from big physics to big biology (such as the Human Genome Project). But with nanotechnology emerging in the first decade of the 2000s and the NIH budget falling off after 2006, if there is a corresponding shift in funding and/or attention away from biotechnology fields to nanotechnology fields, it will mean a shift from more biological science based applications (although of course some biologists are involved in nanotech), to more physics-based technologies. This is a gendered shift, as well. Biological sciences have seen more involvement by women scientists since the 1970s, while physical sciences remain dominated by men and masculinity (e.g., Traweek 1995). Research is needed on this tension between biotech and nanotech, and the areas with more and less representation of women scientists.

If research focused on the contrasts between biotech and nanotech equity and found that the representation of women researchers was greater in biotech, the different field contexts would be an important question for further study. The explanation that women self-select for scientific fields that are more strongly associated with life and living beings is partial at best (Schiebinger 1999). Individual career decisions must be considered within the broader cultural context and the structure of opportunities available. Molfino and Zucco (2008) argue that women's contributions to biotech are facilitated less by preference for "living beings" and more by their experience dealing with "complexity." Emerging technology areas like biotechnology (and presumably nanotechnology) are characterized by complexity. In technoscience settings where simple explanations cannot match the uncertainty, contradiction, and disorder, "women come into the picture as subjects who have had a long experience of dealing with ambivalent situations in which there is no question of choosing one way rather than another.... Unlike men, they are more exposed to the demands of others, and so less individualist in their choices not because less autonomous, but because immersed in relationships" (Molfino and

Zucco 2008, 27). Ridgeway's (2009) theory of gender frames explains that the ways scientific fields are culturally coded send messages to men and women about the appropriateness for their participation and leadership. Engineering and physical sciences have been consistently framed as masculine, while the biological sciences have been more neutral. Ridgeway (2009) argues that this cultural framing, in conjunction with biotech's organizational flexibility identified by Smith-Doerr (2004) and Whittington (Whittington and Smith-Doerr 2008) has opened opportunities for women in the life sciences in contrast to engineering and the physical sciences.

Evidence from computer science is helpful in understanding how nanotech could include broader representation from women researchers, in a non-life science field. In the 1980s, women comprised 30–40% of computer science majors, there was little difference in pay for men and women in computer science, and women were entering computing in large numbers (Wright and Jacobs 1994). When the cultural frame of computing and the institutionalized structure changed—towards having few time/space boundaries for the all consuming work of coding or "hacking," and into a "macho geek" culture (Margolis and Fisher 2002)—then fewer women were included in computing. Rather than focus on attracting women only into nano-bio areas based on a flawed "life science preference" assumption, the trick for equity in nanotech broadly is to avoid machismo and to keep a gender neutral, open-frontier, open-to-comers culture intact along with organizational flexibility.

In addition to its focus on the similarities rather than differences between nanotech and biotech, this chapter perhaps ought to be read with a few other caveats in mind. First, it does not cover all of the literature on women and gender equity in science and engineering. This chapter is more focused on how emerging areas of science and technology may differ from more established settings for technoscientific work. It also focuses on the sociological and STS strands of the gender equity literature on STEM fields. Second, this chapter argues for the value of conceptualizing convergences between two feminist theoretical perspectives—here called liberal and critical feminism—which some social scientists would argue have fundamentally different and irreconcilable assumptions. This chapter does not argue for ignoring those differences, but instead argues that there is value in paying attention to where and how these two feminist perspectives converge (and where they do not) in analysis of equity issues in emerging technologies like nanotech and biotech. Finally, this chapter does not present new empirical data. This review of existing understandings of gender equity in STEM and conceptualization of a research agenda for studying these issues in nanotechnology, however, could be useful prior to launching into data gathering.

There is a fundamental connection between who is included in science-based organizations and the technoscience produced. This assertion is just one way the two feminist perspectives—liberal/equal opportunity and critical/gendered construction of technology views come together. Who is doing science and technology matters for equitable outcomes, and how we study nanotech and gender equity issues also matters. Efforts to develop theoretical perspectives that pay attention to how both structures and cultures are gendered, and the development of research methods to match these theoretical perspectives, would be fruitful. This could take the shape of having feminist research teams that include scholars with both qualitative and quantitative expertise. Such social science endeavors could have positive impacts on nanotechnology, if well integrated into nanotech activities from the beginning (Guston and Sarewitz 2002), to help nano-scientists think about diversity on their research and design teams—whether in countries of the global North or South.

One direct policy implication of this chapter, then, is that nanotechnology research needs to continue to involve social scientists from the beginning of projects. The National Science Fondation-funded Nanoscale Science and Engineering Centers and the Centers for Nanotechnology in Society are a good beginning—but as research in the United States and other countries expands, so should the role of social scientists. This chapter argues that attention must be given to policies for deciding who are the nanotech and social scientists involved—the researchers should include diverse group of feminist and critical scholars who will address equity issues—not only by counting women's participation but also in critical analysis of processes and products of nanotech for equity issues. An indirect policy implication of this chapter for creating equity in nanotech is that variation in organizations should be sought—alternatives to hierarchies, and support for research outside traditional university settings—may create new opportunities for gender equity.

Questions will remain, but the focus on conducting rigorous gender equity research in emerging nanotechnology settings must be a priority. We can learn where to place emphasis on policy changes for halting emerging inequity and/or fostering equality from the inside, and we can also learn if these are social processes that require external social movement pressures on firms and universities. Employing a combination of theoretical perspectives to examine contexts that are experiencing organizational and technoscience changes will lead to better understanding of persistent inequalities and roads to equity. In short, research is needed to discover which social, organizational and technoscience practices in nanotechnology and nanoscience can ensure that the next 30 years for women in STEM careers do not look the same as the last thirty.

Acknowledgements An earlier version of this paper was presented at the workshop organized by Jameson Wetmore and Susan Cozzens on "Nanotechnology, Equity, and Equality," held at Arizona State University, November 2008, Tempe, AZ. Thanks to Jamey and Susan—also this volume's editors—for their helpful comments on drafts of this paper, to two anonymous reviewers for comments, and to Ed Hackett for the "War on Cancer" analogy. Author thanks the National Science Foundation for support in preparation of this work; however, any opinions and conclusions are those of the author and do not necessarily reflect the views of NSF.

References

Bainbridge, William Sims. 2002. Public attitudes toward nanotechnology. *Journal of Nanoparticle Research* 4: 561–570.
Cozzens, Susan E. 2007. Distributive justice in science and technology policy. *Science and Public Policy* 34 (2): 85–94.
Cozzens, Susan E., Isabel Bortagaray, Sonia Gatchair, and Dhanaraj Thakur. 2008. Emerging technologies and social cohesion: Policy options from a comparative study. Paper

presented at the PRIME Latin America Conference, September 24–26, Mexico city. http://prime_mexico2008.xoc.uam.mx/papers/Susan_Cozzens_Emerging_Technologies_a_social_Cohesion.pdf. (accessed July 30, 2010).

Croissant, Jennifer, and Sal Restivo, eds. 2001. *Degrees of compromise: Industrial interests and academic values*. Albany: SUNY.

Ding, Waverly W., Fiona Murray, and Toby E. Stuart. 2006. Gender differences in patenting in the academic life sciences. *Science* 313:665–667.

Ding, Waverly, Toby E. Stuart, Fiona Murray. 2007. Commercial science: A new arena for gender differences in scientific careers? Berkeley, CA: University of California. Unpublished manuscript. (paper specific url: http://www.haas.berkeley.edu/faculty/paper/ding2_gender%20and%20commercial%20science.pdf) Last accessed July 30, 2010.

Eaton, Susan C. 1999. Surprising opportunities: Gender and the structure of work in biotech firms. *Annals of the New York Academy of Sciences* 869: 175–189.

Feynman, Richard P. 1960. There's plenty of room at the bottom: An invitation to enter a new field of physics. *Engineering and Science* (February). http://www.zyvex.com/nanotech/feynman.html. (accessed July 30, 2010).

Firestone, Shulamith. 1970. *The dialectic of sex*. New York, NY: Farrar, Straus and Giroux.

Fox, Mary Frank, Carol Colatrella, David McDowell, and Mary Lynn Realff. 2007. Equity in tenure and promotion: An integrated institutional approach. In *Transforming Science and Engineering*, ed. Abigail J. Stewart, and Jane E. Malley, 170–186. Ann Arbor, MI: University of Michigan Press.

Fox, Mary Frank, and Mohapatra, Sushanta. 2007. Social-organizational characteristics of work and publication productivity among academic scientists in doctoral-granting departments. *Journal of Higher Education* 78 (5): 542–571.

Frietsch, Rainer, Inna Haller, Melanie Funken-Vrohlings, and Hariolf Grupp. 2009. Gender-specific patterns in patenting and publishing. *Research Policy* 38: 590–599.

Ginther, Donna K., and Shulamit Kahn. 2006. Does science promote women? Evidence from academia 1973–2001. *NBER Working Paper Series*, No. W12691.

Guston, David H. and Daniel Sarewitz. 2002. Real-time technology assessment. *Technology in Society* 24:93–109.

Hammonds, Evelynn, and Banu Subramanian. 2003. Conversation on feminist science studies. *Signs* 28 (3): 923–944.

Hans, Asha. 2006. Gender, technology and disability in the south. *Development* 49 (4): 123–127.

Haraway, Donna. 2004. Cyborgs, coyotes, and dogs: A kinship of feminist figurations. In *The Haraway reader*, 321–332. New York, NY: Routledge.

Harcourt, Wendy. 2008. Heading blithely down the garden path? Some entry points into current debates on women and biotechnologies. In *Women in biotechnology*, ed. Francesca Molfino, and Flavia Zucco, 35–69. New York, NY: Springer.

Hartmann, Heidi. 1976. Capitalism, patriarchy, and job segregation by sex. *Signs* 1 (3): 137–169.

Hosek, Susan D., Amy G. Cox, Bonnie Ghosh-Dastidar, Aaron Kofner, Nishal Ramphal, Jon Scott, and Sandra H. Berry. 2005. *Gender differences in major federal external grant programs*. Santa Monica, CA: RAND Corporation. TR-307-NSF-2005.

Jacobs, Jerry, and Scott Frickel. 2009. Interdisciplinarity: A critical assessment. *Annual Review of Sociology* 35: 43–65.

Keller, Evelyn Fox. 1983. *A feeling for the organism: The life and work of Barbara McClintock*. New York, NY: W.H. Freeman.

Keller, Evelyn Fox. 1992. *Secrets of life, secrets of death: Essays on language, gender and science*. New York, NY: Routledge.

Lamont, Michele. 2009. *How professors think: Inside the curious world of academic judgment*. Cambridge, MA: Harvard University Press.

Latour, Bruno. 1996. *Aramis or the love of technology*. Cambridge, MA: Harvard University Press.

Libaers, Dirk, Martin Meyer, and Aldo Geuna. 2006. The role of university spinout companies in an emerging technology: The case of nanotechnology. *Journal of Technology Transfer* 31: 443–450.

Macnaughten, Phil, Matthew Kearns, and Brian Wynne. 2005. Nanotechnology, governance and public deliberation: What role for the social sciences? *Science Communication* 27: 1–24.
Margolis, Jane, and Allan Fisher. 2002. *Unlocking the clubhouse: Women in computing.* Cambridge, MA: MIT.
Mason, Mary Ann, Marc Goulden, and Karie Frasch. 2009. Why graduate students reject the fast track. *Academe Online* (January-February) vol. 95(1) http://www.aaup.org/AAUP/pubsres/academe/2009/JF/Feat/maso.htm. (accessed January 21, 2009).
Massachusetts Institute of Technology. 1999. A study on the status of women faculty at MIT. *The MIT Faculty Newsletter*, XI (4).
McMillan, G. Steven. 2009. Gender differences in patenting activity: An examination of the U.S. biotechnology industry. *Scientometrics* 80 (3): 683–91.
McQuaid, James, Laurel Smith-Doerr, and Daniel J. Monti. 2010. Expanding entrepreneurship: Female and foreign-born founders of New England biotechnology firms. *American Behavioral Scientist* 53(7): 1045–1063.
Metcalfe, Amy Scott, and Sheila Slaughter. 2008. The differential effects of academic capitalism on women in the academy. In *Unfinished agendas: New and continuing gender challenges in higher education*, ed. Judith Glazer-Raymo, 80–111. Baltimore, MD: Johns Hopkins University Press.
Meyer, Alan D., Vibha Gabha, and Kenneth A. Colwell. 2005. Organizing far from equilibrium: Non-linear change in organizational fields. *Organization Science* 16: 456–73.
Mohanty, Chandra Talpade. 2003. *Feminism without borders: Decolonizing theory, practicing solidarity.* Durham, NC: Duke University Press.
Molfino, Francesca, and Flavia Zucco, eds. 2008. *Women in biotechnology.* New York, NY: Springer.
Murray, Fiona, and Leigh Graham. 2007. Buying science and selling science: Gender differences in the market for commercial science. Industrial and Corporate Change 16: 657–689.
National Academies of Science, Engineering and Medicine. 2007. *Beyond bias and barriers: Fulfilling the potential of women in academic science and engineering.* Washington DC, National Academies Press.
NACME. 2008. *Confronting the "new" American dilemma, underrepresented minorities in engineering: A data-based look at diversity.* White Plains, NY: NACME. http://206.67.48.105/NACME_Rep.pdf. (accessed January 25, 2009).
National Science Board. 2008. *Science and engineering indicators 2008.* Arlington, VA: National Science Foundation. http://www.nsf.gov/statistics/seind08/ (accessed January 25, 2009).
Oldenziel, Ruth. 1999. *Making technology masculine: Men, women and modern machines in America 1870–1945.* Amsterdam: Amsterdam University Press.
Powell, Walter W. 1990. Neither market nor hierarchy: Network forms of organization. *Research in Organizational Behavior* 12:295–336.
Powell, Walter W., Kenneth W. Koput, and Laurel Smith-Doerr. 1996. Interorganizational collaboration and the locus of innovation: Networks of learning in biotechnology. *Administrative Science Quarterly* 41: 116–145.
Rayman, Paula M. 2001. *Beyond the bottom Line: The search for dignity at work.* New York, NY: Palgrave.
Reskin, Barbara F., and Debra Branch McBrier. 2000. "Why not ascription? Organizations' employment of male and female managers." *American Sociological Review* 65: 210–233.
Rhoten, Diana, and Stephanie Pfirman. 2007. Women in interdisciplinary science: Exploring preferences and consequences. *Research Policy* 36: 56–75.
Ridgeway, Cecilia L. 2009. Framed before we know it: How gender shapes social relations. *Gender & Society* 23: 145–160.
Rosser, Sue V. 2006. Using the lenses of feminist theory to focus on women and technology. In *Women, gender, and technology*, ed. Mary Frank Fox, Deborah Johnson, and Sue V. Rosser, 13–46. Champaign, IL: University of Illinois Press.
Schiebinger, Londa. 1999. *Has feminism changed science?* Cambridge, MA, Harvard University Press.

Slaughter, Sheila, and Gary Rhoades. 2004. *Academic capitalism and the new economy: Markets, state, and higher education.* Baltimore, MD: Johns Hopkins University Press.

Smith-Doerr, Laurel. 2004. *Women's work: Gender equity v. hierarchy in the life sciences.* Boulder, CO: Lynne Rienner.

Steinpreis, Rhea E., Katie A. Anders, and Dawn Ritzke. 1999. The impact of gender on the review of the curricula vitae of job applicants. *Sex Roles* 41: 509–528.

Suitor, Jill, Dorothy Mecom, and Ilana S. Feld. 2001. Gender, household labor, and scholarly productivity among university professors. *Gender Issues* 19: 50–57.

Thompson, Charis. 2008. Stem cells, women, and the new gender and science. In *Gendered innovations in science and engineering*, ed. Londa Schiebinger. Stanford, CA: Stanford University Press, 109–130.

Tong, Rosemarie. 2008. *Feminist theory: A more comprehensive introduction, third edition.* Boulder, CO: Westview.

Towers, Sherry. 2008. A case study of gender bias at the postdoctoral level in physics, and its resulting impact on the academic career advancement of females. ArXiv Working Paper 0804.2026v3 (April 19), http://arxiv.org/PS_cache/arxiv/pdf/0804/0804.2026v3.pdf. (accessed January 15, 2009).

Traweek, Sharon. 1995. Bodies of evidence: Law and order, sexy machines, and the erotics of fieldwork among physicists. In *Choreographing history*, ed. Susan Leigh Foster, 211–228. Bloomington: Indiana University Press.

Valian, Virginia. 1998. *Why so slow? The advancement of women.* Cambridge, MA: MIT.

Vetter, Betty M. 1976. Women in the natural sciences. *Signs* 1 (3): 713–720.

Wajcman, Judy. 2004. *TechnoFeminism.* Cambridge: Polity.

Whittington, Kjersten Bunker, and Laurel Smith-Doerr. 2005. Gender and commercial science: Women's patenting in the life sciences. *Journal of Technology Transfer* 30: 355–370.

Whittington, Kjersten Bunker, and Laurel Smith-Doerr. 2008. Women inventors in context: Gender disparities in patenting across academia and industry. *Gender & Society* 22 (2): 194–218.

Wenneras, Christine, and Agnes Wold. 1997. Nepotism and peer review in science. *Nature* 387: 341–343.

Wolfinger, Nicholas H., Mary Ann Mason, and Marc Goulden. 2009. Stay in the game: Gender, family formations and alternative trajectories in the academic life course. *Social Forces* 87 (3): 1591–621.

Wright, Rosemary, and Jerry A. Jacobs. 1994. Male flight from computer work: A new look at occupational resegregation and ghettoization. *American Sociological Review* 59 (4):511–536.

Xie, Yu, and Kimberlee A. Shauman. 2003. *Women in science: Career processes and outcomes.* Cambridge, MA: Harvard University Press.

Zucker, Lynne G., Michael R. Darby, and Marilynn B. Brewer. 1998. Intellectual human capital and the birth of U.S. biotechnology enterprises. *American Economic Review* 88: 290–306.

Zuckerman, Harriet, and Jonathan R. Cole. 1975. Women in American science. *Minerva* 13: 82–102.

Zuckerman, Harriet, and Jonathan R. Cole. 1984. The productivity puzzle: Persistence and change in patterns of publication of men and women scientists. *Advances in Motivation and Achievement* 2: 217–258.

Chapter 2
Women and Patenting in Nanotechnology: Scale, Scope and Equity

Yu Meng and Philip Shapira

There are many ways to determine the diversity of the nanotechnology workforce. None of them is perfect, but they can all be telling in some way. In this chapter Yu Meng and Philip Shapira delve into data in one of the areas Smith-Doerr highlights, namely, patenting. Using a comprehensive data set, these authors report the discouraging, but familiar, statistics: very few women are patenting in nanotechnology. Only 17% of the patents in their dataset had only female inventors; twice as many have only male inventors. But the gap between the two figures has gradually been closing over the period studied. Although women's patents are broader in scope than those of male inventors, female patent applicants are concentrated in a few subfields of nanotechnology, especially those with life science connections. The interdisciplinarity of the field, however, is a positive factor in attracting women and it is possible the female representation may continue to increase.—eds.

Y. Meng (✉)
School of Public Policy, Georgia Institute of Technology, Atlanta, GA, USA; Fraunhofer Institute for Systems and Innovations Research, Karlsruhe, Germany
e-mail: mengyu21@gatech.edu

This chapter was peer reviewed.

2.1 Introduction

This paper investigates the scale and scope of nanotechnology patenting by women in the United States. We analyze patent records in nanotechnology by the United States Patent and Trademark Office from 2002 through to mid-2006 to discern the involvement of female inventors and to examine gender differences in patterns of performance, collaboration, and patent technological comprehensiveness. Although research and innovation at the nanoscale is a relatively new interdisciplinary domain of activity, the underlying fields of science and engineering that comprise nanotechnology are long-established. We find that female involvement in nanotechnology patenting is generally low, as in other technological fields. Most women engaged in nanotechnology patenting do so as part of mixed gender inventor teams. But, we hypothesize that women may be able to take advantage of the interdisciplinary and team character of nanotechnology to contribute to a broadening of the scope of technological innovation. The initial findings resulting from our analysis indicate that where women are involved in nanotechnology patenting, patents tend to be more comprehensive in scope, encompassing more technological classes. We consider the implication of this result both in terms of opportunities for further confirming research and for innovation management and policy.

2.2 Context of the Study

2.2.1 Women in Science and Engineering in the United States

Although women are far from reaching overall numerical parity with men in terms of shares of employment in science and engineering (S&E), there has been positive change over the past two decades. In 1973, women held just 7% of all S&E doctoral jobs in U.S. universities and colleges. By 2006, 33% of all academic doctoral positions were held by women. In nonacademic S&E occupations (including those in industry and government), women held 26% of all positions in 2005, up from 12% in 1980. In nonacademic doctoral S&E employment, women held 31% of positions in 2005, up from 23% in 1990 (NSF 2008).

Yet, notwithstanding the increasing involvement of women in the S&E workforce, gender differences remain. The distribution of women by key fields of science and engineering remains uneven. For example, while women occupy 43.4% of S&E occupations in biology and life sciences, they hold only 28.5% and 27.6% of positions respectively in physics and computer science, and just 11.1% of engineering positions (2003 data reported in NSF 2007; see also Fig. 2.1). Female scientists and engineers are also more likely to work in local government, academic, and nonprofit organizations whereas their male counterparts are relatively more employed in industry, self-employed business, and the federal government (Fig. 2.2). In addition to variations by sector and occupations, women scientists and engineers are clustered in jobs associated with less prestige, authority, and pay (Crewson 1995; Kaufman 1995; Lindsey 1997). Across all fields of science and engineering, there are longstanding and widespread concerns about gender inequity

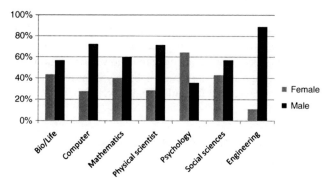

Fig. 2.1 Employed scientists and engineers, by sex and fields, United States, 2003. Source: NSF, Division of Science Resources Statistics, Scientist and Engineers Statistical Data System (SESTAT) 2007

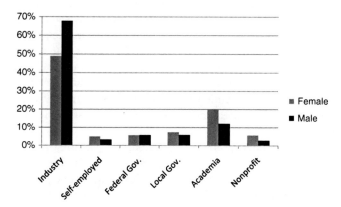

Fig. 2.2 Employed scientists and engineers, by gender and types of employer, United States, 2003. Source: NSF, Division of Science Resources Statistics, Scientist and Engineers Statistical Data System (SESTAT) 2007

including in promotion and career development, in access to research opportunities, and in recognition (see, for example, Reskin 1978a,b; Sorensen 1992; Sonnert and Holton 1995a,b; Evetts 1996; Fox 1999).

Debate is ongoing, not just about the role of and rewards to women in science and engineering, but also about the structures and opportunities which influence their performance (Fox 2001; Fox and Stephan 2001). Attention has focused on women's output in scientific publication, including how women's productivity, co-authorship, and citation patterns compare with those of men (Astin 1969; Reskin 1978a,b; Fox 1983; Cole and Zuckerman 1984; Allison and Long 1987; Levin and Stephan 1991, 1998; Long 1992; Creamer 1998; Xie and Shauman 1998; Long 2001; Prpić 2002; Leahey 2006). Besides ensuring that women are enabled to contribute to the advancement of knowledge and research capability, scientific publication is critical to career development in research universities, hence attention to the relative position and performance of women in scientific publication is understandable and fundamental. However, if we are concerned with how women might (or might not) be influencing the direction of innovation in an emerging technologically-driven field (such as nanotechnology), then an expanded focus (beyond scientific publication) is necessary. We need to investigate not only how and by whom new knowledge is being generated through scientific research, but also how and by whom innovative applications are being developed for and by industry.

2.2.2 Patenting as an Indicator of Engagement in Innovation Activity

Innovation in a technological field involves not just research and development (including drawing on research and development [R&D] in academia and government labs as well as in private industry) but—most crucially—the *application* of new knowledge and discoveries. One of the key pathways for applying new knowledge is through the invention and introduction of new and improved technological products, processes, devices, and systems into the marketplace. Companies (and increasingly universities and government laboratories) often seek to protect (and thus retain control over the real or potential value of) their technological inventions through the creation and ownership of intellectual property rights. This is typically done through patenting (although inventions may also be controlled through maintaining trade secrets and further protected or assigned though contracts and licenses).

In return for intellectual property protection, the patenting process generally requires the public disclosure of inventions—a process that allows others to build on this knowledge to pursue technological developments that might otherwise have not occurred as rapidly, if at all. The study of patenting hence not only measures who is inventing (or, more formally, who is filing for the protection of their inventions), but is also one of several indicators of the broader trajectory of innovation in an emerging field (see Jaffe and Trajtenberg 2002). Inventions that are submitted for patent application undergo independent examination by the patent office to ensure that the invention is novel, non-obvious, and has utility. Since patenting has a cost in terms of money and time, inventors and their assignees generally seek patent protection for inventions that have, or promise, significant value. Generally, this is measured in commercial terms, although patenting is also pursued to protect other core intellectual property, to attract venture capital funding, for other business strategy reasons (for instance, to respond to or undermine competitors), and for status and professional recognition. Although not all patents are subsequently commercialized through innovations, the growth and accumulation of patents in a new area of technology has been viewed as indicating directions for subsequent investments and related product and process innovations (Schmookler 1962; Schmoch 2006).

In recent years, there has been increased scholarly attention devoted to the role of women in, and gender differences related to, patenting. For example, in a study comparing gender differences in patenting in life science, Whittington and Smith-Doerr (2005) found that female researchers are less likely to patent than male researchers and the gap holds across generational cohorts. On average, male researchers produce more patents throughout their careers than female researchers, but the difference shrinks when account is taken only of those who actively patent. In another study on academic life scientists' patenting activity, Ding and colleagues discovered that a statistically significant gender difference remains after holding constant the effects of productivity, networks, field, and employer attributes, with women patenting at only 40% of their equivalent male counterparts (Ding et al. 2006). A salient gender gap is also evident in results from an analysis of a small sample of faculty in biotechnology (Murray and Graham 2007). They found that, in the same sample, men have a higher rate than women in the transition to patenting over time, and a much smaller share of women faculty has ever patented than men (23% vs. 74%). In three other studies that relied upon large-scale patent or patent application data, women's patents only accounted for a minority of total patents, although the range was broad—with patents associated with women varying from 0.8% to 32.9% of totals according to field, year, field, and country (U.S. Department of Commerce 1999; Naldi et al. 2004; Frietsch et al. 2009).

In our study, the focus is also on patenting by women. However, our question is not so much about women's output or productivity performance in patenting compared with men but rather on how women are engaged in patenting and whether women's engagement influences the nature of inventions for which patent protection is sought. The technological arena in which we situate our inquiry is nanotechnology which, as an interdisciplinary domain rapidly making the transition from research discovery to innovative applications, is well suited to provide the topography to explore our line of inquiry.

2.2.3 Nanotechnology Emergence and Applications

Nanotechnology involves the manipulation of molecular-sized materials to create new products and process with novel features due to their nanoscale properties. The development in the early 1980s of scanning tunneling and atomic force microscopy instruments first allowed and stimulated empirical research at the nanoscale (Baird and Shew 2004). Since then, nanotechnology has emerged as an interdisciplinary domain encompassing multiple fields of expertise, including physics, chemistry, materials science, life sciences, and electrical engineering (Porter and Youtie 2009).

The nanotechnology domain has attracted greatly increased governmental and private sector attention around the world, in large part because of its potential to drive new rounds of business and economic growth. In the United States, direct federal investment in nanotechnology R&D in universities, government labs, and private collaborations was more than US$1.6 billion in fiscal year 2010, while estimates of annual U.S. corporate nanotechnology funding approach US$2 billion (Lux Research 2007; Shapira et al. 2010). Applications are occurring (with many more envisaged) across multiple sectors including materials and coatings, electronics, aerospace, medicine and medical devices, energy, environment, construction, packaging, and consumer durables. The United States has a leading role in nanotechnology R&D and in the accelerating shift to the development of commercial applications. About one-quarter of the world's scientific publications in nanotechnology are associated with U.S.-based researchers (Youtie et al. 2008). Estimates by the Georgia Tech Program in Nanotechnology Research and Innovation (Kay and Shapira 2009) suggest that there are upwards of 5,000 corporate establishments in the United States who have entered the nanotechnology domain (about one-third of the total of all the world's corporate establishments identified as active in nanotechnology through publication and patenting). These establishments include large incumbent companies (for whom nanotechnology may as yet be a fractional proportion of their activities), small and mid-sized firms, and new nanotechnology-focused start-up enterprises in high technology and more traditional industries. Of more than 1,000 nanotechnology-based available consumer products catalogued worldwide by the Project on Emerging Nanotechnologies, 53% originated from the United States, including nanotechnology-enabled products in cosmetics, clothing, sporting equipment, electronics, and automotive applications.

These currently available nanotechnology-based consumer products represent just a portion of all the nanotechnology applications available today, and an even smaller sub-set of nanotechnology-enabled applications, devices, and systems that may enter into use in coming years. This is indicated in the large number of patent applications that have been filed to date, which represents an accumulating stock of inventions that can be drawn upon for future use. More than 17,000 nanotechnology-related patent applications were received by the United States Patent and Trademark Office (USPTO) from 1990 to 2008, with more than 12,600 nanotechnology-related patent grants through to the latter year (Georgia Tech global nanotechnology patent database, see Shapira et al. 2010, and also discussion of database later in this chapter). Venture capital (VC) investment in start-up enterprises engaged in nanotechnology totaled about US$590 million in 2006 in the United States (Shapira et al. 2010). In short, the nanotechnology field is making the transition from a focus on research and discovery to the application of results through patents, products, investments in existing companies, and the formation of new enterprises.

The expansion of nanotechnology R&D and the acceleration of industrial activity has raised concerns not only about potential environmental, health and safety implications (which we do not discuss in this chapter) but also about societal consequences, including whether this new technology will reinforce existing social stratification and how economic returns will be distributed (Wood et al. 2003; Zucker and Darby 2005). It is thus appropriate, for multiple reasons, to probe the role of women in inventive activity and innovation in this domain. Examining the number of patents associated with women relative to men in the nanotechnology field is a starting point, since involvement in invention in a leading-edge technology is one of the major ways through which the pattern of subsequent innovation is established and through which opportunities for returns to invention are distributed (via career development, royalties, venture start-up opportunities, professional recognition, and other returns). We do anticipate that women's role in nanotechnology will be much less than that of men, since women hold a minority (and in some instances, a rather marginal share) of occupations in the science and engineering fields most associated with nanotechnology. Beyond this, we seek to understand how women engage in nanotechnology patenting, including their fields of focus and how they are engaged in collaborative inventor teams. We speculate that women may be able to take advantage of the interdisciplinary and team character of nanotechnology to contribute to a broadening of the scope of technological innovation. If this is the case, then the situation reinforces arguments that women should be encouraged to take a greater part in nanotechnology innovation not only for reasons of equity and to ensure their fair share of the rewards and returns that will accrue to successful inventors, but also to further the development of the interdisciplinary and convergent applications that are important if nanotechnology is to leverage its full potential.

2.2.4 Interdisciplinarity, Collaborative Research Behavior, and Gender

Our conjectures about the roles of women in interdisciplinary invention and patenting teams are underpinned by existing contributions to the literature related to both interdisciplinary research and scientific collaboration. For interdisciplinary research, Rhoten and Pfirman (2007) identify several key characteristics, including cross-fertilization, team-collaboration, field-creation, and problem-orientation. They suggest that women's preference for complexity and diversity and their willingness to break with conventional social rules of science and styles of interaction lead women scientists to be more likely to engage in interdisciplinary research. Female scientists may also be attracted to an interdisciplinary field, especially in its early stages of development, because they are less narrowly focused than males on pursuing priority and recognition (Barinaga 1993; Sonnert and Holton 1995).

In terms of collaboration, research has found a strong correlation between productivity and collaboration (Price and Beaver 1966; Zuckerman 1967; Pravdić and Oliuić-Vuković 1986; Durden and Perri 1995). Several elements in collaboration have even been shown to affect productivity, such as complementary skills, intellectual stimulus, access to equipment, and exchange of opportunity information (Lee and Bozeman 2005). Moreover, collaborated papers are found more likely to be accepted in journals (Zuckerman and Merton 1971; Gordon 1980; Presser 1980; Bayer and Smart 1991; Hollis 2001). Some studies argue that women are inclined toward teamwork while men prefer more independent work (Hayes 2001). However, the argument is not consistently supported by empirical evidence. Earlier research found that women scientists were less likely to collaborate (Chubin 1974; Cameron 1978; Cameron and Blackburn 1981; Cole and Zuckerman 1984; Scott 1990; Kyvik and Teigen 1996; Hunter and Leahey 2008) and recent research provided evidence that women scientists are as likely as, but not more than, men to collaborate (Corley 2005).

There is also a relevant literature related to how scientists chose topics to focus upon, whether working within or across disciplines. Rhoten and Pfirman (2007) suggest that women tend to be attracted more towards applications of knowledge, rather than more theoretical aspects. A qualitative study conducted by Elisabeth Piene suggests the existence of gender difference in preference for research questions even in the same discipline—women, compared with men, are less interested in pure technical dimensions of their discipline (cited from Sorensen 1992). Pinker (2005) argues that, compared with men, women are more likely to focus on people-oriented problems. Psychological studies further suggest that women tend to integrate information from across various sources while men like to isolate objects and problems under study (Haier et al. 2005; Science Daily 2005). Rhoten and Pfirman (2007) draw on such evidence to suggest that women scientists' interdisciplinary ways of doing research lead them to a greater focus on cross-fertilization and field-creation.

These insights from this literature about broad gender differences among researchers in their approaches to field selection, interdisciplinarity, and problem choice, suggest that the nanotechnology domain will be a particularly fruitful one to explore questions related to women's roles in inventive activity. However, in exploring trends related to patenting and gender in nanotechnology, we recognize that multiple factors will come into play, some of which are mutually reinforcing while others are counteracting. For example, we note that the wide variety of disciplines engaged in nanotechnology are likely to bring with them, even into a new interdisciplinary domain, inherited ways of working. Some of the contributing disciplines to nanotechnology are more male-dominated (such as chemistry, physics, and engineering) while others have had a relatively greater role for women in recent years (including biology and medicine). It is likely that different fields of patenting will exhibit variations in women's contributions, reflecting these inherited traits. On the other hand, women researchers may be pulled towards inventive activity in the nanotechnology field, even when coming from an established discipline, because of nanotechnology's interdisciplinary character. Moreover, since nanotechnology is a relatively new cross-cutting domain, there may be fresh opportunities for women inventors compared with more established fields. At the same time, success in new interdisciplinary environments may require extensive or additional networking and connections, which might present a disadvantage to women. Similarly, a mix of organizations, large and small, and in the public, private, and non-profit sectors are involved in nanotechnology development and application. Whittington and Smith-Doerr (2005) suggest that women in small firms characterized by flatter and flexible organizational structure are more likely to patent than those in hierarchical organizations, whether in the private or public sector. Whether this finding crosses over into the nanotechnology domain remains to be ascertained. Overall, we note that there is indeed a mix of tendencies and possibilities for the role of women in nanotechnology patenting, some of which may be contradictory. At this still early stage in the development of inventive activity in nanotechnology, it is thus appropriate to initiate exploration of these questions.

2.3 Patent Data Source, Inventor Gender Identification, and Team Classification

2.3.1 Nanotechnology Patent Database

The source of the nanotechnology patent data analyzed in this paper is the Georgia Tech global database of nanotechnology patents. This database is based on searches of patent applications and awards from the United States Patent and Trademark Office (USPTO), the European Patent Office (EPO), the Japan Patent Office, the World Intellectual Property Office, and patents records from other national patents offices for seventy countries. Nanotechnology patents are identified using a two-stage process. In the first stage, searches using validated nanotechnology keywords and strings are combined with searches in nanotechnology patent classifications (e.g., IPC-B82 and USPTO Class 977). Inventions filed with multiple patent authorities are grouped into patent families as appropriate. In the second stage, duplicates are removed and exclusion terms applied (to exclude patents that fall outside the nanotechnology domain, for example those referencing only measurement terms such as nanometer without another substantive combination). The methodology is presented in detail in Porter et al. (2008).

For the period 1990–2006 (mid-year), the Georgia Tech global nanotechnology patent database contains 53,720 patent records. From this database, we extracted a subset of USPTO patent records by application date in the time period 2002–2006 (July). This subset comprises 12,742 records, among which 9,201 (72%) are applications and 3,541 (28%) are awards. Nearly three-fourth of these nanotechnology patents are the result of team collaborations of two or more inventors. Almost half of these patents are assigned to individual inventors, with just over two-fifths to corporate industry assignees, and the balance to academic, governmental, and non-profit institutions. A descriptive summary of the USPTO nanotechnology patents is presented in Table 2.1.

2.3.2 Inventor Gender

Patent applications record the names of inventors. However, inventor gender is not recorded. This requires us to formulate an independent strategy to identify the gender of inventors. Our approach involves developing a first name gender matching algorithm which can be applied to the inventor names in our patent database.

In developing our approach, we drew upon available lists of gender-matched names and resources from other researchers and organizations. One of the most comprehensive recent efforts to identify gender through name matching has been undertaken by Frietsch and colleagues (Frietsch et al. 2009) in their work on patent applications filed with EPO. Their list of gender assignments is helpful, but it is limited primarily to European names. While many first names (of both men and women) in the United States have European origins, the broad diversity of the U.S.

Table 2.1 Summary statistics, U.S. nanotechnology patents, 2002–2006

Patent characteristic		Number	Percent of total
Year granted	2002	1,904	14.9
	2003	2,161	17.0
	2004	2,613	20.5
	2005	3,698	29.0
	2006	2,366	18.6
Inventor team size	1	3,076	24.1
	2	3,333	26.2
	3	2,596	20.4
	4	1,759	13.8
	5 or more	1,978	15.5
International patent classes (IPC)	1	8,009	62.9
	2	3,111	24.4
	3	1,038	8.1
	4	373	2.9
	5 or more	211	1.7
Assignees	Individual	6,239	49.0
	Nonprofit	119	0.9
	Academia	1,155	9.1
	Government	210	1.6
	Industry	5,019	39.4
Total patents		12,742	100

Source: Analysis of patents awards by United States Patent and Trademark Office, 2002–2006, from Georgia Tech global nanotechnology patent database. See Porter et al. (2008), for information on database construction and nanotechnology definition

in terms of ethnicity and national origin and also the variations in U.S. names that have evolved over time mean that other approaches also need to be applied. To extend our list, we thus drew on other sources. We consulted the lists of first names maintained by the U.S. Social Security Administration (SSA).[1] Although patents are filed by inventors of all ages, one recent study finds the mode for first patent application is the early 30s in the United States (Walsh and Nagaoka 2009). To allow for a distribution around this typical age, we extracted the top 1000 male and female U.S. first names recorded by the SSA for the three decades of the 1960s, 1970s, and 1980s. We removed duplicates and also identified same first names which appeared in both the male and female lists. This gave us a list applicable for American first names in three categories: female, male, and both. Additionally, we used two online sources for identifying the gender of Japanese and Indian first names.[2]

As a next step, we matched all the available inventor first names in our database of U.S. nanotechnology patents separately against the four lists noted above. We checked for conflicts (for example, Debra is identified as *female* in the European list but as *both* in the American list) and coded cases of conflicts in first names as matched but not-identified.[3] This gave us a replicable and reasonably consistent first name/gender matching algorithm able to place first names in one of five categories: male, female, both, matched but not-identified, and not-matched and not-identified.

Subsequently, we conducted a verification procedure on the algorithm by using data from a 2007 survey of U.S. inventors and their inventions where first names and confirmed gender information is available.[4] While this allowed us to improve only marginally the correction rate for first names in our non-identified categories, it did validate (to a correspondence rate of 99%) our assignments of names to the male and female categories.

We then created a thesaurus based on our refined algorithm and used data-mining techniques to search through the first names in our database of U.S. nanotechnology patents.[5] Of the 24,331 inventors listed in these patents, we identified 1,789 female inventors (7.4% of the total) and 15,661 male inventors (64.4%). Of the balance of 6,872 (28.2%), most are names of Chinese and Korean origin which cannot be assigned, while some fall into the *both* category. If inventors that cannot accurately be assigned to either female or male are excluded, we find that of the 17,459 inventors that we can gender assign, 10.3% are female and 89.7% are male. This gives a ratio of identified female to identified male nanotechnology patent inventors of about 1:9.

2.3.3 Assignment of Patent Inventor Team Gender Categories

The unit of analysis in this study is the patent record. We are able to assign one of several patent gender categories to the inventors associated with each patent. As noted earlier, the majority of patents are associated with multiple inventors. So, our assignment of patent gender categories is typically at the team level. In other words, the value of *gender* for a given patent is based on the gender combinations of all inventors associated with the patent. The possibilities include: female (if all inventors are women); male (if all inventors are men); and mixed (of the inventors, some are women and some are men). However, there are also additional categories because, although we cannot assign male or female genders to more than one-fourth of the inventors in the database, we frequently can make gender assignments for one or more of the other team members. Hence, we can identify the seven following team gender combinations: (1) all female, (2) female + male, (3) female + unknown (4) female + male + unknown, (5) all male, (6) male + unknown, and (7) unknown.

Since the unit of analysis is the patent record, we cannot directly ascribe the characteristics and behaviors of individual inventors. For example, we do not know (from the available data) the discipline or background of the inventor. However, USPTO assigns patent classifications to each patent which denote the technological area (or areas) covered by the patent. We use the International Patent Classification (IPC) included in the U.S. patent record.[6] Thus, we do know what areas of technology the patent team is focusing upon. Since individual inventors may be associated with more than one patent and our database is defined to comprise nanotechnology patents, it is possible that the inventors included in our database also have patents outside of the domain of nanotechnology. So, while we have a reasonably complete set of all U.S. nanotechnology patents recorded between 2002 and mid-2006, we do not necessarily have every patent produced by inventors included in our database if they have also been awarded non-nanotechnology patents. With these caveats noted, the variables that we can create from the dataset do allow us to analyze the gender combinations of inventor teams in nanotechnology. In turn, this can inform our understanding of the contributions and influence of women in nanotechnology inventor teams.

2.4 Analysis and Findings

2.4.1 Descriptive Results

2.4.1.1 Patent Counts

There are three inventor team categories—female, male, and mixed (female + male)—where we are certain about the composition of the inventor team's gender (see Table 2.2). We find that patents with all male inventor(s) are dominant (38.7%) as a share of all U.S. nanotechnology patent records (2002 through to mid-2006). Teams which combine both men and women inventors are responsible for 6.9% of all these patents, while entirely female inventor teams are responsible for a very small proportion, or just 1%. This is little different from the less than 1% share of

Table 2.2 Gender classification, U.S. nanotechnology patents, 2002–2006

	Year						
	2002	2003	2004	2005	2006	Total	
At least one female inventor							
All female	20	25	29	40	20	134	1.1%
Female + male	147	166	188	214	162	877	6.9%
Female + male + unknown	142	149	188	274	168	921	7.2%
Female + unknown	25	31	49	56	36	197	1.5%
Sub-total	334	371	454	584	386	2,129	16.7%
Percent of column total	17.5%	17.2%	17.4%	15.8%	16.3%	16.7%	
Male inventor(s)							
All male	817	850	1,057	1,345	861	4,930	38.7%
Percent of column total	42.9%	39.3%	40.5%	36.3%	36.4%	38.7%	
Not identified							
Male + unknown	532	665	782	1,259	735	3,973	31.2%
Unknown	221	275	320	510	384	1,710	13.4%
Sub-total	753	940	1,102	1,769	1,119	5,683	44.6%
Percent of column total	39.5%	43.5%	42.2%	47.8%	47.3%	44.6%	
Total	1,904	2,161	2,613	3,698	2,366	12,742	100%

Source: Analysis of U.S. nanotechnology patents granted 2002 through mid-2006 (see Table 2.1)

U.S. patents recorded to women in the later part of the nineteenth century (Khan 2000), although the role for women inventors in mixed gender inventor teams is clearly much larger today, as indicated by our findings for nanotechnology.

2.4.1.2 Research Collaboration

The modal team size for the nanotechnology patents in our database is two inventors (comprising 26.7% of all U.S. recorded nanotechnology patents, 2002 through to mid-2006). (See Table 2.2) However, there is variation by gender. Patents *with at least one female* inventor tend to have larger inventor team size than *male* patents, with the mean size of 4.2 for the former group versus 2.1 for the latter (Table 2.3). Among nanotechnology patents *with at least one female* inventor, only 5.3% were recorded to one inventor but 35.9% recorded to a team with five or more inventors. In contrast, for *male* inventor team patents, about 42.6% were recorded to one inventor while only 5.1% were recorded to teams with five or more male inventors. But the difference is primarily rooted in the fact that a large proportion of the patents involving female inventors is attributed to *mixed* patents involving both female and male inventors. When looking more closely at only *female* and *male* patents (Fig. 2.3), we observed that *female* patents are more likely to be product of individual inventive activity while *male* patents are more likely to be the outcomes of collaborations. This finding suggests that women may still be disproportionately left out of collaborations even in an interdisciplinary domain such as nanotechnology. Inventor teams

Table 2.3 Gender composition and team size, U.S. nanotechnology patents, 2002–2006

	Team size						
	1	2	3	4	5 or more	Total	Mean (S.D.)
At least one female inventor							
All female	113	16	5			134	1.19(0.48)
Female + male		253	249	190	185	877	3.57(1.70)
Female + male + unknown			144	222	555	921	
Female + unknown		99	51	23	24	197	
Sub-total	113	368	449	435	764	2,129	4.23(3.19)
Percent of row total	5.3%	17.3%	21.1%	20.4%	35.9%	100%	
Male inventor(s)							
All male	2,101	1,450	756	372	251	4,930	2.06(1.26)
Percent of row total	42.6%	29.4%	15.3%	7.5%	5.1%	100%	
Not identified							
Male + unknown		1,057	1,175	844	897	3,973	
Unknown	862	458	216	108	66	1,710	
Sub-total	862	1,515	1,391	952	963	5,683	
Percent of row total	15.2%	26.7%	24.5%	16.8%	16.9%	100%	
Total	3,076	3,333	2,596	1,759	1,978	12,742	

Source: Analysis of U.S. nanotechnology patents granted 2002 through mid-2006 (see Table 2.1)

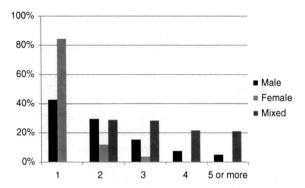

Fig. 2.3 Percentage of U.S. nanotechnology patent records, by gender composition and team size. Source: Analysis of U.S. nanotechnology patent records, 2002 through mid-2006 (see Table 2.1)

including at least one woman were associated with 16.7% of U.S. nanotechnology patents recorded during the period 2002 through to mid-2006 (see Table 2.2). Over the same period, the share of patents recorded to teams comprised entirely of male inventor(s) (38.7%) was more than twice as large. The gap between these two groups has steadily diminished, from 25.4% in 2001 to 20.1% in 2006. Since this measurement might be influenced by changes in the composition of patents where gender is not identified, it is important to check the numerical ratio between these two groups of interest. The ratio of recorded nanotechnology patents where the inventor team included at least one woman to that for all-male inventor teams hardly changed from 1:2.44 in 2002 to 1:2.23 in 2006.

Among other studies on gender disparities in collaboration, Lee and Bozeman (2005) report on collaborations of university professors and researchers in engineering, bioscience, computer science, chemistry, physics, and other science fields. They find that, on average, men have 10.2 scientific collaborators compared with women, who have 3.6 scientific collaborators. Our study, which is of nanotechnology patenting across all sectors (rather than of publication in the academic sector), confirms gender differences again.

2.4.1.3 Technical Focus

Using primary IPC class as an indicator, we summarize the technical subfields for nanotechnology patents. The range of technical fields engaged by nanotechnology is broad, encompassing medicine and medical devices, physical, chemical and mechanical processes, materials, electricity and electronics, computing, optics, paints and coatings, and agricultural products. Our analysis notes where there is a difference in the distribution by IPC classes (N = 110) of at least 0.5% between patents with at least one female inventor and those with male inventors (Table 2.4). The scale of the percentage difference in the distribution for nanotechnology patents with at least one female inventor is largest in the biology-related IPC classes (A61, C08, and C12). While the negative percentages in the two columns

Table 2.4 Field comparison by team gender, U.S. nanotechnology patents, 2002–2006

		Inventor team				
		Male	Female		At least 1 female	
IPC	Class	%	%	Difference from Male (%)	%	Difference from Male (%)
B41	Printing	1.2	12.7	11.5	2.4	1.2
C08	Organic macromolecular compounds	4.5	11.2	6.7	7.8	3.3
C12	Biochemistry; mutation or genetic engineering	2.7	7.5	4.8	6.4	3.7
A61	Medical or veterinary science; Hygiene	13.0	17.2	4.1	16.1	3.1
B08	Cleaning in General; Prevention of fouling in general	0.3	2.2	1.9	0.3	0
C03	Cements; Ceramics	0.9	2.2	1.4	0.9	0
H01	Basic electronic elements	15.8	17.2	1.4	15.5	−0.3
G11	Information storage	4.2	5.3	1.1	3.7	−0.5
C07	Organic chemistry	1.2	2.2	1.0	2.8	1.6
C30	Spinning	0.6	1.5	0.9	0.5	−0.1
G01	Measuring; Testing	9.2	3.0	−6.2	6.6	−2.6
B01	Physical or chemical processes or apparatus in general	5.8	1.5	−4.3	5.5	−0.2
G02	Optics	4.9	2.2	−2.7	3.2	−1.7
G03	Horology	3.3	0.7	−2.6	3.2	−0.1
B29	Plastics	1.6	–	−1.6	0.9	−0.7
C23	Coating metallic material	1.9	0.7	−1.2	1.5	−0.4
B23	Machine tools	1.2	–	−1.2	0.5	−0.7
B32	Layered products	2.7	1.5	−1.2	3.5	0.8
C09	Dyes; paints; polishes; natural resins; adhesives	1.8	0.7	−1.1	2.0	0.2
H05	Basic electric elements	1.0	–	−1.0	0.5	−0.5
	N	4,930	134		2,129	

Source: Analysis of U.S. nanotechnology patents granted 2002 through mid-2006 (see Table 2.1). IPC patent classes ranked (approximately) by largest difference between percentage for patents by male inventors and percentages for patents by all female inventors and at least one female inventor

"Difference from Male" help mark out the IPC classes in which *Male* patents account for a larger share, these IPC classes generally refer to engineering-related fields (e.g. G01, B01, B23). The finding is consistent with literature that indicates that women researchers are more likely to work in such fields as biology and life sciences (NSF 2007; Rhoten and Pfirman 2007; Sorensen 1992). When comparing only *female* inventor with *male* inventor patents, we observe that women are also highly represented in patent classes such as "medical and veterinary science: hygiene" and "printing; lining machines; typewriters; stamps," which include technical fields traditionally occupied by women.

2.4.1.4 Patent Comprehensiveness

To examine our proposition regarding gender differences in ways of undertaking research and innovation, we analyzed the *comprehensiveness* of the nanotechnology patents in our database. This is operationalized as the number of IPC classes assigned to each patent. If a patent has two or more IPC classes assigned, it indicates potential technical applications that range more broadly, across multiple technical areas. We find that patents invented by teams with *at least one female* inventor are more comprehensive than *male* patents (Table 2.5). The mean value of *comprehensiveness* for *male* inventor nanotechnology patents is 1.58 IPC. For *female* inventor patents and patents *with at least one female* inventor the mean values of *comprehensiveness* are 1.71 and 1.62 IPC classes respectively, and these mean differences from the male inventor patents are statistically significant. The share of patents assigned only to one IPC class is 61.0% for patents *with at least one female* inventor, whereas the share for *male* inventor patents is 62.3%. For almost each category along the

Table 2.5 Comprehensiveness of patents, U.S. nanotechnology patents, 2002–2006

	Number of IPC classes per patent						
	1	2	3	4	5 or more	Total	Mean (S.D.)
At least one female inventor							
All female	73	41	10	8	2	134	1.71(1.01)
Female + male	525	240	71	29	12	877	1.61(1.01)
Female + male + unknown	584	215	77	29	16	921	
Female + unknown	117	44	18	13	5	197	1.62 (1.01)
Sub-total	1,299	540	176	79	35	2,129	
Percent of row total	61.0	25.4	8.3	3.7	1.6	100	
Male inventor(s)							
All male	3,070	1,236	398	140	86	4,930	1.58 (0.96)
Percent of row total	62.3	25.1	8.1	2.8	1.7	100	
Not identified							
Male + unknown	2,495	963	348	103	64	3,973	
Unknown	1,145	372	116	51	26	1,710	
Sub-total	3,640	1,335	464	154	90	5,683	
Percent of row total	64.1	23.5	8.2	2.7	1.6	100	
Total	8,009	3,111	1,038	373	211	12,742	

Source: Analysis of U.S. nanotechnology patents granted 2002 through mid-2006 (see Table 2.1)

dimension of number of IPC classes that is more than one, the share of patents *with at least one female* is larger, albeit slightly, than that of *male* patents, with 25.4% vs. 25.1% for two classes, 8.3% vs. 8.1% for three classes, and 3.7% vs. 2.8% for four classes The only exception is that the percentage of patents with *at least one female* in the category "five or more classes" is smaller than that of *male* patents in the same category.

2.4.1.5 Assignee Organization

The assignee of a patent is the person or organization holding the rights or owning the patent. Aided by data-mining techniques, we distinguish among patents owned by individuals (who may also be the inventors) and by corporate industry, academic, government, and non-profit organizations. In most U.S. organizations, the rights to inventions developed by employees belong to the organization, but the organization may or may not file for a patent on the invention. Hence, where the assignee is an individual person, that person may be an independent inventor or an employee of an organization.

Our analysis (see Table 2.6) of the database of recorded U.S. nanotechnology patents shows that *male* inventor patents, compared to patents *with at least one female* inventor, are slightly more likely to be assigned to industry or to individuals (50.3% vs. 48.8% and 40.4% vs. 38.4%). Compared with *male* inventor patents, those *with at least one female inventor* are more likely to be assigned to government

Table 2.6 Gender and type of assignee, U.S. nanotechnology patents, 2002–2006

	Type of assignee					
	Industry	Government	Academic	Nonprofit	Individual	Total
At least one female inventor						
All female	47	3	13	2	69	134
Female + male	333	14	94	4	432	877
Female + male + unknown	373	23	89	4	432	921
Female + unknown	65	5	15	7	105	197
Sub-total	818	45	211	17	1,038	2,129
Percent of row total	38.4	2.1	9.9	0.8	48.8	100
Male inventor(s)						
All male	1,994	77	347	32	2,480	4,930
Percent of row total	40.4	1.6	7.0	0.6	50.3	100
Not identified						
Male + unknown	1,579	66	461	37	1,830	3,973
Unknown	628	22	136	33	891	1,710
Sub-total	2,207	88	597	70	2,721	5,683
Percent of row total	38.8	1.5	10.5	1.2	47.9	100
Total	5,019	210	1,155	119	6,239	12,742

Source: Analysis of U.S. nanotechnology patents granted 2002 through mid-2006 (see Table 2.1)

(1.6% vs. 2.1%), academia (7.0% vs. 9.9%), or nonprofit organizations (0.6% vs. 0.8%). *Mixed* patents (with male + female inventors) are relatively more prevalent in academic organizations and *female* patents in non-profit organizations among the three types of inventor teams, although industry individual invention account for the two largest shares among all types.

2.4.2 Econometric Results

The descriptive statistics presented in the prior section provide an overview of the bivariate relationships found in our database of U.S. nanotechnology patents between the inventor team gender composition and several other variables of interest. To further advance our analysis, we have undertaken an econometric analysis to probe the relationship between a given variable and team gender composition, holding constant the value of other variables. The model we present is limited because we do not account for potential individual-level and other omitted variables that are unavailable in our current dataset. Hence, our results should be viewed as preliminary, subject to the development of a more fully-specified model that will require additional data collection and analysis.

Our model focuses on three types of inventor team compositions where we are assured about the gender composition. There are *female*, *male*, and *mixed* (female + male). We use multinomial logit regression, which is appropriate where there is a nominal dependent variable comprised of more than two classes. The independent variables include: four dummy time variables—*Year2003, Year2004, Year2005, and Year2006*; a dummy variable *Public sector* (where 1 indicates the assignee is an academic, non-profit, or government organization; 0 indicates a corporate industry or individual assignee); and two continuous variables, *Team size* and *Comprehensiveness*. These variables are regressed on the dependent variable *Gender*, where 0 is assigned to *male* inventor patent, 1 to *female* inventor patent, and 2 to *mixed* inventor patent.

Overall, the model contains 5,941 observations and the pseudo R^2 is 0.1396 (Table 2.7). The coefficients of independent variables in the model show that *Public sector, Team size*, and *Comprehensiveness* are good predictors of the *Gender* team inventor composition of a nanotechnology patent. Holding fixed other variables in the model, the team size of *female* patents is significantly smaller than that of *male* inventor patents while the team size of *mixed* inventor patents is significantly larger than that of *male* inventor patents. This is consistent with the bivariate results discussed previously. *Female* inventor patents are more comprehensive than *male* inventor patents, while *mixed* inventor patents seem more comprehensive than *male* inventor patents but the difference is not statistically significant. Holding other variables constant, both *female* and *mixed* inventor patents, compared with *male* inventor patents, are significantly more likely to come from the public sector versus industry or be individually assigned. To check whether the difference in sample size between male and female inventor patents is an issue, we randomly selected 3% of the male patents (to make the size of the two groups similar). We ran the logit regression, and repeated the procedure several times. The

Table 2.7 Multinomial logit regression of team gender composition (all male inventor team patent as reference group)

	Female inventor team	Mixed gender inventor team
Year 2003	0.302	0.159
	(0.31)	(0.14)
Year 2004	0.183	0.021
	(0.30)	(0.13)
Year 2005	0.279	−0.055
	(0.28)	(0.13)
Year 2006	0.012	0.170
	(0.32)	(0.14)
Public sector	0.625**	0.329***
	(0.26)	(0.12)
Team size	−1.450***	0.648***
	(0.19)	(0.03)
Comprehensiveness	0.183**	0.012
	(0.08)	(0.032)
Constant	−1.908***	−3.545***
	(0.35)	(0.12)
Observations	5941	
Pseudo R-square	0.1396	

Source: Analysis of U.S. nanotechnology patent records, 2002 through mid-2006 (see Table 2.1). Standard errors in parentheses *** $p < 0.01$, ** $p < 0.05$, * $p<0.1$

coefficients yielded are slightly different from those presented in Table 2.7, but with one exception the signs and significance remain the same.[7] The exception is that the gender difference for *Public sector* became insignificant when the model was re-run against 3% of the male inventor patents. While our results must still be interpreted with caution, our overall conclusion from these models is that nanotechnology patents invented by female inventor teams are more comprehensive in terms of patent classes yet are developed in smaller teams compared with male patents. When women inventors are engaged in mixed teams, those inventor teams are larger than those for exclusively male invented patents, but there is no statistically significant effect on comprehensiveness. We might tentatively interpret this result to suggest that women have more influence on the technical breadth of nanotechnology patents when they work with other women, but this suggestion needs further research confirmation before it can be upheld. Since all-female teams are a relatively small proportion of all nanotechnology inventor teams, as yet, women probably only exercise a relatively small influence on the overall direction of nanotechnology patenting.

2.5 Conclusion and Discussion

In this study we investigate female involvement in nanotechnology patenting. Although nanotechnology is a new domain of interdisciplinary science, it builds upon a series of long-established disciplines and fields of technical expertise. Several of these disciplines, such as physics or engineering, have long been male-dominated, while others, such as the life sciences, have seen much expanded roles for women in recent years. We expected that male inventors would continue to be prevalent numerically in the nanotechnology field, i.e., that gender disparity would continue. However, we did propose that nanotechnology's interdisciplinary character would present new opportunities for women to influence the character of technological development in the field.

Our analysis of U.S. nanotechnology patent records (2002 through mid-2006) shows that *female* inventor patents and patents *with at least one female* inventor are much fewer than *male* inventor patents. However, the gender gap may be narrower than the general gap discovered in other discipline-based studies. We observe that the gap between the number of patents *with at least one female* inventor compared with *male* inventor patents diminished over our 5-year study period, although the gap between *female* inventor and *male* inventor patents is unchanged. This finding indicates that when women inventors are engaged in nanotechnology patenting, they are increasingly likely to do so as part of *mixed* male and female invention teams. In other words, we are seeing an increase in collaboration between female and male inventors in nanotechnology, although the modal nanotechnology U.S. patent is still one that has been invented by an all-male inventor team.

Female patents tend to be individual inventive products while *male* patents tend to be products of collaboration. *Mixed* inventor patents have a larger team size, which hints, but does not confirm, that women inventors are complementary to men, providing additional expertise, rather than simply substituting. This is consistent with two other results. First is the finding that patents *with at least one female* inventor and *female* inventor patents are prevalent (when compared with male inventor patents) in technical fields involving biology, chemistry, and a few traditional female fields, so where women are included in mixed teams, this tends to be in technical areas where women are more concentrated. Second, patents *with at least one female* inventor or *female* inventor patents are more comprehensive in terms of the breadth of IPC technology classes covered than *male* inventor patents. Our model confirms that this difference is statistically significant ($p < 0.05$) for *female* inventors although not for *mixed* inventor teams. However, again there is a clue that women inventors may be adding to the interdisciplinary scope of the patent teams they are associated with. The observation that *female* patents tend to come from public or quasi-public organizations (based on assignee classifications) hints not so much that conditions in these organizations may be more favorable to women inventors than in private industry, but that women researchers in public-sector organizations might have greater flexibility to develop patents with other women.

These results should be interpreted with caution. Our results relate to U.S. nanotechnology patents recorded over a recent time period. There is no automatic generalizability to patenting in other fields or countries. Second, although we progressed significantly in the development of an algorithm to ascertain an individual's gender based on first name, we succeeded in identifying genders for only 68% of the inventor pool in our database. We judge that our identification is most incomplete for U.S. patent inventors with Asian (especially Chinese and Korean) names. Third, we should be careful not to ascribe too much to assignee organizational type. As noted by Whittington and Smith-Doerr (2005), organizational characteristics and structure are more influential (than organizational type, per se) in female involvement and performance in patenting. Future analysis would be enriched by using variables related to organizational characteristics, as well as those related to individual and team characteristics.

Yet, taking on board these cautions, our analysis does provide insight into female roles, performance, and influence in nanotechnology patenting. In a domain which promises interdisciplinary and technological convergence, women inventors as yet have a small but increasing role. Moreover, where women do have a role in successful nanotechnology invention, it appears to be in ways that reinforce, if not add to, interdisciplinarity. Whether this influence is caused directly by the specific research and innovation strategies that women pursue or because women are more attracted to already interdisciplinary environments and problems (or both) is a question for further study.

Notes

1. www.ssa.gov/OACT/babynames
2. http://www.indianchild.com/indian_baby_names.htm for Indian first names, and http://www.languageisavirus.com/namedatabase/db.cgi?db=default&uid=&ID=&Letter=—&Name=&Gender=—&Origin=Japanese&Meaning=&view_records=Search&nh=1 for Japanese first names.
3. Further details of our first name/gender algorithm are available on request.
4. This survey of U.S. patent inventors was undertaken at Georgia Tech by John Walsh in June-November 2007, and obtained 1,919 responses (see Walsh and Nagaoka (2009), for further details).
5. For datamining, we use VantagePoint, a text-mining software tool (see http://www.thevantagepoint.com/)
6. http://www.wipo.int/export/sites/www/classifications/ipc/en/guide/guide_ipc_2009.pdf
7. Results are not presented here but will be provided on request.

References

Allison, P.D., and S.J. Long. 1987. Inter-university mobility of academic scientists. *American Sociological Review* 52: 643–652.
Astin, H.S. 1969. *The women doctorate in America*. New York, NY: Russell Sage.
Baird, D., and A. Shew. 2004. Probing the history of scanning tunneling microscopy. In *discovering the nanoscale*, ed. D. Baird, A. Nordmann, and J. Schummer, 145–156. Amsterdam: IOS.
Barinaga, M. 1993. Is there a "female style" in Science? *Science* 260: 384–391.
Bayer, A.E., and J.C. Smart. 1991. Career publication patterns and collaborative "style" in American academic science. *The Journal of Higher Education* 62: 613–636.
Cameron, S.W. 1978. *Women faculty in academia: Sponsorship, informal networks, and scholarly success*. Ann Arbor, MI: University of Michigan Press.
Cameron, S.W., and R.T. Blackburn. 1981. Sponsorship and academic career success. *The Journal of Higher Education* 52: 369–377.
Chubin, D.E. 1974. Sociological manpower and womanpower: Sex differences in career patterns of two cohorts of American doctorate scientists. *American Sociologist* 9: 83–92.

Cole, J.R., and H. Zuckerman. 1984. The productivity puzzle: Persistence and change in patterns of publication of men and women scientists. *Advances in Motivation and Achievement* 2: 217–258.

Corley, E. 2005. How do career strategies, gender, and work environment affect faculty productivity in university-based science centers? *Review of Policy Research* 22: 637–655.

Creamer, E.G. 1998. Assessing faculty publication productivity: Issues of equity. *ASCHE-ERIC Higher Education Report* No.26. Washington, DC: ASHE-ERIC/Georgie Washington University.

Crewson, P.E. 1995. A comparative analysis of public and private sector entrant quality. *American Journal of Political Science* 39: 628–639.

Ding, W.W, F. Murray, and T.E. Stuart. 2006. Gender differences in patenting in the academic life science. *Science* 313: 665–667.

Durden, G., and T. Perri. 1995. Coauthorship and publication efficiency. *Atlantic Economic Journal* 23: 69–76.

Evetts, J. 1996. *Gender and career in science and engineering*. London: Taylor and Francis.

Fox, M.F. 1983. Publication Productivity among Scientists: A Critical Review. *Social Studies of Science* 13: 285–305.

Fox, M.F. 1999. Gender, hierarchy, and science. In *Handbook of the sociology of gender*, ed. J.S. Chafetz, 441–457. New York, NY: Kluwer/Plenum.

Fox, M.F. 2001. Women, science, and academia: Graduate education and careers. *Gender and Society* 15: 654–666.

Fox, M.F., and P.E. Stephan. 2001. Careers of young scientists: Preferences, prospects and realities by gender field. *Social Studies of Science* 31: 109–122.

Frietsch, R, I. Haller, M. Vrohlings, and H. Grupp. 2009. Gender-specific patterns in patenting and publishing. *Research Policy* 38: 590–599.

Gordon, M. 1980. A critical reassessment of inferred relations between multiple authorship, scientific collaboration, the production of papers and their acceptance for publication. *Scientometrics* 2: 193–201

Haier, R.J., R.E. Jung, R.A. Yeo, K. Head, and M.T. Alkire. 2005. The neuroanatomy of general intelligence: Sex matters. *NeuroImage* 25: 320–327.

Hayes, E.R. 2001. A new look at women's learning. *New Directions in Adult and Continuing Education* 89: 35–42.

Hollis, A. 2001. Co-authorship and the output of academic economists. *Labour Economics* 8: 503–530.

Hunter, L., and E. Leahey. 2008. Collaborative research in sociology: Trends and contributing factors. *The American Sociologist* 39: 290–306.

Jaffe, A.B., and M. Trajtenberg. 2002. *Patents, citations, and innovations: A window on the knowledge economy*. Cambridge, MA: MIT.

Kaufman, D.R. 1995. Professional women: How real are the recent gains. In *Women: A feminist perspective*. ed. J. Freeman. Mountain View, CA: Mayfield.

Kay, L., and P. Shapira. 2009. Developing nanotechnology in Latin America. *Journal of Nanoparticle Research* 11: 259–278.

Khan, B.Z. 2000. "Not for ornament": Patenting activity by nineteenth-century women inventors. *Journal of Interdisciplinary History* 16: 159–195.

Kyvik, S., and M. Teigen. 1996. Child care, research collaboration, and gender differences in scientific productivity. *Science, Technology, and Human Values* 21: 54–71.

Leahey, E. 2006. Gender differences in productivity: Research specialization as a missing link. *Gender and Society* 20: 754–780.

Lee, S., and B. Bozeman. 2005. The impact of research collaboration on scientific productivity. *Social Studies of Science* 35: 673–702.

Levin, S., and P.E. Stephan. 1991. Research productivity over the life cycle: Evidence for academic scientists. *American Economic Review* 81: 114–132.

Levin, S., and P.E. Stephan. 1998. Gender differences in the rewards to publishing in academe: Science in the 1970s. *Sex Roles: A Journal of Research* 38: 1049–1064.

Lindsey, L.L. 1997. *Gender role: A sociological perspective*. Upper Saddle River, NJ: Prentice Hall.

Long, S.J. 1992. Measures of sex differences in scientific productivity. *Social Forces* 71: 159–178.
Long, S.J. 2001. *From scarcity to visibility: Gender differences in the careers of doctoral scientists and engineers*. Washington, DC: National Academy Press.
Lux Research. 2007. *The nanotechnology report*, 5th ed. New York, NY: Lux Research.
Murray, F., and L. Graham. 2007. Buying science and selling science: Gender differences in the market for commercial science. *Industrial and Corporate Change* 16: 657–689.
Naldi, F., D. Luzi, A. Valente, and I.V. Parenti. 2004. Scientific and technological performance by gender. In *Handbook of quantitative science and technology research*, ed. H.F. Moed, W. Glanzel, and U. Schmoch, 299–314. Boston & London: Kluwer.
NSF. 2007. *Women, minorities, and persons with disabilities in science and engineering*. Arlington, VA: National Science Foundation.
NSF. 2008. *Thirty-three years of women in S&E faculty positions*. Arlington, VA: National Science Foundation, Directorate for Social, Behavioral, and Economic Sciences.
Pinker, S. 2005. *The science of gender and science: A conversation with Steven Pinker and Elizabeth Spelke*. Cambridge, MA: Havard University Press.
Porter, A.L., and J. Youtie. 2009. Where does nanotechnology belong in the map of science? *Nature-Nanotechnology* 4: 534–536.
Porter, A.L., J. Youtie, P. Shapira, and D.J. Schoeneck. 2008. Refining search terms for nanotechnology. *Journal of Nanoparticle Research* 10: 715–728.
Pravdic, N., and V. Oliuic-Vukovic. 1986. Dual approach to multiple authorship in the study of collaborator/scientific output relationship. *Scientometrics* 10: 259–280.
Presser, S. 1980. Collaboration and the quality of research. *Social Studies of Science* 10: 95–101.
Price, D. J., and D. Beaver. 1966. Collaboration in an invisible college. *American Psychologist* 21: 1011–1018.
Pripic, K. 2002. Gender and productivity differentials in science. *Scientometrics* 55: 27–58.
Reskin, B. 1978a. Scientific productivity, sex, and location in the institution of science. *American Journal of Sociology* 83: 1235–1243.
Reskin, B. 1978b. Sex differentiation and the social organization of science. *Sociological Inquiry* 48: 491–504.
Rhoten, D., and S. Pfirman 2007. Women in interdisciplinary science: Exploring preferences and consequences. *Research Policy* 36: 56–75.
Schmoch, U. 2006. Double-boom cycles and the comeback of science-push and market-pull. *Research Policy* 36: 1000–1015.
Schmookler, J. 1962. Changes in industry and in the state of knowledge as determinants of industrial innovation. In *The rate and direction of inventive activity: Economic and social factors*, ed. National Bureau of Economic Research, 195–232. Princeton, NJ: Princeton University Press.
Science Daily. 2005. Intelligence in men and women is a gray and white matter (January 22).
Scott, J. 1990. Disadvantage of women by the ordinary processes of science: The case of informal collaboration. In *Despite the odds: Essays on Canadian women and science*. ed. M. Ainley. Montreal, QC: Vehicule.
Shapira, P., J. Wang, and J. Youtie. 2010. United States. In *Encyclopedia of nanotechnology and society*, ed. D. Guston, and J.G. Golson. New York, NY: Sage.
Sonnert, G., and G. Holton. 1995a. *Gender differences in science careers*. New Brunswick, NJ: Rutgers University Press.
Sonnert, G., and G. Holton. 1995b. *Who succeeds in science? The gender dimension*. New Brunswick, NJ: New Rutgers University Press.
Sorensen, K.H. 1992. Towards a feminized technology? Gendered values in the construction of technology. *Social Studies of Science* 22: 5–31.
U.S. Department of Commerce. 1999. *Buttons to biotech*. Washington, DC: U.S. Department of Commerce.
Walsh, J,. and S. Nagaoka. 2009. Who invents? Evidence from the Japan-U.S. inventor survey. RIETI Discussion Paper Series 09-E-034. http://ideas.repec.org/p/eti/dpaper/09034.html. (accessed October 2009).
Whittington, K.B., and L. Smith-Doerr. 2005. Gender and commercial science: Women's patenting in the life sciences. *Journal of Technology Transfer* 30: 355–370.

Wood, S., R. Jones, and A. Geltard. 2003. *The social and economic challenges of nanotechnology*. Swindon: Economic and Social Research Council.
Xie, Y., and K.A. Shauman. 1998. Sex differences in research productivity: New evidence about an old puzzle. *American Sociological Review* 63: 847–870.
Youtie, J., P. Shapira, and A.L. Porter. 2008. Nanotechnology publications and citations by leading countries and blocs. *Journal of Nanoparticle Research* 10: 981–986.
Zucker, L.G., and M.R. Darby. 2005. *Social-economic impact of nanoscale science: Initial results and nanobank*. Cambridge, MA: National Bureau of Economic Research.
Zuckerman, H. 1967. Nobel laureates in science: Patterns of productivity, collaboration, and authorship. *American Sociological Review* 32: 391–403.
Zuckerman, H., and R.K. Merton. 1971. Patterns of evaluation in science: Institutionalization, structure and functions of the referee system. *Minerva* 9: 66–100.

Chapter 3
Potential Implications for Equity in the Nanotechnology Workforce in the U.S.

Sonia Gatchair

Based on her previous research on employment in high-technology industries, Sonia Gatchair projects that the nanotechnology research and development workforce will be largely male and disproportionately of European or Asian descent. Since the over-representation of white and Asian men is particularly concentrated in high technology industries and remains statistically significant even after controlling for education, Gatchair's analysis suggests that just educating more people of African or Latin descent will not eliminate the problem. Echoing Smith-Doerr's observations, she notes that the workplace itself needs to change; institutions must identify and eliminate processes of differential inclusion and exclusion. This step may be particularly hard to take in nanotechnology, where many jobs are located in traditional industries with established organizational cultures. Like Smith-Doerr, Gatchair calls for research to track differential patterns, so that action can be focused effectively.—eds.

S. Gatchair (✉)
School of Public Policy, Georgia Tech University, Atlanta, GA, USA
e-mail: sgatchair@gatech.edu

This chapter was peer reviewed. It was originally presented at the Workshop on Nanotechnology, Equity, and Equality at Arizona State University on November 21, 2008.

3.1 Introduction

The growth of nanotechnology industries, which represent the application of new, cross-cutting technologies and innovations, may have implications for equity in the workforce and labor markets of areas where growth is strongest. As a result of large sums of money being spent on research and development (R&D) in nanotechnology (NSTC and OSTP 2007), and the possibility that local and regional policymakers will promote clustering or agglomeration around nanotechnologies and complementary technologies (Shapira and Youtie 2008), the potential for growth and creation of new jobs will be enhanced. Well-known examples of areas that have prospered as a result of agglomeration strategies include Silicon Valley in California in electronics and information technologies and the Research Triangle in North Carolina in biotechnologies. The strategies, which for example promote entrepreneurial activities, create start-up firms, and support small and medium enterprises, will result in a mix of jobs, if successful. A small number of the jobs such as those in research and development, process improvements, or technical extension services, will be highly specialized and have higher than average wages. However, the majority will represent the range of skills that characterize the labor market and include production, clerical, sales, marketing, and other service jobs, among others.

As with many other high technology industries, nanotechnology industries are likely to have high levels of R&D relative to sales, and to employ higher than average numbers of scientists and engineers for research and commercialization activities. However, racial and ethnic disparities exist among students taking science and engineering (S&E) subjects in school and in the S&E workforce (NSF 2007). In keeping with observations on the characteristics of employment in high technology industries from a previous study (Gatchair 2007), it is likely that the distribution of employment opportunities resulting from the growth of nanotechnology jobs will not reflect the racial and ethnic diversity of the U.S.

Extrapolating from the study on high technology industries, this paper discusses potential implications for equity in the nanotechnology workforce resulting from differences in the effects of education and other factors on employment and wages among four racial and ethnic groups in the U.S. (Gatchair 2007). In that study, Gatchair compared employment and wage differences among four racial and ethnic groups in S&E occupations with those in other jobs in high technology industries, and with S&E workers outside of the sector. The study is described in greater detail in subsequent sections, and the complete study can be found in *Representation and Reward in High Technology Industries and Occupations: The Influence of Race and Ethnicity* (Gatchair 2007). It is anticipated that the disparities observed in S&E occupations and in other jobs in high technology industries will persist or be even more pronounced in the nanotechnology workforce and, in general, the labor markets where the industries are concentrated.

Although individuals have different views on the usefulness and desirability of equity, some believe that workforce equity, in which the representation of different groups in the workplace is similar to the representation in the population as a whole and rewards are commensurate with qualifications, is important in considerations of fairness and justice. The groups may be defined by religion, gender, race, ethnicity, or other criteria. In addition, equitable access of all groups to well paying S&E employment opportunities has economic implications. Inadequate participation of different racial and ethnic groups will limit the pool from which talented individuals can be drawn (BEST 2004; Pearson and Fechter 1994). Increased diversity and cultural perspectives have the potential to contribute to greater creativity and the development of new knowledge, which can increase productivity and competiveness. Thus moral, social, and economic considerations make it important to determine the extent to which imbalances exist in the racial and ethnic diversity in the workplace, and to take steps to address these imbalances, if they exist.

This chapter discusses potential implications for equity in employment and wages among different racial and ethnic groups in the nanotechnology workforce. The chapter is organized as follows: The first section outlines some of the drivers behind nanotechnology industry growth, and the implications for the demand of skill in the workforce. The second section focuses on issues in the supply of skills, and briefly examines the training of scientists and engineers in the U.S. in general. It discusses racial and ethnic disparities in S&E training and occupations, then more specifically issues in the training of nano-scientists and engineers. The third section outlines the possible composition of the nanotechnology workforce and potential equity issues. This paper argues that the nanotechnology workforce will be dominated by white and Asian males with even less diversity being observed among individuals in S&E occupations. However, although black and Hispanic males are underrepresented in S&E jobs in nanotechnology industries, compared to their representation in the population, those with graduate education are likely to have greater parity in wages with whites and Asians, when compared to differences that exist outside of these industries. Finally, the chapter highlights implications for policy, which could help to provide a more equitable distribution of the rewards from the growth of nanotechnology industries.

3.2 Nanotechnology Industry and Workforce Growth

Nanoscience and nanotechnology represent a range of techniques and approaches based on knowledge from different scientific disciplines including physics, chemistry, and engineering, among others. They involve the manipulation of materials with dimensions in the range of 1–100 nm (NSTC and OSTP 2009). Nanoscale materials possess useful properties that are different from the same materials of larger particle size or in aggregated form. The lure of remarkable properties such as greater strength, electrical conductivity, or enhanced optical characteristics

(Helmus 2006) impelled a flurry of investment in related R&D activities in the U.S. and other countries. In 2008, U.S. federal agencies invested over US$1.55 billion to support nano-related R&D activities (NSTC and OSTP 2009). The amount was more than three times the 2001 funding and 9% more than was spent in 2007. From 2001 to 2008, federal government agencies invested more than US$10 billion in nanotechnology. Hundreds of products containing nano-materials are currently on the market and several estimates suggest that the anticipated market for nano-enabled products will be over US$1,000 billion by the next decade (Helmus 2006; Hullmann 2007). Given the sustained level of public and private sector investments, it is anticipated that nanotechnology industries and the related labor force are poised for substantial growth.

Local or regional policy initiatives that agglomerate or cluster nanotechnology businesses can stimulate growth and have the potential to impact the size and characteristics of the labor market (Saxenian 2002). Areas impacted by the policies and related labor markets are likely to have a higher than average demand for highly skilled individuals, who will drive the creation of new knowledge and products. Although clusters might be centered on universities, which provide skilled workers, if there is considerable industry growth, in-migration of workers may occur to satisfy industry needs. In-migration can lead to shifts in the demographic characteristics of the area where the industries are located (Saxenian 2002). Clustering also facilitates networks of relationships among individuals, firms, universities, and other organizations. As a result, individuals with connections to formal or informal networks are likely to have better access to the mix of high and low skilled jobs available (Granovetter 1983; Saxenian 2002), and individuals from a particular group may become concentrated in the areas where the policies are implemented.

It is difficult to obtain estimates of the number of workers currently employed in nanotechnology because nanotechnology does not represent a single industry; instead, applications are cross-cutting and can be found in numerous industries or sectors. Thus nanotechnologies exhibit some of the characteristics of general purpose technologies (Youtie et al. 2008). General purpose technologies, for example information technologies and biotechnologies, underlie many different industrial activities and spawn a large number of innovative products and processes leading to rapid growth. Since nanotechnology has some of the characteristics of general purpose technologies, many more persons might work with the technology than is apparent from observations of industries directly involved in nano-material research or production. In addition, since the technology is still emerging, it is difficult to estimate the potential demand for skilled workers in the industry and the literature reflects widely varying estimates of workforce needs (Van Horn et al. 2009). One of the most widely quoted is the National Science Foundation (NSF) estimate that more than two million nanotechnology workers will be needed globally by 2015, and approximately half this number will be in the U.S. (National Nanotechnology Initiative 2009).[1]

Although estimates for nanotechnology employment represent a relatively small proportion of the U.S. workforce (considerably less than 1%), the task of meeting the human capital needs of the industries faces unique challenges. Individuals with interdisciplinary S&E skills will be needed to undertake the rapid advancements, and overcome technical challenges associated with R&D, commercialization, and early production activities. Since the products are at the early stage of the product life cycle, production workers also are likely to require greater technical skills when compared to those involved with more mature products in which the processes have become routine. In addition, technicians, sales, marketing, legal, and other workers will be at an advantage if they have skills and knowledge related to nanotechnology. As the applications of nanotechnology become more pervasive an increasing number of individuals, requiring varying levels of knowledge and skill about nanotechnology, will be necessary.

3.3 Issues in Science and Engineering Education in the U.S. and Implications for the Nanotechnology Workforce

3.3.1 Supply and Demand of S&E Workers

The U.S. faces several challenges in its efforts to educate individuals in S&E fields and to develop the S&E workforce, many of which will extend to the nanotechnology workforce. Although the NSF reports that graduate enrollment in S&E fields, with the exception of computer fields, increased between 1999 and 2005, it is uncertain whether the current numbers will meet the demands for increased productivity and competiveness in the U.S. (National Science Board 2008). Approximately 25% of college educated individuals in S&E occupations in the U.S. were foreign-born in 2003 (National Science Board 2008). In 2005, approximately 25% of the graduate students in S&E in U.S. schools of higher education are foreign students on temporary visas (National Science Board 2008). The competition for foreign-born scientists and engineers, many of whom now find increasingly favorable conditions in their home countries (Wadhwa 2009), coupled with the uncertainty in the demand for S&E workers could lead to inadequate numbers of individuals trained in S&E in the U.S. As a consequence, the workforce needs of new nanotechnology industries may not be met.

3.3.2 Racial and Ethnic Disparities in S&E Education and Employment

In addition to the challenges associated with the supply and demand for S&E workers, the U.S. faces issues of inequity in the training of scientists and engineers and in the composition of these occupations. Racial and ethnic disparities exist among S&E graduates and in S&E occupations, despite many programs to recruit and retain minorities in S&E fields. Only about one third of the bachelors' degrees earned by whites, blacks, and Hispanics are in S&E fields, while approximately half of the bachelors' degrees earned by Asians are in S&E fields. White and Asian males are more likely to receive doctoral degrees in science and engineering compared to blacks and Hispanics, and to females (National Science Board 2008). As a result, Asians are disproportionately represented as graduates in S&E fields compared to their representation in the population. Blacks and Hispanics are under-represented compared to their representation in the population, and compared to whites and Asians. The increasing representation of Asian males in the U.S. has reduced the traditional dominance of white males in S&E jobs; however S&E jobs still do not reflect the diversity of nation's population (National Science Board 2008). Scholars argue that the nation is not making the best use of the potential pool of talent in terms of both quality and quantity for S&E (Pearson and Fechter 1994; BEST 2004). The lack of diversity in the S&E workforce is not just a problem in the current make up of those holding the jobs, but given the low representation of blacks and Hispanics in the S&E education pipeline, it is likely that very few members of these groups will find employment in nanotechnology sectors. Employment in the industries will not reflect the diversity of the nation's people.

3.3.3 Nanotechnology Education and Training

In addition to difficulties associated with producing adequate numbers scientists and engineers, and ensuring diversity in the pool of skilled S&E workers, training the nanotechnology workforce is beset with unique problems. According to Fonash (2001), two major challenges constrain the creation and continued growth of a nanotechnology workforce. These are first, to attract sufficient numbers of students to nanotechnology related fields of study, and second, to design suitable curricula to train both scientific and technical personnel for industry. Students, teachers, and the general public are less aware of careers in nanotechnology compared to more visible fields such as computer science and information technology (Fonash 2001). Consequently, the number of students who recognize nanotechnology as a potential career option is likely to be very small. Further, since nanotechnology is based on the intersection of several scientific disciplines or fields, students who can study in different scientific fields will be at an advantage. Therefore training nanotechnology scientists and engineers is even more challenging because fewer students opt for science, technology, engineering, and mathematics (STEM) fields of study as a whole. Since it is important to ensure that the pool of potential talent available for the nanotechnology workforce remains as large as possible, students should not be excluded or eliminated pre-maturely from STEM fields of study because of spurious reasons such as race. If more people from a wider variety of backgrounds see nanotechnology research or application as a viable career option, the number of people available in the pool will likely increase.

Kim (2007) argues that the skills needed for nanotechnology result from the multi-disciplinary integration of knowledge from different areas. Since current approaches to training are pursued primarily along individual disciplinary lines, scientists usually obtain these skills sequentially at different stages of their academic preparation. According to Stephan et al., who focused on a somewhat narrow definition of nanotechnology related training and jobs,[2] students are trained in individual laboratories rather than in specific programs established by universities (Stephan et al. 2007). The additional courses needed to satisfy the multi-disciplinarity of the field results in a demanding academic scenario for students. It means that students will need extended periods to facilitate the acquisition of skills, and nanotechnology training will require even greater resources, which include supervisory expertise, laboratory facilities, and financial support, when compared to training in single disciplines.

The integrated, multi-disciplinary perspective towards training in nanotechnology (Fonash 2001; Kim 2007) can potentially place black and Hispanic minorities at a disadvantage. Blacks and Hispanics already face several constraints, which include poorly equipped schools, low quality of science teachers, discouragement from entering STEM fields of study, as well as disadvantages in socio-economic conditions (Clark 1999; Tang 2000) that contribute to under-representation in S&E fields and occupations. Thus, Blacks and Hispanics are likely to face even greater difficulties pursuing extended training in multi-disciplinary S&E fields. Thus, it is likely that the nanotechnology workforce will reflect even less diversity than other scientific fields.

The small number of blacks and Hispanics in S&E fields of study will not only affect entry into the nanotechnology workforce, but also has the potential to impact the success of those who are employed in the industries, since collaborations of individuals within and across disciplines facilitate endeavors in science and innovation. In a study of the creative potential of scientists working in nanoscience and technology, Heinze and Bauer (2007) found that creativity was greater among scientists who had a larger number of contacts or interactions with groups of scientists from widely diverse disciplines. In addition, networks increase effective communications with colleagues and contribute to knowledge flows, which help to increase productivity, competitiveness, and learning (Cooke 1996, 2007). They also help to establish contacts, which help individuals get jobs (Granovetter 1983). Since the number of black and Hispanic scientists are considerably fewer than whites and Asians, blacks and Hispanics are likely to participate in networks with fewer connections that can facilitate the flow of knowledge as well as getting jobs (Pearson 1985, 2005). Consequently black and Hispanic minorities will have fewer opportunities for employment in nanotechnology industries and will be under-represented in both S&E and non-S&E jobs, such as sales, marketing, and production, among others, in the nanotechnology sector.

3.4 Technology, Technological Changes and Wage Inequality in the U.S.

In addition to contributing to disparities in employment, nanotechnology industry growth potentially can contribute to the rising wage inequalities, which have been observed in the U.S. workforce as a whole since the late 1970s. Wage disparities exist within (vertical inequality) and across (horizontal inequality) groups defined in different ways such as by educational levels, or by race and ethnicity. Studies attribute the disparities to a complex interplay of factors, which include changes in the relative demand and supply of both skilled and unskilled labor, institutional structures, and government policies, among others (Bartel and Sicherman 1999; Card and DiNardo 2002; Galbraith 1998; Goldin and Katz 2008).

Technology, technological changes, and the growth of technology industries have been implicated in wage disparities in a number of different ways. Technological changes, which increase the demand for skills (typically referred to as "skill biased technological change") and rising wage differences between the college educated and those who do not attend college (increasing returns to education) are considered to be among the chief causes of the growing wage inequality (Goldin and Katz 2008; Juhn et al. 1993). In studies which focus specifically on the effects of technological changes, capital intensity, and skill differences on wages and wage inequalities, Aghion et al. argue that technologies characterized as general purpose technologies have an even greater propensity to give rise to increased wage inequality (Aghion et al. 2002). They argue that the effects of technology are self-reinforcing and drive increased demand for workers with greater skill or adaptability; skilled workers in turn produce technologies that require greater skill (Aghion et al. 2002; Bartel and Sicherman 1999). In addition, the growth of technology-based industries, which gain greater profits as a result of monopolistic or oligopolistic activities and pass these on to employees in the form of higher wages, can contribute to greater wage disparities between workers inside and outside of these industries (Galbraith 1998). Nanotechnologies, with their fast growth, pervasiveness, and cross-cutting characteristics, already show features typical of general purpose technologies, which suggest that the industries are likely to exacerbate current wage inequalities.

3.5 Potential Labor Market Effects

The number of individuals in the nanotechnology workforce and the racial and ethnic makeup at the present time and in the long term are not known. However, extrapolations from data on high technology industries can provide a good idea of the potential composition. This paper extrapolates from a study that systematically compared the effects of education on employment and wage differences among different racial and ethnic groups in high technology industries, and more specifically in S&E occupations within and outside of these industries (Gatchair 2007). The groups of jobs compared in the study include: (1) S&E occupations in high technology industries (referred to subsequently as high tech S&E jobs); (2) high technology industries jobs, which are not S&E (other high technology jobs); (3) S&E jobs outside of the high technology sector—those mainly in academia, utility companies, government—(other S&E jobs); (4) all other jobs (non-technology jobs). The groups studied (non-Hispanic whites, non-Hispanic blacks, Asians, and Hispanics) correspond to the categories that the Federal government uses for data collection and analyses of racial and ethnic differences in the U.S. The primary data used in the study came from the March Annual Demographic Survey of the Current Population Survey for the period 1992 to 2002 (Bureau of Labor Statistics 1992–2002).

Since the study faced a number of limitations because the dataset used did not contain sufficient information to ensure that all statistical assumptions were met, for example information needed to overcome endogeneity issues,[3] several methods were used in the analyses. These include multinomial regression analyses of different models of the likelihood of employment in different industry-occupation groups and ordinary least squares, t-tests, and non-parametric graphical comparisons for the analyses of wage differences. The different regression techniques pointed to the bias in estimations of the magnitude of coefficients on the variables but supported the overall conclusions on the direction and significance of the effects of the variables. As a result, the ensuing discussions focus on inferences drawn from the direction and significance of relevant variables, rather than on the magnitude of the differences.

Another limitation was that the number of women in S&E fields among the different racial and ethnic groups and other groups in the sample were insufficient to provide reliable regression estimations for women. As a result, the extrapolations from this study focus on male workers. The study was also limited by the absence of direct information on the fields of study of individuals at different educational levels, with the result that it was not known whether individuals studied in S&E fields of study or not. As an approximation, the study focuses on the occupations of individuals and assumes that for individuals with graduate education, the correspondence between the individuals' occupation and field of study will be more closely related.[4]

3.5.1 Employment in High Technology Industries

Employment patterns in high technology industries can provide insights on potential characteristics of the nanotechnology workforce such as the size, growth rate, and composition. Depending on the definition used, high technology industry employment represents less than 12% of employment outside of the agricultural sector in the U.S. (Hecker 2005). Table 3.1, which shows information on the distribution of employment among the industry/occupational groups, as well as other characteristics of the sample used in the study, also supports this observation (Gatchair 2007). However, employment growth rates for sub-sectors vary considerably, with some sectors, such as aerospace manufacturing, showing a decline in employment levels; while others, such as the information and communications technologies services sector, are growing faster than the national average (Hecker 2005). Given considerable expenditures on R&D, and efforts to stimulate industry growth, the nanotechnology sector may be among those in which employment growth is above the national average. However, in keeping with observations for high technology industries in general, it is anticipated that relatively few jobs will be created, and the

Table 3.1 Comparison of means and standard deviations of characteristics of male and female full, part-time and non-workers

Variable	Female		Male	
	Mean	S.D.	Mean	S.D.
Race/ethnicity				
White	0.725	0.447	0.737	0.440
Black	0.131	0.338	0.114	0.318
Asian	0.038	0.191	0.037	0.188
Latino	0.106	0.307	0.113	0.316
Education				
Without high school	0.180	0.384	0.198	0.399
High school	0.528	0.499	0.502	0.500
College	0.292	0.454	0.299	0.458
Industry/occupation				
High technology employment	0.044	0.206	0.097	0.296
Science & engineering occupations	0.012	0.108	0.043	0.202
High technology / S & E	0.005	0.072	0.024	0.152
Non-high technology / S & E	0.007	0.081	0.019	0.137
High technology / non-S & E	0.039	0.193	0.073	0.261
Non-high technology / non-S & E	0.949	0.220	0.884	0.320
Other characteristics				
Age	38.356	13.319	37.913	13.247
Marital status	0.733	0.442	0.664	0.472
Children present	0.474	0.499	0.413	0.492
Home ownership	0.677	0.468	0.685	0.465
Foreign born	0.130	0.337	0.137	0.344
Self employed	0.052	0.223	0.103	0.303
Full time worker	0.374	0.484	0.533	0.499
Union member or coverage	0.021	0.142	0.030	0.170
Average income $	14,591	20,486	28,069	36,351
Region/locality				
New England	0.051	0.220	0.052	0.221
Mid east	0.172	0.377	0.167	0.373
Great lakes	0.164	0.371	0.164	0.370
Plains	0.068	0.252	0.069	0.254
South east	0.244	0.429	0.239	0.426
South west	0.105	0.307	0.106	0.307
Rocky mountains	0.031	0.172	0.032	0.177
Far west	0.165	0.371	0.172	0.377
Central city	0.249	0.433	0.245	0.430
Other urban area	0.414	0.493	0.418	0.493
Rural	0.190	0.392	0.191	0.393
Annual unemployment rate	5.432	1.487	5.433	1.496
Proportion high technology firms (1996)	0.050	0.012	0.051	0.012
Proportion high technology employment (1996)	0.059	0.023	0.059	0.024
N	5,21,917		4,88,707	

number of new jobs created will not be as great as those associated with older manufacturing industries. Skilled or better educated workers will be in a better position to take advantage of these jobs even for non-S&E jobs. In addition, similar to other high technology industries, nanotechnology industries are likely to favor younger workers, who have not yet had children (Gatchair 2007).

3.5.2 Effects of Education on Employment

Gatchair (2007) found that education had the greatest effect on determining employment in S&E jobs in high technology industries as well as those outside of high technology industries, when compared to other variables in the study (Table 3.2). Other variables in the study included the number of years of potential experience, race, marital status, region of residence, urban status, and proportion of high technology businesses, among others. In keeping with this observation, it is anticipated that education will be the most important factor in determining employment in nanotechnology industries. When the effects of education are examined in greater detail (see: Table 3.2, Model 2), the results show that graduate education increased the odds of employment in S&E occupations even more than the effects of bachelors' degrees (Gatchair 2007). Thus it is expected that individuals with the highest levels of skills (masters or doctoral degrees) will have the advantage and gain employment. Anecdotal evidence, which suggests that low skilled jobs in the high technology sector are either part-time are outsourced, supports the view that low skilled individuals, for example those without high school diplomas, will find only limited employment in these industries. Thus the nanotechnology workforce will comprise highly educated individuals, with the proportion of scientists and engineers being greater than the average for all industries in general.

Surprisingly, the effects of education on employment were greater in the high technology S&E jobs compared to other S&E jobs, which are mainly in academia, government, or the utility companies (Gatchair 2007). The larger effects of education on employment in S&E jobs in the high technology industries may reflect a demand for higher levels of skills, which can drive creativity, productivity, and competitiveness in industry; and that the industries favor employment of more highly educated graduates. Industry appears to have the advantage in the competition with academia and the utilities for the scarce S&E graduates. Similar pressures on the demand for S&E skills are expected to prevail in the case of nanotechnology, which further supports the view that the skill levels of the nanotechnology workforce, will, on average, be above that of other industries.

In keeping with current trends, which show a steady increase in employment in S&E occupations since the 1980s (National Science Board 2008), nanotechnology industry growth likely will contribute to the growing proportion of scientists and engineers in the workforce. Thus the findings on the importance of education reinforce the current imperative, which emphasizes efforts to attract and retain students in S&E fields of studies. Although additional skills acquired through experience and on-the job training will also be important, skills as measured through years of educational attainment will be an important determinant of employment.

Table 3.2 Comparison of odds ratios of employment from multinomial logit models with all variables included and college education as single variable and separated into bachelors and post graduate degrees for male full, part-time and non-workers

Variables	Model 1			Model 2		
	High technology/ science and engineering	High technology/ non-science and engineering	Non-high technology/ science and engineering	High technology/ science and engineering	High technology/ non-science and engineering	Non-high technology/ science and engineering
High school	7.8355*	1.6880*	6.2629*	7.8767*	1.6855*	6.2676*
	(1.6516)	(0.0594)	(1.1543)	(1.6604)	(0.0593)	(1.1553)
College	55.5015*	2.1436*	33.3930*			
	(11.6691)	(0.0788)	(6.1203)			
Bachelors degree				54.0375*	2.2203*	33.4999*
				(11.3695)	(0.0832)	(6.1456)
Post graduate degree				61.3064*	1.9252*	33.3291*
				(13.0127)	(0.0836)	(6.2151)
Experience	0.9884**	1.0467*	1.0110**	0.9893**	1.0469*	1.0118**
	(0.0048)	(0.0027)	(0.0051)	(0.0048)	(0.0027)	(0.0051)
Experience (squared)	0.9997**	0.9990*	0.9995*	0.9997**	0.9990*	0.9995*
	(0.0001)	(0.0001)	(0.0001)	(0.0001)	(0.0001)	(0.0001)
Black	0.1252***	0.7297***	0.3145**	0.1245***	0.7302***	0.3134**
	(0.1009)	(0.0723)	(0.1599)	(0.1004)	(0.0723)	(0.1593)
Latino	0.0460*	0.7150*	0.1520*	0.0459*	0.7151*	0.1514*
	(0.0344)	(0.0398)	(0.0650)	(0.0344)	(0.0398)	(0.0647)
Asian	0.5806***	0.9792	0.8043	0.5917**	0.9770	0.8124
	(0.1223)	(0.0703)	(0.1500)	(0.1245)	(0.0702)	(0.1514)
Black × high school	3.2983	1.2604**	1.4095	3.3000	1.2604**	1.4105
	(2.6972)	(0.1312)	(0.7368)	(2.6986)	(0.1312)	(0.7373)
Black × College	4.7179	1.0373	2.4901			
	(3.8049)	(0.1197)	(1.2801)			
Black × Bachelors				4.6610	1.0261	2.3917

Table 3.2 (continued)

	Model 1			Model 2		
Variables	High technology/ science and engineering	High technology/ non-science and engineering	Non-high technology/ science and engineering	High technology/ science and engineering	High technology/ non-science and engineering	Non-high technology/ science and engineering
Black × Post graduate				(3.7596) 5.2154**	(0.1209) 1.0261	(1.2337) 2.9430**
Latino × High School	6.6924**	1.0482	3.0143**	(4.3066) 6.6825**	(0.1903) 1.0487	(1.5737) 3.0178**
Latino × College	(5.0736) 13.0549*	(0.0666) 1.1354	(1.3288) 4.2624*	(5.0660)	(0.0666)	(1.3304)
Latino × Bachelors	(9.8211)	(0.0852)	(1.8503)	13.1685* (9.9158)	1.0952 (0.0870)	4.3064* (1.8746)
Latino × Post graduate				13.1622*	1.2569	4.1033***
Asian × College	2.3288* (0.4871)	1.0681 (0.0913)	1.8186*** (0.3460)	(10.0806)	(0.1552)	(1.9012)
Asian × Bachelors				1.7269** (0.3700)	1.0739 (0.0987)	1.3250 (0.2622)
Asian × Post graduate				3.4904*	1.0761	2.9471*
Married	1.3489* (0.0544)	1.3566* (0.0328)	1.2139* (0.0527)	(0.7551) 1.3052* (0.0528)	(0.1311) 1.3638* (0.0330)	(0.5923) 1.1946* (0.0519)
Own child in household	0.9042*	0.9050*	0.9287**	.9063***	0.9038*	0.9296**

Table 3.2 (continued)

	Model 1			Model 2		
Variables	High technology/ science and engineering	High technology/ non-science and engineering	Non-high technology/ science and engineering	High technology/ science and engineering	High technology/ non-science and engineering	Non-high technology/ science and engineering
Buying/own House	(0.0272) 1.1796* (0.0387)	(0.0159) 1.2706* (0.0238)	(0.0312) 1.2246* (0.0420)	(0.0273) 1.1782* (0.0388)	(0.0159) 1.2739* (0.0239)	(0.0313) 1.2282* (0.0423)
Full-time worker	2.8769* (0.0956)	2.6537* (0.0487)	2.8091* (0.0972)	2.8711* (0.0953)	2.6513* (0.0487)	2.8069* (0.0971)
Self-employed	0.4732* (0.0228)	0.4334* (0.0131)	0.1085* (0.0103)	0.4655* (0.0226)	0.4355* (0.0132)	0.1078* (0.0102)
Member/covered by union	0.1789*	0.8136*	0.5187*	0.1793*	0.8131*	0.5190*
	(0.0241)	(0.0329)	(0.0482)	(0.0241)	(0.0328)	(0.0482)
Year 1994–1996	0.5283* (0.0252)	0.5730* (0.0147)	0.5038* (0.0245)	0.5282* (0.0252)	0.5731* (0.0147)	0.5029* (0.0245)
Year 1997–1999	0.6407* (0.0358)	0.6565* (0.0208)	0.5593* (0.0327)	0.6380* (0.0358)	0.6565* (0.0208)	0.5562* (0.0326)
Year 2000–2002	0.6438* (0.0345)	0.6005* (0.0179)	0.5289* (0.0300)	0.6406* (0.0344)	0.6008* (0.0179)	0.5260* (0.0299)
Live in central city	1.5787* (0.0684)	1.1663* (0.0277)	1.2176* (0.0527)	1.5643* (0.0679)	1.1707* (0.0279)	1.2183* (0.0526)
Live in other urban area	2.0467* (0.0726)	1.3675* (0.0261)	1.3251* (0.0479)	2.0330* (0.0721)	1.3703* (0.0262)	1.3232* (0.0478)
New England	1.6782* (0.0907)	1.5418* (0.0518)	0.9709 (0.0586)	1.6629* (0.0901)	1.5468* (0.0520)	0.9677 (0.0583)
Mid-eastern region	1.0233 (0.0493)	1.0095 (0.0282)	1.0458 (0.0513)	1.0182 (0.0492)	1.0115 (0.0283)	1.0435 (0.0512)

Table 3.2 (continued)

Variables	Model 1			Model 2		
	High technology/ science and engineering	High technology/ non-science and engineering	Non-high technology/ science and engineering	High technology/ science and engineering	High technology/ non-science and engineering	Non-high technology/ science and engineering
Great lakes	1.2025*	1.7055*	0.9638	1.2041*	1.7052*	0.9631
	(0.0573)	(0.0452)	(0.0485)	(0.0576)	(0.0451)	(0.0485)
Plains	0.9142	1.1851*	1.0218	0.9133	1.1843*	1.0173
	(0.0626)	(0.0456)	(0.0651)	(0.0626)	(0.0455)	(0.0648)
South west	1.4091*	1.2876*	1.1867***	1.4158*	1.2864*	1.1888***
	(0.0782)	(0.0407)	(0.0697)	(0.0785)	(0.0406)	(0.0698)
Rocky mountains	1.3254*	1.0490	1.1144	1.3420*	1.0480	1.1209
	(0.0895)	(0.0442)	(0.0754)	(0.0907)	(0.0442)	(0.0759)
Far west	1.4490*	1.1754*	1.0206	1.4721*	1.1743*	1.0363
	(0.0761)	(0.0369)	(0.0577)	(0.0773)	(0.0369)	(0.0584)
Unemployment rate	0.9872	1.0267*	0.9700**	0.9858	1.0269*	0.9688**
	(0.0135)	(0.0081)	(0.0147)	(0.0135)	(0.0081)	(0.0147)
Proportion S & E degrees	15.0904*	3.4573*	8.7378*	12.7306*	3.5152*	7.8579*
	(4.6878)	(0.6235)	(2.7931)	(3.9762)	(0.6347)	(2.5182)
Pseudo R-square	0.1172			0.1179		
chi2	21000			22000		
p	0.0000			0.0000		

Note: Numbers in parentheses are standard errors
Significance Levels *p<0.000; ** p<0.05; *** p<0.01

3.5.3 Effects of Education and Race on Employment

The patterns in the effects of education on employment suggest that other factors that have implications for equity are also important in determining employment. In Gatchair (2007), the odds of employment of minorities in S&E jobs increased significantly more than the odds of similar whites when the level of educational attainment changed from bachelors to graduate degrees. Hispanics gained the most, followed by Asians, then blacks. While employment prospects for minorities improve greatly with additional education for all jobs, the increases were greatest in high technology S&E jobs. Despite larger gains however, compared to similar whites, college educated blacks and Hispanics still had significantly lower odds of employment in high technology S&E jobs relative to non-technology jobs (that is, non-S&E jobs outside of high technology industries), and to S&E occupations outside of high technology industries (Gatchair 2007). For blacks and whites with graduate education, there was no significant difference between the odds of employment in other S&E jobs (those outside of the high technology industries) relative to other non-technology jobs. Unlike blacks however, Hispanics had significantly lower odds of employment in both types of S&E jobs relative to non-technology jobs compared to similar whites (Gatchair 2007).

Since Hispanics show lower odds of employment and Asians show higher odds of employment in both types of S&E jobs relative to non-technology jobs compared to whites (Gatchair 2007), it is possible that these results reflect differences in the composition of the educational pipeline. That is, the results are indicative of the lower proportion of Hispanic graduates in S&E education, and the higher proportion of Asian students. However, the mixed results obtained for blacks suggest that it is not just the racial and ethnic composition of the S&E educational pipeline that determines employment.

These findings indicate that it is possible that blacks and Hispanics may have difficulty getting jobs as scientists and engineers in nanotechnology industries despite having graduate level training in the sciences (Gatchair 2007). Under-representation of blacks and Hispanics in S&E fields of study and occupations (National Science Board 2008), as well as greater difficulties getting S&E jobs in industry, will result in predominantly white and Asian males being employed in nanotechnology industry S&E jobs. In keeping with observations on employment in high technology industries (Gatchair 2007), it is anticipated that Asians will have disproportionately higher representation compared to their representation in the population. As a result, the racial and ethnic make-up of scientists and engineers in the nanotechnology workforce will not reflect the diversity in the U.S. population.

For individuals with graduate education, no significant differences were found among the odds of employment of similar blacks, Hispanics, Asians and whites in other high technology jobs (non-S&E) relative to non-technology jobs (Gatchair 2007). The absence of significant differences in odds of employment of the four racial and ethnic groups does not mean that no differences exist, but that if any differences exist, they are the same in two groups of jobs compared. Asians with bachelors' degrees have significantly higher odds of being employed in other high technology industry jobs compared to any other racial group. Several explanations could contribute to this observation. It may reflect the advantage of having large numbers in the S&E education pipeline, since technology based industries hire individuals with S&E backgrounds or skills for non-S&E positions such as sales and production (RESULTAR 2008). Alternatively, it may represent stronger networks among Asians that provide information on available jobs. Or it may just reflect the belief of those who control the employment process that Asians are better suited to work in technology industries. Thus it is possible that there will be disproportionate representation of Asians in non-S&E jobs in nanotechnology industries, which further supports the view that the nanotechnology industry workforce will not reflect the racial and ethnic diversity of the U.S. population.

Although under-representation of blacks and Hispanics in nanotechnology industries could be due to exclusion, differences in employment among racial and ethnic groups could be due to skill differences, which are not reflected in educational attainment through years of study. Black and Hispanic minorities, even with graduate degrees, may have greater difficulties acquiring interdisciplinary scientific training and skills, which are required for nanotechnology. In addition, individuals may differ in the so called soft skills (communication, leadership, problem solving, etc.). Thus under-representation of blacks and Hispanics may arise in part from the absence of specific skills in nanotechnology or from more general skills.

In addition, economists also argue that employment differences may arise because individuals may choose to be in one type of job more than the other. Yet, it is difficult to rationalize why, on average, non-Asian minority scientists deliberately choose lower paying academic jobs in preference to higher paying industry jobs. However, racial/ethnic differences observed in the patterns of employment for both groups of S&E jobs and non-S&E jobs in high technology industries suggest that race and ethnicity are included among the factors used to determine employment.

A complex interplay of social, cultural, historic, and economic factors contributes to the effects of race and ethnicity on employment in the U.S. context and leads to situations in which workforce equity is absent. While some non-Asian minorities with the requisite skills will gain employment in nanotechnology industries, exclusion will contribute to fewer opportunities. Sociologists and economists have attempted to explain the process of exclusion in different ways. For instance, Becker (1971) argues that employers might have a "taste for discrimination," that is they prefer to employ individuals from a particular group, even if it means a reduction in profits. In a competitive market discriminators may eventually be eliminated, but in a noncompetitive market they may persist indefinitely. In another explanation, termed "statistical discrimination," employers with limited information about the productivity of prospective employees use their perceptions about productivity of the group as a whole in deciding who to hire. Asians are often perceived as being "good" or "better" at science and engineering compared to other minorities, so they have an advantage in getting these jobs (Tang 1997, 2000).

In yet another explanation of the disparities between minorities and whites, Tomaskovic-Devey (1993) argues that "status based social closure" could also be used as a mechanism to exclude certain groups from jobs. Thus dominant groups, in this case white and Asian males, could restrict the employment of others in lucrative, highly paid jobs in industry using criteria such as licensing or credentialing, race, or gender because different factors can be used as closure mechanisms. As a result, the racial and ethnic diversity in the sector will not reflect that of the population. Blacks with graduate education, unlike Hispanics are likely to have better access to S&E jobs outside of high technology industries (Gatchair 2007), which is in keeping with previous observations that affirmative action policies have worked better in academia than in industry.

3.5.4 Potential Implications for Wages

When considering issues of workforce equity, it also is important to examine the distribution of wages, as well as the distribution of jobs, because patterns in the allocation of these two outcomes may be different. Gatchair (2007) found that the wages of blacks and Hispanics with graduate education in high technology S&E jobs were systematically lower than comparable whites, although these differences were not statistically significant for all sub-groups matched according to levels of experience. Nevertheless, the differences of approximately US$100 per week or US$5000 per annum, could be economically meaningful over the lifetime of individuals. For example, in the U.S. context, it could affect where individuals buy homes, which in turn affects the schools that children attend. The wage differences could have potential implications for the prospects of individuals and their families, over the long term.

For both types of S&E jobs (those in high technology industries and those outside), Asians and whites with graduate education did not have significantly different wages at most of the experience levels evaluated. However, older blacks and Hispanics with graduate education had significantly lower wages than similar whites in S&E jobs outside of high technology industries; but wages were not significantly different from whites for younger workers. Since the four racial/ethnic groups had greater parity in wages in high technology S&E jobs compared to the other groups of jobs, it is anticipated that qualified minorities, who get nanotechnology S&E jobs will earn wages that are close to similar whites. Based on observations in Gatchair (2007), racial and ethnic differences in wages for non-S&E jobs in nanotechnology industries are less likely for individuals with graduate level education, when compared to those with bachelors' level education. Wage disparities for individuals with bachelors' level education are likely to be very similar to the disparities that prevail in the workforce in general.

While it is anticipated that blacks and Hispanics will have greater difficulty getting employment in nanotechnology S&E jobs, if wage inequalities are present they may not be along racial and ethnic divisions. However since nanotechnologies have some of the characteristics of general purpose technologies, they have the potential to exacerbate wage differences, within and across groups of individuals because of their effects on the demand for skills. Given the challenging S&E skills requirements, and inadequate supplies of skilled individuals in the S&E

education pipeline, the wages of the nanotechnology workforce are likely to be higher than the average for other industries. In keeping with studies that show increasing returns to education (Lemieux 2006), wage differences between qualified and less qualified individuals within nano-industries may be on average greater than in the rest of the economy. Further, if productivity increases and profitability are high, and the gains are passed to workers as observed with previous technological changes, then nanotechnologies will contribute to the growing inequality observed in the U.S.

3.6 Conclusions

The growth of nanotechnology will lead to the creation of some new jobs; however it is likely that the number of new jobs will not be as great as the numbers observed in older manufacturing industries. Highly skilled individuals, in particular those with S&E backgrounds, will be favored, even for jobs that are not designated as S&E jobs. White and Asian males, who are disproportionately represented in the S&E education pipeline, and who might have more extensive networks that provide information and contacts, will fill these jobs. White and Asian males may even have an additional edge in the employment process because of statistical discrimination or closure mechanisms. Thus it is likely that the nanotechnology workforce will not reflect the racial and ethnic diversity of the U.S.

Asians are likely to have higher representation in nanotechnology jobs compared to their representation in the population. Non-Asian minorities on the other hand, will have a lower representation. Education can contribute to a reduction of the gap in the likelihood of employment of blacks and Hispanics with whites and Asians in nanotechnology jobs. However, it will not eliminate the disparities completely, since the effects of education vary with the level of education, industry/occupation, and racial group under consideration. Racial and ethnic differences in employment and wages will depend on whether individuals have graduate or bachelors' level education; work within or outside the nanotechnology sector; or work in S&E occupations. Under-representation of blacks will not be due solely to differences in the number of students pursuing S&E education, but will be due to other factors. Although the concentration of particular groups in industries or occupations is not unusual, depending on one's views, this may be undesirable. Economists argue that differences in representation may be due in part to choices that individuals make based on where they perceive that they will obtain the best returns for their efforts. However, differences may also be due to systematic practices, for example discrimination or closure practice, that result in barriers for some racial groups.

Although non-Asian minorities are under-represented, wage disparities between minorities and whites with graduate education are likely to be less for S&E jobs, when compared to other jobs, and least for high technology S&E jobs. However, based on previously observed effects of technological changes, nanotechnology industry growth is likely to contribute to vertical wage disparities.

Since education by itself will not lead to an equitable distribution of employment among different racial and ethnic groups, policies which specifically aim to reduce disparities, such as affirmative action and support for research that provides information on employment and wage patterns among racial groups, are necessary and should be continued. For example, additional resources could be provided to gather data that specifically relates to the nanotechnology workforce so that the implications suggested here could be substantiated or refuted.

In order to reduce potential under-representation of non-Asian minorities, efforts should be continued to attract and retain a broad cross-section of students to science and engineering fields of study. Special efforts should be continued to attract non-Asian minorities in order to ensure that the nation makes the best use of the potential talent. The policies will contribute to a nanotechnology workforce that reflects the diversity of the nation, as well as helps to provide a larger pool of individuals to undertake S&E activities.

Since racial and ethnic disparities in the workforce can extend to the labor market area, if the concentration of industries is large enough, economic developers need to address the issue of diversity in their recruitment strategies that aim to increase the number of businesses and create jobs. Thus recruitment strategies should make an effort to encourage individuals of different origins to come to the area in order to increase the diversity, rather than build a concentration of a single group. In addition, in order to create demand for different levels of skills, which might contribute to a more diverse workforce, the strategies should encourage complementary businesses and those that cater to the needs of different ethnic groups. If diversity issues are not addressed, in-migration of specific groups can lead to shifts in the demographics of a particular area, and tensions can develop between new and existing groups, whether defined by socio-economic status, age, race, ethnicity, or other characteristics. Individuals who are included may find this quite advantageous, but for the excluded the situation may be less desirable since they are unable to reap the benefits of employment and wages created by the new industries.

This paper argues that racial and ethnic disparities in S&E education and in the treatment of workers can contribute to potential disparities in the nanotechnology workforce. Efforts to grow diversity in the pool of S&T workers are important, and can increase the diversity of the workplace and regions. If these suggestions are taken into consideration, then a more equitable distribution of benefits from the new industries could be achieved among different groups and the nation as a whole.

Notes

1. National Nanotechnology Initiative, Education and Workforce Needs, retrieved June 23, 2009 from http://www.nano.gov/html/society/Education.html. Although other estimates of the nanotechnology workforce exist, this paper uses that adopted by the NNI, which represent multiple U.S. government agencies involved in nanotechnology.

2. The researchers arrived at available jobs by using a search strategy, which included the keywords of "nano" and "mems" in the announcement. Although the authors argue that the strategy was likely to capture most of the available jobs, given the variety of nanoproducts, techniques, and applications, it is likely that the approach used could have missed a number of jobs that did not explicitly use these terms.
3. Endogeneity arises because independent variables e.g. the levels of educational attainment in the model are affected by unobserved variables such as family background, non-cognitive skills, which also affect the dependent variables. As a result, estimates of the effects of the independent variables are biased, that is they do not reflect the true value of the effects.
4. Studies use different approaches to define the S&E workforce (National Science Board 2008). Gatchair (2007) defined the S&E workforce using a carefully selected set of occupations identified in the CPS dataset. This approximation was adopted in the study because of the absence of information on S&E fields of study in the CPS dataset.

References

Aghion, Philippe, Peter Howitt, and Giovanni L. Violante. 2002. General purpose technology and wage inequality. *Journal of Economic Growth* 7 (4): 315–345.
Bartel, Ann P., and Nachum Sicherman. 1999. Technological change and wages: An interindustry analysis. *Journal of Political Economy* 107 (2): 285–325.
Becker, Gary Stanley. 1971. The economics of discrimination. *Economics research studies of the Economics Research Center of the University of Chicago*. 2nd ed. Chicago: University of Chicago Press.
BEST. 2004. The talent imperative: Diversifying America's science and engineering workforce. ed. Building Engineering and Science Talent. San Diego, CA.
Bureau of Labor Statistics. 1992–2002. Current population survey. Bureau of Labor Statistics.
Card, David, and John E. DiNardo. 2002. Skill-biased technological change and rising wage inequality: Some problems and puzzles. *Journal of Labor Economics* 20 (4): 733–783.
Clark, Julia V. 1999. Minorities in science and math. *ERIC Digest*, http://chiron.gsu.edu/cgi-bin/homepage.cgi?link=zoer&style=&_id=803d513f-1165344073-8437&_cc=1. (accessed July 31, 2010).
Cooke, Philip. 1996. The new wave of regional innovation networks: Analysis, characteristics and strategy. *Small Business Economics* 8 (2): 159–171.
Cooke, Philip. 2007. How benchmarking can lever cluster competitiveness. *International Journal of Technology Management* 38 (3): 292–320.
Fonash, Stephen J. 2001. Education and training of the nanotechnology workforce. *Journal of Nanoparticle Research* 3 (1): 79–82.
Galbraith, James K. 1998. *Created unequal: The crisis in American pay*. New York, NY: Free.
Gatchair, Sonia Denise. 2007. Representation and reward in high technology industries and occupations the influence of race and ethnicity. Atlanta, GA: Georgia Institute of Technology. http://hdl.handle.net/1853/19770. (accessed July 30, 2010).
Goldin, Claudia Dale, and Lawrence F. Katz. 2008. *The race between education and technology*. Cambridge, MA: Belknap Press of Harvard University Press.
Granovetter, Mark. 1983. The strength of weak ties: A network theory revisited. *Sociological Theory* 1: 201–233.
Hecker, Daniel. 2005. High-technology employment: A NAICS-based update. *Monthly Labor Review* 128 (7): 57–72.
Heinze, Thomas, and Gerrit Bauer. 2007. Characterizing creative scientists in nano-S&T: Productivity, multidisciplinarity, and network brokerage in a longitudinal perspective. *Scientometrics* 70 (3): 811–830.

Helmus, Michael N. 2006. How to commercialize nanotechnology. *Nature Nanotechnology* 1 (3): 157–158.
Hullmann, Angela. 2007. Measuring and assessing the development of nanotechnology. *Scientometrics* 70 (3): 739–758.
Juhn, Chinhui, Kevin M. Murphy, and Brooks Pierce. 1993. Wage inequality and the rise in the returns to skill. *Journal of Political Economy* 101 (3): 410–442.
Kim, Kelly Y. 2007. Research training and academic disciplines at the convergence of nanotechnology and biomedicine in the United States. *Nature Biotechnology* 25 (3): 359–361.
Lemieux, Thomas. 2006. Postsecondary education and increasing wage inequality. *American Economic Review* 96 (2): 195–199.
National Nanotechnology Initiative. 2009. *Education and workforce needs*. http://www.nano.gov/html/society/Education.html. (accessed June 23, 2009).
National Science Board. 2008. Science and engineering indicators 2008. Arlington, VA: National Science Foundation.
NSF. 2007. Women, minorities and persons with disabilities in science and engineering: 2007 (NSF 07-315). ed. D. o. S. R. S. National Science Foundation. Arlington, VA.
NSTC, and OSTP. 2007. The national nanotechnology initiative: Supplement to the Presidents FY2008 budget. ed. National Science and Technology Council and Office of Science and Technology Policy. Washington, DC.
NSTC, and OSTP. 2009. The national nanotechnology initiative: Supplement to the Presidents FY2010 budget. ed. National Science and Technology Council and Office of Science and Technology Policy. Washington, DC.
Pearson, Willie. 1985. *Black scientists, white society, and colorless science: A study of universalism in American science*. Millwood, NY: Associated Faculty.
Pearson, Willie. 2005. *Beyond small numbers: Voices of African American PhD chemists, diversity in higher education*. Amsterdam: Elsevier.
Pearson, Willie, and Alan Fechter, eds. 1994. *Who will do science? Educating the next generation*. Illie Baltimore, MD: Johns Hopkins University Press.
RESULTAR. 2008. Insulin case study. In *Distributional consequences of emerging technologies* Atlanta, GA: Georgia Institute of Technology.
Saxenian, Annalee. 2002. Silicon valley's new immigrant high-growth entrepreneurs. *Economic Development Quarterly* 16 (1): 20–31.
Shapira, Philip, and Jan Youtie. 2008. Emergence of nanodistricts in the United States. *Economic Development Quarterly* 22 (3): 187–199.
Stephan, Paula, Grant C. Black, and Tanwin Chang. 2007. The small size of the small scale market: The early-stage labor market for highly skilled nanotechnology workers. *Research Policy* 36 (6): 887–892.
Tang, Joyce. 1997. Evidence for and against the 'double penalty' thesis in the science and engineering fields. *Population Research and Policy Review* 16 (4): 337–362.
Tang, Joyce. 2000. *Doing engineering: The career attainment and mobility of caucasian, black and Asian-American engineers*. Lanham, MD: Rowman & Littlefield.
Tomaskovic-Devey, Donald. 1993. *Gender and racial inequality at work: The sources and consequences of job segregation, Cornells studies in industrial and labor relations*. Ithaca, NY: ILR.
Van Horn, Carl, Jennifer Cleary, and Aaron Fichter. 2009. The workforce needs of pharmacuetical companies in New Jersey that use nanotechnology: Preliminary findings. New Brunswick, NJ: John J. Heldrich Center for Workforce Development, Rutgers, The State University of New Jersey.
Wadhwa, Vivek. 2009. A reverse brain drain. *Issues in Science & Technology* 25 (3):45–52.
Youtie, Jan, Maurizio Iacopetta, and Stuart Graham. 2008. Assessing the nature of nanotechnology: Can we uncover an emerging general purpose technology? *Journal of Technology Transfer* 33 (3): 315–329.

Chapter 4
Exploring Societal Impact of Nanomedicine Using Public Value Mapping

Catherine P. Slade

Scholars looking to promote the idea that public values, like equity, should guide scientific research often run into a tricky problem: who decides which values are most important? It can be a bit presumptuous for individual scholars to claim that they know what is best for the world and which values should be pursued. One recent technique developed to deal with this dilemma is Public Value Mapping or PVM. The basic idea behind PVM is that while deciding which values should be pursued by scientific institutions can open a can of worms in regards to representation and ethics, at the very least institutions should be held accountable for the values they public claim they are pursuing. The first step of PVM is to map the values publicly espoused by the organizations involved in the development of science and engineering research projects, and then analyze the effects these organizations have on the world to see if there is a correlation. PVM scholars routinely find that the grand social visions promised by scientific institutions rarely become reality. In this chapter Catherine Slade turns the PVM lens on nanomedicine and the organizations who claim they are working for an equitable distribution of the benefits of the field. Unfortunately Slade finds the organizations are falling well short of their stated goals in terms of both actions and outcomes, but her work shows that PVM can be a useful tool in the effort to ensure that statements about equity are not simply lip service.—eds.

C.P. Slade (✉)
Consortium for Science, Policy and Outcomes; Arizona State University, Tempe, AZ, USA
e-mail: Slade3534@comcast.net

Originally presented at the Workshop on Nanotechnology, Equity, and Equality at Arizona State University on November 21, 2008.

4.1 Introduction

Public policies and the discourse surrounding the United States' scientific and technological enterprise manifest a wide array of values, including values ranging from security and leadership, to health and quality of life. Although some of these values are inherently economic or can be reduced to economic surrogates (Murashov and Howard 2008), economic values rarely serve as preferred end-state values for public policies, especially those that relate to health and well-being. When economic values serve as surrogates for social or public values (e.g. the economic value of conservation of habitat or the marginal cost-benefit of improved health) these indirect economic indicators often prove problematic (Anderson 2006; Norton and Noonan 2007).

Despite the fact that a range of values motivating investments into science and technology in general—and nanotechnology in particular—cannot conveniently be evaluated in economic terms, there exists a long-standing tendency to resort to market-oriented measures or even metaphors in assessing the activities that occur in and around United States science and technology laboratories (Bozeman et al. 2007). In part this practice is owing to the traditional economic growth rationale for public investment in science (Nelson 2000; Bozeman and Sarewitz 2005). But, just as important, non-economic analysis presents a set of analytical tools useful, in some cases, for measuring policy response to public and social choices and values. When decision-makers wish to consider "the public interest" or social values for nanotechnology, available concepts and theories tend to be unfocused, imprecise and underdeveloped (Scheufele and Lewenstein 2005).

The idea of public value mapping (PVM) of science policy was initially developed by researchers at the Consortium for Science, Policy and Outcomes (CSPO) at Arizona State University as part of a grant from the Rockefeller Foundation and, more recently, has been supported by the U.S. National Science Foundation's "Science of Science Policy" program in a grant entitled, "Public Value Mapping: Developing a Non-Economic Model of the Social Value of Science and Innovation Policy" (NSF #0738203) (Bozeman 2007). From its beginnings, the primary rationales for the public value mapping of science are that: (1) the focus of science policy should be on end-state social goals and public values, and (2) current research evaluation and science policy analysis methods and techniques may be useful in many important respects but are not entirely sufficient for analyzing the impacts of research on public values. PVM is not and does not aspire to be a unified method; rather it is a logical model using a variety of qualitative and quantitative methods aimed at understanding the role of public values in science and research investments and their societal impacts and outcomes.. While PVM has heretofore provided little precision, its core assumptions have changed little since its genesis, including that PVM is not an analytical technique but a model that includes a guiding theoretical framework that is context specific (Bozeman 2007; Bozeman and Sarewitz 2005).

Accordingly this case study using PVM focuses on nanomedicine for cancer and the intrinsic value of equity. The National Cancer Institute (NCI) policies and initiatives, including nanomedicine for detection and treatment of cancer, demonstrate high hopes for the technology and its societal impact, especially with respect to fair and equitable access to emerging and promising technologies (NCI 2006).

Nanomedicine policy in this and other venues encourages multidisciplinary collaborations between social scientists, biomedical scientists and health scientists (Kahn 2008). NCI has thus embraced the idea that social science and societal impact, such as reducing health disparities, are important to nanomedicine research policy and funding priorities (De Melo-Martin 2009), making it a prime candidate for PVM analysis.

This paper uses the evolving theory of public value mapping (Bozeman 2007) to examine the discourse on nanomedicine and its expected social imperatives, such as equity in access to these promising and novel treatments, and how the discourse relates to science policy analysis and development. The focus of the study is on analysis of public values as represented by public value statements in the discourse of stakeholders concerned with nanomedicine in cancer. In the PVM model, and conceptually with respect to values theory (Kahneman and Tversky 2000), there are of two types of public values of interest: intrinsic and instrumental. Intrinsic values are those that are ends in themselves; once they are achieved, they represent an end state of preference. Instrumental values are instead concerned with means and with process; they have no value in themselves but are valued in relation to an intrinsic value. The interplay between proposed intrinsic and instrumental values in a case study comprises its public value logic. In this case the logic describes how nanomedicine in cancer will meet the intrinsic value of equity primarily, and how science policy related to nanomedicine will address equity through stated instrumental values. I chose the intrinsic public value of equity and the instrumental values related to minority access to nanomedicine for cancer because of their currency in the health science policy arena (Resnik and Tinkle 2007).

Why such a narrow focus on equity in and minority access to nanomedicine given the broad expectations for nanotechnology science policy development? While PVM was conceived with the goal of developing instruments and applications to broader research evaluation, its early and current proof-of-concept applications have focused on and benefited from specific case studies with important policy imperatives. In Bozeman's (2007) treatise on PVM he notes the work of Gaughan and Bozeman (2002) that applied PVM to breast cancer research. In the same book Gupta (2002) is credited with using PVM to analyze genetically modified crops. In both cases, the analyses were preceded by developing public policy statements and initiatives similar to policy statements of NCI for nanomedicine equity. Other applications, also rooted in case studies, have focused on public values failure in climate change (Bozeman and Sarewitz 2005), and influenza vaccine research (Feeney and Bozeman 2007). Recently, as part of the NSF-sponsored phase of the PVM project, inter-linked case studies have been developed looking at technology transfer mechanisms, cancer disparities, and a more in-depth view (compared to earlier PVM research) of climate change. Thus, the best and most promising applications of PVM are grounded in specific case studies where science policy has dramatic societal implications for specific intrinsic public values.

The chief purpose of this paper is to develop a PVM case study by evaluating the current discourse on public and social values for nanotechnology science and innovation policy, focused on the case of nanomedicine and equity. This is a health science policy issue characterized by multiple, competing and complex criteria and stakeholders (Best and Khushf 2006; Hede and Huilgol 2006) in which only using economic rationales clearly under-serves social and public values (Bozeman 2007). When we also consider that nanomedicine policy is a moving target, with a churn of competing policy aspirations, innovation objectives and implementation strategies pursued by a diverse set of research and technology performers, it is easy to see that understanding the discourse of public values and public value mapping is useful here (Bozeman 2003, 2007; Slade forthcoming).

The analysis begins with an identification of the primary intrinsic and instrumental public values that nanomedicine in cancer promises to address. Using the public value mapping model, public value statements of participants in the nanomedicine discourse are identified. How they predict impacts of nanomedicine on society, especially intrinsic values for equity in access to these promising technologies, is explored.

A secondary purpose of this study is to inform the scientific and policy communities representing numerous organizations and agencies about the importance of relating developing nanomedicine in cancer to those values held publicly relevant and important much earlier in the research and development process. Gaps in the value logic model and/or disconnects in value statements between interrelated groups addressing nanoscience and engineering policy are addressed. Understanding these gaps and disconnects should result in better policy development, including information for setting public funding and research priorities.

4.2 Public Values and Emerging Nanotechnology

With regard to nanotechnology science and engineering policy (NSE), the relationship between public values and public policy is especially interesting given the underlying ambivalence of perceptions of the field (Goorden et al. 2008) and also given that the motivations behind nanotechnology policy prescriptions are themselves potentially contradictory (Fisher and Mahajan 2006). Currently, more is unknown than known about the relationship between nanotechnology and societal impacts (Sarewitz and Woodhouse 2003). This is not a situation unique to nanotechnology as a science discipline. Guston (2000, 2008) has written extensively about the importance of the discourse among science policy actors to ensure "the production of the public good of basic research, and the investment in future prosperity that is research" (Guston 2000, 32). Guston and others (see Guston and Sarewitz 2002) promote an agenda for science policy, and especially NSE policy, which ensures that science productivity is valued in greater terms than generation of intellectual property or adherence to standards for scientific integrity and ethics. Indeed, this is the foundation for the "science of science policy," an emerging field of interdisciplinary research to assess the impacts of emerging scientific and engineering technologies and innovations (National Science and Technology Council 2008).

Several aspects of nanotechnology policy make it an especially high priority for public value mapping, as evidenced by the growing body of literature on its sociocultural meaning (Bainbridge 2004). Since public perceptions of and values for emerging technology drive policy decisions, then science that is more readily grasped and put in context by laypersons usually gets more traction in the policy arena; with its inherent complexity, nanotechnology needs special attention (Scheufele et al. 2009). Thus there could be a disconnect between expert NSE opinions of scientists and engineers and policy-makers that would be bridged by revealing shared assumptions of public perceptions and values (Bainbridge 2004; Roco and Bainbridge 2005; Scheufele and Lewenstein 2005).

Why the focus on nanomedicine? Some justification comes from the growing hype about its prospects for its societal impact, both beneficial and detrimental (Lenk and Biller-Andorno 2007; Resnik and Tinkle 2007). Further, there is the tendency to oversimplify what is commonly considered one technology, like nanomedicine for cancer diagnosis, but is in actuality multiple nanomedicine technologies that involve manipulation of matter and devices at the nanoscale (Slade forthcoming; Sparrow 2009). Finally, the general public acceptance of nanomedicine is based on growing and sustained confidence in the technology; a confidence that comes from alignment of research and development priorities with public values and priorities (Agres 2004; Guston and Sarewitz 2002).

The situation of rapidly emerging nanomedicine accompanied by ambivalence about priorities for science policy provides an opportunity for prospective research like PVM. PVM can be used to clarify the baseline of non-economic expectations, goals, and promises that appear to animate the major investments being made in nanoscience—before those expectations are affected by knowledge of impacts and outcomes and in order to ground future policy assessments (Bozeman 2007). One of many challenges for analyzing public values in nanomedicine policy is the sheer volume of participants in the discourse, and PVM is especially designed for this situation. Discourse participants are motivated by a wide divergence of agendas and values drawn from a range of industrial, government, academic, and civil society actors (e.g., Fisher et al. 2008). Further, the discourse participants are disciplinarily more diverse than what might be seen in other realms of health science and innovation policy; a situation that accords with the notable nanomedicine policy emphasis on interdisciplinarity evident in U.S. legislation and federal funding initiatives (Romig et al. 2007).

To understand the role of PVM in nanomedicine policy analysis specifically requires understanding the role of PVM in public policy development and Science and Innovation Policy (SIP) in general. Figure 4.1 below shows the PVM logic model in simplified form. The key assumption is that policy that focuses on market or science success alone could be ignoring public values (Bozeman 2007). The importance of PVM is demonstrated in its case study approach and its enhancement of policy analysis. According to the model, typical policy development and analysis can ignore public values. The steps leading to policy recommendations using PVM include: (1) producing an inventory and categorization of salient public value statements, (2) refining a model for values aggregation, (3) creating a policy imperative diagnostic based on a more detailed public values failure analysis, and (4) describing the relationship between market value failure and public value failure. The current study uses this model to demonstrate that analysis of public values statements and analysis of failure to meet core public values is illuminating with respect to SIP development, especially for nanomedicine in cancer.

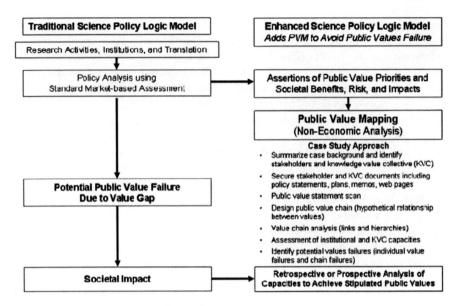

Fig. 4.1 Public value mapping logic model

4.3 The PVM Nanomedicine Case Study: A Focus on Equity

The emerging field of nanomedicine as a subset of nanotechnology holds great promise for transforming both cancer research and clinical approaches to cancer care (Heath and Davis 2008). Nanomedicine is currently being used to: (1) detect cancer at its earliest stages, (2) pinpoint the location of cancerous cells within the body, (3) deliver cancer-fighting drugs to specific cells, and (4) assess the effectiveness of treatment. Nanomedicine approaches to drug delivery especially are touted as the breakthrough technology for curing cancer in all its forms. The significant and often fatal problems with current cancer diagnosis and treatment technology, for example current chemotherapy toxicity and lack of specificity to tumor cells, may be resolved by nanomedicine but not without as yet unknown risks (Best and Khushf 2006; Hede and Huilgol 2006). Many of the social impact issues related to the scientific and technological progress of nanomedicine in cancer are not unique to this discipline and are similar to those related to the development and application of other new biomedical procedures (Best and Khushf 2006; Faunce and Shats 2007; Lenk and Biller-Andorno 2007; Mehta 2004; Resnik and Tinkle 2007). But unlike some prior cancer technology advances, including those now considered "gold standards" of treatment, nanomedicine is developing at an unprecedented rate especially with respect to funding and implementation of clinical research (Wonglimpiyarat 2005). Thus nanomedicine for cancer is well positioned to provide clinical breakthroughs and societal impact, potentially positive and negative. And that makes it a prime candidate for PVM analysis.

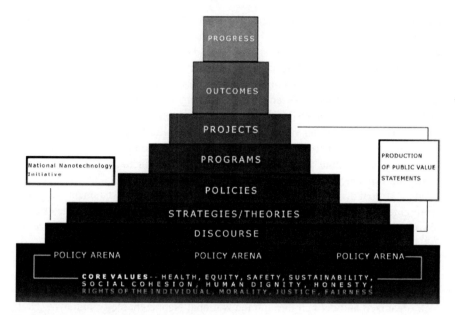

Fig. 4.2 Identifying sources of public values statements

To narrow the focus and make it a better case study candidate for PVM analysis my research focuses on the intrinsic value of equity and the instrumental values associated with minority access to this promising group of technologies (Ford et al. 2008). Figure 4.2 below shows my own model of the expectations for the discourse surrounding nanomedicine in cancer from the broad and varied stakeholder group engaged in policy development. Note my predication that discussion of new SIP such as nanomedicine in cancer (the looming technological advances) lacks a smooth entree to the policy discourse and my contention that a public value mapping effort, including analysis of public value statements of the stakeholders (on the right), should enhance communication between the stakeholders and subsequent policy analysis and development of this type, even for newcomers and outliers of the policy discourse.

4.4 The Method for Analysis of Value Discourse in Nanomedicine Stakeholder Communities

In this case study of nanomedicine and equity as an intrinsic value and access for minorities in various forms as instrumental values, we have longstanding discourse on a variety of interrelated issues of medical technology development and health disparities, with looming nanomedicine innovation driving some of the current policy initiatives. Thus the universe of stakeholder groups (providers, advocates, funding agencies, research consortia, and academic institutions for example) working with the topic is broad and large and their discourse (the source of public value statements to be analyzed) is exponentially larger. The approach to assessing the potential for public failure began with establishing the most important stakeholder groups through their common topical policy interests as noted in public documents on their websites and then analyzing their production and treatment of "public value statements" specific to nanomedicine and equity through access for minorities. The public value statements are specific language in official documents and the anchor for PVM. In this case, and most other similar cases, public value statements come from a wide range of sources including: (1) official and legitimated statements of policy goals; (2) goals in less well-articulated policy statements; (3) goal statements in strategic plans; (4) aggregated statements of values represented in opinion polls; (5) official policy statements by policy actors; and (6) official policy statements by relevant non-governmental policy actors (Bozeman 2003).

The stakeholders for nanomedicine in cancer are complex and composed of some members with long term collaborations and others with only intermittent engagements. Relevant stakeholders in nanotechnology development seem to be fairly well organized and defined through the National Nanotechnology Initiative (NNI), managed within the framework of the National Science and Technology Council (NSTC), a council with direct communication with the President's Cabinet and by which the administration coordinates science, space, and technology policies across the Federal Government. Conversely, the stakeholders engaged in policy to address equity issues that might be affected by nanomedicine are many in number at the federal, state, or local level, less centrally coordinated, and constantly changing (see for example the constituents of the Department of Health and Human Services Office of Minority Health).

Table 4.1 below summarizes the sources for public values statements for this case study with a brief description of their role in this policy discourse. To identify stakeholders, the analytic method was to start at the websites for the two seemingly most disparate stakeholder groups in the discourse: the National Nanotechnology Initiative (NNI) and the United States Department of Health and Human Services Office of Minority Health (OMH). The NNI is new to, but needed in the discourse on, health disparities and the OMH is responsible for driving the strategic framework for improving racial/ethnic minority health. On the websites of these two entities combinations of the terms "nano*", "clinical research", "emerging technology", "cancer" and "health disparities" were searched to identify potential stakeholders and relationships between stakeholders in working groups and within this policy arena. The stakeholder scan was nonlinear, and in many cases circular, meaning that stakeholder documents were searched several times as new terms were identified. The intent was less to have all of the potential sources of statements than to have stakeholders of all government levels and as many non-governmental organizational types as possible, as well as any prominent individuals and organizations represented. One finding from this part of the analysis is that the stakeholder interests and proliferation of related public values statements were quite broad.

Table 4.1 Sources of public value statements in emerging nanomedicine

Key	Source
AHRQ	Agency for Healthcare Research and Quality
AACR	American Association for Cancer Research
ACS	American Cancer Society
AMA	American Medical Association
AMHPS	Association of Minority Health Professions Schools
CDCOMH	CDC Offices of Minority Health and Health Disparities and Office of Public Health Research
CEHDMSM	Center of Excellence on Health Disparities—Morehouse School of Medicine
CF	The Commonwealth Fund
CHDPRG	Cancer Health Disparities Progress Review Group
DSMD	David Satcher, M.D., Ph.D.
DCHMHAC	State of Georgia Department of Community Health's Office of Health Improvement and Minority Health Advisory Council
DHHSOMH	DHHS Office of Minority Health
EPANCER	Environmental Protection Agency National Center for Environmental Research
HDC	Health Disparities Collaborative
KFF	Kaiser Family Foundation
KKI	Kennedy Krieger Institute
MCPHI	Morehouse College Public Health Institute—Research Center on Health Disparities
NCICRCHD	National Cancer Institute Center to Reduce Cancer Health Disparities
NCIANC	NCI Alliance for Nanotechnology in Cancer
NCMHHD	National Center on Minority Health and Health Disparities
NIEHS	National Institute of Environmental Health Science
NMA	National Medical Association
REACH	REACH US: Racial and Ethnic Approaches to Community Health Across the US
RWJ	Robert Wood Johnson Foundation Finding Answers
WKK	WK Kellogg Foundation

The analysis included 25 stakeholders and over 1,100 pages of documentation on publicly accessible websites. Table 4.2 below shows examples of related public value statements.

In the PVM model, and conceptually with respect to values theory (Bozeman 2007), there are of two types of public values of interest: intrinsic and instrumental. Intrinsic values are core values and ends in themselves; once they are achieved they represent an end state of preference. Instrumental values are instead concerned with means and with process; they have no value in themselves but are valued in relation to an intrinsic value. The interplay between proposed intrinsic and instrumental values in a case study comprises its public value logic. Given the focus on the intrinsic value of equity for nanomedicine by policy makers and the public, it was surprising to find no mention of reduced or eliminated health disparities among the stakeholders of the basic research enterprise. Is it possible that their greater emphasis on economic development (e.g. competitiveness, rationalized resources, expandable capacity) signals that basic researchers are not yet as cognizant of their influence on public values as needed? This is an important finding for this case, given that societal priorities for nanomedicine in cancer are currently in the hands of the basic research enterprise (Agres 2004; Emerich and Thanos 2003).

Table 4.2 Examples of related public value statements

Value statements	Examples
Access	The Community Networks Program of the National Cancer Institute's Center to Reduce Cancer Health Disparities ... aims to increase access to and use of beneficial cancer interventions, such as proven approaches for ... early detection and treatment of breast, cervical, prostate and colorectal cancers. (www.crchd.cancer.gov/about/overview.html)
Collaboration	The American Cancer Society's international mission concentrates on capacity building in developing cancer societies and on collaboration with other cancer-related organizations throughout the world in carrying out shared strategic objectives (www.cancer.org/ACS Mission Statement).
Consumer empowerment	The specific aims of the community partnership development resource of the Center of Excellence on Health Disparities of the Morehouse School of Medicine are to ... create consumer participant groups .. that will act as advisors to academic researchers on community based projects (www.web.msm.edu/EXPORT/cpd.html).
Data and analytical methods	The call to action for the Trans-HHS Cancer Health Disparities Progress Review Group includes establishing "new approaches for data collection and sharing to aid in the study of the effects of cancer and their relationship to variables such as race, ethnicity, and socioeconomic status" (www.hhs.gov/chdprg/about/).
Education	The AMHPS is intended to ... build and strengthen institutional infrastructure supporting the development and implementation of effective programs to advance professional development, education, and research training for racial and ethnic minorities (www.cdc.gov/omhd/CoopAgree/BAA.html)
Effective research and interventions	Center of Excellence on Health Disparities at the Morehouse School of Medicine's goal Is to increase the critical mass of talented research faculty pursuing health disparities research training in biomedicine and provide the research infrastructure needed for investigators to pursue excellence in the twenty-first century (www.web.msm.edu/EXPORT/history.html).
Minority participation in clinical trials	The AHRQ concludes that "clinical investigators need effective strategies to improve participation of underrepresented populations in cancer clinical trials" (www.ahrq.gov/clinic/tp/recruittp.htm)

Table 4.2 (continued)

Value statements	Examples
Systematic change	The Strategic Framework for Improving Racial/Ethnic Minority Health and Eliminating Racial/Ethnic Health Disparities presents a vision—and provides the basis—for a "systems approach" to addressing racial/ethnic minority health problems within and outside of HHS. A systems approach implies that all parties engaged ... are themselves part of system or nested systems (www.omhrc.gov/npa/framework/)
Translational research	The Translational research core [of the CEHD of Morehouse School of Medicine] will offer a centralized location where researchers can share and collect biological information that will aid in leading-edge basic science technologies to probe the biological basis of health disparities. This integration of knowledge from lab bench to bedside to community is another fulfillment of the "three dimensional" approach to health disparities research.(www.web.msm.edu/EXPORT/translational.html)

Table 4.3 shows the summary results of the categorization using PVM methodology language concerning public value chain analysis (Bozeman 2007). Consistent with the model in Fig. 4.2 above, the translation of core or intrinsic values likely to be relevant to any public policy issue can be categorized in this case into instrumental discourse categories of stakeholder groups using public value statements.

4.5 The Public Values Failure Assessment

A core component of PVM is the analysis of potential public values failures from science policy. Public values failure occurs when the market and the public sector fail to provide goods and services required to achieve the core values of society. Bozeman (2007) has demonstrated the value of this analytic technique and its use in case studies like this. Table 4.3 below summarizes my diagnostic of public failure for a broad range of intrinsic values related to fairness and equity for a specific aspect of nanomedicine in cancer, the clinical trial. In the following paragraphs I explain the failure definition and illustration in more detail.

4.5.1 Interest Aggregation or Articulation Criterion

The PVM model suggests that public failure can result from an inability to aggregate interests. In this case, the National Institutes of Health (NIH) especially has detailed policies and requirements for including minorities in clinical trials, yet with little

Table 4.3 Public values failure model applied

Public failure	Failure definition	Illustration
Interest articulation or aggregation	Political processes and social cohesion insufficient to ensure effective communication and processing of public values	Even within NIH, there is no consistent or effective policy/strategy to increase minority participation in clinical trials. NIH requirements for minority participation in sponsored research dating back to 1993 have been ineffective in increasing proportion of minorities in trials.
Imperfect monopolies	Private provision of goods and services permitted even though government monopoly deemed in the public interest	Minorities, especially low income persons in minority groups, tend to receive their health care in private community settings that are least likely to have physicians with access to or an interest in participation in sponsored research including clinical trials.
Benefit hoarding	Public commodities and services have been captured, limiting distribution to the population	Lack of diversity in potential study populations (those with access to participating physicians or centers) results in inequitable distribution of clinical trials (often life-saving) resources. Most trials limit co-morbid conditions that are more prevalent in minority populations.
Scarcity of providers	Despite the recognition of a public value and agreement on the public provision of goods and services, they are not provided because of the unavailability of providers	In addition to a lack of minority physicians in general, only 3–4% of board-certified minority physicians participate in clinical trials (compared to several times that for white physicians).

Table 4.3 (continued)

Public failure	Failure definition	Illustration
Short time horizon	A short time horizon is employed when a longer term view shows that a set of actions is counter to public value	Healthy People 2010 and 2020s short term goals for cures for cancer and elimination of health disparities are inconsistent with timeframes for nanotechnology development. This could result in some technologies being favored for sponsored research and clinical trials in the short term (to get a quick bang for the buck) to the exclusion of those with better long term benefits and efficacy, especially for minorities.
Conservation of resources	Policies focus on substitutability (or indemnification) even in cases when there is no satisfactory substitute	Health care and health science resources are not equitably distributed and not easily redistributed.
Threats to human dignity and subsistence	The core value of subsistence is violated	Results of clinical trials often have limited generalizability to the population as a whole and with even less generalizability to minority groups that may experience different biological responses to drugs and devices than most study participants. The result could be greater risk to minorities of the "unintended consequences" of nanotechnology.

impact on reducing health disparities so far (Bruner et al. 2006). Conversely, NIH has limited specific policies for translational research; at this point it is only a roadmap. Garber and Arnold (2006) suggest that the admirable plans of NIH are often distorted through their policies, perhaps because the arguments minorities receive for clinical trials participation are not compelling or tailored to their needs or desire for novel treatment. Unfortunately little has changed in the NIH policy to better align the minority participation policies with the public values they are designed to address. This is a clear example of lack of aggregation of interests among key stakeholder groups.

4.5.2 Imperfect Monopolies Criterion

Imperfect monopolies occur when the government's lack of protection of competition and free enterprise leads to public value failure. Health care in the United States is a public-private venture, fraught with challenges. Cicatiello (2000) describes the myriad of government policies, programs and processes that nanomedicine must be integrated into if it is to be of societal benefit. Government "owns" the regulatory process for medicine and research. Yet, as Long et al. (2004) suggest, a scan of the recent health disparities literature shows few if any new interventions; policy research is almost entirely devoted to documenting continuing disparities. That the government has failed to create insurance programs to reduce disparities in access to cancer clinical trials (Colon-Otero et al. 2008) suggests that government is overconfident in the private provision of insurance coverage of clinical research.

4.5.3 Benefit Hoarding Criterion

Health disparities, especially in minority access to clinical trials, are not entirely the result of financing. Murthy et al. (2004) demonstrated that accession and participation disparities often result from preexisting condition and co-morbidity exclusion criteria that are more prevalent in minority populations. The NIH requires justification for explicitly excluding minorities from clinical trials, suggesting that approved trials have access that is then fair and equitable. Yet the research design itself may subtly limit access to minorities resulting in the benefits of novel treatments being captured by certain population groups.

4.5.4 Scarcity of Providers Criterion

Cancer research is geographically widely distributed. However, limited progress has been made to ensure that community physicians can participate. Pinto et al. (2000) documented the myriad of reasons why community physicians are challenged to participate in clinical trials to the benefit of their patients. Regardless of the source of barriers to community physician participation (Swanson and Ward 1995), suffice it to say that limited access for community physicians, the primary source of health care for minorities, will result in a scarcity of providers to accession minority patients to trials of novel treatments like nanotechnology.

4.5.5 Short Time Horizon Criterion

Public value failure can occur when a short-term solution is applied to a situation clearly requiring longer term action. In this case, the charge to cure cancer by 2015 (see Healthy People 2010), could supersede equitable distribution of the benefits of novel treatments. Lacking minority participation in cancer trails, the research community could come to the conclusion that the War on Cancer is won, even when there is no evidence that effective treatment for non-minorities works for minorities as well (Yancy 2008).

4.5.6 Conservation of Resources Criterion

Conservation of resources deals with the issue of substitutability. In this case this would mean access to novel treatments would not deplete key resources, such as funding for health care delivery or the use of health care facilities and services. The funding for clinical trials of novel treatments is often misconstrued. Clinical trials in many respects are financed with private insurance coverage just the same as standard treatment. It has been well-documented that the insured are more likely to participate in research than the uninsured and that minority populations are less likely to be insured (Garber and Arnold 2006). The NIH Revitalization Act of 1993 mandates that NIH create programs and policies to not allow cost to be an acceptable exclusion from clinical research. So long as the treatment aspect of clinical research is essentially financed through health insurance programs, then disparities in access will continue to occur. Since health care resources and access to services due to insurance constraints are not equitable, then these resources are being depleted by some well-insured persons at the detriment of others who have more limited access.

4.5.7 Threats to Human Dignity and Subsistence Criterion

The failure to make cancer clinical trials (even those related to nanotechnology) accessible to all has resulted in less generalizability to all groups within the population. In the case of nanotechnology, the results of this incomplete information could be unintended consequences for minorities who are less studied.

4.6 Conclusion

The public values statements analysis shows that the current discourse on nanotechnology, equity, and health disparities is long on rhetoric and short on translation to policy and program interventions. Working groups and more long-term stakeholder entities are producing symbolic reports developed by some of the same participants and/or qualified by mostly the same very shallow (at best) scholarly support. Advocates for new technological advances, such as nanotechnology in cancer, may have difficulty finding and claiming a seat at the health disparities policy development and implementation table because they are late entries into the discourse with no specific mechanism (and perhaps little incentive) for joining. I suggest that Public Value Mapping early in the development of science and technology, specifically at the basic research level, creates the sensitivity needed to affect societal benefit in the long term. For example, and given that they have the same general sources of extramural funding, why not include statements of public values considerations in federal grant applications by basic researchers similar to some of the impact statements required in grants for human subjects research?

At this point the results of this analysis are more descriptive than prescriptive but at the very least they help explain the source of much frustration with a resource intensive and emotionally charged aspect of health science policy. My investigations thus far suggest that translational research efforts of NCI, especially as they relate to the implications of emerging technologies such as nanotechnology on health disparities, lack a unifying framework for discourse in the translation between the various basic, developmental and clinical research groups. My research in public value mapping indicates that it holds promise in this respect, yet it needs further testing and development, especially with respect to identifying and analyzing public values statements in frequency and context and in the way public values change over time.

Potential public value failure of nanotechnology in cancer is not entirely a government policy/program problem, which is why I believe it lends itself especially well to the PVM analytic framework. For example, in addition to the short term inequitable access of minorities to nanotechnology trials, the long-term effects could be additive and cumulative as a result of the current system for technology development and commercialization in medicine (Hede and Huilgol 2006). Government may be able to encourage basic research in areas that are more equitable in impact even if less profitable and it may encourage minority participation in trials. However, equitable access to the technology depends on private funding of commercialization initiatives and insurance coverage and cannot be assured without significant changes in national health care policy that are largely unrelated to research policies. A case in point is that the 1993 National Institutes of Health Revitalization Act has done little to increase black participation in clinical trials, due at least in part to private health care cost and access issues faced by minority populations in particular (Murthy et al. 2004).

The public value implications roll out in other ways as well. The impact of emerging technologies such as nanotechnology on health disparities have been the subject of much discussion due to unknown potential for biological differences between ethnic/racial groups and their influence on the effectiveness of certain treatments. The research community expects that nanotechnology will improve both the understanding of physiologic markers of disease and drug responsiveness that is individual and personalized, regardless of race and ethnicity. Until such technology is available and liberally tested on all subpopulations, the contentious racial differentiation in medical research (i.e. is race physiologically differentiated or socially constructed or both?) will continue to be a distraction from other important scientific issues (Yancy 2008).

The means of access to emerging and promising cancer technologies is no different for minorities than for the rest of the population: insurance coverage, adequate financial resources, proximity to providers, and the like. With portals to early access to emerging and promising technologies, like clinical trials, facing logical and empirically-evidenced barriers, clearly there are many aspects of policies for emerging technologies that need greater attention in terms of social imperatives. Success in the development of health technologies is as much a societal issue as a health system institutional issue because individual health has societal impact. This is yet another reason for PVM treatment. For example, too little is known about what compels minorities to pursue or agree to experimental treatment or accessing novel technologies (Garber and Arnold 2006). The reasons for disparities in access to emerging technology are not completely clear either, but rarely is the success or failure of a clinical trial primarily focused on societal implications like these (Getz and Faden 2008; Garber and Arnold 2006; Pinto et al. 2000). Rather, inequities are described in terms of barriers as opposed to failures (Ford et al. 2008). A public value mapping treatment would bring a more organized approach to consideration of social implication and values, irrespective of or even in conjunction with discussion of market efficiency.

An emerging body of theoretical and empirical work suggests that greater awareness of social contexts among basic researchers can influence choices that the researchers make (e.g., Guston and Sarewitz 2002). I suggest that there is a need for better understanding of basic researcher awareness of public values, such as equity in access to novel medical technologies. If basic researcher concern about equity is well-established then the PVM framework can be used to understand how these values are translated from the earlier stages in the science process to clinical applications. If the basic researcher concerns about public values like equity are more limited, then the PVM framework might be used to incorporate these values earlier in the scientific process. Either way, and given that scientists have many choices available to them about what research paths to follow and how to follow them, better understanding of the ways that basic researchers regard public values such as equity and ensuing challenges such as health disparities could be beneficial in research funding policy development by tracking how research agendas evolve with greater or lesser likelihood of public value success. In addition, because PVM counters conventional economic thinking, better understanding about the ways that public values factor into the related research incentives, and strategies that facilitate scientist success in the technology transfer process, could be beneficial in designing policy that balances market-driven concerns against the need to address social problems such as health disparities.

References

Agres, T. 2004. Opportunity awaits small thinkers. *Drug Discovery and Development* 7 (2): 15.

Anderson, M.M. 2006. Summer conference. Embryonic innovation: Path creation in nanotechnology. Paper presented at the DRUID Conference on Knowledge, Innovation and Competitiveness; Dynamics of Firms, Networks, Regions and Institutions, Copenhagen, Denmark.

Bainbridge, W.S. 2004. Sociocultural meanings of nanotechnology: Research methodologies. *Journal of Nanoparticle Research* 6: 285–299.

Best, R., and G. Khushf. 2006. The social conditions for nanomedicine: Disruption, systems, and lock-in. *Journal of Law, Medicine and Ethics* 34 (4): 733–740.

Bozeman, B. 2003. Public value mapping of science outcomes: Theory and method. In *Knowledge flows and knowledge collectives: Understanding the role of science and technology policies in development*, ed. Daniel Sarewitz, 3–48. Center for Science, Policy and Outcomes. Vol 2.

Bozeman, B. 2007. *Public values and public interest: Counterbalancing economic individualism.* Washington, DC: Georgetown University Press.

Bozeman, B., and M. Gaughan. 2002. Public value mapping: The case of breast cancer research. New York, NY: Center for Science, Policy and Outcomes—Report to the Rockefeller Foundation.

Bozeman, B., P. Laredo, and V. Mangematin. 2007. Understanding the emergence and deployment of 'nano' SandT: Introduction. *Research Policy* 36 (6): 807–812.

Bozeman, B., and D. Sarewitz. 2005. Public values and public failure in US science policy. *Science and Public Policy* 32 (2): 119–136.

Bruner, D.W., M. Jones, D. Buchanan, and J. Russo. 2006. Reducing cancer disparities for minorities: A multidisciplinary research agenda to improve patient access to health systems, clinical trials, and effective cancer therapy. *Journal of Clinical Oncology* 24 (14): 2209–2215.

Cicatiello, J.S. 2000. A perspective of health care in the past: Insights and challenges for a health care system in the new millennium. *Nursing Administration Quarterly*, 25 (1): 18–23.

Colon-Otero, G., R.C. Smallridge, L.A. Solberg Jr., T.D. Keith, T.A. Woodward, F. B. Willis, et al. 2008. Disparities in participation in cancer clinical trials in the United States. *Cancer* 112 (3): 447–454.

De Melo-Martin, I. 2009. Creating reflective Spaces: Interactions between philosophers and biomedical scientists. *Perspectives in Biology and Medicine* 52 (1): 39–47.

Emerich, D.F., and C.G. Thanos. 2003. Nanotechnology and medicine. *Expert Opinion On Biological Therapy* 3 (4): 655–657.

Faunce, T., and K. Shats. 2007. Researching safety and cost-effectiveness in the life cycle of nanomedicine. *Journal Of Law And Medicine* 15 (1): 128.

Feeney, M.K., and B. Bozeman. 2007. Public values and public failure. *Public Integrity* 9 (2): 175–190.

Fisher, E., and R.L. Mahajan. 2006. Contradictory intent? US federal legislation on integrating societal concerns into nanotechnology research and development. *Science and Public Policy* 33: 5–16.

Fisher, E., C. Selin, and J. Wetmore, eds. 2008. *The yearbook of nanotechnology in society, Vol. 1: Presenting Futures.* New York, NY: Springer Science and Business Media.

Ford, J.G., M.W. Howerton, G.Y. Lai, T.L. Gary, S. Bolen, M.C. Gibbons, et al. 2008. Barriers to recruiting underrepresented populations to cancer clinical trials: A systematic review. *Cancer* 112 (2): 228–242.

Garber, M., and R. Arnold. 2006. Promoting the participation of minorities in research. *American Journal of Bioethics* 6 (3): W14–W20.

Getz, K., and L. Faden. 2008. Racial disparities among clinical research investigators. *American Journal of Therapeutics* 15 (1): 3–11.

Goorden, L., M. Van Oudheusden, J. Evers, and M. Deblonde. 2008. Nanotechnologies for tomorrow's society: A case for reflective action research in Flanders, Belgium, In *The Yearbook of*

Nanotechnology in Society, Vol. 1: Presenting Futures, ed. E. Fisher, C. Selin, and J. Wetmore. New York, NY: Springer Science and Business Media.

Gupta, A. 2002. Ensuring 'safe use' of biotechnology in India: Key challenges. *Economic and Political Weekly* 6: 2762–2769.

Guston, D.H. 2000. *Between politics and science: Assuring the integrity and productivity of research.* New York, NY: Cambridge University Press.

Guston, D.H. 2008. Innovation policy: Not just a jumbo shrimp. *Nature* 454 (7207): 940–941.

Guston, D.H., and Sarewitz, D. 2002. Real-time technology Assessment. *Technology in Society* 24: 93–109.

Healthy People 2010 and United States Department of Health and Human Services. 2000. *Healthy people 2010: Understanding and improving health.* Washington, DC: U.S. Department of Health and Human Services.

Heath, J.R., and M.E. Davis. 2008. Nanotechnology and cancer. *Annual Review of Medicine* 59: 251–265.

Hede, S., and N. Huilgol. 2006. 'Nano': The new nemesis of cancer. *Journal of Cancer Research and Therapeutics* 2 (4): 186–195.

Kahn, K.L. 2008. Moving research from bench to bedside to community: There is still more to do. *Journal of Clinical Oncology* 26 (4): 523–526.

Kahneman, D., and A. Tversky, eds. 2000. *Choices, values and frames.* New York, NY: Cambridge University Press.

Lenk, C., and N. Biller-Andorno. 2007. Nanomedicine-emerging or re-emerging ethical issues? A discussion of four ethical themes. *Medicine, Health Care, And Philosophy* 10 (2): 173–184.

Long, J.A., V.W. Chang, S.A. Ibrahim, and D.A. Asch. 2004. Update on the health disparities literature. *Annals of Internal Medicine* 141: 805–812.

Mehta, M.D. 2004. The future of nanomedicine looks promising, but only if we learn from the past. *Health Law Review* 13: 16–18.

Murashov, V., and J. Howard. 2008. The U.S. must help set international standards for nanotechnology. *Nature Nanotechnology* 3: 635–636.

Murthy, V.H., H.M. Krumholz, and C.P. Gross. 2004. Participation in cancer clinical trials: Race-, sex-, and age-based disparities. *JAMA: Journal of the American Medical Association* 291 (22): 2720–2726.

National Cancer Institute. 2006. *The NCI strategic plan for leading the nation to eliminate the suffering and death due to cancer* (No. NIH Publication No. 06-5773). Bethesda, MD: Department of Health and Human Services.

National Science and Technology Council. 2008. The science of science policy: A federal research roadmap. Report on the Science of Science Policy to the National Science and Technology Council, Office of Science and Technology Policy. November.

Nelson, R. 2000. *The sources of economic growth.* Cambridge: Harvard University Press.

Norton, B., and D. Noonan. 2007. Ecology and valuation: Big changes needed. *Ecological Economics* 63 (4) 664–675.

Pinto, H.A., W. McCaskill-Stevens, P. Wolfe, and A.C. Marcus. 2000. Physician perspectives on increasing minorities in cancer clinical trials: An Eastern Cooperative Oncology Group (ECOG) initiative. *Annals of Epidemiology,* 10 (8 Suppl): S78–284.

Resnik, D.B., and S.S. Tinkle. 2007. Ethical issues in clinical trials involving nanomedicine. *Contemporary Clinical Trials* 28: 433–441.

Roco, M.C., and W.S. Bainbridge. 2005. Societal implications of nanoscience and nanotechnology: Maximizing human benefit. *Journal of Nanoparticle Research* 7, 1–13.

Romig Jr, A.D., A.B. Baker, J. Johannes, T. Zipperian, K. Eijkel, B. Kirchhoff, H.S. Mani, C.N.R. Rao, and S. Walsh. 2007. An introduction to nanotechnology policy: Opportunities and constraints for emerging and established economies. *Technological Forecasting and Social Change* 74: 1634–1642.

Sarewitz, D., and E. Woodhouse. 2003. Small is powerful. In *Living with the genie: Essays on technology and the quest for human mastery,* ed. A.P. Lightman, D.R. Sarewitz, and C. Desser, 307–336. Washington, DC: Island.

Scheufele, D.A., E.A. Corley, T. Shih, K. Dallrymple, and S. Ho. 2009. Religious beliefs and public attitudes toward nanotechnology in Europe and the U.S. *Nature Nanotechnology* 4 (2): 91–94.

Scheufele, D.A. and B.V. Lewenstein. 2005. The public and nanotechnology: How citizens make sense of emerging technologies. *Journal of Nanoparticle Research* 7:651–667.

Slade, C. Forthcoming. Public value mapping of equity in emerging nanomedicine. *Minerva*.

Sparrow, R. 2009. The social implications of nanotechnology: An ethical and political analysis, *Journal of Bioethical Inquiry* 6 (1): 13–23.

Swanson, G.M., and A.J. Ward. 1995. Recruiting minorities into clinical trials toward a participant-friendly system. *Journal of the National Cancer Institute* 87: 1747–1759.

Wonglimpiyarat, J. 2005. The nano-revolution of Schumpeter's Kondratieff cycle. *Technovation* 25: 1349–1354.

Yancy, C.W. 2008. Race-based therapeutics. *Current Hypertension Reports* 10: 276–285.

Chapter 5
Ableism and Favoritism for Abilities Governance, Ethics and Studies: New Tools for Nanoscale and Nanoscale-enabled Science and Technology Governance

Gregor Wolbring

The values that shape what nanotechnologies are developed are wide and varied. Some hope to remedy environmental problems, others desire to cure cancer, and still others are looking for the next "indispensible" consumer product; but underlying these goals are deeper values that we rarely think about. Gregor Wolbring argues that many of these goals are shaped by our vision of which abilities are desirable and which are to be avoided. Wolbring calls this moral judgment of abilities "ableism," and he uses it to show how even people with the best of intentions can help to create an increasingly inequitable world. He explores human enhancement technologies—a favorite theme of futurist portrayals of nanotechnology—to show how developing nanotechnologies that further certain abilities can create new inequities. He argues that the development of nano-enhanced bodies may make those deemed to have physically perfect bodies today the disabled of the future. This may be a wake up call for all of us—forcing us to realize that even if we presently have power and status, if we don't consider the ways in which nanotechnology may help to create inequities, we may find ourselves on the other side of a nano-divide.—eds.

G. Wolbring (✉)
Department of Community Health Science, Program in Community Rehabilitation and Disability Studies, Faculty of Medicine, University of Calgary, Calgary, AB, Canada
e-mail: Gregor.wolbring@asu.edu

Originally presented at the Workshop on Nanotechnology, Equity, and Equality at Arizona State University on November 21, 2008.

5.1 Introduction

Nanoscale S&T[1] has huge positive potential. However, bringing the positive potential to fruition depends on the right social environment and foresight to identify societal and other problems and the willingness to address them. The research, development, and use of nanoscale S&T embodies and shapes the perspectives, purposes, prejudices, particular objectives, and cultural, economical, ethical, moral, spiritual, and political frameworks of different social groups and society at large. Nanoscale S&T influences and is influenced by various discourses and areas of action. Ableism is one concept that impacts the direction, vision, and application of nanoscale S&T, and vice versa.

Ableism is a set of beliefs, processes, and practices that produce, based on one's abilities, a particular kind of understanding of one's self, one's body, and one's relationship with others of one's species, other species, and one's environment, and includes one being judged by others (Wolbring 2008a). Ableism reflects the sentiment of certain individuals, households, communities, groups, sectors, regions, countries, and cultures to cherish and promote certain abilities over others. For instance, in much of the Western world abilities such as productivity and competitiveness are seen as more valuable than abilities such as empathy, compassion, and kindness (Wolbring 2008a).

Advances in nanoscale S&T, and nanoscale enabled S&T including molecular manufacturing, biotechnology, information technology, cognitive science, neuro-engineering, synthetic biology, and geo-engineering, increasingly make possible a new transhumanized form of ableism. The ability of science and technology research and development (R&D) products to modify the appearance and functioning of biological structures, including the human body, beyond existing norms and species-typical boundaries makes it possible for people to be judged for having (or not having) abilities that were not previously available to human beings. This transhumanized form of ableism (Wolbring 2008a) is a set of beliefs, processes, and practices based on the idea that: (a) the improvement of functioning of biological structures beyond typical boundaries is essential; and (b) species-typical biological structures are limited, defective, and in need of constant improvement beyond biological structure-typical boundaries (Wolbring 2008a).

Existing discourses on nanotechnology do address equity, equality, human security, social cohesion, medical health and environmental safety, and distributive justice. However, the ableism angle to nanotechnology in general and to these issues in particular is rarely, if ever, covered despite the impact that ableism has on the direction and vision of nanoscale S&T. The purpose of this paper is to introduce the reader to: (a) the concept of ableism; (b) ableism and ability inequity, inequality, insecurity and injustice; (c) the transhumanization of ableism; and (d) four new fields of academic and non-academic inquiry—ableism ethics, ableism governance, ableism foresight and ableism studies—that might be additional tools usable in nanoscale S&T governance. The paper contends that there is a pressing need for society to deal with ableism in all of its forms and consequences, including its impact on science and technology discourses in general and nanoscale S&T

discourses in particular. To set the stage it first outlines the definition of nanotechnology that it will use, the type of ethicist that is needed to understand the ethical implications of nanotechnology and other emerging technologies, and a more full description of what ableism is and how it works.

5.2 Setting the Stage

5.2.1 What Meaning of Nanotechnology Does This Paper Use?

Various definitions of nanotechnology are employed by different people. The author's understanding is in sync with the definition used by the International Organization for Standardization Technical Committee 229 on Nanotechnology (ISO/TC229); the organization that is charged to produce standards for classification, terminology and nomenclature, basic metrology, and calibration (International Organization for Standardization (ISO) 2008). This definition is based on the concept of nanoscale and covers: (a) many different nanoscale science and technology products and processes; (b) any science and technology fields with nanoscale aspects; and (c) nanoscale-enabled science and technology fields, product and processes. For the author, nano "scale" is the important aspect. In the same way that the move from millimeter to micrometer allowed for many new products and processed to be developed, so will the move towards the nanoscale lead to many new applications, products and processes? Nanoscale products and processes will be ubiquitous and will impact every facets of life on earth.

5.2.2 Creating Ethicists that Could Inform Policy

Today's scientific news seems to become yesterday's news, fast replaced by even more astonishing news. One field of science is chased by another at an ever-increasing speed. It seems every major science and technology field has ethics attached to it. Some examples are information technology ethics, bioethics, medical ethics, neuroethics, gene-ethics, nanoethics, and synthetic biology ethics. Some question the validity of nanoethics. For instance bioethicist Holm (quoted in (Althoff and Lin 2008)) rejects the notion of a nanoethics field with the argument that no "ethical issues are raised by nanoscale science and technologies that are not already raised by other kinds of technologies." Others believe that the nanoscale synthetic biology work on a minimal genome does not cross any moral boundaries (Cho et al. 1999). But the question is whether a science and technology driven ethics field is still useful and can give guidance as how to use science and technology.

Ever since the 2001 workshop on "Nanotechnology, Biotechnology, Information technology and Cognitive science (NBIC): Converging Technologies for Improving Human Performance," organized by the U.S. National Science Foundation and the

U.S. Department of Commerce (Roco and Bainbridge 2003), the convergence of different science and technologies seems to be the buzz. If this buzz has a real foundation one might argue we need a "converging science and technology" ethicist, an ethicist who can cover numerous science and technology fields. The nanoscale ethicist might be best positioned to be just such a converging science and technology ethicist (CT ethicist). He or she would have the ability to look at ethical issues arising from the convergence of different science and technologies, especially the convergences enabled by nanoscale characteristics.

If this was achieved, however, one could argue that it is still not enough. Arguably the NBIC report (Roco and Bainbridge 2003) that resulted from this workshop had two main impacts. First, it moved the focus away from any given science and technology towards a set of them. And second, it moved the focus away from any one science and technology towards a goal driven focus for which different science and technologies are needed. We might, therefore, need goal based ethicists who understand and are aware of all the existing and envisioned science and technologies as they relate to the goal. Independent of how we brand the ethicist, if goals really are the driver it might make sense to look at what drives the generation of goals and to compliment science and technology governance with goal governance. Changing the purview of ethicists from focusing on the ethics in one science and technology field to the ethics of establishing goals for converging science and technology might also make sense in light of the fact that it seems new science and technology fields are being generated at ever increasing speeds. This would also open the door for ethicists to study ableism as it is one concept that shapes the goals people put forward.

5.2.3 Ableism and Favoritism for Abilities

Ableism in its general form is a set of beliefs, processes, and practices that produce, based on one's abilities, a particular kind of understanding of one's self, one's body, and one's relationship with others of one's species, other species, and one's environment, and includes one being judged by others. Ableism is a favoritism for certain abilities that are projected as essential, while at the same time labeling real or perceived deviation from or lack of these essential abilities as a diminished state of being. Those deemed as not having the proper abilities often experience disablism (Miller et al. 2004), the discriminatory, oppressive, or abusive behavior arising from the belief that people without these "essential" abilities are inferior to others. Ableism is one of the most societally entrenched and accepted "isms" and it exists in many forms such as biological structure based ableism, cognition based ableism, ableism inherent to a given economic system, and social structure based ableism (Wolbring 2008a).

Ableism based on biological structure is a set of beliefs, processes and practices that favors species-typical normative body structure based abilities and labels subnormative species-typical biological structures as: deficient, as not able to perform as required, as being in need of fixing, and/or as a diminished state of being (seen as the healthy state). This idea, or prejudice, has a long history and is often linked to the discourse around health, disease, and medicine. People who are seen as subnormative are labeled as being in ill health and are the main target of medical interventions. The discourse around deafness and Deaf Culture would be one example. While many hearing people expect the ability to hear and believe deafness is a deficiency to be treated through medical means, many deaf people see hearing as an irrelevant ability and do not perceive themselves as ill. This form of ableism which labels subnormative as a state of ill health is critiqued within the disabled people rights discourse and by the academic field of disability studies (Campbell 2001; Carlson 2001; Overboe 2007). These fields question the assumption of deficiency intrinsic to non-normative sub species-typical body abilities and the favoritism for normative species-typical body abilities.

However ableism is much more pervasive and not limited to areas where species-typical abilities are expected and sub species-typical abilities are seen as deficiencies. The favouritism of abilities is rampant today in many forms such as biological structure based ableism, cognition based ableism, social structure based ableism, or ableism inherent to a given economic system.

Ableism has been used historically and still is used by various social groups to justify their elevated level of rights and status in relation to other social groups, other species, and to the environment they live in (Wolbring 2008a,b,d). The favoritism of abilities is rampant today and is inherent in, or contributes to, other "isms" such as:

Racism—It is often stated that the favored race has superior cognitive abilities over other races;

Sexism—At the end of the nineteenth century women were viewed as biologically fragile (lacking the preferred ability of strength) and emotional (thereby possessing an undesirable ability), and thus incapable of bearing the responsibility of voting, owning property, and retaining custody of their own children (Wasserman et al. 1998; Wolbring 2003b);

Caste-ism—As Gail Omvedt (2001) said in the *Hindu*, a daily newspaper in India: "For caste, like race, is based on the notion that socially defined groups of people have inherent, natural qualities or 'essences' that assign them to social positions, make them fit for specific duties and occupations." I submit that the natural inherent qualities are "abilities" which make them fit for specific duties and occupations.

Ageism—Missing the abilities one has as a youth;

Speciesism—The elevated status of the species *homo sapiens* is often justified by stating that the homo sapiens has superior cognitive abilities;

Anti-Environmentalism—The disregard for nature reflects another form of ableism: humans are here to use nature as they see fit as they see themselves as superior to nature because of their abilities. We might see a climate change-driven appeal for a transhuman version of ableism where the transhumanisation of humans is seen as a solution for coping with climate change. This could become especially popular if we reach a "point of no return," where severe climate change consequences can no longer be prevented.

GDP-ism—The ability of a country to produce; and

Consumerism—The ability to consume whatever one wants.

A quick glance at this list demonstrates that ableism is the basis of and permeates many of the preferences that have shaped society in the past and will likely shape the future.

5.3 The Challenges Today and to Come: The World of Ableism

5.3.1 Ableism and Nanoscale Science and Technologies

Science and technology, including nanoscale S&T research and development, and different forms of ableism have always been and will continue to be inter-related. The desire and expectations for certain abilities have led and will continue to lead to the support of science and technology research and development that promises the fulfillment of these desires and expectations. Science and technology research and development has led and will continue to lead to products that enable new abilities, and expectations and desires for new forms of abilities, making possible new forms of ableism. Science and technology developed in the future—in particular nanoscale enabled S&T—will continue to enable the modification of appearance and functioning of non-biological structures and increasingly enable the modification of appearance and functioning of biological structures, including the human body and the bodies of other species, beyond existing norms and boundaries within and between species. Unless there is a radical shift in how we judge abilities, these new technologies will lead to a changed understanding of one's self, one's body, and one's relationship with others of one's species, other species, and one's environment. Accompanying these changes will be a change in anticipated, desired, and rejected abilities, and the transhumanization of the ableism (Wolbring 2008a).

The desire to change human abilities is not just implicit in this process, but explicit. The report that resulted from the landmark 2001 NSF NBIC workshop (Roco and Bainbridge 2003), which stressed the idea of the convergence of various sciences and technologies under the umbrella of nanoscale, mentions the term "productivity" 60 times, "efficiency" 54 times, and "competitive" 26 times (Wolbring 2008a). The report and the workshop reflect the belief that human performance enhancement is a way to fulfill the goals of productivity, efficiency and competitiveness. The focus on human performance enhancement changes the character of ableism. Under given social dynamics, the increasing availability of human performance enhancements enables the transhumanization of ableism.

The challenge created by these new abilities will be the new types of discrimination that will likely come with them. If the expectation becomes widespread that enhancement of body abilities beyond species-typical boundaries is needed to achieve certain desired goals it will come with consequences for those who do not have access to or refuse to adopt these enhancements. This group, which I have called the "techno-poor impaired and disabled people," (Wolbring 2007; Wolbring 2008a–d) will likely suffer from a variety of forms of disablism. In the same way that today's "normal" impaired and disabled people experience disablism due to the fact that a form of ableism cherishes species-typical functioning and sees and treats people not fitting that norm negatively, in the future one can expect that under a transhumanized form of ableism the techno-poor impaired and disabled (those who can't afford the latest body "upgrades") will be perceived as people in a diminished state of being human and will experience disablism (Miller et al. 2004) accordingly.

This section has looked at the intersection between ableism and science and technology. The next two sections look at the human security and the equality and equity discourses, discourses that are central to the medical and social wellbeing of disabled people and others who are on the negative side of the ableism sentiments, and argue that an ability angle can make a positive contribution to these discourses.

5.3.2 Ability Security and Self-Identity Security

Within the human security discourse one reads a lot about economic security, food security, health security, environmental security, personal security, community security, and political security. As much as nanoscale S&T advances impact these facets of human security, and as much as this area is underserved within the nanodiscourses (Wolbring 2010), the author submits here that emerging and envisioned science and technology and nanoscale S&T processes and products pose additional challenges to human security and generate the need to expand the list of human security aspects to include ability security (Wolbring 2006) and self-identity security (Wolbring 2006), concepts not typically covered in the human security discourse.

"Ability security" means that one is able to live a decent life with whatever set of abilities one has, and that one will not be forced to have a prescribed set of abilities to live a secure life (e.g. even if one doesn't have the ability to walk, he or she should be able to secure employment) (Wolbring 2006). "Self-identity security" could be seen as a subset of personal security and means that one is accepted with one's set of abilities and that one should not be forced (physically or by circumstance) to accept a perception of oneself one does not agree with (e.g. one is not expected to have the ability to walk or is seen as a "deficient product" if one cannot walk) (Wolbring 2006).

Although "ability security" and "self-identity security" are not really used as such in the disabled people rights discourse, many of the demands from disabled people could be collected under these terms. The United Nations Convention on the Rights of Persons with Disabilities, for instance, can be seen as an international document that provides a harmonized framework for, among other things, ability security of disabled people as it is calls for "their participation in the civil, political, economic, social, and cultural spheres with equal opportunities, in both developing and developed countries" (United Nations 2007). The basic idea of self-identity security can also be seen in numerous other actions and writings such as the slogan "the right to be different," the different disability pride events, the concept of Deaf culture, and numerous academic and non-academic articles that denounce the medical model's labeling of people as "disabled" even if they do not agree with such label (see Wolbring 2003b for many references). Many of these works conclude that self-identity security for disabled people does not exist, but should.

Although ability and self-identity security are topics of concern for disabled people they are also of concern to many other people as well, at least in some point of their lives. For instance, ability security is very important to the elderly who sometimes feel they are hindered in having a decent life due to lacking certain abilities of the youth. Self-identity security is of importance, for instance, to women and ethnic and cultural oppressed groups who are punished simply for who they are. Women, simply for being women, are treated badly in many corners of the world. Indigenous people were and are still in many places hindered in their attempt to live out their traditional knowledge and their way of seeing the world. Their knowledge is ignored and seen as inferior to other knowledge.

As we are able to further enhance the abilities of the body through science and technology, ability and self-identity security will increasingly be of importance to every person who does not want to modify their body according to the wishes of others or who cannot afford these modifications. People who currently see themselves as the norm, as non-impaired or "species-typical," will be seen by others as impaired or deficient if they do not enhance themselves according to "new norms," or the wishes of powerful others.

We have seen for decades, if not centuries, that when science and technology develops new tools that are external to the body, people often have to learn how to use these devices so that they can capitalize on the new abilities if they want to keep their job or get a new job. Although these non-impaired species-typical labeled people might not see it as such, they also have no ability security. They are forced to adopt the latest advances or they will be unemployable. For instance as the typewriter diffused across the United States, secretaries with perfect penmanship (a long valued ability) were replaced by secretaries who could type at high speeds. If predictions of new and emerging nanoscale S&T products and processes come true, and if we focus a lot on the ability of competitiveness and consumerism, we will see this dynamic not only speed up we might see a move towards the transhumanization of ableism. Those people now considered non-impaired, species-typical, will no longer have a competitive set of abilities to be considered fit workers. They will have to decide whether they want a job badly enough to modify their body so they meet the new expected types of abilities.

At first this statement may seem startling, but we as a society have already begun to accept body modification as a routine part of maximizing our abilities so that we can be productive members of society. A recent survey of 1,427 readers of the journal *Nature* revealed that 79% felt that healthy people should be allowed to take cognitive enhancers (Maher 2008). When asked whether healthy children under the age of sixteen should be restricted from taking these drugs, unsurprisingly most respondents (86%) said that they should. One-third of respondents said they would feel pressure to give cognition-enhancing drugs to their children if other children at school were taking them. Some believe that drugs that can modify the cognition and emotions of people will move into the mainstream between 2010 and 2040 and "competitive advantage will come not just from managing knowledge generated within your company, but by cogniceutically managing the ability of your employees to learn, think, be creative" (Hind 2005).

5.3.3 Ability Equality and Equity

The impact of nanoscale S&T onto equality and equity and the existence of inequity and inequality is the theme of this Yearbook. Many definitions exist for equity and equality. However none of them have an ability angle. In order to properly address the issues outlined above, this needs to be remedied. The author submits that: (a) definitions for ability equity, equality, inequity and inequality are essential for linking the Ableism discourse to the equity and equality discourses and (b) that an ability angle is essential to look at in the equity and equality discourse. Many definitions for equity and equality exist. In Table 5.1 below and in the paragraphs that follow, I cite a number of different framings and generate a transmuted version of them with an ability angle.

As most people are not familiar with the concept of ability inequity and inequality the author submits here some examples. For both ability inequity and ability inequality two subgroups exist. One group is linked to intrinsic bodily abilities and the other group is linked to external abilities, abilities generated by human interventions that impact humans. These two subgroups of internal and external ability inequities and inequality are quite distinct in their effects and discourse dynamics, involved stakeholders and goals.

Definition: *Ability inequality* is a descriptive term denoting any uneven distribution of access to and protection from abilities generated through human interventions, right or wrong (modified from Cozzens 2007).

Situations that meet this definition of ability inequality are plentiful. For example the majority of blind people do not have access to the content of webpages as most webpages are not designed with their access in mind. If specific jobs and education require such increased expectations on abilities, disabled people who do not have access to them will very likely lose out. When the telephone was developed by those who privileged hearing (i.e. those who can hear), it led to uneven distribution of access. The deaf were not able to use the new technology, and when it became increasingly essential for work and basic socialization to communicate quickly over distances, their ability to be productive and active members of society was lessened. When the walking found ways to build stairs that suited them they added floors to buildings. The disabled ones who could not go up the stairs were immediately denied the ability to partake in everything happening on the floors accessible by stairs.

Ability inequalities also are experienced by so called species-typical people. Eating certain food leads to better abilities, but not everyone has access to this food. Clean water leads to better abilities, but not everyone has access to it. And when some modify their bodies and add to their abilities not everyone will be able to follow suit. "Enabling" enhancements will lead to ability inequalities for those who do not have access to them or who choose not to modify their bodies.

Definition: *Ability inequality* is a descriptive term denoting any uneven judgment of abilities intrinsic to biological structures such as the human body, right or wrong (modified from Cozzens 2007).

The author covered elsewhere (Wolbring 2003b) many references that highlight the negative judgments made of disabled people and their intrinsic set of abilities.

Table 5.1 Transmutation versions of equity and equality definitions

Original definition	Definition where the underscored is replaced by an ability angle
Gender equality, equality between *men and women*, entails the concept that all human beings, both men and women, are free to develop their personal abilities and make choices without the limitations set by stereotypes, rigid *gender roles* and prejudices. *Gender* equality means that the different behavior, aspirations and needs of *women and men* are considered, valued and favored equally. It does not mean that *women and men* have to become the same, but that their rights, responsibilities and opportunities will not depend on *whether they are born male or female* (Unit for the Promotion of the Status of Women and Gender Equality UNESCO 2000)	*Ability* equality, equality between *people of different abilities*, entails the concept that all human beings, both men and women, are free to develop their personal abilities and make choices without the limitations set by stereotypes, rigid *ability expectations* and prejudices. *Ability* equality means that the different behavior, aspirations and needs of *people of different abilities* are considered, valued and favored equally. It does not mean that *people with different abilities* have to become the same, but that their rights, responsibilities and opportunities will not depend on *their abilities*
Gender equity means fairness of treatment for *women and men* according to their respective needs. This may include equal treatment or treatment that is different but which is considered equivalent in terms of rights, benefits, obligations and opportunities. (Unit for the Promotion of the Status of Women and Gender Equality UNESCO 2000).	*Ability* equity means fairness of treatment for *people with different abilities*, according to their respective needs. This may include equal treatment or treatment that is different but which is considered equivalent in terms of rights, benefits, obligations and opportunities.
Equality, a descriptive term denoting any even distribution, right or wrong. (Cozzens 2007)	Ability equality has two aspects. One aspect deals with non-bodily abilities and the other is linked to bodily abilities: *Ability* equality, a descriptive term denoting any even distribution *of access to and protection from abilities generated through human interventions,* right or wrong. *Ability* equality a descriptive term denoting any *factual judgment of abilities intrinsic to biological structures such as the human body,* right or wrong
Equity, a normative term denoting a just or fair distribution (Cozzens 2007)	Ability equity has two aspects. One aspect deals with non-bodily abilities and the other is linked to bodily abilities: *Ability* equity, a normative term denoting a just or fair distribution *of access to and protection from abilities generated through human interventions* *Ability* equity, a normative term denoting a just or fair judgment *of abilities intrinsic to biological structures such as the human body*

Table 5.1 (continued)

Original definition	Definition where the underscored is replaced by an ability angle
Inequity, a normative term denoting an unjust or unfair distribution (Cozzens 2007)	Ability inequity has two aspects. One aspect deals with non-bodily abilities and the other is linked to bodily abilities: *Ability* inequity, a normative term denoting an unjust or unfair distribution of *access to and protection from abilities generated through human interventions* *Ability* inequity, a normative term denoting an unjust or unfair *judgment of abilities intrinsic to biological structures such as the human body*
Inequality a descriptive term denoting any uneven distribution right or wrong (Cozzens 2007)	Ability inequality has two aspects. One aspect deals with non-bodily abilities and the other is linked to bodily abilities: *Ability* inequality a descriptive term denoting any uneven distribution *of access to and protection from abilities generated through human interventions*, right or wrong *Ability* inequality a descriptive term denoting any *uneven judgment of abilities intrinsic to biological structures such as the human body*, right or wrong
Inequality is the unequal distribution of something people value (Cozzens et al. 2008)	Ability inequality is the unequal distribution of access to and protection from something people value based on difference in people's abilities: *Ability* inequality is the unequal distribution of *access to and protection from abilities generated through human intervention that* people value. *Ability* inequality is the unequal distribution of *access to and protection from something people value such as abilities generated through human interventions whereby some are intrinsic abilities of a modified biological entity (i.e. the ability of the human eye to see infrared) whereas others are abilities generated by human activities (i.e. ability to use a car).*

Many are cemented into law such as the inequality related to anti-genetic discrimination laws which only protect people who are identified as having a gene that might lead in the future to the loss of abilities favored by the norm, but do not protect the people who have actually lost the abilities demanded by the norm (Wolbring 2004). The negative judgments made of disabled people and their intrinsic set of abilities is also linked to science and technology advances such as genetic tests which exist to detect subnormative genetic and body structure compositions that are linked to subnormative abilities (Wolbring 2003a; The International Sub-Committee of BCODP 2000).

Definition: *Ability inequity* is a normative term denoting an unjust or unfair distribution of access to and protection from abilities generated through human interventions (modified from Cozzens 2007).

One could say that one of the purposes of the United Nations Convention on the Rights of Persons with Disabilities (United Nations 2007) was to highlight which ability inequities are unjust and to prescribe some remedies for them. As the report of the convention states: "The purpose of the convention is to promote, protect and ensure the full and equal enjoyment of all human rights by persons with disabilities" (United Nations 2007, article 1). It covers a number of key areas such as accessibility, personal mobility, health, education, employment, habilitation and rehabilitation, participation in political life, and equality and non-discrimination. The convention marked a shift in thinking about disability from a social welfare concern to a human rights issue, which acknowledges that societal barriers and prejudices are themselves disabling (United Nations 2007, FAQ).

Definition: *Ability inequity* is a normative term denoting an unjust or unfair judgment of abilities intrinsic to biological structures such as the human body (modified from Cozzens 2007).

The United Nations Convention on the Rights of Persons with Disabilities (United Nations 2007) negates the negative judgment linked to the abilities or perceived lack thereof of disabled people when it discusses in article 1 the respect for the inherent dignity of disabled people: "The purpose of the present Convention is to promote, protect and ensure the full and equal enjoyment of all human rights and fundamental freedoms by all persons with disabilities, and to promote respect for their inherent dignity" (United Nations 2007).

5.4 Favoritism for Abilities and Ableism: The way Forward

Thus far this paper has shown how the goals and visions for science and technology, including nanoscale S&T, are linked to the favoritism of abilities and the ableism one accepts. In turn, this favoritism can generate ability insecurity, inequality, inequity and injustice. Despite this alarming situation, there has been hardly any discourse about ableism (outside of the disabled people community), no in depth analysis of the problem, and no one has developed any policy recommendations. With such an immense problem, what can be done? The author contends that the first step must be to develop a better understanding of the issues and mechanisms. Therefore ableism ethics, ableism governance, ableism foresight and ableism and ability studies should be developed into academic and non-academic fields of inquiry to help with science and technology and nanoscale S&T governance and direction.

5.4.1 Ableism Ethics

Ableism ethics is a framework of standards and values that: (a) guides beliefs, processes and practices of ableism; (b) guides the favouritism for certain abilities and how one decides which abilities to favour over others; and (c) guides the reactions (disablism) towards humans and other biological entities that are seen—whether real or perceived—to lack these essential abilities.

A comprehensive study of ableism ethics should also include: (a) the perspectives of many different groups, especially of the people labelled as lacking certain "essential" abilities; (b) impact assessment of different forms of ableism onto different ethical theories and ethical principles including health ethics theories and their use to govern science and technology and health research, care, and policy; and (c) identification of ethical actions that flow from a favouritism for certain abilities.

It is believed that many negative consequences of S&T for humankind could be avoided by using ethical principles to govern them. As many of the negative consequences are linked to which abilities one favors and what ableism one adheres to, the author submits that in the same way that bioethics, nanoethics, neuroethics, and converging technology ethics are important so is ableism and favoritism for abilities ethics. Developing guidelines around which ableisms and favoritisms of abilities are ethical, e.g. which form and shape of competitiveness might be ethical and which might not, can be a useful tool for the governance of S&T.

5.4.2 Ability Studies

Favoritism for abilities and ableism studies (ability studies, for short) is about the investigation of: (a) the social, cultural, legal, political, ethical and other considerations by which any given ability may be judged and which lead to favoring one ability over another; (b) the impact and consequence of favoring certain abilities and rejecting others; (c) the consequences of ableism in its different forms, and its relationship with and impact on other isms; (d) the impact of new and emerging technologies on ableism and consequent favoritism towards certain abilities and rejection of others; and (e) identification of the abilities that would lead to the most beneficial scenario for the maximum number of people in the world.

5.4.3 Ableism Foresight

The ability to anticipate and understand shifting social dynamics enabled by advancing sciences and technologies is an essential tool for governance. With the ever faster changing advances in science and technologies and changes in social values it is increasingly important to reflect on potential scenarios so one is prepared for actions needed. To understand how ableism might change in the future is one essential area of foresight. Changing ableisms are an indicator for changing social values and social dynamics.

5.4.4 Ableism Governance

In the same way that we govern science and technology in order to maximise the positive potential of S&T and to minimize the negative aspects so do we have to govern favoritism for abilities and ableism to maximise the positive potential of the choice of abilities and to minimize the negative aspects ableism and the disablism linked to it.

5.5 Conclusion

If forms of ableism that favor productivity, efficiency, and competitiveness are main drivers, if not the main drivers, for envisioning and directing science and technology in general and nanoscale S&T products and processes in particular, one can expect product developments that further these forms of ableism. As long as these goals, these abilities, remain the standard by which institutions and individuals are measured, S&T ideas and projects that claim to be able to further these abilities will get considerably more attention than the concerns ableism scholars raise about them.

There may be some discourses currently playing out in the debates over nanotechnology that do get in the way of the push for greater productivity, efficiency, and competitiveness. For instance environmental and medical health safety issues are covered in the nano discourse and these may slow down the achievement of greater productivity, efficiency, and competitiveness. But they only slow that process down and in the end might benefit the ability to be productive, efficient, and competitive. Other prevalent issues linked to nano advances such as those discussed in this article receive much less attention. This is possibly because they simply are not noticed or those concerned about them have very little voice. Or it may in the end be because to rectify inequity, inequality, insecurity, injustice in general, and ableism and ability inequity, inequality, insecurity, and injustice in particular might impair the ability of nano to fulfill the productivity, efficiency, and competitiveness goals. To fully address and incorporate these values into the nano discourse might require that one abolishes competitiveness as a goal and replaces it with other goals such as maximum distribution.

The direction and governance of science and technology in general and nanoscale S&T in particular is linked to the ableism one exhibits and cherishes. There is a need to address the nearly unconscious acceptance of the desire for certain abilities and the emerging transhumanization of these desires. There is a need to look in a coherent fashion at ableism and favoritism for abilities. The author submits that four new academic and policy inquiry fields of favoritism for abilities and ableism ethics, governance, studies and foresight are useful tools for the governance of science and technology in general and nanoscale S&T in particular.

Note

1. As for the rest of the paper when I write nanoscale S&T I mean two areas; one being nanoscale science and technologies products and processes and the other being science and technology products and processes that are enabled by some nanoscale component but are in the end a mixture of nano and non-nano components.

References

Althoff, Fritz, and Patrick Lin. 2008. What's so special about nanotechnology and nanoethics? *International Journal of Applied Philosophy* 20 (2): 179–190.
Campbell, Fiona A.K. 2001. Inciting legal fictions: 'Disability's' date with ontology and the ableist body of the law. *Griffith Law Review* 10 (1): 42.
Carlson, L. 2001. Cognitive Ableism and Disability Studies: Feminist Reflections on the History of Mental Retardation. *Hypatia* 16(4):124–146.
Cho, Mildred K., David Magnus, Arthur L. Caplan, and Daniel McGee. 1999. Policy forum: Genetics. Ethical considerations in synthesizing a minimal genome. *Science* 286 (5447) (October 12): 2087–2090.
Cozzens, Susan E. 2007. Distributive justice in science and technology policy. *Science and Public Policy* 34 (2): 85–94.
Cozzens, Susan E., Isabel Bortagaray, Sonia Gatchair, and Dhanaraj Thakur. 2008. Emerging technologies and social cohesion: Policy options from a comparative study. Paper presented at the PRIME Latin America Conference, September 24–26, 2008. http://prime_mexico2008.xoc.uam.mx/papers/Susan_Cozzens_Emerging_Technologies_a_social_Cohesion.pdf. (accessed August 4, 2010).
Hind, John. 2005. What's the word: Cogniceuticals n. medicines for saving and increasing cognition. *The Observer*. July 24. http://www.guardian.co.uk/theobserver/2005/jul/24/features.magazine97. (accessed August 4, 2010).
International Organization for Standardization (ISO) 2008. Business plan ISO/TC 229 Nanotechnologies. International Organization for Standardization, http://isotc.iso.org/livelink/livelink/fetch/2000/2122/4191900/4192161/TC_229_BP_2007-2008.pdf?nodeid=6356960&vernum=0
International Sub-Committee of BCODP. 2000. The new genetics and disabled people.
Miller, Paul, Sophia Parker, and Sarah Gillinson. 2004. Disablism: How to tackle the last prejudice. http://www.demos.co.uk/files/disablism.pdf. (accessed August 26, 2009).
Maher, Brendan. 2008. Poll results: Look who's doping. *Nature* 452: 674–675.
Omvedt, Gail. 2001. The U.N., racism and caste – II Opinion. *The Hindu*, April 10.
Overboe, James. 2007. Vitalism: Subjectivity exceeding racism, sexism, and (psychiatric) ableism. *Wagadu: A Journal of Transnational Women's and Gender Studies* 4.
Roco, Mihail, and William Bainbridge. 2003. *Converging technologies for improving human performance: Nanotechnology, biotechnology, information technology and cognitive science.* Dordrecht: Kluwer.
Unit for the Promotion of the Status of Women and Gender Equality UNESCO. 2000. *Gender equality and equity* UNESCO. http://unesdoc.unesco.org/images/0012/001211/121145e.pdf. (accessed August 4, 2010).
United Nations. Convention on the rights of persons with disabilities. 2007. http://www.un.org/disabilities/default.asp?id=259. (accessed August 4, 2010).
Wasserman, Anita, David Mahowald, Mary B. Becker, and Lawrence C. Silvers. 1998. *Disability, difference, discrimination: Perspective on justice in bioethics and public policy.* Lanham, MD: Rowman & Littlefield.

Wolbring, Gregor. 2003a. Disability rights approach towards bioethics. *Journal of Disability Studies* 14 (3): 154–180.

Wolbring, Gregor. 2003b. Science and technology and the triple D (disease, disability, defect). In *Converging technologies for improving human performance: Nanotechnology, biotechnology, information technology and cognitive science,* ed. Mihail C. Roco, and William Sims Bainbridge, 232–243. Dordrecht: Kluwer. http://www.bioethicsanddisability.org/nbic.html. (accessed August 4, 2010).

Wolbring, Gregor. 2004. Disability rights approach to genetic discrimination. In *Society and genetic information. Codes and laws in the genetic era.* ed. J. Sandor, 161–187. Budapest: Central European University Press.

Wolbring, Gregor. 2006. Human security and NBICS. Innovationwatch.com. http://www.innovationwatch.com/choiceisyours/choiceisyours.2006.12.30.htm. (accessed August 4, 2010).

Wolbring, Gregor. 2007. Glossary for the 21st century. International Center for Bioethics, Culture and Disability. http://www.bioethicsanddisability.org/glossary.htm. (accessed August 4, 2010).

Wolbring, Gregor. 2008a. Why NBIC? Why human performance enhancement? *Innovation; The European Journal of Social Science Research* 21 (1): 25–40.

Wolbring, Gregor. 2008b. Is there an end to out-able? Is there an end to the rat race for abilities? *Media and Culture* 11 (3). http://journal.media-culture.org.au/index.php/mcjournal/article/viewArticle/57. (accessed August 4, 2010).

Wolbring, Gregor. 2008c. Ableism, enhancement medicine and the techno poor disabled. In *Unnatural selection: The challenges of engineering tomorrow's people*. Chapter 24. ed. Peter Healey, and Steve Rayner. London: Earthscan.

Wolbring, Gregor. 2008d. The politics of ableism. *Development* 51 (2): 252–258. http://www.palgrave-journals.com/development/journal/v51/n2/index.html. (accessed August 4, 2010).

Wolbring, Gregor. 2010. Nanotechnology and social cohesion. *International Journal of Nanotechnology* 7(2/3): 155–172.

Chapter 6
i Will Go Further

Yonex Corporation USA

A quick examination of the Woodrow Wilson Center's Project on Emerging Nanotechnologies Consumer Product Inventory reveals that the majority of nano-enhanced consumer products on the market today—including cosmetics, athletic apparel, and teddy bears impregnated with nanosilver—are targeted at wealthy Western customers. The strategy makes economic sense. Adding nano to an existing product usually increases the costs of production so companies are seeking out customers willing to pay a premium to own the latest and greatest. In some cases, especially cosmetics, corporations, don't label their products as containing nanotechnology because they worry that concerns over health and safety will hurt their sales. But in other areas there is a strong belief that the "nano" label moves product. There are few areas where strategies like this work as well as in the sports equipment industry. In many sports the difference between winning and losing can be the slimmest of margins. Many athletes hope that more technologically advanced equipment can provide the edge they need. The Nanospeed i golf club by Yonex is one of many products that uses nanotechnology to convey a sense that the product is made with the most cutting edge technology available. According to the manufacturer the club employs nanotechnology in two ways: it uses carbon nanotubes for a stronger shaft without adding weight, and an elastic nano titanium alloy so that the shaft snaps back quicker for increased clubhead speed. This advertisement is a reminder of who is buying nano-enhanced products today and that the goal of buying such products is often to get a leg up on the competition.—eds.

This advertisement is reproduced with permission from the Yonex Corporation USA.

Part II
Uneven Structures

Chapter 7
Nanotechnology and the Extension and Transformation of Inequity

Georgia Miller and Gyorgy Scrinis

While the authors that have contributed to this book believe that furthering the cause of equity is a laudable goal, there are many people who benefit from existing unequal political arrangements. In this chapter, Georgia Miller and Gyorgy Scrinis argue that many of those currently directing the future of nanotechnology have a strong incentive to maintain these patterns of unequal distribution. They note that nanotechnology is arising from actions that align it with powerful economic and political interests in the Global North. Despite paying lip service to studying the "ethical, legal, and social implications" of nanotechnology, those who are driving the rapid expansion of nanotechnology have not shown any genuine commitment to reorienting the enterprise to human needs or a more equal society. Given the power disparities between nano advocates and critics, Miller and Scrinis find it improbable that there will be any fundamental realignment. In a sense Miller and Scrinis offer a challenge to all the authors in the volume to find ways to break through the barriers to equity.—eds.

G. Miller (✉)
Friends of the Earth Australia, Melbourne, VIC, Australia
e-mail: georgia.miller@foe.org.au

7.1 Introduction

Governments in the European Union, the United States, Australia, and elsewhere are acting slowly to address the new health and environment risks associated with nano-ingredients now used in hundreds of products world-wide. Non-governmental organizations (NGOs), social scientists, and members of the public involved in early stage engagement activities have emphasised that governments need also to address nanotechnology's social dimensions alongside its new safety risks. But governments have shown little interest in supporting critical reflection about the interactions between nanotechnology, science, and society, or in implementing measures to address equity concerns at an early stage of nanotechnology development.

Discussion of nanotechnology's societal dimensions remains largely divorced from questions of innovation policy, research funding, and governance. Where social, ethical, or equity issues are acknowledged, they tend to be peripheral to the "main game" of technical research and industry development. United Kingdom think tank Demos (Stilgoe 2007, 16) suggests that for many proponents: "Social and ethical concerns have become an obligatory footnote to nanotechnology's technological promise." However a number of NGOs have warned that based on their experiences with past technologies, and given nanotechnology's development trajectory to date, nanotechnological development is likely to widen existing inequalities between and within countries (e.g. ETC Group 2005a–2005c; Friends of the Earth Australia 2006; NanoAction 2007).

Nanotechnological innovation may further entrench or deepen a number of forms of existing inequalities. This includes inequalities of wealth and income; unequal access to employment, to the means of production, and to other social goods, such as health care (Invernizzi et al. 2008); the further concentration of economic and corporate power; the further loss of privacy and the aggregation of information collected on the citizenry (Cribb 2007); greater inequity in exposure to hazardous chemicals and wastes in the workplace and in the environment; and greater instability and insecurity for war-affected regions as a result of nano-weaponry and new means of destruction—casualties may increasingly be borne by the technologically inferior side (Woodhouse and Sarewitz 2007). Where nanotechnology is applied in the quest to "eliminate" disabilities or different biological realities, it could further marginalise disabled people (Cabrera 2009; Wolbring 2002). However, nanotechnology may not merely extend existing socio-economic relations and forms of inequality, but also re-shape and transform them, such that these inequalities and imbalances in wealth and power may take new and novel forms in the contemporary era. This may include new forms of exclusion, disadvantage, dispossession, exploitation, and control, and these may combine with or re-frame existing forms of inequalities and power imbalances. The emerging field of "human enhancement," for example, could even create new elite minorities of people whose cognitive or physical capacities have been extended beyond species-typical boundaries (Roco and Bainbridge 2002).

7.2 Acknowledging the Social and Economic Values Shaping Nanotechnology's Development Trajectory

Scientific practices and technological development are often viewed as existing outside of social processes. It is common scientific and public policy practice to frame social dimensions of technology development as external "risks" or "impacts"— something to be considered as *secondary effects* rather than as *core aspects* of technology development that require attention during each stage of the innovation cycle (Kearnes et al. 2006a,b; Macnaghten et al. 2005; Mohr 2007). Woodhouse and Sarewitz (2007, 140) observe that "unequal outcomes associated with science and technology are [not] usually interpreted as emerging from... the structure of the research and development (R&D) enterprise itself." The need to identify and interrogate the unacknowledged political, cultural and economic forces shaping development of new technologies has therefore been emphasised by social scientists (Irwin 2006; Mohr 2007; Rogers-Hayden et al. 2007; Wynne 1993, 2007).

Political and economic pressures, the assumptions and aspirations of researchers, industry groups and government decision makers, the membership of decision making bodies, institutional cultures, and the allocation of research and industry development funding have the potential to shape nanotechnology development. Growing financial pressures on scientists in universities and public research institutions mean that the innovation priorities of corporate sector research partners are increasingly influential (Woodhouse and Sarewitz 2007). Such factors influence the scope and direction of research, the regulatory context of nanotechnology commercialisation, and the extent of government support—financial and political—for industry development. They also affect the likelihood that the views of less privileged actors will be sought—and their interests incorporated or ignored—in nanotechnology oversight. These factors all influence the extent to which equity considerations will be perceived as legitimate and the priority which they will be accorded.

Nanotechnology's early development was strongly driven by public funding; as late as 2003 public money constituted half of nanotechnology research and development funding world-wide (Lawrence 2005), and in 2004 it was still a full third (Hullman 2006). Yet governments have largely failed to acknowledge that its developmental trajectory is mutable, and could—and should—be shaped to maximise social utility, and better reflect community preference (Sparrow 2007). To improve nanotechnology governance, to facilitate proper evaluation of nanotechnologies in society, and to reduce the likelihood that nanotechnology will deepen existing or create new forms of inequities, it is essential to: open up nanotechnology assumptions, institutions, funding, and governance to critical scrutiny and debate; to undertake early and mid-stage technology assessment to inform the allocation of research funding, development of innovation strategies, and governance; to investigate and implement measures that will prevent or mitigate a "nanotechnology divide" which magnifies existing global socio-economic inequities, including potential reform of existing intellectual property and patenting systems; and to support public participation in decision making in each of these areas, including of marginalised groups.

7.3 The Economic, Social and Political Context of Nanotechnology Development

Nanotechnology development has been aggressively funded and promoted by national governments; "governments everywhere grasp that they have already entered a nanotechnology race" (Whitman 2007, 277). However there are also interlocking non-government institutional interests that are deeply committed to supporting nanotechnology's rapid development. Business, academics, industrialists, the research community, and military interests all view nanotechnology as essential to maintaining economic, scientific and military competitiveness, and are therefore also strong proponents (Whitman 2007).

The strong network of financially and technologically interested groups committed to nanotechnology development has significant implications for equity. Those most closely engaged in techno-scientific policy deliberations tend to come from privileged classes and nations, and to have a particularly optimistic view of the social, economic, and environmental benefits of technological innovation (Woodhouse and Sarewitz 2007). It makes it likely that public concerns of a fundamental or precautionary nature will carry little political weight (Whitman 2007; Woodhouse and Sarewitz 2007). The ready access that financially and technologically interested groups have to the decision making process, and the central role of governments as nanotechnology proponents, public policy developers, regulators, educators, and facilitators of public engagement, is also an impediment to effective governance (FoEA 2009).

The quest for economic and military competitiveness that motivates nanotechnology development shapes research agendas and research cultures, and the kinds of knowledge that nanotechnology produces. Private sector investment in techno-scientific research is traditionally oriented towards delivering products for potential customers with wealth and access, rather than the needs of the poor and disenfranchised (Woodhouse and Sarewitz 2007). But even within public research institutions and universities, there is strong pressure on scientists to produce commercially useful research and to pursue intellectual copyright. Jamison (2009) argues that the links between researchers and industry have become so intimate that science has entered a new, market-oriented mode of knowledge-making, where profit-making is central. He suggests that this diminishes the possibility that nanotechnology will be developed for altruistic or public interest purposes, and results in willful neglect of its social, cultural, and environmental implications. Similarly, Invernizzi et al. (2008, 136) observe that the argument that nanotechnology products will help the poor is belied by its development trajectory to date: "Since nanotechnology's development is essentially guided by corporations' search for profits, a majority of innovations are directed to Northern, affluent societies."

At a time of unprecedented food, ecological, and climate crises, nanotechnology's most important equity issues arguably relate to whether or not it will: further concentrate Northern corporations' control of trade; magnify existing socio-economic inequities between and within countries; further jeopardise the livelihoods and resilience of poor people; add to their pollution burden; and further undermine the ability of communities to retain local control and ownership of food production (ETC Group 2005a,2005b; Invernizzi and Foladori 2005; Invernizzi et al. 2008; Mooney 2006; Nyéléni 2007; Scrinis and Lyons 2010).

There is ongoing debate about the role of technology in causing or deepening inequality at a global scale. Many observers suggest that technology deepens existing inequality, even where it is not the main force creating it; Woodhouse and Sarewitz (2007) caution that new technoscientific capacities introduced into a non-egalitarian society tend disproportionately to benefit already privileged people. Others point to the complex dynamics of inequality and suggest that in some contexts emerging technologies could reduce rather than increase inequalities (see Cozzens et al. 2006). Despite ongoing disagreement about technology's role in deepening inequity, our experience in recent decades demonstrates conclusively that technological innovation alone will not redress inequity. During the last 30 years, a period of significant technological progress and innovation in which microelectronics, information technologies, medical treatments, and telecommunications were developed, the gap between the global rich and the global poor has widened.[1] When global inequality has increased during the expansion of such powerful technologies over recent decades, the obvious question is "why would it be any different for nanotechnologies?" (Invernizzi et al. 2008).

7.4 Potential for Nanotechnology to Exacerbate Existing Inequity

Proponents predict that nanotechnology will deliver breakthroughs in medicine, energy, agriculture, and communications. Yet these breakthroughs—as with previous technical breakthroughs—may be inaccessible to poor or marginalised groups (Royal Society and Royal Academy of Engineering 2004). The availability of technologies does not guarantee access to those who have most need of them. In many instances, efficient and relatively cheap technologies already exist to address public health, sanitation, medical, energy, and agricultural needs of poor people and even these are often not accessible (Invernizzi et al. 2008). Furthermore, it is possible that by concentrating ownership and control of essential platform techniques, processes, and products, nanotechnology may exacerbate existing inequity (Shand and Wetter 2006).

The ETC Group's Shand and Wetter suggest that: "With applications spanning all industry sectors, technological convergence at the nanoscale is poised to become the strategic platform for global control of manufacturing, food, agriculture, and health in the immediate years ahead" (Shand and Wetter 2006, 80). Should predictions of nanotechnology's potential as a platform technology prove accurate, countries and companies which are making early investments, patenting aggressively, and can afford to defend patent claims, are likely to cement and expand their control of key industries and trade (Corporate Watch 2005; ETC Group 2001, 2005a,2005b, 2008; FoEA 2006). In an analysis of nanotechnology patent grants up to 2003, Hullman

(2006) found that Northern countries were well ahead of Southern countries in registering nano-patents; the United States was the most active nation in the world for registering patents, followed by Japan, Germany, the United Kingdom and France. There is a wide disparity among Southern countries in nanotechnology investment, development and patenting. In recent years patent grants have grown in high-growth emerging economies (Liu et al. 2009). In particular, the patent growth rate in China has been remarkable; since 2005 China has held the largest number of nanotechnology patents internationally (Preschitschek and Bresser 2010). Nonetheless, the majority of patents world-wide are still held by Northern countries, and the majority of Southern countries hold few nanotechnology patents. Patenting trends therefore reflect not only a North-South but also a South-South divide.

Nanotechnologies may enable corporations to extend their control over markets and other producers, via proprietary control of essential platform techniques and products of nanotechnology (ETC Group 2005a,b). Bowman (2007, 313) notes that: "Of particular concern is the progressive blurring of the invention/discovery interface under Article 27 [of the Agreement on Trade-Related Aspects of Intellectual Property Rights (TRIPS Agreement)] that may produce uncertainty over the types of nano-products that can be patented... wide interpretation of Article 27(1) may result in the monopolisation of fundamental molecules and compounds." Forero-Pineda (2006) observes that strong protection of scientific and technological intellectual property, including the patenting of research tools, can constrain the capacity of scientists in Southern countries to carry out their own research and development. Without active international cooperation, Southern countries must exert considerable energy to access scientific results and information.

In addition to consolidating the domination of technological intellectual property by corporations and governments based largely in Northern or high-growth emerging countries, nanotechnology may disrupt markets on which many Southern countries' economies depend. Novel nanomaterials and nano-innovations may disrupt or displace the markets for existing products, commodities, services, and technologies. This could have a disproportionate impact on Southern economies which are heavily reliant on commodity trade, and which may lack the capacity for rapid transformation in the face of new economic circumstances (ETC Group 2005a,b; NanoAction 2007).

A range of nanotechnological innovations have the potential for displacing workers and the demand for labour in a range of industries. This would be consistent with previous technological innovations and revolutions, yet in the present techno-economic context, there is the potential for the displacement and redundancy of workers on an unprecedented scale. Examples here include the ability to further automate factory production, and the displacement of agricultural workers through innovations in agricultural mechanisation, chemical input applications or precision farming systems (Scrinis and Lyons 2007). Workers in both Northern and Southern countries will be vulnerable to displacement as a result of nano-automation and any gains in efficiencies, particularly manual labourers. Nonetheless, Southern workers may be most adversely affected, with limited capacity for government support and fewer alternative employment opportunities.

Nanotechnology may also exacerbate existing environmental injustices, such as the exposure of poorer communities to toxic substances and wastes in their workplaces or neighbourhoods. Again, this is likely to affect poor and marginalized communities in both Northern and Southern countries. Further, Southern countries may find themselves disproportionately shouldering nano-risks by becoming manufacturing centres for nano-products Northern workers would prefer not to handle, or else as dumping grounds for nano waste. Since Southern countries usually have weaker environmental regulations, it is possible that international companies will choose to locate plants and waste disposal sites in these countries, leaving local communities exposed to greater risks (Invernizzi et al. 2008).

Beyond the potential for exacerbation of economic inequity and environmental injustice, nanotechnology presents threats to privacy and to accepted human freedoms. Cribb (2007) suggests that the data storage, fusion, mining, and analytic capacity of quantum computing—advances that may be achieved within a generation—will enable round the clock surveillance of every aspect of a person's life. He suggests that this is "no less than the enabling technologies for the global police state, though no-one is admitting as much" (Cribb 2007, 4). Further, he suggests that this will have a key bearing on future human culture: "Like the *observer principle* in quantum physics where the mere act of observation changes the event being observed, people who know they are, or may be, under surveillance around the clock are bound to modify their natural behaviour" (Cribb 2007, 9). In addition to their role in political surveillance or law enforcement, nano-enabled remote sensing and surveillance technologies may also be used by corporations and governments to enforce proprietary rights and contract compliance on farmers and other users of nano-products (Shand 2005).

Next generation nanotechnology applications in the field of therapeutic or human "enhancement" are predicted to alter people's cognitive and physical capacities. NGOs and bioethicists have warned that nanotechnology "has the potential to challenge our understanding of what it means to be human, what it means to have impairments, to differ from the norm or to be different" (Cabrera 2009, 1) and to expand social inequalities (ETC Group 2004; FoEA 2006; Wolbring 2002, 2008). Human enhancement could create new elite minorities of wealthy citizens who have access to the technology, and a new majority of people who are seen as "impaired" or "disabled" because their "performance" has not been nanotechnologically "enhanced" (Wolbring 2002, 2008; Chapter 5).

7.5 Nanotechnology Decision Making and Policy Development Entrenches Existing Inequities

R.E. Sclove (1995), then director of the Loka Institute, argued that the extent to which democratic involvement should be sought in oversight of a given technology "should correspond roughly to the degree to which it promises, fundamentally or enduringly, to affect social life." Governments in Australia, the United States, and elsewhere have predicted that nanotechnology will transform every aspect of our lives (DITR 2006; NSTC 2000). The APEC Centre for Technology Foresight observes that major breakthroughs associated with nanoscale convergent technologies will inevitably be associated with large-scale social upheaval:

> If nanotechnology is going to revolutionise manufacturing, health care, energy supply, communications and probably defence, then it will transform labour and the workplace, the medical system, the transportation and power infrastructures and the military. None of these latter will be changed without significant social disruption. (APEC 2002)

Given the scale of anticipated global nanotechnology-driven social, economic, and ecological change, NGOs have argued for wide-ranging public involvement in decision making to ensure that nanotechnology is managed in the interests of wider publics, not just that of the emerging industry. The Loka Institute (2007) has argued that: "the general public of every nation, their children, and their children's children [are] the key stakeholders in this potential revolution." Friends of the Earth Australia (2006, 8) has urged that "It is essential that civil society has an informed debate about whether or not it actually wants the changes that nanotechnology will bring, and has the opportunity to be involved in decision making about public policy and regulatory development." Greenpeace UK's Doug Parr (cited in Regaldo 2003) has cautioned that: "What we want to avoid is the situation where a small group of financially and technologically interested people develop something and thrust it on the rest of the world." Yet it seems clear that "financially and technologically interested people" remain at the centre of nanotechnology decision making worldwide, while the rest of the global population is ignored, or at best given a tokenistic opportunity to take part in dialogue that has no capacity to affect outcomes.

Nanotechnology decision making is concentrated in the hands of the emerging industry, based largely in Northern countries, and in the hands of governments. Governments' principal international policy forum is the OECD, whose membership is exclusively Northern and dominated by European nations (for membership see OECD undated) although some Southern countries are invited to participate as observers.[2] The activities of the OECD's Working Party on Nanotechnology and Working Party on Manufactured Nanomaterials are conducted primarily in English, with French translation assistance offered for meetings. This is a clear barrier to wider participation of many countries. The OECD is not a decision making forum, but its joint research initiatives, policy forums, and workshops are extremely influential. The absence of Southern countries from OECD nanotechnology activities means that this central forum is guided by, is responsive to, and advocates for the interests of Northern countries. Other key international nanotechnology regulatory forums and conferences and even international consumer-interest meetings also tend to be trans-Atlantic (focusing on Europe and the United States), rather than transnational in character. A frequent assumption is that once a trans-Atlantic regulatory agreement is negotiated (presumably on terms acceptable to both the United States and Europe), it could be the blueprint for a future global agreement (e.g. Inside US Trade 2009)

The exclusion of Southern countries' interests from United Nations forums is also problematic. For example, in mid 2009 the United Nations Food and

Agriculture Organization (FAO) met jointly with the World Health Organization to consider nanotechnology's use in food and agriculture for the first time (WHO 2009). The meeting took place at the tail end of an unprecedented global food crisis. Nonetheless, the United Nations' key food policy institution restricted its agenda to consideration of safety issues (of key sensitivity in Northern countries) with no consideration for the broader implications for food sovereignty and food security (of vital importance for the South). The Nyéléni Forum for Food Sovereignty had earlier warned that nanotechnology's use in food and agriculture would further undermine the capacity of small-scale farmers to meet their own food needs (Nyéléni 2007). NGOs had also been critical of the large-scale, input and capital-intensive, export-oriented, and corporately-controlled paradigm of food production which nanotechnologies are primarily being used to support and extend (ETC Group 2004; FoE 2008). Yet despite the FAO being the United Nation's key forum for food policy, all socio-economic and equity aspects of nanotechnology's use in food and agriculture were excluded from discussion. Opportunities to present to the meeting were limited to scientific "experts" with expertise in the technical risks of nanotoxicology; small scale farmers and international farmers' advocacy networks such as La Via Campesina were excluded. In 2010 the FAO held a conference to identify and promote applications of nanotechnology in food and agriculture that could benefit Southern farmers. However even this meeting was overwhelmingly dominated by technical nano-scientists. In a field of 300+ participants there were one or two social scientists and only 3 people representing community NGOs; neither farmers nor farmers' representative groups were represented. Although held in Brazil, the meeting was conducted exclusively in English. There was little acknowledgement that nanotechnology may intensify economic or other pressures on small farmers.

"Financially and technologically interested people" also firmly control nanotechnology decision-making at a national level—even where efforts are made to give the impression that the outcomes of public engagement exercises will inform policy development. Friends of the Earth Australia (2009) has pointed out that economic pressures, and the unacknowledged role of governments as key technology proponents, can fatally constrain and compromise the capacity of public engagement to affect the decision making process. As Whitman (2007, 279) asks: "When set against the political, institutional and financial backing already driving nanotechnology and its projected growth, how much purchase is deliberative democracy likely to have?"

Nanotechnology marks one of the first instances where the need for "upstream engagement" has become part of the "master narratives of public policies" in many OECD countries (CIPAST 2008; Joly and Kaufmann 2008). However this does not reflect an acknowledgement by governments that wider publics have the right to be involved in decision making that will affect them, or recognition that public involvement will result in better decisions. There has yet to be a public dialogue with explicit links to decision making within government, industry, or the scientific community. Instead, governments' interest in "engaging" their publics on nanotechnology is largely explained by a wish to avoid a repeat of the backlash that greeted genetically engineered foods. The stated objective of many countries' public engagement programs is to build public acceptance of nanotechnology (CIPAST 2008). In its survey of seventy international public engagement initiatives on nanotechnology, CIPAST (2008) notes that many rate poorly on Arnstein's (1969) "ladder of citizen participation." That is, using Arnstein's ladder, nanotechnology engagement efforts are more accurately described as "manipulation," "therapy," or "informing." Rather than offering "citizen power," nanotechnology engagement generally constitutes "non-participation" or "tokenism." Evaluating recent public engagement activities in Australia, Lyons and Whelan (2009) conclude that: "industry interests have captured policy makers and regulators, dissenting voices have been excluded from engagement processes, and engagement processes have not connected with actual policy making activities."

7.6 Discordant Standards Between Innovation and Regulatory Policies Support Industry Development, While Leaving the Public Exposed to Risks/Costs

NGOs and social scientists have raised critical questions related to equity and nanotechnology development. They have questioned: the scope, direction, and purpose of nanotechnology research and commercial development; the assumptions of government, industry, and scientists; which groups, institutions, and individuals are entitled to participate in decision making; whose interests nanotechnology is managed in; the social distribution of benefits and costs; and the mutability and controllability of nanotechnology's development trajectory (Hepburn 2006; FoEA 2007, 2009; Kearnes et al. 2006a,b; Loka Institute 2003, 2007; Macnaghten et al. 2005; Mohr 2007; Sparrow 2007; Stilgoe 2007). Yet equity issues are excluded entirely from innovation and regulatory policy.

The fact that governments are both principal proponents and facilitators of nanotechnology, as well as principal agents for securing and framing governance arrangements, is a key obstacle to appropriate governance (FoEA 2009; Whitman 2007). Government and industry proponents have claimed wide-ranging economic, social and environmental benefits of nanotechnological innovations (e.g. DIISR 2009; DITR 2002; IFRI 2008). But they have largely failed to acknowledge and assess the potential for economic, social, and environmental "costs" or detrimental consequences of nanotechnology development, or to explore the more complicated issues associated with intellectual property and questions of ownership and access. Potential "downsides" of nanotechnology development are largely ignored, or narrowly defined—primarily as toxicological risks. Discordant evidentiary standards are also applied to nanotechnology innovation and regulatory policy.

Innovation policy, including generous government support for nanotechnology research, and industry development and promotion, is underpinned by widely claimed, but poorly scrutinized predictions of economic, social, and broader benefits. The perceived value of these benefits underpins practical and financial government support for rapid nanotechnology commercialization, and forestalls precautionary scientific risk management. Yet claimed benefits remain largely unexamined and outside the scope of any systematic assessment; the inevitability of these benefits is assumed.

Conversely, *regulation* is considered legitimate only to address proven examples of toxicological risk. Contrary to the lax evidentiary standards applied to claims of benefits, risks must be definitely proven and quantified before regulation will be enacted to protect public health and safety, and even before nano-specific safety assessment of new products will be required. In short, publicly funded support for industry development is assured, whereas basic precautions to ensure public safety are stalled.

This "benefits versus risks" framing of innovation and regulatory policy is extremely problematic (Miller and Scrinis, forthcoming). It ignores broader costs, challenges, social and equity dimensions of new technologies, and privileges narrowly defined technical risk as the only legitimate basis for new technologies' regulation. This reinforces the tendency of scientists and decision-makers "to see themselves as purveyors of objective risk assessment, and... to view public concerns as subjective perceptions of risk that are thus marginal to the decision-making process" (Ross 2007, 215).

J. Clarence Davies, former United States (U.S.) Environmental Protection Agency official and fellow of the U.S. Woodrow Wilson Center's Project on Emerging Nanotechnologies asserts that: "what is needed is a capability to consider the overall impacts of major new technologies and to do so while there is still time to deal with the impacts" (Davies 2009, 31). Yet governments have largely been unwilling—or unable—to undertake systematic technology forecasting and assessment of nanotechnology's social dimensions as a part of the governance process. In some instances this is a reflection of the loss of technology assessment capabilities. The places where technology assessment was once carried out in countries such as Denmark and the United States have been significantly reduced in size and shape during the last 15 years (Jamison 2009).

7.7 Intrinsic Properties of Nanotechnology Make it More Likely to Expand Inequity

In evaluating the structural and systemic implications of nanotechnological development, in addition to acknowledging the realities of the socio-economic context in which nanotechnological innovations are being developed and deployed, it is important to consider some of the more or less intrinsic characteristics of this technological platform. The intrinsic characteristics of nanotechnology are in many cases common to other emerging technological systems of the twenty-first century. For example, nanotechnological instruments and systems are often capital-intensive, require highly specialised knowledge, will be controlled by patents, and will enable the closer integration of a range of technological systems. It is therefore well-resourced corporations that are better placed to control and even monopolise the development and commercialisation of nanotechnological systems.

Where nanotechnological innovations increase the productivity and efficiency of manufacturing or agricultural systems, they may similarly facilitate the growth and concentration of market share of large-scale producers, and thereby undermine the economic viability of smaller-scale producers (Foladori and Invernizzi 2008; Scrinis and Lyons 2007; FOE 2008). The surveillance and monitoring capabilities of nano-scale technologies are also likely to be utilised by, and of most benefit to, large and powerful corporations and governments, at the expense of the liberties and autonomy of workers and citizens. The control of the global market for genetically-modified seeds by a handful of agri-biotech corporations may represent the future pattern of nanotechnological development in a range of industries.

The potential benefits of nanotechnological innovation for poor or disadvantaged social groups, communities, or countries are often discussed in terms of identifying individual beneficial applications. For example, the development of cheap water filtration technologies, cheap pharmaceuticals or medical diagnostic kits, and decentralised energy generation systems, have been used to demonstrate the broad-based potential benefits of nanotechnological applications for poor communities (Salamanca-Buentello et al. 2005). Particular nano-applications may indeed offer benefits to their users - especially where existing technologies or management systems are inadequate or expensive. However nano-products and systems do not always offer more effective services or treatments than existing technologies, nor necessarily represent the best value-for-money investment for resource-poor communities and governments. Further, important questions regarding the extent to which nano-products establish a relationship of dependence are often ignored (Stilgoe 2007). It is questionable whether recipient communities will have control over the future manufacture, maintenance, and distribution of such nano-products, and at what cost. Due to their highly technical and capital-intensive nature, manufacturing or maintaining nano-products may be outside the skills base or economic affordability of recipient communities. Should nanotechnology create dependency on ongoing "technological charity" by foreign companies or governments, it may be of limited long-term benefit to recipient communities. Similarly, if communities become reliant on products manufactured far away, they may be vulnerable to fluctuations of nano-product price or availability. If nanotechnology applications displace alternative, community-controlled solutions there could be a loss of traditional knowledge that comes at a high social cost.

Importantly, individual beneficial nano-applications do not challenge or displace the broader socio-economic structures which create, entrench, and extend existing inequalities and power imbalances, and which frame the deployment and use of these individual applications (Invernizzi et al. 2008). At the same time, the public focus on individual applications ignores the ways in which nanotechnological systems may reinforce, extend, and transform broader socio-economic structures, and in ways that may deepen and create new forms of inequality, disadvantage, exclusion, dispossession, and power imbalances. Prominent promotion of examples of "technological charity" may disguise the extent to which nanotechnology perpetuates or exacerbates existing inequity, poor industry practice, environmental pollution, unjust intellectual property regimes, etc. It may also give the incorrect impression that public good applications for poor communities are a major focus of nanotechnology research and development. In fact, in the United States, the world's largest government funder of nanotechnology R&D, military applications receive the greatest proportion of public funding (U.S. National Nanotechnology Initiative 2005), while the private sector is focused on developing consumer items for wealthy and comparatively healthy people in the Global North. The structure and focus of the nanotechnology research and development enterprise itself therefore reflects and perpetuates broader social and economic inequities.

7.8 Conclusion

Governments and other nanotechnology proponents have shown little interest in supporting critical reflection about nanotechnology's social dimensions. Proponents have been keen to promote the potential for individual nanotechnology applications to meet social or environmental needs. However, they have largely avoided the question of whether or not nanotechnology innovation as a whole will exacerbate existing inequity. Whereas social and ethical concerns have to some extent become an "obligatory footnote to nanotechnology's technological promise," such issues remain marginal to the principal business of industry commercialisation. Nanotechnology research, development, and commercialization to date demonstrate clearly that it is driven by a quest for scientific, economic, and military competitiveness, rather than a desire to overcome inequity.

The potential for nanotechnology to reduce existing inequities, rather than exacerbate them, is limited on a number of fronts. Nanotechnology research is expensive and scientists face strong pressure to develop profitable products for a wealthy clientele. Addressing concerns relating to the potential for further concentration of corporate ownership of potential future platform technologies would pose significant direct challenges to intellectual property and patenting regimes internationally. Nanotechnology is a highly technical field and those with the greatest understanding of its risks and challenges have a professional or financial interest in its development. The members of specialist groups most closely involved in technoscientific policy development come from privileged backgrounds, and often hold an overly optimistic view of the potential for technological innovation to be of wider benefit. Participatory processes to seek input in policy and decision making processes from wider publics and marginalised groups remain tokenistic, and are sidelined from the main business of industry development. Regulatory systems are lagging well behind commercial research and development for practical as well as political reasons. There are inherent uncertainties in the technology and its applications, as well as significant knowledge gaps in its implications for human health and the environment. The capacity to detect and monitor particles is extremely low compared to their widespread use and environmental diffusion. Regimes to control military applications face enormous obstacles in the face of an unacknowledged emerging nano-arms race, and political pressure from technologically advanced nations to reduce their soldiers' exposure to conflict.

For these reasons, there may be limits to the extent to which—through better regulation and more democratic control—nanotechnological development can simply be directed towards equitable and just goals and applications, or can be used to redress existing forms of inequalities and power imbalances. This is not a reason not to pursue efforts to increase the levels of democratic participation in policy development, to delay action to protect public health and the environment, or to forestall measures to prevent or mitigate greater inequities. However given these limits, it is legitimate to question the societal benefits of nanotechnological innovation as a whole, and to expose the embedded interests of the broader technological and economic paradigms that will shape the development and deployment of this technological platform.

Notes

1. The gap between the global rich and the global poor is growing, although by some measures economic inequality between countries is decreasing. Milanovic (2005; cited in Cozzens et al. 2008) has examined global data, and concludes that inequality *between* countries' gross domestic product (GDP) per capita is rising. However, if GDP is weighted by population, inequality between countries is declining. Nonetheless, data analysed by Milanovic and others demonstrate that inequality *within* countries is increasing.
2. Interested NGOs and the nanotechnology industry may also send observers to meetings.

References

APEC. 2002. *Nanotechnology: The technology for the 21st Century. vol. 2: The full report*. Bangkok: The APEC Center for Technology Foresight, National Science and Technology Development Agency

Arnstein, S. 1969. A ladder of citizen participation, *Journal of the American Planning Association* 35 (4), 216–224.

Bowman, D. 2007. Patently obvious: Intellectual property rights and nanotechnology. *Technology in Society* 29 (3): 307–315.

Cabrera, L. 2009. Nanotechnology: Changing the disability paradigm. *International Journal of Disability, Community & Rehabilitation* 8 (2). http://www.ijdcr.ca/VOL08_02/articles/cabrera.shtml (accessed August 03, 2010).

Citizens Participation in Science and Technology.2008. Nanotechnology and society: Where do we stand in the ladder of citizen participation? *CIPAST Newsletter Nanotechnology* March 08. www.cipast.org/download/CIPAST%20Newsletter%20Nano.pdf (accessed August 03, 2010).

Corporate Watch. 2005. *Nanotechnology: What it is and how corporations are using it*, London: Corporate Watch. www.corporatewatch.org.uk/?lid=2147 (accessed August 03, 2010).

Cozzens, S.E., S. Gatchair, E. Harari, and D. Thakur. 2006. *Distributional assessment of emerging technologies: A framework for analysis*. http://www.cds.edu/globelics/susan%20E%20cozzens.pdf (accessed August 03, 2010).

Cozzens, Susan E., Isabel Bortagaray, Sonia Gatchair, and Dhanaraj Thakur. 2008. Emerging technologies and social cohesion: Policy options from a comparative study. Paper presented at the PRIME Latin America Conference, September 24–26, 2008. http://prime_mexico2008.xoc.uam.mx/papers/Susan_Cozzens_Emerging_Technologies_a_social_Cohesion.pdf(accessed August 03, 2010).

Cribb, J. 2007. The dwarf lords: Tiny devices, tiny minds and the new enslavement: The Governance of Science and Technology. A Joint GovNet/CAPPE/UNESCO Conference. August 9–10, 2007, Australian National University. http://www.onlineopinion.com.au/view.asp?article=6323 (accessed August 03, 2010).

Davies, J.C. 2009. *Oversight of next generation nanotechnology*, Washington, DC: Project on Emerging Nanotechnologies. www.nanotechproject.org/publications/archive/pen18/ (accessed August 03, 2010).

DIISR. 2009. *National enabling technologies strategy – Discussion Paper*, Department of Innovation, Industry Science and Research, Canberra.

DITR. 2002. *Smaller, cleaner, cheaper, faster, smarter: Nanotechnology applications and opportunities for Australian industry*. A Report for the Commonwealth Department of Industry, Tourism & Resources, Canberra.

DITR. 2006. *Options for a national nanotechnology strategy*. June 2006. Australian Government, Department of Industry, Tourism and Resources, Canberra.

ETC Group. 2001. *New enclosures: Alternative mechanisms to enhance corporate monopoly and bioserfdom in the 21st century, ETC Group Communiqué No.73*, Ottawa: ETC Group.

ETC Group. 2004. *Down on the farm: The impact of nano-scale technologies on food and agriculture*, Ottawa, ON: ETC Group. www.etcgroup.org/en/materials/publications.html?pub_id=80 (accessed August 05, 2010).

ETC Group. 2005a. *Nanotech's "second nature" patents: Implications for the global south*, Ottawa, ON: ETC Group. www.etcgroup.org/en/materials/publications.html?pub_id=54 (accessed August 05, 2010).

ETC Group. 2005b. The potential impacts of nano-scale technologies on commodity markets: The implications for commodity dependent developing countries. South Centre research paper No.4., Ottawa, ON: ETC Group. http://etcgroup.org/en/node/45 (accessed August 05, 2010).

ETC Group. 2005c. *Nanogeopolitics, ETC Group Communiqué No.89*, Ottawa, ON: ETC Group. http://etcgroup.org/en/node/51 (accessed August 05, 2010).

ETC Group. 2008. Who owns nature? Corporate power and the final frontier in the commodification of life, Ottawa, ON: ETC Group http://www.etcgroup.org/en/node/707. (accessed 05 August 2010).

Foladori, G., and N. Invernizzi. 2008, The workers' push to democratize nanotechnology, Chapter 2. In *The yearbook of nanotechnology in society vol. 1: Presenting futures*. ed. E. Fisher, C. Selin, and J. Wetmore, 23–36. Springer.

Forero-Pineda, C. 2006. The impact of stronger intellectual property rights on science and technology in developing countries. *Research Policy* 35(6):808–824.

Friends of the Earth Australia. 2006. *The disruptive social impacts of nanotechnology*, Melbourne, VIC: FoEA. www.nano.foe.org.au/node/152 (accessed August 03, 2010)

Friends of the Earth Australia. 2007. *Who's afraid of the precautionary principle?* Melbourne, VIC: FoEA. www.nano.foe.org.au/node/186 (accessed August 03, 2010)

Friends of the Earth. 2008. *Out of the laboratory and on to our plates: Nanotechnology in food and agriculture*, Sydney, NSW: FoE Australia, Europe and U.S. http://nano.foe.org.au/sites/default/files/Nanotechnology%20in%20food%20and%20agriculture%20-%20web%20resolution.pdf (accessed August 03, 2010).

Friends of the Earth Australia. 2009. *Questioning government's role as chief nanotechnology proponent – a biased adjudicator?* Melbourne, VIC: FoEA. www.nano.foe.org.au/node/307 (accessed August 03, 2010).

Hepburn, J. 2006. Technology, risk and values: From genetic engineering to nanotechnology. *Chain Reaction* 97: 40–41.

Hullman, A. 2006. Who is winning the global nanorace? *Nature Nanotechnology* 1:81–83.
IFRI. 2008. *Nanotechnology, food, agriculture and development*. IFPRI Policy Seminar, June 18, 2008. http://www.ifpri.org/event/nanotechnology-food-agriculture-and-development (accessed August 03, 2010).
Inside U.S. Trade. 2009. *U.S., EU differ on product safety for nanomaterials, trade fight looms*. October 9.
Invernizzi, N., and G. Foladori. 2005. Nanotechnology and the developing world: Will nanotechnology overcome poverty or widen disparities? *Nanotech Law & Business* 2 (3): 101–110.
Invernizzi, N., G. Foladori, and D. Maclurcan. 2008. Nanotechnology's controversial role for the south. *Science Technology Society* 13 (1): 123–148.
Irwin, A. 2006. The politics of talk: Coming to terms with the 'new' scientific governance. *Social Studies of Science* 36 (2): 299–320.
Jamison, A. 2009. Can nanotechnology be just? On nanotechnology and the emerging movement for global justice. *Nanoethics* 3:129–136.
Joly, P.B., and A. Kaufmann. 2008. Lost in translation? The need for 'upstream engagement' with nanotechnology on trial. *Science as Culture* 17 (3): 225–247.
Kearnes, M., P. Macnaughten, and J. Wilsdon. 2006a. *Governing at the nanoscale: People, policies and emerging technologies*. Demos, London.
Kearnes, M., R. Grove-White, P. Macnaughten, J. Wilsdon, and B. Wynne. 2006b. From bio to nano: Learning lessons from the UK agricultural biotechnology controversy. *Science as Culture* 15 (4): 291–307.
Lawrence, S. 2005. Nanotech grows up. *Technology Review* 108 (6): 31.
Liu, X., P. Zhang, X. Li, H. Chen, Y. Dang, C. Larson, M. Roco, and X. Wang. 2009. Trends for nanotechnology development in China, Russia, and India. *Journal of Nanoparticle Research* 11:1845-1866.
Loka Institute. 2003. *Langdon Winner's testimony to the Committee on Science of the U.S. House of Representatives on the societal implications of nanotechnology, Wednesday, April 9*. www.loka.org/Documents/Winner_nano_testimony.pdf (accessed August 03, 2010).
Loka Institute. 2007. *Precaution, participation and nanotechnology*, Loka Nanotechnology Group. August 2007. http://www.loka.org/FedNanoPolicy.html (accessed August 03, 2010).
Lyons, K., and J. Whelan. 2009. Community engagement to facilitate, legitimize and accelerate the advancement of nanotechnologies in Australia. *Nanoethics*. doi:10.1007/s11569-009-0070-2.
Macnaghten, P., M. Kearnes, and B. Wynne. 2005. Nanotechnology, governance, and public deliberation: What role for the social sciences? *Science Communication* 27 (2): 1–24.
Miller, G., and G. Scrinis. Forthcoming 2010. The role of NGOs in governing nanotechnologies: Challenging the 'benefits versus risks' framing of nanotech innovation. Chapter 3. In *International Handbook on Regulating Nanotechnologies*, ed. G. Hodge, D. Bowman, and A. Maynard. London: Edward Elgar.
Mohr, A. 2007. Against the stream: Moving public engagement on nanotechnologies upstream. In *Risk and the public acceptance of new technologies*, ed. R. Flynn, and P. Bellaby, 107–125. New York, NY: Palgrave Macmillan.
Mooney, P. 2006. Hype and hope: A past and future perspective on new technologies for development. *Development* 49 (4): 16–22.
NanoAction. 2007. *Principles for the oversight of nanotechnologies and nanomaterials*, Washington, DC: International Center for Technology Assessment. www.nanoaction.org/nanoaction/page.cfm?id=223 (accessed August 03, 2010).
National Science and Technology Council Interagency Working Group on Nanoscience, Engineering and Technology. 2000. *National nanotechnology initiative: Leading to the next industrial revolution*. Washington, DC: NSTC.
Nyéléni. 2007. *Nyéléni 2007 Forum for Food Sovereignty, Sélingué Mali*, February 23–27. www.foei.org/en/publications/pdfs/nyeleni-forum-for-food-sovereignty (accessed August 03, 2010).
OECD. n.d. *Ratification of the convention on the OECD*, http://www.oecd.org/document/58/0,3343,en_2649_201185_1889402_1_1_1_1,00.html. (accessed August 03, 2010).

Parr, D. 2003. Without a reality check, claims of nanotech's benefits are a con. *Small Times*, September 26. http://www.smalltimes.com/articles/stm_print_screen.cfm?ARTICLE_ID=268999 (accessed September 2, 2009).

Preschitschek, N. and D. Bresser. 2010. Nanotechnology patenting in China and Germany - a comparison of patent landscapes by bibliographic analyses. Journal of Business Chemistry. 7(1):3–13.

Regaldo, A. 2003. Greenpeace warns of pollutants derived from nanotechnology. *Wall Street Journal.* July 25. http://www.mindfully.org/Technology/2003/Pollutants-From-Nanotechnology25jul03.htm. (accessed September 2, 2009).

Roco, M., and W. Bainbridge. 2002. Converging technologies for improving human performance: Nanotechnology, biotechnology, information technology and cognitive science (NBIC). NSF/DOC-sponsored report. http://www.wtec.org/ConvergingTechnologies/Report/ (accessed August 05, 2010).

Rogers-Hayden T., A. Mohr, and N. Pidgeon. 2007. Introduction: Engaging with nanotechnologies – engaging differently? *NanoEthics* 1:123–130.

Ross, K. 2007. Providing "thoughtful feedback": Public participation in the regulation of Australia's first genetically modified food crop. *Science and Public Policy* 34 (3) 213–225.

Royal Society, and Royal Academy of Engineering. 2004. *Nanoscience and nanotechnologies: Opportunities and uncertainties.* London: RS-RAE.

Salamanca-Buentello, F., D. Persad, E. Court, D. Martin, A. Daar, and P. Singer. 2005. Nanotechnology and the Developing World. *PLoS Med* 2(5), e97 doi 10.1371/journal.pmed.0020097.

Sclove, R. 1995. *Democracy and technology.* New York, NY: The Guildford Press.

Scrinis, G., and K. Lyons. 2007. The emerging nano-corporate paradigm: Nanotechnology and the transformation of nature, food and agri-food systems. *International Journal for the Sociology of Food and Agriculture* 15 (2): 22–44.

Scrinis, G., and K. Lyons. 2010. Nanotechnology and the techno-corporate agri-food paradigm. Chapter 16: In *Food security, nutrition and sustainability,* ed. G. Lawrence, K. Lyons, and T. Wallington 252–270. London: Earthscan.

Shand, H. 2005. New enclosures: Why civil society and governments should look beyond life patents. In *rights and liberties in the biotech age: Why we need a genetic bill of rights,* ed. Sheldon Krimsky, and Peter Shorett, 40–48. Rowman & Littlefield.

Shand, H., and K. Wetter. 2006. Shrinking science: An introduction to nanotechnology. Chapter 5. In *State of the world 2006: Special focus, China and India.* The Worldwatch Institute. 78–95 New York, NY: WW Norton.

Sparrow, R. 2007. Revolutionary and familiar, inevitable and precarious: Rhetorical contradictions in enthusiasm for nanotechnology. *NanoEthics* 1 (1): 57–68.

Stilgoe, J. 2007. *Nanodialogues: Experiments in public engagement with science.* London: Demos.

US National Nanotechnology Initiative. 2005. Research and development leading to a revolution in technology and industry. *Supplement to the President's 2006 budget, nanoscale science, engineering, and technology subcommittee on technology,* National Science and Technology Council. http://www.nano.gov/NNI_06Budget.pdf (accessed August 05, 2010).

Whitman, J. 2007. The governance of nanotechnology *Science and Public Policy* 34 (4): 273–283.

WHO. 2009. *Joint FAO/WHO expert meeting on the application of nanotechnologies in the food and agriculture sectors: Potential food safety implications.* June 1–5 2009. http://www.who.int/foodsafety/fs_management/meetings/nano_june09/en/index.html (accessed August 05, 2010).

Wolbring, G. 2002. Science and technology and the triple d (disease, disability, defect). In *Converging technologies for improving human performance,* ed. Mihail.C. Roco, and William S. Bainbridge, 232–243. Arlington: NSF.

Wolbring, G. 2008. Why NBIC? Why human enhancement? *European Journal of Social Science Research* 21 (1): 25–40.

Woodhouse, E., and D. Sarewitz. 2007. Science policies for reducing societal inequities. *Science and Public Policy* 34 (2): 139–150.
Wynne, B. 1993. Public uptake of science: A case for institutional reflexivity. *Public Understanding of Science* 2: 321–337.
Wynne, B. 2007. Public participation in science and technology: Performing and obscuring a political–conceptual category mistake. *East Asian Science and Technology Society: an International Journal* 1: 99–110.

Chapter 8
Nanotechnology and the Sixth Technological Revolution

Mark Knell

One of the major driving forces, if not the major driving force, behind nanotechnology is economics. The argument that nanotechnology will radically alter the world as we know it is often made by people who stand to profit from the changes. They expect that there will be new winners and losers, and they are trying to be among the winners. In this chapter, Mark Knell reviews the thinking of evolutionary economists on the relationship between economic inequalities and long waves of technological change. According to this line of thought, when a revolutionary new technology is first introduced it disrupts social relationships, including the institutions developed under the old technological regime for redistribution of wealth to reduce inequality. Until new institutions are developed that match the new regime, inequality increases. If nanotechnology actually represents a sixth technological revolution—and Knell leaves this question open—societies will have to invent new forms of redistribution to accompany it.—eds.

M. Knell (✉)
Norwegian Institute for Studies in Innovation, Research and Education, Oslo, Norway
e-mail: Mark.knell@nifustep.no

This chapter was peer reviewed. Originally presented at the Workshop on Nanotechnology, Equity, and Equality at Arizona State University on November 22, 2008.

8.1 Introduction

There have been five successive technological revolutions starting from about the time Adam Smith published the *Wealth of Nations* in 1776, according to Freeman and Perez (1988). Arkwright's development and placement of a mechanical spinning machine into the first water-powered mill in 1771 in Cromford, England marks the first industrial revolution. The technology essentially mechanized the cotton industry, but it also lead to the development of a new transport and communications infrastructure. Waterpower, turnpikes, canals, and other waterways were essential to this revolution. Around 1830 steam engines and machinery made of iron triggered the second industrial revolution, with railways, steam ships, telegraphs, and a universal postal service providing the means to network the economy. Three more industrial revolutions occurred since then, at roughly 40–60 year intervals. Each of them involves the introduction of new technologies and new and improved communications and transport infrastructures.

Freeman and Perez (1988) maintain that we are currently in the middle of the fifth industrial revolution, or what might be described as the information and communication technological (ICT) revolution. The starting point can be traced back to 1971 when Intel introduced the first commercially viable microprocessor, which made it possible to incorporate all of the functions of a central processing unit (CPU) onto a single integrated circuit. This technology led to the development of personal computers, digital control instruments, software, and application of integrated circuits in a wide variety of products and services. Semiconductors also make it easier to access certain renewable energy sources. It also made it possible to develop a global digital telecommunications network and the Internet, together with electronic mail and other e-services, as new ways to communicate and network the economy. Computing performance per unit cost has roughly doubled every 2 years since their introduction. The number of the transistors per integrated circuit has increased from 29,000 transistors on the Intel 8086/8088 microprocessors introduced in the late 1970s, to over 2 billion transistors on the Intel Tukwila microprocessor planned for production in early 2010.

How might nanotechnology fit into this story? There is an active debate over whether it will form the basis for another long wave. A study by the National Academy of Sciences (2002) suggests, "the present state of nanodevices and nanotechnology resembles that of semiconductor and electronics technology in 1947." This date is important in the history of technology, and for the ICT Revolution in particular, because it was the year that three scientists at Bell Laboratories first demonstrated the point-contact transistor amplifier. Over the next 13 years, Bell Labs developed several different types of transistors, including the silicon transistor and the MOS transistor. Texas Instruments produced the first commercially viable silicon transistor in 1954, and an industrial cluster emerged in the Santa Clara Valley, which culminated in the introduction of the first Intel microprocessor in 1971 (Lécuyer 2006). While it is not easy to compare the way nanotechnology might evolve over time with the development and evolution of the microprocessor, the parallel does point to the idea of Perez (2002) that the origin of one industrial revolution can be found in the preceding revolution. This provides some basis from which to contemplate the possibility that nanotechnology has what it takes to be an important driving force in the sixth industrial revolution.

Long-wave theorists have suggested that the cycle of techno-economic paradigms are connected to the dynamics of inequality in society. Creative destruction brings new business opportunities but destroys jobs in older industries. The redistributive mechanisms of the older paradigm period may not operate effectively in the new structural configuration. Inequality will decrease only when new employment opportunities are generated and new institutions developed to address the new structure (Freeman 2000). Under conditions of globalization, job destruction and creation may happen in different places, effectively blocking traditional forms of redistribution.

This chapter addresses the possible connections between nanotechnology and these distributional dynamics. The next section of the chapter explores the idea of technological revolutions and techno-economic paradigms and goes into greater detail on the five technological revolutions that are already known. Section Three identifies some of the common trends across the different techno-economic paradigms. These trends are then used in Section Four to explore the possibility that nanotechnology could be an important driving force in the sixth revolution. Section Five examines the distributional consequences of the emergence of a new techno-economic paradigm as applied to the current realities of developments based on nanotechnologies. The chapter closes with questions for further study.

8.2 Technological Revolutions and Techno-economic Paradigms

Economists have long been interested in the idea that business cycles may have a regular pattern that can span over 50 years or more. Sometimes referred to as long-wave theory, the first notable contribution was by Nikolai Kondratieff. Using statistics on price behavior, including wages, interest rates, raw material prices, foreign trade, and bank deposits, Kondratieff (1928) identified two full cycles of expansion, stagnation, and recession that took place from 1790 to 1849 and from 1850 to 1896. Since what Kondratieff described was mainly a statistical phenomena and not a theoretical one, most academic economists were not convinced by the argument. However, Joseph Schumpeter took up some of these ideas and incorporated them in his theory of innovation and the business cycle.

Schumpeter is best known for his theory of innovation. In his book on business cycles, Schumpeter (1939) defined innovation as the "setting up of a new production function," which included the introduction of a new product, new process, and new forms of organization. Throughout his life, he focused on the theory of production and how changes in technology could affect the structure of production. In contrast to Kondratieff, Schumpeter (1939, 178) described long waves in terms of production and dated them according to the broadly defined technology that characterized the period: the industrial revolution for the period between 1780s and 1842; the age of steam and steel for the period between 1842 and 1897; and the Kondratieff era of electricity, chemistry, and motors, which begins in 1898. The empirical data indicate that innovations tend to appear in bunches and often depend on clusters of entrepreneurs located in the same general location. Schumpeter also noted that even given the extensive statistical analysis, there is considerable doubt as to the dating of each cycle.

An extensive literature developed out of Schumpeter's interpretation of the Kondratieff wave, some of it questioning the validity of long wave theory and Kondratieff's own description of long waves as sinusoidal-like cycles, and others developing the idea that technological revolutions are essential to each long wave cycle. Freeman and Louçã (2001) and Perez (2002) provide the basis for much of the recent discussion on the existence of long waves. These authors consider each cycle or long wave to represent not only a technological revolution but also a change in the techno-economic paradigm. Each technological revolution represents a kind of Kuhnian paradigm shift (Kuhn 1962) and is based on Schumpeter's (1939) idea that major or radical innovations initiate a fundamental change in the way things are produced, the types of products being produced, how a firm is organized, and the way people transport things and communicate. Each long wave involves not only the appearance and evolution of a new General Purpose Technology (GPT), but also a change in the way people relate to the technology. The introduction of the microprocessor, for example, not only transformed the way economic growth is generated, but also how people interact. Effective innovation and public procurement policies can also induce a paradigm shift, as was the case in the ICT revolution, when the needs of the military helped drive the development of the transistor in the 1950s (Lécuyer 2006), and the Internet in the 1960s and 1970s. Yet, social uses of the Internet eventually led to further development of the technology behind the backs of the military, which then helped shape the paradigm and the way the people related to each other.

Perez (2002, 8) defines a technological revolution as "a powerful and highly visible cluster of new and dynamic technologies, products and industries, capable of bringing about an upheaval in the whole fabric of the economy." It is a cluster of several interrelated radical breakthroughs, forming a major constellation of interdependent technologies. A GPT will be at the core of the cluster of new technologies and will be essential in driving down the cost of production over time. These costs will be lowered mainly through the development of new sources of energy and the development of new materials that are vital to the development of new products and processes, as well as through the improvement of the transport and communications infrastructure. A GPT can also be thought of as a basic commodity in Sraffa's (1960) sense, in that it is used either directly or indirectly in the production of all other goods or services. This theory suggests that the radical innovations underlying a technological revolution will spread far beyond the sector where it was developed. Long-term economic growth, therefore, is not confined only to the products and processes directly related to the major technological innovations, but to the economy as a whole. As Fig. 8.1 illustrates, the new GPT will become the new engine of economic growth, which through the creation, use, and diffusion of the new technologies across all sectors and industries in the economy, will create the potential for long-term productivity growth in the economy as a whole.

8 Nanotechnology and the Sixth Technological Revolution

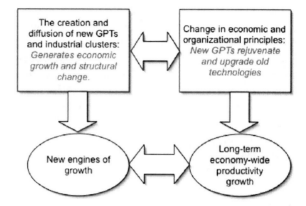

Fig. 8.1 The nature of technological revolutions. Source: Based on Perez (2002)

Each technological revolution appears as a cycle, lasting up to 60 years. The cycle does not appear smooth and continuous, but contains many upheavals as new enterprises, industries, and technologies displace the old and mature ones. Schumpeter (1942) described this process as one of creative destruction. Perez (2002) identifies four distinct phases in the techno-economic paradigm: (1) irruption, when the new technology is introduced; (2) frenzy, or the period of intense exploration; (3) synergy, when the technology is diffused throughout the economy; and (4) maturity, as the diffusion process becomes complete. Figure 8.2 illustrates the technology

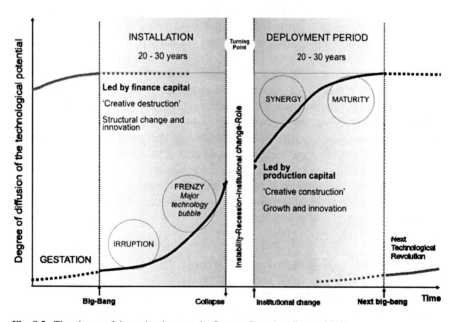

Fig. 8.2 The phases of the technology cycle. Source: Based on Perez (2002)

cycle. Both stagnation and dynamic growth appear in the irruption stage, as old technologies mature and new technologies have not diffused through the economy. During the frenzy stage, many new opportunities to apply the new technology open up, leading to the creation of new markets and the revival of old industries. Dynamic expansion, economies of scale, and diffusion are most common during the synergy phase, when producers tend to dominate. Perez (2002, 53) describes this phase as the "golden age" when economic growth is harmonious and social cohesiveness becomes an imperative. In the last phase complacency appears as the technology reaches maturity and diffuses through the economy.

Freeman and Louçã (2001, 146) used the term technology system to describe how the Schumpeterian clusters are formed and the dynamic interrelatedness that develops within them. Their long-wave contains six phases, of which phases two to five roughly correspond to the technology cycle depicted in Fig. 8.2. The first and last phases overlap in the sense that both the new and old technology systems coexist around the time of the big-bang. A new technology system will begin a laboratory-invention phase, with prototypes, patents, and early applications. This phase could be seen as the gestation period of the new technology and may present a challenge to the dominant technology system. Establishing the feasibility of the new technology by demonstrating its potential application to products and processes roughly corresponds to the first part of the irruption phase. This phase is followed by the explosive take off, which is also marked by significant and sometimes turbulent changes to industrial structure and the regulatory regime. In the fourth phase, stable long-term growth occurs as the new technology system asserts itself as the dominant system in the countries on the technological frontier. This period corresponds to the steep upward slope of the technology life cycle in Fig. 8.2. The technology system then enters a period of maturity, when it first experiences a slowdown and erosion of profitability and then becomes increasingly challenged by the new technologies that will drive the next technology system.

A novel feature of the technology life cycle in Perez (2002) is that each long wave is marked by a turning point when financial capital is supplanted by production capital. This idea follows from Schumpeter's recognition of the entrepreneur and financier as two independent agents that drive the innovation process. The financier dominates in the first two phases of the cycle and the entrepreneur dominates in the second two phases. Financial bubbles are also common in the second phase as confidence in the financial system to support the new technology gains momentum. There is also a tendency for free market policies to dominate in the first two phases and a re-evaluation of the governance systems and institutional arrangements of the economy in the second half of the cycle.

8.3 Technological Revolutions Through Time

Freeman and Louçã (2001) and Perez (2002) have identified five successive technological revolutions that have occurred since the 1770s. Table 8.1 summarizes the five revolutions, based on Perez (2002). All three authors date the start of the

Table 8.1 Five successive technological revolutions, from the 1770s until today

Technological revolution	Popular name	Initial technology	Core input (GPT)	New technologies	New infrastructures
First (1771)	Industrial revolution	Arkwright's mill (Cromford, Britain)	Iron, raw cotton, coal	Mechanization, factories, and wrought iron machinery	Canals, turnpike roads, sailing ships and water power
Second (1829)	Age of steam and railways	The Rocket steam engine (Manchester, Liverpool, Britain)	Iron, coal	Agglomeration, standard parts, construction and steam engines and machinery	Steam power, steam ships, railway, telegraph and ports
Third (1875)	Age of steel, electricity and engineering	The Carnegie Bessemer steel plant (Pittsburgh, USA)	Electricity, steel, copper, metal alloys	Steel, copper, chemistry, electrical equipment, science (R&D), and electric equipment	Steel construction (steamships, railway, bridges, giant structures), telephone, and standardization
Fourth (1908)	Age of oil, the automobile (motorized vehicles) and mass production	Ford Model T moving assembly line (Detroit, USA)	Oil, gas, synthetic materials	Mass production, economies of scale, standardization of products, refining, synthetics, and home appliances	Airlines and airports, motorways, universal electricity, cheap oil, radio and analogue telecommunications
Fifth (1971)	Age of information and telecommunications	Intel micro-processors (Santa Clara, USA)	Integrated circuits	Information intensity, computers, software, control instruments	Digital communications (internet, web), high-speed travel

Source: Based on Perez (2002) and Freeman and Louçã (2001)

industrial revolution around 1771 when the first water-powered cotton mill was opened in Cromford, England. This event, or big-bang, marked the beginning of large-scale factory manufacturing with mechanized production, entrepreneurship, and a new proletariat. Water became the main source of power as well as a way to transport goods over long distances, together with roads and turnpikes. Iron, raw cotton, and coal became the key inputs (GPTs) into the production system, with iron being applied in virtually every industry through the development of new machinery and equipment that replaced wooden ones, and as various inputs in the production process. Labor productivity growth also began to take off, with much of it attributed to time-saving management and specialization in tasks (von Tunzelmann 1995). There was also a turning point in the 1790s when canal mania created a technology bubble, followed by a collapse in the British financial system in 1793 (Perez 2009).

This technology paradigm was followed by a period where steam became the main source of power that propelled both the mechanization of industry and transport. The railway, telegraph, and steamship became central to the transport and communications infrastructure. While coal and iron remained key inputs into the production system, steam engines, machinery tools, and the engineering industries emerged as the new leading sectors in the second technology revolution. Steam engines were installed and working in commercial enterprises as early as 1776, but they were not applied widely because of high costs and the lack of technical feasibility. This changed when a sequence of innovations in the machine tool industry and in precision engineering from 1800 to 1830 made it possible to design and construct high-pressure engines (Freeman and Louçã 2001). These developments then triggered the emergence of the steamships and railways, along with the telegraph. Steam made it possible to provide energy anywhere, and the railroad and telegraph made it possible to create a national network that not only included all of the industrial cities, but also created a truly national market. It was essential for the United States to agglomerate its various markets, and to catch up with Britain in terms of productivity per worker (Chandler 1977). Large joint stock companies with hundreds of employees on average, were created during this time, creating the need for principled management, but it was also a period when repression of the working class was at its strongest and inequality very extreme (see both Karl Marx and Charles Dickens). In the mid 1840s railway mania set in, with massive amounts of financial capital being funneled into the industry, which eventually led to a financial panic in Britain during 1847 and revolutions throughout Europe in 1848 (Perez 2002).

The 1850s marked the golden age of the second technological revolution, and the period leading up to the early 1870s marked the period of maturity, when many countries in the global economies experienced significant nationwide economies of scale. Eventually the declining scale economies, together with the emergence of new industries and technologies; namely the discovery of electricity, steel, and heavy chemical and civil engineering; led to the third technological revolution. Electricity as an energy source and the availability of relatively inexpensive steel were the most important GPTs developed during this techno-economic paradigm, which together led to the development of an electrical equipment industry, and new ways of packaging things, especially foods. It also led to the further development of a global infrastructure, which not only included the creation of a world wide telegraph and telephone network, but also the further development of shipping (steamships), railways, and great bridges and tunnels. Economies of scale were located mainly within the plant, and enterprises became larger, depending increasingly on a managerial bureaucracy using principles of "scientific management" (Taylorism). At the same time, science became a productive force through the creation of industrial research laboratories. Edison's Menlo Park (New Jersey) laboratory set up in 1876 with the specific purpose of producing a steady stream of new products for the market, including the phonograph, microphones, electric lighting, and a system for electrical distribution, as well as other goods. Excessive loans to the railway industry set off a series of bank failures that led to the Panic of 1893 and the bankruptcy of many enterprises including three large railroads.

Motorized vehicles and oil were core inputs in the fourth technological revolution. Both existed long before Henry Ford introduced the moving assembly line in 1913 to build the Model T, but it was this application of mass production techniques, including making use of machines and presses to stamp out parts and insure interchangeability, that led to the relative cheapness of large-scale production and the emergence of mass consumption (Hounshell 1984). Many other industries emerged using mass production techniques, including automotive components, tractors, aircraft, consumer durables, and synthetic materials, and to insure that mass consumption continued, consumer credit innovations were essential (Freeman and Louçã 2001). The rapid growth of the consumer markets generated large economies of scale, but it also created large corporations that required new ways of managing diverse operations, including several different brand names, consumer credit operations, and an internal research laboratory (Chandler 1977). Oil became the key source of energy, mainly because of its abundance and relative cheapness. Just as in the previous technological revolutions, frenzy developed in the 1920s, which led to the financial crash of 1929 and the subsequent (Great) depression. This time the crisis was deeper and required more drastic measures in the form of the New Deal, including banking reforms, some of which were designed to control speculation (Glass-Steagall Act of 1933). By the end of the 1930s, the world was in turmoil, as much of the world was involved in the Second World War.

After World War II, mass production fueled the economy, resulting in high growth for long periods of time and a vast array of new product innovations. But by the end of the 1960s, the U.S. economy entered a period of stagnation that lasted through the 1970s. During this time, however, the ICT technological revolution gained momentum, culminating in the introduction of the microprocessor. Perez (2002) dates the "big-bang" to 1971, when the first microprocessor was produced, with the three subsequent decades comprising the installation period. Computing, software, and digital communications were in their infancy prior to 1971, with products such as the IBM 360 and mobile radiotelephones making an appearance. Many new technologies and products appeared in the market after the microprocessor was invented, often developed by small and medium sized entrepreneurial firms, which initially depended on venture capital and other forms of external finance, including public procurement by the U.S. government for use in the military and space.

Apple, Cisco Systems, and Microsoft are three examples of small entrepreneurial enterprises that emerged during the early part of this period that became large corporations in a relatively short period of time. The so-called "dot-com bubble" of 2000 marked the peak of the frenzy period, which culminated in the financial collapse of 2008. More than one hundred banks failed in the United States in the first 10 months of 2009 (CNN 10-03-09). The techno-economic paradigm may be going through the turning point at the moment this chapter is being written, as new global agreements negotiated in 2009 appear to be shifting power away from the financiers and toward the entrepreneurs. Nevertheless, the current global recession may continue for some years, because, as Galbraith (1955) pointed out many years ago, the great crash of 1929 was the financial event that preceded the Great Depression. Past history also suggests that we are far from the end of the ICT revolution, and that the best period of economic growth has yet to come. And it is still not clear what role the ICT revolution will play in the resurgence of renewable energy as an inexpensive source of energy.

If we are in the middle of the ICT Revolution as Perez (2002) suggests, then the ICT revolution as it is currently evolving will enter its golden age in time. The next revolution will begin to assert itself as this one stagnates when profitable opportunities vanish and silicon microprocessors reach their limit. Financial capital, which will generally be driven by expected profits during the deployment period, will search for new profitable opportunities as the core technologies reach maturity. To help push down the costs of production, capital will move to the more marginalized sectors, sometimes off-shoring production to less costly locations. Some of the more risky capital may search for new opportunities in the new industries. Between revolutions there is a period when two paradigms may coexist, or what Perez (2002, 90) describes as a "bifurcation of the production structure." During this period, which may last two or three decades, the two technology systems compete with each other. This is a period of economic and social turmoil, when the long-term unemployment rate increases and the new techno-economic paradigm challenges the old institutional arrangements. Eventually, there will be a new financial frenzy, where financial capital will gravitate to the new technology, eventually leading to another financial collapse.

8.4 Imagining the Future of Nanotechnology

The potential application of nanotechnology to a wide variety of uses brings up a widely-debated issue: whether nanotechnology is or will be a General Purpose Technology (GPT). As the preceding sections have described, all of the core technologies and products in each of the five successive technological revolutions were GPTs in that they affected the entire economy both at the national and global levels. Society also experienced considerable change and upheaval as the world economy moved through one technological revolution to another. While nanotechnology may have the potential to be a GPT, it is too early to determine whether it will become pervasive in the economy. The idea of a GPT however, can provide an interesting lens for examining the distributional effects of nanotechnology that are the focus of this chapter.

As Perez (2002, 13) points out, biotechnology, bioelectronics, and nanotechnology are all at "a stage equivalent to the oil industry and the automobile at the end of the nineteenth century or to electronics in the 1940s and 1950s, with vacuum-TV, radar and analog control equipment and telecommunications." She continues, "The breakthrough that would make it cheap to harness the forces of life and the power hidden in the infinitely small is still unpredictable." This will depend partly on "when the current information revolution approaches limits to its wealth-generating power." According to Perez (2009), the dot-com bubble and the financial crisis marks the turning point within the ICT technological revolution. After an institutional rearrangement of the global economy, the cycle should enter the deployment period or what Perez describes as the "golden age" of the ICT revolution. As the revolution reaches maturity, the time will be ripe for a sixth wave.

Nanotechnology appears to have many of the characteristics that could drive the next techno-economic paradigm, but whether it will do so is still highly uncertain. The Organization for Economic Cooperation and Development (OECD) emphasizes that nanotechnology is in an early laboratory-invention phase, with prototypes, patents, small-scale demonstrations, and early applications (Palmberg et al. 2009). And while the future is highly uncertain, there have been many early applications of nanotechnology, particularly in health care, electronics, and energy production. Some of the most novel nano-scale devices are intended to assist scientists to learn more about the unique properties of matter at both the nanoscale and mesoscale. But scientists have a long way to go before they fully understand the kind of physics that govern matter at this scale. The lack of any widespread interaction between science, technology, and industry (Meyer 2007), or of any substantial cost savings, suggests that nanotechnology will not challenge the prevailing technology system based on ICT until some time in the future. Finally, the markets for early products using nanotechnology are highly fragmented and show no sign of coalescing into a cluster of markets that have the power to drive a new techno-economic paradigm or technology system. Freeman and Louçã (2001) emphasize that it took more than a half-century of science and invention before the ICT technologies could challenge the techno-economic paradigm of oil, automobile, and mass production (see also Lécuyer 2006).

Roukes (2007) affirms that there are many opportunities for the practical application of nanotechnology, but that scientists still have much to learn about the physics that govern matter on a nanoscale. He point out that most nanotechnology research starts by generating a pattern on a larger scale and then reduces its lateral dimensions before cutting out nano-scale devices. Very little is known about how to build up from atoms and molecules to nanostructures, an idea that Richard Feynman described in a lecture given more than 50 years ago at CalTech titled, "There's Plenty of Room at the Bottom" (Feynman 1959; Drexler 2004). In this lecture

Feynman speculates on the ability of scientists to manipulate matter on an atomic scale, and what this might mean for the development on computer circuitry and other specialized optical equipment. He suggests that nanotechnology is a platform of ideas that may have quantum mechanical effects influenced by micromechanics forces such as elastic, surface tension, electrostatic, electromagnetic, thermal, and piezoelectric forces. Its application consists of the processing of separation, consolidation, and deformation of materials at the atomic or molecular levels.

Virtually all manufacturing of nano-materials and nano-scale devices today use the top-down approach, which depends on larger, externally controlled devices to direct their assembly. A big down side of this manufacturing process is that it starts with large pieces of materials and ends with the need to dispose of significant excess material, a process that tends to keep production costs relatively high (Roukes 2007). The bottom-up approach, as proposed by Feynman, starts from the atomic and molecular levels to build more complex assemblies. Innovations arising from nanoscience are more likely to be commercialized as greater control over atom-by-atom and molecule-by-molecule construction improves. The great advantage of the bottom up approach is that when molecules are put together, they will self-assemble into ordered structures, leaving little or no waste. It also makes it possible to create organic molecules and structures that have never existed in nature. This approach is more likely to yield significant cost savings and increase the opportunity for practical application. It also provides the most convincing argument supporting the idea of convergent technologies.

There has been considerable interest in applying these approaches to existing products and processes. The development of nano-scale instrumentation, which Meyer (2007) suggests will provide a common platform for the various nanotechnology fields, is perhaps the most important application. In 1981 the Scanning Tunneling Microscope (STM) was developed at the IBM Research Laboratory in Zurich, enabling atomic-scale imaging of various surfaces. The same laboratory then developed the Atomic Force Microscope (AFM) in 1986, which significantly increased the range of the STM. This discovery earned the inventors a Nobel Prize. Since then, developments have resulted from key discoveries of nanotechnology-enabled new materials, or nanomaterials, and resulted in several innovations in already available products and processes, as well as the creation of some new ones. One key discovery made in 1985 using the instrumentation was the Fullerene, a nanomolecule composed entirely of carbon that can take on the form of a hollow sphere, ellipsoid, or tube (also known as buckyballs and buckytubes, named after Buckminster Fuller). Both mechanically strong and flexible, the nanomolecule was applied in several existing products, including solar cells, various coatings, sports equipment, memory chips, etc.

The commercial success of medical devices using nanotechnology suggests that some combination of nano and biotechnologies might be the new core. While nanotechnology is an enabling technology that stretches across many disciplines, its most promising applications have been in biotechnology. The highly interdisciplinary nature of both technologies (Rafols and Meyer 2007) as well as the interdependencies of the technologies, combined with the gradual reduction in the size of the transistor, suggests that there might be a convergence of technologies over the next decades. What might appear to be two or three independent technological systems could converge into one technology system, which would then trigger the explosive take-off into the sixth technological revolution. If Meyer (2007) is correct in arguing that nanotechnology is not one but several fields of technology, then it is easy to support the idea of Roco and Bainbridge (2002) that nanotechnology, biotechnology, information technology, and cognitive sciences are converging emerging technologies. The rate of knowledge growth is very high in these technologies, mainly because it relies so heavily on basic and applied research in the scientific community (universities as well as public and private laboratories).

Despite the uncertainties in application, there is considerable interest around the world in nanotechnology. In 2007, total research and development (R&D) investment in nanotechnology was US$13.8 billion, of which US$7.3 billion came from private corporations and venture capital funds and US$6.5 came from various public sources (Roco 2009). Final production goods incorporating nanotechnology are an estimated US$200 billion in the global economy, with a growth rate of approximately 25%, but these figures greatly overestimate the real contribution of nanotechnology, as it makes up only small fraction of total value-added (Palmberg et al. 2009).

8.5 Inequality in Changing Technological Revolutions

Inequality is a key tension in each technological revolution. Within each cycle, income distribution tends to widen as new technologies create new financial opportunities. This is especially the case when finance capital dominates over production capital. Perez (2002) uses the Schumpeterian term "creative destruction" to characterize the first half of the cycle and "creative construction" to characterize the second half. One consequence of creative destruction during periods of structural adjustment is the rapid loss of jobs, particular in the older more mature industries (Freeman 2000). This process will continue through the cycle, but during the financial frenzy, many new jobs are created. Some will be high-skilled jobs, but most will be low-skilled production and service-oriented jobs.

Economists going back to at least to the time of Adam Smith were concerned with the employment effects of technological change (Freeman et al. 1982). Virtually all of them agreed that the introduction of a new technology most often resulted in unemployment over the short run, either directly in the industry where it was introduced or in other industries related to it. But there was not universal agreement on the long-term effect of the new technology. Over time, increases in demand for the re-employment of labor in other industries would increase employment and decrease the unemployment rate. During a technological revolution employment tends to grow most rapidly in the carrier branches that are directly related to the new core inputs, and it tends to decline in those industries from the previous cycle that have reached maturity. Inequality resulting from structural change can also appear as a mismatch of skills and qualifications, which will be reflected in the earnings dispersion, particularly between high-skilled and low-skilled jobs (see Valdivia's discussion in Chapter 9). Education, training, and other learning processes can help resolve this mismatch, as well as facilitate the movement of labor between industries to help insure that unemployment does not become long-term. In essence, the technological cycle over time reduces the inequalities between those that are part of the new technological revolution and those that were left behind in the previous one.

Freeman (2000) maintained that in previous technological revolutions, inequality became most extreme across countries. In 1820, just a few years before the first technological revolution gave way to the second one, the interregional spread of gross domestic product (GDP) per capita was just under 3–1, whereas in 2003, it was over 18–1 (Maddison 2007, 70). Over the course of the cycle, economic growth tends to be uneven and differentiated, sometimes resulting in some countries catching-up, but most ending up falling behind. The diffusion of technology to the least developed economies tends to be slow and costly, and is heavily influenced by difficulties in learning and adaptation. But during the deployment period, technology tends to diffuse more rapidly, perhaps because of the logic of the production system itself.

Large inequalities also appear as production becomes increasingly fragmented, with many of the low-skilled production jobs and some of the service-oriented jobs moving to less developed regions. Since financiers have the most power in the first half of the cycle, they and those holding stocks and bonds in successful enterprises in the new industries will gain most of the new wealth. The process of creative destruction also ensures that some entrepreneurs will be winners, while others will be losers. One issue that makes the distributional consequences more complicated is that the entrepreneur, financier, manager, and worker are all functions or activities, and individuals can perform more than one function. For example, a worker with a retirement fund invested in the stock market also performs the function of a financier, albeit a small one.

Since nanotechnology is at the very beginning of the gestation period, inequality mainly appears as a lack of competence in carrying out R&D activity, especially in the application of new ideas to the emerging technology, or a lack in capabilities, namely the ability to learn and acquire new knowledge and to apply this knowledge in specific social and economic contexts (von Tunzelmann and Wang 2003). Statistics on the number and distribution of publications and patents indicate that nanotechnology-related R&D activities are concentrated in the United States, Japan, Germany, France, the United Kingdom, and some of the smaller OECD member states, along with the small but growing research activity in India and China (Palmberg et al. 2009). As it progresses through this period and moves toward challenging the dominant ICT paradigm, access to, and control over, essential resources such as information, knowledge, skills, and finance will become increasingly more important.

8.6 Concluding Comments on Technological Revolutions and Nanotechnology

The National Academy of Sciences (2006) in the U.S. maintains, "nanotechnology is an enabling technology that promises to contribute at many frontiers of science and technology." It promises to be an essential process-based technology that can lead to many different product innovations in the future, including the potential development of a seemingly endless number of new materials. If this promise is fulfilled, then nanotechnology could become a key input into the global economy. It has the potential to bring flexibility and variety to the economy in a way that no other technology revolution has brought. Some, including IBM, suggest that new materials developed through the use of nanotechnology and biotechnology could replace the current silicon-based microprocessor as the industry reaches maturity.

The claim that that nanotechnology is an emerging General Purpose Technology (GPT), however, is premature. Bresnahan and Trajtenberg (1995) and Helpman (1998) define GPTs as having three characteristics: pervasiveness, technological dynamism, and innovational complementarities. These are all characteristics of the techno-economic paradigm, which is a broader concept that covers the entire economic system, including the institutional arrangements of the society. But all of these characteristics require a well functioning market economy, driven by innovative entrepreneurs in search of economic profit. The idea that nanotechnology is a GPT relies on patent and patent citation databases, which captures the inventive stage of the technology cycle (essential in the gestation period) and not the innovative stage (essential in the frenzy and golden age periods). Pervasiveness in the market occurs because of the diffusion of the many complementary innovations after the new techno-economic paradigm is already in place. Arguments based on patent and patent citation databases, such as Youtie et al. (2007), describe the interdisciplinary nature of nanotechnology and confirm the idea that it is being applied in several fields of technology, but they do not capture the dynamic productivity effects that appear after the GPT asserts itself during a technological revolution. (This criticism applies equally to Moser and Nicholas (2004), who use a patent citation database to claim that electricity is not a GPT.)

The story told in this chapter does not explore the implications that nanotechnology might have on society in any detail. It does describe how inequality might be generated during the course of the long wave, but not how the potential risks to the environment, health, and safety might steer the development of nanotechnology in the future. The reason may be that long-wave theorists tend to focus on the long-term benefits of technology central to the techno-economic paradigm and do not discuss the costs or implications of the technology. Nevertheless, these theorists recognize that social tensions are an important part of the gestation period, or the period prior to the installation of the new techno-economic paradigm. National Nanotechnology Initiative agencies in the United States currently are trying to smooth these tensions by fostering both international R&D collaboration that explore nano-materials and nano-scale devices and to promote the "responsible development of nanotechnology." Nightingale et al. (2008) makes it clear that the risk and uncertainty associated with the implications of nano-materials and the global nature of the technology, suggest a need for precautionary regulations to protect against "possible" hazards that involve international organizations. Yet social unrest is one of the best sign posts in long-wave theory, as it often signals the emergence of a new technology, massive structural change as the economy adapts to the new technology, and the subsequent inequality generated by the change over time. Throughout history, major social unrest tends to dominate any national initiatives or attempts to steer the development of a new, potentially dominant, technology.

Virtually all of the evidence suggests that nanotechnology is an emerging technology that is at the beginning of the gestation period of the sixth technology revolution. But it is all but impossible to predict what role this technology will play there. Despite creating new knowledge about nanotechnology at an astounding rate, collaborative efforts between scientists and scientists and industry are fragmented and punctuated at best. Nevertheless, there is a strong hint that the bottom-up approach, where molecules to self-assemble into ordered structures, not only provides a promising way to reduce costs and waste, but also provides a basis for a convergence of emerging technologies. Perez (2009) emphasized, "no matter how important and dynamic a set of new technologies may be, it only merits the term revolution if it has the power to bring about a transformation across the board." Nanotechnology will become the sixth techno-economic paradigm if it has a widespread impact in all parts of the economy and if the socio-institutional structure changes along with it.

Acknowledgements The author wishes to thank Susan Cozzens and an anonymous reviewer for their helpful comments on an earlier draft.

References

Bresnahan T., and M. Trajtenberg. 1995. General purpose technologies: 'Engines of growth?'. *Journal of Econometrics* 65: 83–108.
Chandler, A.D., Jr. 1977. *The visible hand*. Cambridge: Harvard University Press.
Drexler, K.E. 2004. Nanotechnology: From Feynman to funding. *Bulletin of Science, Technology and Society* 24: 21–27.
Feynman, R.P. 1959. There's plenty of room at the bottom. Lecture given at the annual meeting of the American Physical Society at CalTech, December 29. http://www.zyvex.com/nanotech/feynman.html.
Freeman, C. 2000. Social inequality, technology and economic growth. In *Technology and in/equality: questioning the information society*, ed. S. Wyatt, F. Henwood, N. Miller, and P. Senker, 149–171. London: Routledge.
Freeman, C., J. Clark, and L. Soete, eds. 1982. *Unemployment and technical innovation: A study of long waves in economic development*. London: Frances Pinter.
Freeman, C., and F. Louçã. 2001. *As time goes by. From the industrial revolutions to the information revolution*. Oxford: Oxford University Press.
Freeman, C., and C. Perez, 1988. Structural crisis of adjustment, business cycles and investment behaviour. In *Technical change and economic theory*, ed. G. Dosi, C. Freeman, R. Nelson, G. Silverberg, and L. Soete, 38–66. London: Pinter.

Galbraith, J.K. 1955. *The great crash 1929.* London: Hanish Hamilton.
Helpman, E., ed. 1998. *General purpose technologies and economic growth*, Cambridge: MIT.
Hounshell, D.A. 1984. *From the American system to mass production, 1800 to 1932.* Baltimore: Johns Hopkins University Press.
Kondratieff, N. 1928 (1999). The long wave cycle. Reprinted in *Foundations of long wave theory, vol. 1*, ed. F. Louçã, and J. Reijnders, 25–138. Cheltenham: Edward Elgar.
Kuhn, T.S. 1962. *The structure of scientific revolutions.* Chicago: University of Chicago Press.
Lécuyer, C. 2006. *Making silicon valley: Innovation and the growth of high tech, 1930–1970.* Cambridge: MIT.
Maddison, A. 2007. *Contours of the world economy, 1–2030 AD*, Oxford: Oxford University Press.
Meyer, M. 2007. What do we know about innovation in nanotechnology? Some propositions about an emerging field between hype and path-dependency. *Scientometrics* 70: 779–810.
Moser, P., and T. Nicholas. 2004. Was electricity a general purpose technology? Evidence from historical patent citations. *The American Economic Review* 94(2): 388–394.
National Academy of Sciences. 2002. *Small wonders, endless frontiers: A review of the national nanotechnology initiative.* Washington, DC: National Academies Press.
National Academy of Sciences. 2006. *A matter of size: Triennial review of the National Nanotechnology Initiative.* Washington DC: National Academies Press.
Nightingale, P.M. Morgan, I. Rafols, and P. van Zwanenberg. 2008. *Nanomaterials innovation systems: Their structure, dynamics and regulation.* UK: SPRU mimeo.
Perez, C. 2002. *Technological revolutions and finance capital: The dynamics of bubbles and golden ages.* Cheltenham: Edward Elgar.
Perez, C. 2009. The double bubble at the turn of the century: Technological roots and structural implications, *Cambridge Journal of Economics* 33: 779–805.
Palmberg, C., H. Dernis, and C. Miguet 2009. Nanotechnology: An overview based on indicators and statistics, *STI Working Paper 2009/7*. Paris: OECD.
Rafols, I., and M. Meyer, 2007. How cross-disciplinary is bionanotechnology? Explorations in the specialty of molecular motors. *Scientometrics*, 70: 633–650.
Roco, M.C. 2009. Global development and governance of nanotechnology, The Frank Howard distinguished lecture, GWU, October 12, 2009.
Roco, M.C., and W.S. Bainbridge. 2002. *Converging technologies for improving human performance: Nanotechnology, biotechnology,iInformation technology, and cognitive science.* NSF/DOC-sponsored report, Arlington, Virginia.
Roukes, M. 2007. Plenty of room indeed. Scientific American Reports: Special edition on nanotechnology. *Scientific American* 17: 4–11.
Schumpeter, J.A. 1939. *Business cycles: A theoretical, historical, and statistical analysis of the capitalist process.* New York, NY: McGraw-Hill.
Schumpeter, J.A. 1942. *Capitalism, socialism and democracy,* New York, NY: Harper and Row.
Sraffa, P. 1960. *Production of commodities by means of commodities: Prelude to a critique of economic theory.* Cambridge: Cambridge University Press.
Von Tunzelmann, G.N. 1995. *Technology and industrial progress: The foundations of economic growth.* Cheltenham: Edward Elgar.
Von Tunzelmann, G.N., and Q. Wang. 2003. An evolutionary view of dynamic capabilities. *Economie Appliquée* 6: 33–64.
Youtie, J., P. Shapira, A. Urmanbetova, and J. Wang. 2007. A brief history of the future of manufacturing: U.S. manufacturing technology forecasts in retrospective, 1950-present. *International Journal of Foresight and Innovation Policy* 3: 311–331.

Chapter 9
Innovation, Growth, and Inequality: Plausible Scenarios of Wage Disparities in a World with Nanotechnologies

Walter D. Valdivia

Walter Valdivia provides another economic analysis of the patterns nanotechnology is weaving. Although much growth theory neglects distributional issues, one version considers the income dynamics connected to the diffusion of a new general purpose technology (GPT). A GPT is a technology that causes widespread change because it radically affects the productivity of many other technologies. Because nanotechnology research is so pervasive in the sciences and engineering, many observers are working with the assumption that it will be as important a GPT as computing technology. Using a model proposed by Philippe Aghion to explain the relationship between skill and the diffusion of a GPT, Valdivia analyzes ways wage inequality might appear in several possible paths for the development of nanotechnologies. Since new growth theories leave more room for policy creativity than their predecessors, they can be used more productively in the process of anticipatory governance to create plausible scenarios. Valdivia advocates doing this, thus linking this approach to governance to economic theory.—eds.

W.D. Valdivia (✉)
Arizona State University, Tempe, AZ, USA
e-mail: Walter.valdivia@gmail.com

What will be the impact of nanotechnologies on wage inequality? What can we do about it? There are of course several avenues to answer these two questions. Here I take a somewhat heterodox avenue drawing from *economic growth theory*, which has been strongly influential in shaping economic and innovation policy, and from *anticipatory governance*, a framework in the governance of emerging technologies that is rapidly gaining currency.

A brief word is in order to justify the blend of theories put forth in this chapter. The Keynesian school of economics dominated macroeconomic theory and defined economic policy for three decades in the post-WWII period.[1] One of the fundamental tenets of Keynesian economics, not only as theorized but also as contrasted against empirical observations, was the trade-off between unemployment and inflation. During the economic crisis of the early 1970s, many economists found themselves unable to explain the pervasiveness of simultaneous high inflation and high unemployment and could not make policy prescriptions congruent with their theories. At the same time, a host of theoretical challenges rooted in the neoclassical school were coming to maturity; prominent among them were rational expectations and economic growth theory. Modeling of long-term growth had made headway since the seminal work of Robert Solow (1956) and was poised at the end of the 1970s and early in the 1980s to become the new paradigm in macroeconomics.

The model that I bring to bear on the emergence of nanotechnologies and wage disparities belongs to this second family of economic models, and more specifically to the sub-family of endogenous growth models.[2] As these models began to hold sway in economic schools, their influence in academic, entrepreneurial, and policy circles increased. In fact, this family of models has significantly shaped economic and innovation policy during the last three decades.[3] In addition to the pull and standing of the neoclassical paradigm, the model that threads the argument in this chapter was a deliberate choice for three reasons:

(i) *Income Distribution:* Income generation and distribution are determined simultaneously. Nevertheless, growth models have generally focused on growth and neglected income distribution, suggesting a strong conviction on the part of modelers that "all boats rise with the tide." In fact, many of the policy prescriptions deduced from neoclassical growth models are unconcerned about their distributive effects. What is more, innovation policy could be renamed innovation-for-growth policy because of its exclusive attention to accelerating innovation to boost growth (via productivity gains)—sometimes using competitiveness of the national industry as a surrogate for economic growth. There are, however, a minority of economists working within the neoclassical paradigm who have shown a symmetrical concern for income distribution and economic growth (see surveys in Saint-Paul 2008; Grossman 2001; Aghion et al. 1999). In Section 9.1, I bring to bear one of these voices—a model posited by Philippe Aghion (2002)—to show that the analytical tools of this paradigm can be put to use in understanding wage inequality.

(ii) *Major Technological Change:* The standard economic explanation that links a new technology with wage disparities is referred to as *skill biased technological change* (SBTC). However, SBTC cannot be reconciled with certain observed trends, particularly, the simultaneous increase in wage inequality and decline of productivity growth. Thus, the model considered here uses the concept of major technical change—formalized in the literature as *general purpose technology* (GPT)—to provide an explanation of wage disparity that fares better than SBTC with the observed data. In Section 9.2, I discuss how nanotechnologies could be relevant to this model.

(iii) *End of Prediction:* The neoclassical revival of the late twentieth century not only revamped economic theory substantively and methodologically, but also reconsidered its epistemological basis. Perhaps the most radical aspect of this new paradigm was that it shifted the aim of theory from prediction to explanation. Consequently, growth theory does not and cannot be used to predict future states of the economy; instead, it is used to understand certain critical relationships, such as those between the emergence of a technology and wage disparities. The model proposed in this chapter is admittedly one among many plausible explanations, but the fact that it illuminates possible trajectories in the emergence of GPT-related inequality may prove useful for policy design. In the last section of the chapter, I further link current economic growth theories to the questions this volume addresses by proposing a role for models in the anticipatory governance of nanotechnologies.

9.1 Economic Theory, Technological Change, and Wage Disparities

For more than three decades now, income inequality in the United States has dramatically increased. Table 9.1 below compares the cumulative rate of income growth of five income groups for three decades before and after 1974. During the period following WWII, income growth is somewhat similar across all income groups. Conversely, in the last three decades, the cumulative growth of the 95th percentile is six times that of the 20th percentile.

Four hypotheses have been put forward to explain these increasing wage disparities.[4] The first is deunionization and is supported by evidence at the firm level showing that unionized firms have a wage variance 25% lower than their non-unionized peers (Freeman 1991). At the aggregate level, however, this explanation

Table 9.1 Cumulative income growth

Percentile	1947–1974	1974–2005
20th	97.5	10.3
40th	100	18.6
60th	102.9	30.8
80th	97.6	42.9
95th	89.1	62.9

Source: Based on US Census Bureau data tabulated by Bartels (2008)

is not entirely satisfactory because deunionization in the United States started in the 1950s and wage inequality did not start to increase until the mid 1970s. The second hypothesis is the expansion in the supply of unskilled labor—as a result of the incorporation of female workers or immigrants into the labor force. This idea has been dismissed because the inflow of unskilled labor was more than offset by the fast increase in skilled labor due to efforts in education policy.

The third is the opening up of world economies to international trade. Under this theory, trade liberalization should expand the relative demand for skilled labor in developed countries and for unskilled labor in developing countries—where they are relatively cheap, respectively—thus increasing the relative wage of skilled labor in the former and of unskilled labor in the latter, widening the wage gap in industrialized economies, and decreasing the gap in lower income countries. Evidence, however, does not support this hypothesis. In the first place, United States trade with non-OECD countries amounts to only 2% of its GDP (gross domestic product); second, relative prices of skill-intensive goods have not fallen according to this theory; and third, only 20% of the shift from blue-collar to white-collar jobs has been observed across industries (the remaining 80% has been within-industry shifts).

The fourth hypothesis is "skill biased technical change." SBTC rests on the idea that technological change increases the productivity of skilled labor more than it does unskilled labor. When wages reflect factor productivity, the emergence of a new technology should increase the *skill premium*—that is, the additional wage paid for education, experience, or other forms of skill. SBTC hinges on the complementarity of technology and skill. Nelson and Phelps (1966, 70) observed, "educated people make good innovators, so that education speeds the process of technological diffusion," and concluded that "the rate of return to education is greater the more technologically progressive is the economy" (Nelson and Phelps 1966, 75). In addition, the relative price of capital equipment in the United States has persistently declined in time; a notorious recent case is the relative price of equipment in information technologies, which has dropped at an annual rate of 10%. Because equipment is a complement of skilled labor and a substitute of unskilled labor, cheaper equipment displaces the demand for unskilled labor, widening the wage gap. Furthermore, the complementarity of information technologies and skilled labor, argue Milgrom and Roberts (1990), reduce the need for monitoring and supervision and increase communication, and thus lead to flatter organizational hierarchies. In this logic, technical change leads to organizational changes that benefit, in terms of promotion and remuneration, the more educated (compare with the superstars wage model of Rosen 1981).

In other words, the logic underlying SBTC is that the expansion in the relative demand of skilled labor outpaces the expansion in the relative supply (think of the relative supply of skilled labor as the proportion of college graduates with respect to the labor force). This claim is hard to reconcile with the fact that the relative supply of skilled labor has not remained constant; rather, during the past 60 years it has consistently increased. This is particularly perplexing given that the last 30 years have also seen a rapid increase in college wage premium. This empirical contradiction is

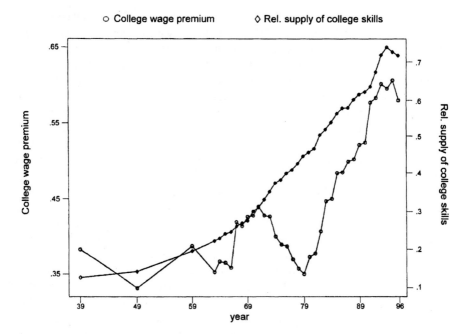

Fig. 9.1 Relative supply of college skills and college premium
Source: Acemoglu (2002)

clearly illustrated by Acemoglu's (2002) telling comparison (reproduced in Fig. 9.1 below) between the college premium and the relative supply of college skills.

Acemoglu (1998) explains this contradiction by observing that an expanding supply of skilled labor creates an incentive for firms to invest in new equipment that those better-trained workers will be able to exploit. This expands the market for new equipment, particularly for new equipment that utilizes high skill levels, signaling innovators that it is this particular type of technological change (i.e., skilled biased) where they need to aim their efforts. It is this *market size effect* that drives SBTC and that keeps increasing the productivity of skilled labor and, with it, the skill premium. At its core, this explanation suggests that innovation, and particularly the skill-biased type of innovation, is partly driven by the supply of skilled labor. If the supply is somewhat determined by the levels of educational achievement, it would seem that an unintended consequence of more education is a wider wage gap. I will return to this point below to characterize it better.

A second empirical contradiction in the SBTC hypothesis is the underlying assumption (at least under endogenous growth models[5]) that the rate of growth of the economy should be proportional to the level of resources devoted to research and development (R&D). Were this to be the case, an increase in the relative supply of skilled labor (as observed in Fig. 9.1) should accelerate the economy's productivity (output per hour unit of labor). However, as Jones (1995) has shown, significant R&D investments have not paid off in terms of productivity growth in the United

States. In fact productivity growth has slowed to a historic low, an annual compound rate from 1970 to 2008 of only 1.7%.[6] This slow down cannot be explained solely by decreasing returns of the R&D sector that feeds skill biased technologies into the economy (Acemoglu 2002) because the other sectors of the economy need not exhibit decreasing returns. This could be the case if innovations in the other sectors were subsidiary to a major technological breakthrough that exhibits decreasing returns. For this reason Aghion (2002) proposed a model that explicitly incorporated the notion of an economy-wide technological change—the model I now describe.

In economic literature, the notion of major technological change has been formalized in the concept of *general purpose technology* (GPT). In Section 9.3, I will discuss the salient aspects of the GPT concept. Here is sufficient to note that: (i) GPTs are technological developments that restructure the production of all sectors in the economy—such as the steam engine, the dynamo, or the computer; and that (ii) the costs and uncertainties of migrating the production platform form an old to a new technology vary across sector and across firms, and therefore the diffusion of a GPT throughout the economy is neither simultaneous nor linear.

At first, only a few risk-taking firms will experiment with the new technology, while the risk-averse majority keeps its existing technological platform. As further developments in the GPT reduce the uncertainties about its profitability, increasing number of firms start experimenting. At some point there should be enough firms with successful stories of adoption of the GPT that the rest of firms consider transition *en masse*. Some techniques or procedures of these initial successful stories become *templates* of innovation for the large number of firms now shifting to the GPT. At this point, firms shift to the GPT in larger numbers because everyone else is shifting and otherwise they confront being stuck with the obsolete platform and driven out of business. In this second phase the pace of adoption slows down, as only the laggard few that remain in operation complete the transition. As described, the adoption of a GPT is similar to the epidemiological curve of the spread of a contagious disease, as illustrated in Fig. 9.2.

The model also needs to incorporate a monotonic increase in the relative supply of skilled labor in order to account for the observed increase in the share of college graduates in the labor force. Now Aghion's model can explain the first contradiction without resorting to the market size effect. At the beginning of the GPT adoption cycle, the relative supply of skilled labor keeps increasing, yet only a few firms are shifting to the new GPT. These few firms can hire only a portion of those workers fresh out of college; the rest will have to find jobs in pre-transition firms where they are not compensated according to their skill level. Consequently, during the first phase of the GPT adoption cycle the skill premium should decline. This condition coincides with the observed decline in the skill premium during the 1970s (see Fig. 9.1), which arguably was the first phase of adoption of a new GPT of information and communication technologies (ICTs). In the acceleration phase of the GPT adoption cycle the relative demand for skilled labor shoots up because firms need many better-educated and more versatile workers to adapt the GPT to their needs (using the *templates* from the pioneers). The skill premium should continue to increase as long as the relative demand for skilled labor outpaces the relative

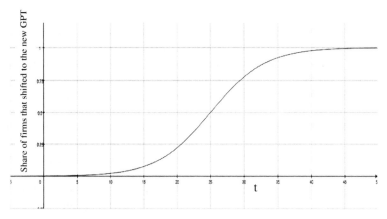

Fig. 9.2 Hypothesized nonlinear diffusion of a GPT
Source: Author's own rendering

supply; and when the transition craze has elapsed, the skill premium should adjust again, assuming the relative supply of skilled labor keeps a vigorous pace.

At this point I would like to return to the role of the education system that I first mentioned with respect to the market size effect. In contrast to the earlier conclusion, this model now shows that efforts to increase consistently the share of college graduates in the labor force may sometimes widen and sometimes close the wage gap between skilled and unskilled labor. Education serves, in this light, as a braking system for wage inequality, which rises only during the acceleration in the GPT adoption cycle; before and after the acceleration, the relative demand for skilled labor does not outpace the relative supply. The more vigorous the pace of growth of the relative supply of skilled labor, the higher the threshold that the relative demand must cross in order for the wage premium to increase. This is to say that a strong education system helps to shorten periods where the wage gap expands.

Aghion's model also fares well against the second empirical contradiction noted above; that of the observed deceleration of productivity. While firms are engaged in the GPT adoption cycle, their productivity should fall because the transition decisions are made based on the expected profitability, not the actual profitability, of shifting to the new technological platform. At the beginning, the costs of equipment, learning, and additional R&D efforts, will all come at the expense of the firm's productivity, market share, and profitability. Only in time will firms return to positive profits; however, the transition becomes increasingly necessary for them to stay in business. This situation is related to what Schumpeter (1942) observed as innovation coming at the cost of disuse of productive capacity, and the shutting down of businesses that have become obsolete, a process he called "creative destruction." For this reason, Aghion designates his work a Schumpeterian growth model.

It should be noted that Aghion (2002) has also used the GPT argument to explain within-group wage inequality by introducing a stochastic process of reallocation of workers to firms that at any given time are at different stages in the transition to the new GPT. Skilled workers have higher transferability of skills, which is to say they can shift across sectors and across vintages of equipment. This ability to transfer serves them well when they are placed in a firm at the forefront of the transition; their productivity is higher and so is their wage level. The other skilled workers, placed in backward firms, will not make use of their potential, so their wages will fall behind the wage level of their educational cohort. The problem of within-group inequality, also called *residual wage inequality*, is significant; nevertheless, the current understanding of it in relation to technological change remains thin. Perhaps this problem has little to do with the changes in productivity and the relative supply and demand of skilled labor, but has more to do with cultural norms (class status and access to social networks), institution of the labor market (unions, structures of work contracts), and idiosyncrasies of each industry (why do doctors earn so much more than school teachers?).

Notwithstanding the limitations of Aghion's model, it is possible now to understand certain important links between the emergence of a major technological change and wage inequality. In the next section I discuss how nanotechnologies might fit this model.

9.2 General Purpose Technologies and Nanotechnologies

It is now apparent that the lynchpin connecting nanotechnologies and wage inequality in this model is the concept of general purpose technology. The question as to whether nanotechnology is showing signs of becoming a GPT has received some attention. For Shea (2005) and Palmberg and Nikulainen (2006), the answer is in the affirmative. The former, however, does not follow the canonical definition of GPT (more on this below) and focuses instead on the disruptive character of nanotechnology for firm's management and business strategy. The latter do use the standard definition of GPT but offers only a broad discussion, not an empirical test. In turn, looking at patent data, Youtie et al. (2008) found comparable generality between nanotechnology and the computer sector. However, only one of the three standard quantitative tests was possible with available data (see below for the three GPT characteristics) and even when their qualitative analysis supported the GPT hypothesis, they remained cautious in their conclusions.

Surely, caution in declaring that nanotechnology is a GPT is warranted. The first and most obvious reason is that both concepts, a technology at the nano-scale and a technology of a general purpose, are not always defined in the same way, and as a consequence, they are used inconsistently. Because the contours of each concept are not precisely demarcated, a second reason for caution is that both concepts may be wrought and fitted together in more than one way. The particular manner in which they ought to be linked is subject to the explanatory framework in which these concepts are deployed. Let me clarify, then, the use of these concepts as they apply to the model described in the previous section.

Before Bresnahan and Trajtenberg (1992) formalized the concept of *general purpose technology*, other students of economic history had observed the importance of certain technologies in determining long-term growth and income distribution. Simon (1987) had compared the significance of the computer against the steam engine, and David (1989, 1990), in his seminal study of the dynamo and the computer, suggested that entire "techno-economic regimes formed around general purpose engines" (see also Mokyr 1990). These studies were followed by a wave of empirical investigations that sought to refine the GPT concept. Bresnahan and Trajtenberg (1992, 4) introduced the concept in these terms:

> We think of the technologies prevalent in any given period as structured in a hierarchical pattern (i.e. as forming a sort of "technological tree"), which in the simplest case would consist just of two levels: a handful of "basic" technologies at the top (perhaps just one), and a large number of product classes or sectors that make use of the former at the bottom. Those at the top are characterized first of all by their *general purposeness*, that is, by their performing some generic function that is vital to the functioning of a large segment of existing or potential products and production systems. Such a generic function would be, for example, "continuous rotary motion," performed at first by the steam engine and later on by electrical motors; "binary logic" would the corresponding generic function for electronics, the obvious candidate GPT of our times [emphasis in original].

In addition to *generality of purpose*, they argued for two additional characteristic of a GPT: *innovational complementarities* with downstream sectors of application (e.g.; the mutual complementarity of semiconductors and computer architecture), and *technological dynamism* that refers to the "continuous innovational efforts, as well as learning, [that] increase over time the efficiency with which the generic function is performed" (Bresnahan and Trajtenberg 1995, 5). This dynamism is owed to the broad scope for improvement and further elaboration of the GPT. These three characteristics constitute the canonical definition of a GPT (Bresnahan and Trajtenberg 1995).

Not surprisingly, many technologies fit this definition. Subsequent studies found that, in retrospect, the waterwheel, steam engine, dynamo, and internal combustion engine were GPTs of power generation and distribution, and the railways and motor vehicles were GPTs in transportation (Lipsey et al. 1998). Moreover, Ruttan (2006) examined the technologies spurred by military procurement and found aircrafts, nuclear energy, the computer, the Internet, and missiles and satellites fitting the definition of GPTs. Going even further, Ruttan (2006) argued that an important GPT was the system of mass production by means of interchangeable parts—introduced in the national armories of Springfield, Massachusetts, circa 1830; and considered the precursor of the Fordist system. Likewise, Rosenberg (1998) reasoned that the introduction of chemical engineering in the United States was itself a GPT.

There is irony in the successful identification of multiple GPTs because the power of the concept to explain long-term changes in productivity and income distribution is diluted. For this reason, David and Wright take the position that "some of these sweeping innovations should be better viewed as sub-categories of deeper conceptual breakthroughs in a hierarchical structure" (1999, 10) and they "prefer to regard the foregoing streams of technical development as subsidiary, or perhaps 'tributary'" (1999, 11) of higher order GPTs. David and Wright identify two GPTs in the twentieth century: the electric dynamo and information and telecommunication technologies (ICTs). This conceptualization of a GPT is the most adequate for the model discussed in this chapter because wage inequality in the economy, taken as a whole, is explained as an effect of one primary GPT and not several secondary innovations. Whether nanotechnology may become the next higher order GPT or simply ride the ICT wave is too early to tell. In either case, nanotechnology should be defined such that it interlocks with this view of a GPT.

The growth model described above conceives of technology simply as a type of object that is new to the economic system, as innovation that is embodied in new vintages of physical capital. Putting emphasis on the materiality of nanotechnology exposes the possible inadequacy of speaking of it in the singular. As apples and oranges are distinct fruits, so are different nano-things distinct, such as composites produced with nano-particles and drugs engineered at the molecular level.[7] Considering only technical aspects, it would be more appropriate to speak of nanotechnologies, in the plural, to reflect the heterogeneity of things studied, researched, and produced that can be measured in nanometers.

This narrower conceptualization of nanotechnology excludes a number of things generally designated by the concept. The most obvious exclusion is the set of technologies that do not directly enhance physical capital. For instance, advances in medicine may reduce the number of work hours lost to disease, or human enhancement promises to make workers more productive; neither, however, directly enhances physical capital. Other exclusions under the model are final products, because a GPT must be an input to manufacturing sectors, and an input of wide applicability.

These wide ranging inputs are also referred to as nano-materials (Lux Research 2007) and I illustrate below in Table 9.2 some examples of likely GPT applications.

To get a perspective on all sectors in the economy and those where a GPT in nanotechnology is likely to originate, I juxtaposed the applications suggested above

Table 9.2 Nano-materials and likely GPT applications

Nano-materials	Possible GPT areas of application
Nano-tubes.	Fuel cells, industrial composites, memory chips
Ceramic and metal nano-particles (particularly metal oxide particles).	Solar cells, catalysts for fuel cells.
Fullerenes (caged molecules)	Catalysts, memory chips, and industrial composites.
Nanostructured metals	Coatings and solar cells.
Nanowires	Solar cells and memory chips

Source: Lux Research (2007)

over the NBER patent classification system. Thus, a GPT may emerge from nanotechnologies in chemicals, coating and resins; in computers, microprocessors and information storage; and in electronics, power systems, and semiconductor devices (Table 9.3 below).

Admittedly, all these nanotechnologies may emerge and yet none become a GPT with the transformative power of the dynamo or the computer. This might occur if a nanotechnology emerges to ride the ICT cycle. For example, any new and more efficient microprocessor or memory chip will simply be plugged into the existing architecture of hardware and software. In this case, nanotechnology cannot be modeled as GPT but the model could be adapted to consider that boost in the

Table 9.3 NBER classification of patents

Category		Sub-category		Possible GPT nanotechnology
1	Chemical	11	Agriculture, food, textiles	
		12	Coating	Fuel cells and thin solar panels.
		13	Gas	
		14	Organic compounds	
		15	Resins	Additives for low emissions combustion
2	Computers and communications	21	Communications	
		22	Computer hardwr. and softwr.	Higher transistor density chips
		23	Computer peripherals	Carbon nano-tubes storage surfaces
		24	Information storage	
3	Drugs & medical	31	Drugs	
		32	Surgery and medical instruments	
		33	Biotechnology	
4	Electrical & electronic	41	Electrical devices	
		42	Electrical lighting	
		43	Measuring and testing	
		44	Nuclear and X-rays	
		45	Power systems	Thin solar panels and fuel cells.
		46	Semiconductor devices	High efficiency fiber optic
5	Mechanical	51	Materials processing and handling	
		52	Metal working	
		53	Motors, engines and parts	
		54	Optics	
		55	Transportation	

Excluded from this table are the miscellaneous sub-categories: for chemical 19, drugs 39, electrical 49, mechanical 59. Also excluded is the entire category Other 6, that includes agriculture, husbandry, food 61, amusement devices 62, apparel and textile 63, earth working and wells 64, furniture and house fixtures 65, heating 66, pipes and joints 67, receptacles 68, and its own miscellaneous
Source: Hall et al. (2001)

adoption and diffusion of ICTs. Another situation when a non-GPT nanotechnology may emerge might occur when significant technical improvements are achieved simultaneously, all looking like a GPT but neither of higher order than the others. For instance, consider the hypothetical case where several new technologies, all viable and technically different from each other, change the economics of energy production and distribution—say, harnessing wind, geothermal, biomass, and solar energy. Even if a given nanotechnology (thin solar panels) is part of this energy revolution, the model suggested here will not adequately depict the link between nanotechnology and wage inequality. If these circumstances are at all possible, how far can the model be stretched to guide the governance of nanotechnology? In the final section of this chapter, I offer a general perspective on the use of models in governance, and a more specific one deploying Aghion's model in these two cases of non-GPT nanotechnologies.

9.3 Implications for Governance

I opened this chapter asking: How might nanotechnologies impact wage inequality? And, what can be done about it? To answer the first question, I have imported from economic theory a reasoned speculation of how a GPT might impact wage inequality and have discussed the extent to which a nanotechnology could become such a GPT. The answer to the second question is less straightforward for several reasons. First, one model, or even a family of them, is only one among many explanations of a complex relationship—in the case at hand, between technology and inequality—and it would be presumptuous to assume that policy prescriptions inferred from that one model are definitive. Second, even if we had produced sound policy prescriptions, their political feasibility is condition *sine qua non* for their implementation. More specifically, the suggested use of policy instruments (related to training programs, education system, and the timing of policy) may clash with political interests that are simply not concerned about wage inequality; some may want to privilege a faster pace of innovation even if it comes at the expense of higher inequality. Third, the question is strictly normative; thus, an adequate answer is supported by some concept of legitimacy in policy making. The notion of governance comprises these three functions: the reasoned production of policy prescriptions, the political negotiation of what course of action is feasible, and the implicit requirements of legitimate public policy. Thus, the answer to the second question must be developed within a particular approach to governance, and I suggest here applying the approach called anticipatory governance.

9.3.1 Anticipatory, Not Predictive Governance

Anticipatory governance has been discussed from various perspectives: from the debates in technology assessment (Guston and Sarewitz 2002; Smits et al. 2009), from its implications for political theory (Karinen and Guston 2009), and from the literature in reflexive governance (Barben et al. 2008). I will introduce here a somewhat alternative perspective that, in keeping with the standard formulation of anticipatory governance, will nevertheless permit me to enlist economic models into its analytical arsenal. This perspective emphasizes the response of anticipatory governance to two common practices of administration, namely: forecasting and contingency planning.

Forecasting is the consequence of conceiving social systems[9] in much the same way as conceiving the economic system as a home appliance. If we believe that scientific study can deliver a user's manual of the social system that tells us what button serves what function and what policy produces desired consequences, then governance becomes a problem of prediction: take action A to achieve goal G, or do this much of A to achieve a given level of G. Forecasting is pervasive in business management and public administration. For instance, the popular doctrine of management-by-objectives[10] asks employees and operational units to set goals and evaluates their performance against these goals. This is evidently an exercise in forecasting, because it presumes that the achievement of desired outcomes is fully determined by effort and dedication. Even administrators who see greater contingency and irreducible uncertainty in the functioning of their organizations may chose to require such forecasts because the practice provides them with an easy system of accountability. Likewise, policy makers who understand the stubborn complexity of the social system may nevertheless use prediction to bask in the popular faith in scientific laws to legitimate their preferred policies.

Anticipatory governance starts from a different conception of the social system, not so much denying that certain regularities may hold, at least for a while, but highlighting the contingent character of the system that reconfigures itself constantly by the effect of changes in its institutions, knowledge, and technologies. It further recognizes that, when these changes are significant, uncertainty and contingency are even more acute. If the transformative promise of nanotechnologies is realized, not even basic system regularities could be worked out for prediction, much less a plausible forecast of the social system for policy makers. It is plain that the practice of forecasting in the face of radical technological change is problematic, and may even be futile. Without dismissing all the uses of forecasting, anticipatory governance recognizes that the possibility of useful prediction is directly related to the maturity of the technology and inversely related to the degree of originality of the technical innovation in question. Still, one cannot help but wonder how often it happens that regularities are identified with sufficient confidence for useful prediction only after the technology in question has fallen in disuse.

Contingency planning is the second common disposition among administrators. We all saw it at work in our respective workplaces before the arrival of the year 2000. Although servers, interfaces, and personal computers were debugged, the interdependence of electronic systems prompted fears that the failure of a single sub-system could precipitate a massive chain reaction. What many organizations did was to put in place contingency plans. Task forces identified critical operations in their organizations and figured out how to assure continuity of these operations if any of the sub-systems of support failed. These teams quickly figured out that the difficulty of designing a contingency plan is directly related to the size and complexity of the organization (or system) as well as the magnitude of the disruption expected in sub-systems of support. One can imagine how much more difficult it would be to design a contingency plan if nanotechnologies affect as many sub-systems of the economy (application sectors) as it has been suggested and if they are as technically revolutionary as some commentators believe.

The literature of organizational behavior gives us one first area of difficulty in envisioning contingency plans. At least two basic conditions have been suggested for effective contingency plans: well coordinated operational networks and high adaptability to environmental shocks (see on coordination Boin and 't Hart 2003; cf. Roberts and Wargo 1994). Yet, networks of a free-market democracy do not naturally cooperate (prevalence of the prisoner's dilemma) nor is the adaptability and resilience of the economic system automatic (as we were reminded during the last financial crisis). A second problem with contingency planning is the mustering of political support for the heavy investments required to implement them (see discussion in Zimmerman 1985). While organizations may be able to afford to invest in redundant sub-systems to secure operational continuity, the creation of redundancy at the social scale could impose a cost on taxpayers that would not be tolerated in the ballot box. A third difficulty with contingency planning is its reactive character. Whether the reaction is to a crisis, an emergency, or a sub-system malfunction, the assumption is that something bad has happened. However, in dealing with the most serious threats from emerging technologies (as once was the case with a nuclear holocaust), it may be prudent to dedicate more resources to prevention (e.g., détente) than to contingency planning.

Anticipatory governance contemplates three central functions: foresight, integration, and engagement (Barben et al. 2008). These functions, indirectly address the aforementioned difficulties of contingency planning. A likely side effect of integrating societal implications to the work and routines of scientific researchers is an increased coordination of social networks and greater adaptability of the entire system to exogenous shocks. Engaging the various publics affected by the emergence of nanotechnologies is, ostensibly, an effort to democratize the co-production of science, technology and social order. In a subtle way, it also is a negotiation of political feasibility of a specific course of action—with the expectation that such a course is set by an inclusive and symmetrical deliberation. Still, the most important distinction lies in the attitude towards prevention. Notice that contingency planning forces prediction because specific actions are conceived in response to accurately described worst-case scenarios; these scenarios are themselves predictions of future states of affairs. Conversely, foresight exercises within anticipatory governance do not seek to predict possible futures; instead, they consider a wide range of possible situations without committing to take any of them as an accurate depiction of the future. The point of anticipation is not to run drills on well-predicted futures, but to enable society to deal with a wider variety of possible situations. Capacity building has a double aim: first, to improve the ability of the system to cope with unexpected situations; and second, to prevent the most perilous ones.

9.3.2 Modeling, Anticipation, and Governance

With expanding computational capacity in the postwar period, modeling had evolved into increasingly complex representations of the economic system (reduced-form equations) that sought to predict accurately future economic conditions. Growth models, in contrast, consist of minimalist representations of the critical operations of the system and resort to abstractions grounded in probability theory to explain economic behavior (e.g., the representative agent). The emphasis shifted to formulating plausible hypothesis about the dynamics of decision-making of consumers and firms and the effect of these decisions in aggregate output. Whereas the previous models conceived of the system as an impenetrable black box and focused efforts on replicating output data, the primary concern of growth models is to produce sound speculations about a single critical operation within the black box. The old school sought accuracy in prediction even at the expense of incorporating ad-hoc elements into their models; conversely, the new school is mostly concerned with the logic and internal consistency of the model. Prediction fell into disuse among the growth theorists.

It could even be suggested that the most radical aspect of the neoclassical wave was to shift the aim of modeling from prediction to explanation. Undoubtedly, the substantive contributions to growth theory as well as the methodological ones[11] had enormous merit in their own right, but abandoning prediction as the purpose of modeling is as radical a change as refusing to use forecasts to organize governance. This commonality of convictions with respect to prediction will permit me to enlist economic modeling as a tool of anticipatory governance. From this perspective, models should offer more than policy prescriptions; they can be used to sketch plausible scenarios that guide efforts in capacity building.

In effect, the practice of using economic models to create scenarios is commonplace. When economists introduce a model, they generally follow the mathematical analysis with a description of the "intuition behind the model." This latter step appeals to our familiarity with actors and situations of real life where we can see the logic of the models at work. This pedagogical technique is similar to the philosopher's thought experiment. A thought experiment is the deliberate construction of a scenario that, however extreme, can be located in the world as we know it. The extremity of such a scenario reveals a paradox, a fundamental contradiction or distinction that the philosopher wants to describe, understand, and resolve. Likewise, models give a rough idea of the range of possible states that could result from varying conditions. Like thought experiments, economic models allow us to envision scenarios against the backdrop of real life.

The strength of scenarios elaborated from economic models is that these models are built from a set of standard primitives, i.e., well defined concepts and well characterized assumptions. Because these concepts and assumptions are standard, they more easily percolate into public discourse. Because their relationships are explicit, the models seem open to challenge, and thus, they may be construed as an honest form of discourse. The perception of familiarity and validity, not to mention their internal consistency, makes these scenarios plausible and consequently useful for mobilizing resources to develop capacities in the social system.

9.3.3 Towards Scenarios of Wage Inequality

I will close this discussion by hinting very briefly at the way scenarios can be constructed using the model from Section 9.1. I address here the challenges presented at the end of Section 9.2 regarding the emergence of two non-GPT nanotechnologies.

Consider a situation in which advances in crystallography further expand the technical capacity of microprocessors and open up the door to a new generation of cost reductions in the production of chip-based electronics. We could expect that these technologies elongate the wave of adoption of ICTs. Considering the pervasive character of microprocessors (from mobile phones to toasters), a significant reduction in production costs would improve the profit margin of leading application sectors, and with it, an increase in the marginal productivity of skilled labor. As explained in the model, this could increase the skill premium only while demand outpaces the expanding relative supply of skilled labor. By virtue of the current pervasiveness of computers and the fact that we are at an advanced stage in the ICT adoption cycle, nano-chips would only have a marginal effect on wage inequality. This condition would hold only insofar as the education system continues to expand the relative supply of computer skills. Less significant than the temporal increase in the skill premium may be the loss of jobs if the few white-collar jobs that require no intense use of computers are phased out. In turn, this development should not directly affect salaries in blue-collar jobs.

Consider now the possibility that thin solar panels developed with nano-scale coatings of cadmium telluride (CdTe) cross the cost-efficiency threshold of 20% and become competitive with coal-based electric plants. In addition, within a short period of time, other energy sources also become cost-efficient too, including geothermal and wind energy. Nano-solar panels would then only be part, not the whole of a revolution in the energy sector. The transition to the new technologies, however, could significantly affect the economy in the way described by Aghion's model. This portfolio of new energy technologies could behave as a GPT. The model suggests that wage inequality will increase if the relative supply of skilled labor lags behind the increase in the demand for skilled labor of the new sector. This is a fertile area for policy intervention from government and from incumbent firms in the energy sector. As I explained, overall productivity could be expected to decline as a result of diverting resources away from production towards experimentation and implementation of the portfolio of new technologies. Then, incumbent firms could justify heavy investments in training their personnel to facilitate the transition to the retooled production platforms. Government could also justify and encourage these training programs to boost the supply of specific skills. If the education system reacts with a lag to the new demand of skills, it will be training programs and not formal education that shorten the period of growing wage inequality.

In both sketches, the model is not the explanatory core of the scenario; it is only the kernel of a well specified set of causal attributions—GPT adoption cycle, relative supply of skilled labor, and wage inequality. Scenarios are in fact polycentric and allow different logics to explain parcels of them. Economic modeling can thus be used in this way, as part of the integrative effort of anticipatory governance to build the social capacities necessary to cope with the advent of major technological change.

9.4 Concluding Remarks

In conclusion: endogenous growth models explain wage disparities by introducing a stylized characterization of major technological change, or GPT, in a non-linear adoption cycle. While some nanotechnologies may become a GPT, many may emerge without having the transformative effect of electricity or ICTs. These nanotechnologies, however, could either ride the ICT adoption cycle or be part of a new GPT. In both cases the model is useful to explain effects on wage inequality; and in a general sense, the model illustrates the anticipatory value of emphasizing training in information technologies and the use of computers as well as other skills relevant to the dominant GPT. In a still more general sense, economic models that do not seek prediction are compatible with foresight efforts in anticipatory governance.

Notes

1. The theoretical advances in the Keynesian vein that dominated the field for nearly four decades are often referred to as neo-Keynesian economics or the neoclassical synthesis. These advances, pioneered by Hicks (1937) and Modigliani (1944), adapted some elements from the neoclassical school (of the turn of the twentieth century) into Keynes' general theory. Because generations of economists were trained with Samuelson's (1947) articulation of the synthesis, and because of his no small contributions to economic theory, the paradigm is also associated with his name.
2. While it is customary to refer to all models in the paradigm as neoclassical (see, for instance, Nelson 2005), authors working within the paradigm reserve the adjective "neoclassical" only for the first generation of growth models (Solow 1956), and distinguish these from AK models (Frankel 1962) and endogenous models (Romer 1990; Aghion and Howitt 1992). The argument in this chapter is unaffected by these distinctions; the reader interested in them is referred to Parente (2001) and Aghion and Howitt (2006).
3. This is not to say that other paradigms such as evolutionary economics have not been influential. See Verspagen (2005) for a comparison of neoclassical and evolutionary growth theory.
4. This discussion draws heavily from Aghion (2002) and Saint-Paul (2008).
5. In Solow's (1956) model, R&D investment is an exogenous variable; in contrast, in endogenous growth models (Romer 1986, 1990; Grossman and Helpman 1991; Aghion and Howitt 1992) the level of R&D is determined within the model as a decision of firms who respond to economic incentives, specifically, the profitability of their investment. Notice that the market size effect of Acemoglu (2000) assumes this endogeneity.

6. It should be noted that the period 1995–2008 has seen a modest improvement in productivity growth with an annual compound rate of 2.1%.
7. Material engineering and biomedicine may have common interests—for instance, crystallography is applied to the study of proteins and semiconductors respectively—but the example stands for the great many different domains of research in these fields.
8. Lux Research (2007) lists ten nano-materials. Those in their classification without clear GPT applications are: dendrimers (spherical branched polymers) and polyhedral oligomeric silsesquioxanes (similar to fullerenes), used for medical diagnostics and treatments; nanoporous materials used for drug delivery systems; nano-scale encapsulation, used to increase solubility of active components in pharmaceuticals; and quantum dots used for biological testing and brighter displays.
9. Some may prefer the term socio-technical systems in order to emphasize the inextricability of the technical and from the social and vice versa. I do not object to that view, but I prefer to refer to socio-technical systems as plainly social.
10. Introduced more than half a century ago by Peter Drucker (1954).
11. The end of prediction implied that the research strategy shifted from verification to falsification; which is put in evidence by the widespread use of calibration (input values for model parameters such that modeled variables behave in accordance with historic data). In addition, another important improvement was the introduction of methodological individualism in macroeconomics; models build upwards from individual behavior. New methods were adapted from mathematics, in particular taking the analysis from static to the dynamic equilibrium.

References

Acemoglu, Daron. 1998. Why do new technologies complement skills? Directed technical change and wage inequality. *Quarterly Journal of Economics* CXIII: 1055–1090.
Acemoglu, Daron. 2002. Technical change, inequality, and the labor market. *Journal of Economic Literature* 40: 7–72.
Aghion, Philippe. 2002. Schumpeterian growth theory and the dynamics of income inequality. *Econometrica* 70: 855–882.
Aghion, Philippe, Eve Caroli, and Cecilia Garcia-Peñalosa. 1999. Inequality and economic growth: The perspective of the new growth theories. *Journal of Economic Literature* 37: 1615–1660.
Aghion, Philippe, and Peter Howitt. 1992. A model of growth through creative destruction. *Econometrica* 60: 323–351.
Aghion, Philippe, and Peter Howitt. 2006. Joseph Schumpeter Lecture, appropriate growth policy: A unifying framework. *Journal of the European Economic Association* 4: 269–314.
Barben, Daniel, Erik Fisher, Cynthia Selin, and David H. Guston. 2008. Anticipatory governance of nanotechnology: Foresight, engagement, and integration. In *The handbook of science and technology studies,* 3rd ed., ed. Edward J. Hackett, Olga Amsterdamska, Michael Lynch, and Judy Wajcman. Cambridge: MIT.
Bartels, Larry M. 2008. *Unequal democracy: The political economy of the new Gilded Age.* Princeton, NJ: Princeton University Press.
Bresnahan, Timothy F., and Manuel Trajtenberg. 1992. *General purpose technologies: Engines of growth?* NBER working paper 4148. Cambridge: National Bureau of Economic Research.
Bresnahan, Timothy F., and Manuel Trajtenberg. 1995. General purpose technologies: engines of growth? *Journal of Econometrics* 65: 83–108.
Boin, Arjen, and Paul 't Hart. 2003. Public leadership in times of crisis: Mission impossible? *Public Administration Review* 63: 544–553.
David, Paul A. 1989. *Computer and dynamo: The modern productivity paradox in a not-too-distant mirror.* Center for Economic Policy Research, No. 172. Palo Alto: Stanford University.
David, Paul A. 1990. The dynamo and the computer: An historical perspective on the modern productivity paradox. *American Economic Review* 80: 355–361.

David, Paul A., and Gavin Wright. 1999. *General purpose technologies and surges in productivity: Historical reflections on the future of the ICT revolution.* Discussion papers in economic and social history. Oxford: University of Oxford.

Drucker, Peter. 1954. *The practice of management.* New York, NY: Harper.

Frankel, Marvin. 1962. The production function in allocation and growth: A synthesis. *American Economic Review* 52: 995–1022.

Freeman, Richard B. 1991. *How much has de-unionisation contributed to the rise in male earnings inequality?* NBER working paper 3826. Cambridge: National Bureau of Economic Research.

Grossman, Gene M. and Elhanan Helpman. 1991. *Innovation and growth in the global economy.* Cambridge: MIT.

Grossmann, Volker. 2001. *Inequality, economic growth, and technological change: New aspects in an old debate.* Heidelberg: Physica-Verlag.

Guston, David H., and Daniel Sarewitz. 2002. Real-time technology assessment. *Technology in Society* 24: 93–109.

Jones, Charles I. 1995. R&D-based models of economic growth. *The Journal of Political Economy* 103: 759–784.

Hall, Bronwyn H., Adam B. Jaffe, and Manuel Trajtenberg. 2001. *The NBER patent citations data file: Lessons, insights and methodological tools.* NBER working paper 8498. Cambridge: National Bureau of Economic Research.

Hicks, John R. 1937. Mr. Keynes and the "classics": A suggested interpretation. *Econometrica* 5: 147–159.

Jones, Charles I. 1995. R&D-based models of economic growth. *Journal of Political Economy* 103: 759–784.

Karinen, Risto, and David H. Guston. 2009. Toward anticipatory governance; The experience with nanotechnology. In *Governing future technologies: Nanotechnology and the rise of an assessment regime*, ed. Mario Kaiser, Monika Kurath, Sabine Maasen, and Cristoph Rehmann-Sutter, 217–232. Dordrecht: Springer.

Lipsey, Richard G., Cliff Bekar, and Kenneth Carlaw. 1998. What requires explanation? In *General purpose technologies and economic growth*, ed. Elhanan Helpman, 15–54. Cambridge: MIT.

Lux Research, Inc. 2007. *The nanotech report^{TM}. Investment overview and market research for nanotechnology*, 5th ed. New York, NY: Lux Research Inc.

Milgrom, Paul, and John Roberts. 1990. The economics of modern manufacturing: Technology, strategy, and organization. *The American Economic Review* 80: 511–528.

Modigliani, Franco. 1944. Liquidity preference and the theory of interest and money. *Econometrica* 12: 45–88.

Mokyr, Joel. 1990. *The lever of riches: Technological creativity and economic progress.* New York, NY: Oxford University Press.

Nelson, Richard R. 2005. *Technology, institutions, and economic growth.* Cambridge: Harvard University Press.

Nelson, Richard R., and Edmund S. Phelps. 1966. Investments in humans, technological diffusion, and economic growth. *American Economic Review* 56: 69–75.

Palmberg, Christopher, and Tuomo Nikulainen. 2006. *Industrial renewal and growth through nanotechnology? An overview with focus on Finland.* ETLA discussion paper 1020. Helsinki: ETLA Elinkeinoelämän Tutkimuslaitos, The Research Institute of the Finnish Economy.

Parente, Stephen L. 2001. The failure of endogenous growth. *Knowledge, Technology and Policy* 13: 49–58.

Roberts, Nancy C., and Linda Wargo. 1994. The dilemma of planning in large scale public organizations: The case of the United States Navy. *Journal of Public Administration Research and Theory* 4: 469–491.

Romer, Paul. 1986. Increasing returns and long-run growth. *The Journal of Political Economy* 94: 1002–1037.

Romer, Paul. 1990. Endogenous technological change. *The Journal of Political Economy,* Part 2: The Problem of Development: A conference of the Institute for the Free Enterprise Systems 98: S71–S102.

Rosenberg, Nathan. 1998. Chemical Engineering as a general purpose technology. In *General purpose technologies and economic growth*, ed. Elhanan Helpman, 15–54. Cambridge: MIT.
Rosen, Sherwin. 1981. The economics of superstarts. *The American Economic Review* 71: 845–858.
Ruttan, Vernon W. 2006. *Is war necessary for economic growth? Military procurement and technology development*. New York, NY: Oxford University Press.
Saint-Paul, Gilles. 2008. *Innovation and inequality: How does technical progress affects workers?* Princeton, NJ: Princeton University Press.
Samuelson, Paul A. 1947. *Foundations of economic analysis*. Cambridge, MA: Harvard University Press.
Schumpeter, Joseph A. 1942. *Capitalism, socialism, and democracy*. New York, NY: Harper and Brothers.
Shea, Christine M. 2005. Future management research directions in nanotechnology: A case study. *Journal of Engineering and Technology Management* 22: 185–200.
Simon, Herbert A. 1987. The steam engine and the computer: What makes technology revolutionary. *EDUCOM Bulletin* 22: 2–5.
Smits, Ruud, Rutgerg van Merkerk, David H. Guston, and Daniel Sarewitz. 2009. *Strategic intelligence; The role of TA in systemic innovation policy*. Utrecht University, Innovation Studies Utrecht, Working Paper 08.01. http://www.geo.uu.nl/isu/pdf/isu0801.pdf. (accessed November 1, 2009).
Solow, Robert M. 1956. A contribution to the theory of economic growth. *The Quarterly Journal of Economics* 70: 65–94.
Verspagen, Bart. 2005. Innovation and economic growth. In *The Oxford handbook of innovation*, ed. Jan Fagerberg, David Mowery, and Richard R. Nelson, 487–513. Oxford: Oxford University Press.
Youtie, Jan, Maurizio Iacopetta, and Stuart Graham. 2008. Assessing the nature of nanotechnology: Can we uncover an emerging general purpose technology? *Journal of Technology Transfer* 33: 315–329.
Zimmerman, Rae. 1985. The relationship of emergency management to governmental policies on man-made technological disasters. *Public Administration Review* 45: 29–39.

Chapter 10
Metropolitan Development of Nanotechnology: Concentration or Dispersion?

Jan Youtie and Philip Shapira

In this chapter, Jan Youtie and Philip Shapira leave the world of economic theory and plunge into the economic realities of the regional distribution of nanotechnology activities today in the United States. Some emerging technologies in the past have developed in specific locations, with the best known being Silicon Valley in California, Research Triangle Park in North Carolina, and Route 128 near Boston. Such "technology districts" can be a great boon for local economies, but they inherently open up inequalities with other areas. Youtie and Shapira explore how current nanotechnology research is distributed among regions in the United States. They find that publications in nanotechnology tend to appear from places that are somewhat more dispersed than in earlier emerging technologies. Still, there are "nanodistricts" in the United States where the emerging activity is concentrated—especially when activity is measured in terms of patents instead of publications. The pattern they describe may not leave much leeway for localities to attract new nano-based business—or at least not as much as they might like.—eds.

J. Youtie (✉)
Enterprise Innovation Institute, Georgia Institute of Technology, Atlanta, GA, USA
e-mail: jan.youtie@innovate.gatech.edu

Originally presented at the Workshop on Nanotechnology, Equity, and Equality at Arizona State University on November 21, 2008.

10.1 Introduction

Nanotechnology is expected to be a new driver of technology-oriented business and economic growth with broad implications in redefining products, industries, skills, and places (Lux 2007). While there are likely to be many benefits associated with the development of nanotechnologies, there are potential risks—not only related to health and safety but also in terms of economic and societal impacts. In this chapter, we explore some of the emerging spatial consequences associated with the development of nanotechnology at the regional level. We focus on metropolitan areas in the United States, identifying those areas where research, development, and early innovation agglomerations in nanotechnology are developing. This analysis is important because geographical distribution will be one of the influences on the division of the economic and social consequences of nanotechnology. We probe whether nanotechnology development is reinforcing the position of already leading technological centers such as Northern California's Silicon Valley or Boston, Massachusetts, or whether new metropolitan trajectories are emerging which may change the distribution of research and technology complexes in the United States We also examine the organizational makeup of these areas to understand the institutional mechanisms that underlie the distribution of nanotechnology. We draw on this evidence to develop propositions about the multiple ways in which patterns of concentration and distribution at the metropolitan level may be influenced in future years by the development of nanotechnology research and innovation.

10.2 The Spatial Distribution of Nanotechnology: Similar or Different from Other Technologies?

Novel technologies promise new benefits to users and consumers in terms of greater efficiency, performance enhancement, and added capabilities. Proponents also point to job and business opportunities and societal benefits. At the same time, such new technologies often have costs, for example, in jobs displaced through efficiency or obsolescence, as well as health or environmental risks. Moreover, where there are net economic benefits, these are generally not distributed equally. Yet, as Street observes, "there are no easy generalisations to be made about how technology affects the choices available to groups and individuals, except to say that each new technology changes the pattern of possibilities." (Street 1992, 101) What we can predict is that the distribution of effects of an emerging technology will largely be context dependent relative to the attributes of that technology; its positioning relative to other technologies; the current distribution of economic, social, and human capital; the broader institutional framework; and the policies and conditions which govern use of and access to the technology.

Given this context, can we discern features already associated with the emerging landscape of nanotechnology research and commercialization which can offer signals as to the ways in which nanotechnology is developing? At the global level, analyses have already been undertaken, usually in the context of international competitiveness, asking which nations are assuming leadership in nanotechnology publications and patenting (Huang et al. 2003, 2005; Kostoff et al. 2006). For example, Youtie et al. (2008) find that Europe, the United States, and Japan, as might be expected, are prominent in terms of the number of nanotechnology publications. However, nanotechnology publication in several other Asian countries is growing at rapid rate, especially in China, which is now the world's second largest producer of nanotechnology research publications after the United States (Shapira and Wang 2009). The rise of China in the new domain of nanotechnology represents a significant change in the global technology development landscape, especially as institutional, regulatory, commercialization, and socio-economic frameworks differ in China from those typically found in fully developed economies. Nanotechnology research and development (R&D) is also emerging in selected other developing countries, including in Latin America (Kay and Shapira 2009), although generally most developing countries have limited capabilities not only to undertake R&D in nanotechnology but also to manage and regulate its deployment (Bürgi and Pradeep 2006).

The national level is an appropriate level to consider how the promise of nanotechnology will be fulfilled and to consider policy issues and strategies related to its deployment. In addition, the national level is the source of the bulk of nanotechnology public R&D funding. In the case of the United States, funding through the National Nanotechnology Initiative for nanotechnology R&D grew by more than 200% from 2001 to 2008, exceeding US$1.5 billion in the 2009 budget. In contrast, U.S. state level funding is nearly an order of magnitude lower, estimated at about US$400 million in 2004 (Lux 2005). However, from an economic development perspective, concentrations of emerging technologies are invariably located at the sub-national regional and metropolitan level. Moreover, new technology development tends to be rather unevenly positioned across geographical space (Feldman and Florida 1994; Malecki 1997). Theories of clustering and agglomeration tell us that there are internal and external economies of scale and scope in regional co-location. These concepts have been highlighted since Alfred Marshall's (1890) work on industrial districts and are prominent in models of high tech regions (Saxenian 1994; Porter 1990).There is a body of related work that examines how knowledge spillovers and exchanges, along with scientific and technical capital and complementary and supporting industries, produce geographic hot spots for high-technology research and commercial activity (Jaffe et al. 1993; Krugman 1991). These factors make for a "spiky" world, suggests Florida (2005) in his description of how certain localities advance development of emerging technologies to a greater extent than most others.

While the *use* of nanotechnologies is likely to be widespread, we would expect the *development* of new nanotechnologies to be more spatially concentrated (as in the prior rounds of technology development observed by the authors mentioned above). But what is the pattern of spatial concentration likely to be? And are there opportunities for new locations to enter as significant players in U.S. nanotechnology so as to create a less uneven distribution of nanotechnology research and commercialization activities at the sub-national level? One reason why the distribution of research and commercialization assets in new technologies is skewed to a small number of locations is because these locations are able to build on capabilities developed in prior rounds of technological development (Fuchs and Shapira 2005). For example, Zucker and Darby (2005) find that there is significant overlap in the United States in the concentration of established biotechnology centers and emerging nanotechnology locations. Any new technological round starts not with a "clean slate" but in a landscape already preconfigured for research and commercialization. This influence of prior R&D is what Zucker et al. (2007) found in their initial study relating nanotechnology article production at the regional level to prior non-nanotechnology article knowledge stocks. However, the correspondence between the geographic concentrations of prior technologies and nanotechnology is not perfect. Shapira et al. (2010) compare nanotechnology to biotechnology, using the biotechnology center classification of U.S. metropolitan areas by Cortright and Mayer (2002) and find both similarities and differences. For example, Boston, San Francisco, New York, and Washington, DC, are in leading positions in both technologies. Conversely, major R&D centers in biotechnology such as Raleigh-Durham, Seattle, and Philadelphia are not as dominant in nanotechnology; while places like Atlanta, Austin, Phoenix—considered average or not significant in the biotechnology study—are more prominent in nanotechnology; and places with large government laboratories with major nanotechnology programs such as Knoxville/Oak Ridge are not even listed as potential biotechnology centers. Thus, while up-and-coming technologies often take root in well-established and prominent technology locations, thereby reinforcing cumulative advantage, the correspondence is not perfect and there is at least some scope for new regions to emerge—and old regions to reposition themselves.

The factors affecting regional emergence and positioning through nanotechnology could be expected to include the multidisciplinarity of the field, involvement of incumbent firms, and the nature of commercialization. Early studies of the multidisciplinarity of nanotechnology find that scientific and technological publications in the field encompass a range of disciplines, as measured by the Institute for Scientific Information (ISI) journal subject categories (SC)—more than 150 of the 175 science and technology SCs—and draw on knowledge in cited references from a wide range of SCs, although with some concentration in fields related to materials science (Schummer 2004; Porter and Youtie 2008). So, what might be the distributional consequences of the multidisciplinary nature of nanotechnology? Case studies by Meyer and Rafols (2009) suggest that needs for instrumentation, special materials, and knowledge sharing underlie multidisciplinary activities in nanotechnology. A range of localities have the potential to fulfill these needs if a single location lacks sufficient resources to address them intra-institutionally; and therefore investigators seek collaborative, cross institutional, and sometimes even cross-regional, relationships.

Another factor with the potential to affect a broader distribution of nanotechnology R&D is its use and commercialization patterns. As noted, the use of nanotechnologies is expected to be widespread. Early evidence observes that nanotechnology has the characteristics of a general purpose technology that is pervasive, innovation spawning, and offers scope for improvement (Youtie et al. 2008; Graham and Iacopetta 2008). Indeed there are already over 1,000 consumer products on the market which are enabled or augmented by nanotechnologies, ranging from electronics, computers and appliances, to paints, cleaners, cosmetics, other chemicals, clothing and sports equipment (Project on Emerging Nanotechnologies 2009). To the extent that nanotechnology is pervasive and user-driven, nanotechnology R&D may be located closer to markets, hence stretching out and opening up the spatial distribution of performers and locations. However, we need to keep in mind the complexity and range of disciplines required for advanced nanotechnology development. It could well be that this will encourage R&D concentrations in locations which already have (or can develop) appropriate technological capabilities and deep talent pools. Indeed, an analysis of global nanotechnology patenting by leading U.S. multinationals indicates that inventive activity is associated with the technological breadth and science capabilities of host locations rather than by market specific factors (Fernandez-Ribas and Shapira 2009).

The enterprise characteristics and industrial structures associated with nanotechnology development are also likely to influence its spatial arrangements. If nanotechnology is viewed not as a new industry but as a cross-cutting process technology, then its development may engage not only many industries but also large firms engaged in volume production. Areas with existing industries might thus be favored, including those with traditional materials-based industries as well as those developing high-technology electronics and medical devices. Indeed Laredo (2008) and Rothaermel and Thursby (2007) find that a distinctive attribute of nanotechnology is the involvement of large incumbent firms in its emergence, which somewhat diverges from the biotechnology paradigm of innovation being driven in small startups often with a university relationship. Laredo (2008, 15) maintains that there is "not a new nano industry, but nano as a problem-solving resource for existing actors ('nano-enabled' products)." Examples of large multinational companies that are prominent in nanotechnology patenting include IBM (information technology), AMD (semiconductors), Motorola (information technology), Boeing (aerospace), General Electric (electronics and other products), 3M (adhesives), Dow/Rohm and Haas (chemicals), DuPont (chemicals), Kimberly Clark (paper), and PPG Industries (glass and coatings). Some of these large companies are headquartered in traditional high technology locations in California and New York, while others have head offices in traditional Midwestern manufacturing locations. Yet, this is not to say that there will not be a high level of venture start-up and spin-off activity in nanotechnology. Wang (2007) identified some 230 new nanotechnology-based venture start-ups formed in the United States through to 2005, with about one-half being companies that had spun-out from universities. Fernandez-Ribas (2008) reports that the role of small and medium-sized enterprises (SMEs), relative to that of large corporations, has grown over the past 10 years. SMEs account for one-third of all U.S. companies with World Intellectual Property Organization (WIPO) Patent Cooperation Treaty (PCT) filings in nanotechnology domains by 2006 compared to less than 20% in 1997.

Nanotechnology is thus intrinsically a diverse domain, involving large firms and SMEs across multiple industries and providing opportunities both for incumbents and new ventures. The technology is emerging during a period of heightened globalization and attention to user-driven and open innovation. At the same time, nanotechnology is characterized by its science-driven complexity and multidisciplinary nature, which require high levels of technological capability and coordination to master. Local agglomeration economies are likely to remain significant, and places which already have capabilities in the development and financing of advanced technology will surely have advantages. Yet, the characteristics of nanotechnology—and, perhaps we should add, the availability of funding driven by policy and local economic development interest in the field—suggest that multiple developmental locations will emerge. Some of these "new" nanotechnology locations may well be "old" locations of mature industries that have been regenerated, where nanotechnology is adapted to existing products and processes as well as to fresh applications. Other sites may offer opportunities for new entrants based on technology and knowledge transfer from universities and public research institutions. To sort this through, and to identify what locational trajectories are emerging in nanotechnology, requires empirical examination. We have undertaken such work, and report our methods and findings in the next section.

10.3 Nanodistricts: Measurement and Results

In this section, we use the concept of "nanodistrict" to designate a geographical cluster of nanotechnology research and commercialization activities (Mangematin 2006). Our focus is on the development of nanodistricts in the United States, which is the world's top producer of nanotechnology articles and patents and continues to be a leader as measured by the citation quality of its research output (Youtie et al. 2008). To assess the growth of nanodistricts in the United States, we use the standard nomenclature of combined statistical areas (CSA) and metropolitan statistical areas not within a CSA (Office of Management and Budget 2006), hereafter termed metropolitan areas. We identify the top 100 metropolitan areas, based on the number of nanotechnology articles assigned to the metropolitan area according to the city and state of the institutions of the authors of a nanotechnology article. Nanotechnology articles used in this analysis come from a database of more than 100,000 U.S. nanotechnology publications (and more than 400,000 global nanotechnology publications) developed at Georgia Tech from the Web of Science's Science Citation Index (WOS-SCI) from 1990 to mid-year 2006. (see Porter et al. (2008) for the details of development of this database.) A similar process, drawing from a set of 54,000 global nanotechnology patents that were extracted from the Micropatents database, was used to aggregate nanotechnology-related patents at the metropolitan level. The number of publications in the top 100 metropolitan areas ranges from 115 (Reno-Sparks, Nevada) to 9612 (New York) over this time period. For patents, the range is from 3 (El Paso, Texas) to nearly 3600 (San Jose-San Francisco-Oakland, California). (see Fig. 10.1.)

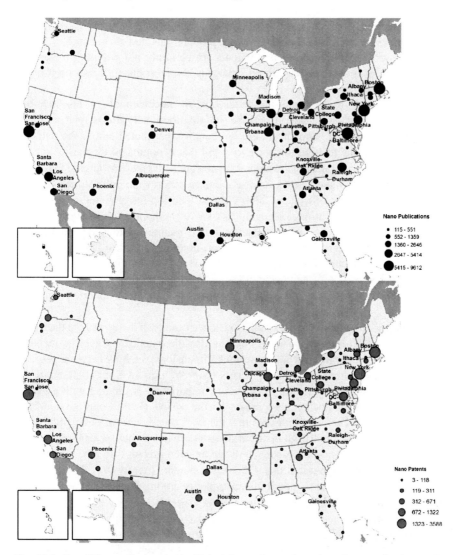

Fig. 10.1 Top 100 nanodistricts in the United States, by total number of nanotechnology publications (*top map*) and Patents (*bottom map*) from 1990 to 2005. Source: Author analysis of Georgia Tech databases of nanotechnology publications and patents (see Porter et al. 2008) *Top 30 nanodistricts based on total number of publications from 1990–2005 are labeled.

Why is it significant to consider the distribution of publications and patents with respect to the likely emergence of nanotechnology in the economic domain? Previous research identifies publications and patents in technology areas as early indicators of economic activity. For example, Shapira et al. (2003) report that metropolitan areas' corporate and university publications in the 1980s were early indicators of employment in information and communication technology industries, and patents also performed this function in the 1990s. With respect to nanotechnology, Shapira et al. (2010) found a relationship between U.S. nanodistricts' corporate and total nanotechnology publications and their nanotechnology patenting activity. Publications and patents are certainly not without problems in representing economic activity in an emerging domain, but they can perform an important early indicator role.

We first seek to examine the landscape of nanotechnology research and commercialization over time to probe its distributional effects and changes. Nanopublications and nanopatents are not equally distributed across the database (see Fig. 10.2). The distribution has a long tail that is populated by New York, San Jose-San Francisco-Oakland, the greater Boston area, and the Washington D.C. area. Indeed, the top six nanodistricts represent nearly 40% of the publications but only 24% of the population as of the 2000 Census. The top thirty nanodistricts represent 84% of the publications compared to 44% of the population in 2000 (Shapira and Youtie 2008). This heavy concentration of nanotechnology development in urban areas suggests a corresponding limited potential for rural areas (as well as for smaller cities) with less concentrated assets and capabilities to play a leading role in the emergence of nanotechnology, even as the promise of nanotechnology for addressing agriculture and natural resource has significant importance to rural economies.

It is important not only to look at the existence of overall differences in the spatial concentration of nanotechnology R&D, but also the extent to which the inequalities in the distribution differ along various dimensions, such as between research and commercialization activity, and across time. We employ Gini coefficients and Atkinson inequality indexes to probe these comparisons, beginning with a nanopublication versus nanopatent comparison for the period 1990 through to mid-2006 (see Table 10.1). The table indicates that the Gini coefficient for nanopublications across the top 100 nanodistrict distribution is less than 0.6, but for nanopatents it is greater than 0.7. Atkinson measures indicate an even greater difference in inequality between publications (less than 0.5) versus patents (nearly 0.7). These indexes suggest that there is greater distributional equality among the top 100 nanodistricts in terms of nanopublishing than nanopatenting.

To examine changes over time, Table 10.1 also presents inequality measures calculated for nanopublications in each of the top 100 nanodistricts for three time periods: 1990–1995, 1996–2000, and 2001–2006 (midyear). The results suggest a decline in the concentration of the distribution of nanopublications from the first to the last time period under analysis. The Gini coefficient drops from 0.65 to below 0.6 from the first to the last time period and the Atkinson measure shows an even larger decline, from nearly 0.6 in the first time period to 0.47 in the more recent time period. These findings indicate that there may be a broadening in more recent years of research participation and publication across U.S. regions.

We next explore a set of factors characteristic of the development of these nanodistricts. This exploration involves an examination of a set of prototype nanodistrict groups that were developed in a previous study conducted by the authors, which focused on the top thirty nanodistricts in terms of nanopublication counts (Shapira and Youtie 2008). These districts are examined based on groupings of

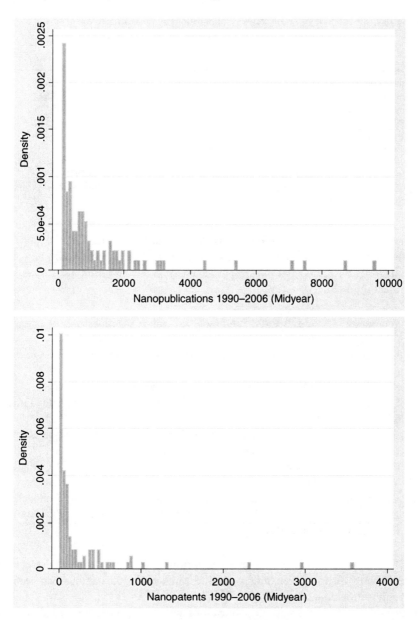

Fig. 10.2 Distribution of nanotechnology Publications and patents across the top 100 U.S. nanodistricts. Source: Author analysis of Georgia Tech databases of nanotechnology publications and patents (see Porter et al. 2008)

Table 10.1 Distribution of nanotechnology publications and patents across the top 100 nanodistricts, 1990 through mid-2006: measures of inequality

	Gini coefficient	Atkinson[a]
Nanopublications	0.594	0.485
Nanopatents	0.723	0.694
Nanopublications 1990–1995	0.645	0.576
Nanopublications 1996–2000	0.606	0.510
Nanopublications 2001–2006	0.585	0.474

[a]The Atkinson measure of inequality has similar theoretical properties to the Gini coefficient but also includes a concentration-aversion weight (Atkinson 1970)
Source: Author analysis of Georgia Tech databases of nanotechnology publications and patents (see Porter et al. 2008)

districts, rather than the districts themselves, because the groupings demonstrate shared factors reflecting common measurable attributes that characterize how these nanodistricts have emerged. Using cluster analysis, seven prototype groups were identified in this study:

(1) Technology leaders (Boston, San Francisco–San Jose (Bay Area), Washington, D.C., and Chicago);
(2) New York (which emerged as a standalone metropolitan technology agglomeration);
(3) Southern California nano-bio districts (San Diego, Los Angeles);
(4) University-dominated districts (Santa Barbara, Ithaca, Lafayette Indiana, Champaign-Urbana);
(5) Government-dominated districts (Knoxville-Oak Ridge, Albuquerque, Denver);
(6) Focused medium-performing districts (Austin, Phoenix; State College, Pennsylvania; Albany, New York; Atlanta; Gainesville, Florida; Madison, Wisconsin; and Seattle); and
(7) Diversified medium-performing districts (Dallas–Fort Worth; Philadelphia; Detroit; Research Triangle Park, North Carolina; Pittsburgh; Houston; Cleveland, Ohio; and Minneapolis).

These clusters were analyzed based on factors characteristic of the development of technology districts, including: overall research performance (total nanotechnology publication counts), early entry (publications 1990–1995), late entry (publications 2001–2006), percentage of publications authored by government laboratories, universities, corporations, and other types of institutions, organizational diversity based on the Herfindahl Index, sectoral focus (share of publications in nanobio), knowledge sharing via co-authorships in and out of the immediate region, and research quality (based on citations). We also analyzed nanotechnology patenting in these districts (see Table 10.2).

We note that all nanodistricts in the seven prototype groups had relatively high levels of cross-regional co-authorships. The average nanodistrict has about half of

Table 10.2 Nanodistrict prototypes based on cluster analysis of top 30 U.S. Metropolitan nanodistricts

District type	Key characteristics	Metropolitan membership
Technology leaders	Many publications, early publications, low Herfindahl, high corporate publications, high patenting	Boston, MA Chicago, IL San Francisco-San Jose, CA Washington DC-Baltimore, MD
New York	Most publications, corporate publications, nanoelectronics	New York
Southern California Nanobio	Nanobio, high citations, high networking	Los Angeles, CA San Diego, CA
University	University-dominated (high H), early publications, fewest patents	Champaign, IL Ithaca, NY Lafayette, IN Santa Barbara, CA
Government	Government lab dominated, high networking	Albuquerque, NM Denver, CO Knoxville, TN
Focused	Late entrants—more university, higher Herfindahl, nanomaterials, later publications, fewer patents	Albany, NY Atlanta, GA Austin, TX Gainesville, FL Madison, WI Phoenix, AZ Seattle, WA State College, PA
Diversified	Diverse—medium publication performance, mid Herfindahl, more corporate publications, nanobio, patents	Cleveland, OH Dallas, TX Detroit, MI Houston, TX Minneapolis, MN Philadelphia, PA Pittsburgh, PA Research Triangle, NC

Note: Based on cluster analysis of top 30 U.S. publishing nanodistricts (for full details, see Shapira and Youtie 2008)

its publications involving an out-of-area author and this percentage is relatively constant across prototype nanodistricts. (The only outlier is Ithaca, in which nearly all of its nanotechnology publications are authored by researchers in the Ithaca area, largely at Cornell University.) This extent of co-authorship may be considered an indicator of sharing of information and nanotechnology developments across intra-regional boundaries (Lewenstein 2005). In other words, author-based cross-regional connections provide an opportunity for investigators in less prominent areas to take part in the emergence of nanotechnology R&D. We do not know the details surrounding the cross-regional author-based linkages, but presumably they involve the need to address gaps in instrumentation, special materials, and knowledge that have the potential to build up nanotechnology R&D in nontraditional locations. Moreover, the extent of sharing suggests that a significant role in nanodistrict development is played not just by within-district factors but also by non-proximate knowledge sharing and exchange, and that this inter-district sharing likely accelerates the diffusion rate of the technology.

These attributes and their implications for nanodistrict development, have consequences for the distributional concentration or dispersion of nanotechnology R&D. We do indeed see that our prototypes include metropolitan complexes that have long been in leadership positions in prior rounds of new technology emergence, particularly Boston and the San Francisco Bay Area. However, as anticipated, we also see the presence of cities in the Midwest and Sunbelt that are not traditionally thought of as playing major roles in the emergence of new technologies.

We also find that our emerging nanotechnology districts differ from the standpoint of their organizational diversity. We observe that the most well-developed prototypes—New York, Technology Leaders, and Southern California—have the highest number of publications and had prominent concentrations of publications in the early stages of nanotechnology development (1990–1995). These three prototypes have a high diversity of organizational participants with multiple companies, universities, and government laboratories and non-profit research institutes significantly participating in nanotechnology R&D.

In contrast, the other clusters of nanodistrict prototypes—especially the Focused, University, and Government—have more monocentric approaches to nanotechnology R&D around a single government laboratory or university. On the one hand, it appears that the effort of these regions without a long history of prominence in new technology emergence involved focusing research resources on a single institution. This level of focus appeared to be important in raising the profile of the regions in the near-term. On the other hand, this monocentric strategy does pose longer-term issues, such as the potential for crowding out other institutions—e.g., other universities, including minority serving institutions, or possibly even private sector firms—with capabilities or specializations in areas not preferred in the dominant institution. Strengthened participation of these other institutions offers the opportunity for developing and invigorating the regional cluster. Moreover, these three monocentric prototype regions have comparatively weak levels of commercialization as measured by number of nanotechnology-related patents. Addressing the relatively weak levels of commercialization requires the involvement of additional organizations, especially local companies that conduct R&D as well as local companies with capabilities to stimulate demand- or user-led business innovation in nanotechnology.

10.4 Conclusions

This work has examined the distribution of nanotechnology publications and patents across metropolitan areas within the United States. These metrics are used as prospective indicators of emerging regional capability. The results have shown that there are considerable inequalities in concentrations of nanotechnology R&D, but the gap between the leading technology regions and other nanodistricts is lessening over time. One caveat is that nanotechnology is still at an early stage of development, with just a generation elapsed since the invention (in the 1980s) and subsequent diffusion of the scanning, tunneling, and atomic force microscopes which enabled the practical investigation and development of nanotechnology. Moreover, much of nanotechnology R&D is in the first of the four nanotechnology development generations proposed by Mihail Roco (2004): passive nanostructures (including aerosols, coatings); active nanostructures (including targeted drugs, adaptive structures); systems of nanosystems (including guided assembly); and molecular nanosystems (including molecular devices by design). It may be that as nanotechnology proceeds through later generational progressions, this development path will affect the regional distribution of nanotechnology R&D, although the direction (more or less inequality) is unclear.

As yet, we have not examined internal features within nanodistricts such as the racial, ethnic, and income attributes of those participating in nanotechnology within each nanodistrict. The process of developing a new technology is usually not a mass participation activity; rather, leading-edge work is conducted by a subset of well-educated (typically doctoral level) scientists and engineers in selected research universities, specialized corporate units, and large government laboratories. That said, Richard Florida (2002) finds that places where new technology develops also have strong and well-educated talent bases, and environments to allow for cultural and racial diversity. However, this does not mean that we expect nanotechnology development by itself to foster a reduction of inequality. Indeed, we have found that nanodistricts have quite different organizational configurations. This implies that nanodistricts vary by the employment opportunities that they present to the labor force. For example, universities and government laboratories, which have been found to provide more equal pay in high-technology careers across racial and ethnic groups than private sector firms (Gatchair 2007), may offer greater opportunities for greater equality in higher paying employment. Although detailed empirical work has yet to be undertaken, we might expect differences in employment equity between public-sector dominated mono nanodistricts and multi-centric nanodistricts where labor market structures include those of multiple private companies as well as universities and government laboratories.

The findings of this study have implications for the fair regional distribution of nanotechnology along four dimensions: (1) advantage reinforcement, (2) technology positioning, (3) cross regional connection, and (4) intra-city structures. Regarding advantage reinforcement, or the extent to which nanotechnology reinforces the position of traditional technology centers, this study has shown that this is the case. Regions that led the emergence of developments in prior rounds of new technologies are also leading in nanotechnology R&D publications and patenting. These technology-leading regions have the highest share of publications. Even more, the leading regions maintain a higher share of patents, suggesting that they may be in a preferential position to take advantage of long-term nanotechnology developments (Florida 2005).

While nanotechnology seems to be subject to relatively skewed allocation of research and especially commercialization from a geographic standpoint, there are signs that nanotechnology also offers the potential for greater geographic dispersion and new technology positioning. Nanotechnology research is observed to be starting to spread to other locations as the field develops over time. This spread includes both traditional technology leaders as well as new areas in the Midwest and Sunbelt without a strong history of prominence in the emergence of new technologies. The difference between the traditional technology leaders and the non-traditional technology regions is changing with respect to publications, though not with respect to patenting, nor are rural areas likely to play a major role at least on the nanotechnology research front.

Cross regional diffusion and linkage is important in enabling "have not" or next tier regions to have access to the latest nanotechnology R&D. This research used a proxy for diffusion and linkage, namely, the high rate of out-of-region co-authorships. As these metropolitan areas begin to try to leverage the research developed within their boundaries to achieve commercial benefits, at the same time, much cross-country sharing of nanotechnology R&D is taking place among the researchers themselves. There a tacit sharing of R&D within the nanotechnology research community, which may fuel a more equal distribution of research if not eventual economic development benefits over the long run.

Intra-city structures influence the future ability of next tier regions to emerge more strongly in the nanotechnology R&D enterprise. The finding that next tier regions tend to adopt monocentric organizational approaches may limit the participation of other universities, government facilities, or businesses in nanotechnology opportunities.

This analysis was set within the U.S. nanotechnology R&D context. While each country has specific features, in general we would also expect to find patterns of intra-country regional inequalities, with the emergence of a small group of dominant city-regions in terms of nanotechnology publication and patenting activity. This expectation is born out in several studies of Europe (Shapira et al. 2009), Latin America (Kay and Shapira 2009), and China (Tang and Shapira 2007). From a global perspective, Laredo (2008) finds that 44% of all nanotechnology articles are found in twelve city-regions, including three each in Japan and the United States, two in China, and one each in France, Russia, Korea, and Singapore. The issue is not whether these clusters exist; they do. More to the point, now that we can identify them, we need to probe how their capabilities can be leveraged so that broadly-shared benefits from nanotechnology can be experienced.

Acknowledgements This study uses data from the large-scale global nanotechnology publication and patent datasets developed by the group on Nanotechnology Research and Innovation Systems at Georgia Institute of Technology—a component of the Center for Nanotechnology in Society (CNS-ASU). Support for the research was provided through CNS-ASU with sponsorship from the National Science Foundation (Award No. 0531194). The findings and observations contained in this paper are those of the authors and do not necessarily reflect the views of the National Science Foundation.

References

Atkinson, Anthony. 1970. On the measurement of inequality. *Journal of Economic Theory* 2: 244–263.
Bürgi, Birgit, and T. Pradeep. 2006. Societal implications of nanoscience and nanotechnology in developing countries. *Current Science* 90(5): 645–658.
Feldman, Maryann P., and Richard Florida. 1994. The geographic sources of innovation: Technological infrastructure and product innovation in the United States. *Annals of the Association of American Geographers* 84(2): 210–229.
Fernandez-Ribas, Andrea. 2008. Analysis of small businesses international patent strategies: Preliminary results. Presented at The Center for Nanotechnology in Society, Tempe, Arizona, January 14–16, 2009.
Fernandez-Ribas, Andrea, and Philip Shapira. 2009. Technological diversity, scientific excellence and the location of inventive activities abroad: The case of nanotechnology. *Journal of Technology Transfer* 34(3): 286–303.
Florida, Richard. 2002. *The rise of the creative Class*. New York, NY: Basic Books.
Florida, Richard. 2005. The world is spiky. *The Atlantic Monthly* October: 48–51.
Fuchs, Gerhard and Philip Shapira, eds. 2005. *Rethinking regional innovation and change. Path dependency or regional breakthrough?* Boston, MA: Springer.
Gatchair, Sonia. 2007. Representation and reward in high technology industries and occupations: The influence of race and ethnicity. Doctoral Dissertation, Atlanta, GA: Georgia Institute of Technology.
Graham, Stuart, and Maurizio Iacopetta. 2008. Nanotechnology and the emergence of a general purpose technology. Paper presented at the NBER Conference on Emerging Industries: Nanotechnology and NanoIndicators, May 1–2, 2008, Cambridge, MA.
Huang, Zan, Hsinchun Chen, Lijun Yan and Mihail C. Roco. 2005. Longitudinal nanotechnology development (1991–2002): National Science Foundation funding and its impact on patents. *Journal of Nanoparticle Research* 7: 343–376.
Huang Zan, Hsinchun Chen, Alan Yip, Gavin Ng, Fei Guo, Zhi-Kai Chen and Mihail C. Roco. 2003. Longitudinal patent analysis for nanoscale science and engineering: Country, institution and technology field. *Journal of Nanoparticle Research*. 5: 333–363.
Jaffe, Adam, Manuel Trajtenberg, and Rebecca Henderson. 1993. Geographic localization of knowledge spillovers as evidenced by patent citations. *Quarterly Journal of Economics* 108 (3): 557–598.
Kay, Luciano, and Philip Shapira. 2009. Developing nanotechnology in Latin America. *Journal of Nanoparticle Research* 11, 259–278.
Kostoff, Ronald N., Jesse A. Stump, Dustin Johnson, James S. Murday, Clifford G.Y. Lau and William M. Tolles. 2006. The structure and infrastructure of global nanotechnology literature. *Journal of Nanoparticle Research* 8: 301–321.
Krugman, Paul. 1991. Increasing returns and economic geography. *Journal of Political Economy* 99(3): 483–499.
Laredo, Philippe. 2008. Positioning the work done on nano S&T associated to PRIME. Paper presented at Nanotechnology Science Mapping and Innovation Trajectories, Manchester, UK, September 9, 2008.
Lewenstein, Bruce. 2005. What counts as a 'social and ethical issue' in nanotechnology? *HYLE—International Journal for Philosophy of Chemistry* 11(1): 5–18.
Lux, 2005. *Benchmarking U.S. states for economic development from nanotechnology*. New York, NY: Lux Research.
Lux, 2007. *The nanotech report. Investment overview and market research for nanotechnology*, 5th ed. New York, NY: Lux Research.
Malecki, Edward J. 1997. *Technology and Economic Development*, 2nd ed. Harlow: Addison Wesley Longman.

Mangematin, Vincent. 2006. Emergence of science districts and divergent technology: The case of nanotechnologies. Paper presented at workshop on Mapping the Emergence of Nanotechnologies and Understanding the Engine of Growth and Development, Grenoble, France, March 1–3, 2006.

Marshall, Alfred. 1890. *Principles of economics*. London: Macmillan.

Office of Management and Budget. 2006. *Update of statistical area definitions and guidance on their uses* (OMB Bulletin No. 07–01). Washington, DC: Executive Office of the President.

Porter, Alan L., and Jan Youtie. 2008. How interdisciplinary is nanotechnology? *Journal of Nanoparticle Research* 11, 1023–1041.

Porter, Alan L., Jan Youtie, Philip Shapira, and Dave Schoeneck. 2008. Refining search terms for nanotechnology. *Journal of Nanoparticle Research* 10: 715–728.

Porter, Michael. 1990. *The competitive advantage of nations*. New York, NY: Free.

Project on Emerging Nanotechnologies. 2009. Consumer products, Washington, DC: Woodrow Wilson International Center for Scholars. http://www.nanotechproject.org/topics/consumer_products/. (accessed December 2009).

Rafols, Ismael, and Martin Meyer. 2009. Diversity and network coherence as indicators of interdisciplinarity: Case studies in bionanoscience. *Scientometrics* 81 (2), Online First.

Roco, Mihail C. 2004. Nanoscale science and engineering: Unifying and transforming tools. *AIChE Journal* 50(5): 890–897.

Rothaermel, Frank, and Marie Thursby. 2007. The nanotech versus the biotech revolution: Sources of productivity in incumbent firm research. *Research Policy* 36(6): 832–849.

Saxenian, Annalee. 1994. *Regional advantage*. Cambridge: Harvard.

Schummer, Joachim. 2004, Multidisciplinarity, interdisciplinarity, and patterns of research collaboration in nanoscience and nanotechnology. *Scientometrics* 59: 425–465.

Shapira, Philip, and Jan Youtie. 2008. Emergence of nanodistricts in the United States: Path dependency or new opportunities? *Economic Development Quarterly* 22(3): 187–199.

Shapira, Philip, Jan Youtie, and Stephen Carley. 2009. Prototypes of emerging nanodistricts in the US and Europe. *Les Annales d'Economie et de Statistique*. In Press.

Shapira, Philip, Jan Youtie, and Sushanta Mohapatra. 2003. Linking research production and development outcomes at the regional level. *Research Evaluation*, 12(1): 105–116.

Shapira, Philip, and Jue Wang. 2009. From lab to market: Strategies and issues in the commercialization of nanotechnology in China. *Journal of Asian Business Management* 8(4): 461–485.

Street, Paul. 1992. *Politics and technology*. New York, NY: Guilford.

Tang, Li, and Philip Shapira. 2007. Networks of research collaboration in China: Evidence from nanotechnology publication activities, 1990–2006. Working Paper. Program on Nanotechnology Research and Innovations Assessment, Atlanta, GA: Georgia Institute of Technology.

Youtie, Jan, Maurizio Iacopetta, and Stuart Graham. 2008. Assessing the nature of nanotechnology: Can we uncover an emerging general purpose technology? *Journal of Technology Transfer* 32 (6): 123–130.

Youtie, Jan, Philip Shapira, and Alan Porter. 2008. Nanotechnology publications and citations by leading countries and blocs. *Journal of Nanotechnology Research*, 10(6): 981–986.

Wang, Jue. 2007. Resource spillover from academia to high tech industry: Evidence from New nanotechnology-based firms in the U.S. Doctoral Dissertation, Atlanta, GA: Georgia Institute of Technology.

Zucker, Lynne G., and Michael R. Darby. 2005. *Socio-economic impact of nanoscale science: Initial results and nanobank*, (Working Paper 11181). Cambridge, MA: National Bureau of Economic Research.

Zucker, Lynne, Michael Darby, Jonathan Furner, Robert Lieu, and Hongyan Ma. 2007. Minerva unbound: Knowledge stocks, knowledge flows and new knowledge production. *Research Policy* 36: 850–863.

Chapter 11
The Role of Organized Workers in the Regulation of Nanotechnologies

Guillermo Foladori and Edgar Zayago Lau

While many of the chapters in this book examine the economic differences between countries, the economy also structures unequal relationships within the workplace. Guillermo Foladori and Edgar Zayago Lau call attention to one such historical relationship that forms an important part of the context of the regulation of nanotechnology: the relationship between workers and managers. Past experience has taught workers not to trust management to watch out for their best interests when it comes to risk. Workers also have been relatively neglected in the technology assessment process, even though they are more likely than the general public to be exposed to new substances and production processes at a time when risk is unknown. Foladori and Zayago Lau point to the important role of nongovernmental organizations in calling attention to risks. Their analysis suggests that action by unions is another possible route to the more even, and therefore fairer, distribution of risks from nanotechnology.—eds.

G. Foladori (✉)
Development Studies Program, Universidad Autónoma de Zacatecas, Zacatecas, Mexico
e-mail: gfoladori@gmail.com

Originally presented at the Workshop on Nanotechnology, Equity, and Equality at Arizona State University on November 21, 2008. This chapter was peer-reviewed.

11.1 Introduction

The regulation of nanotechnologies is a subject that has been on the discussion agendas of governments, international organizations, and other social sectors. Several governments have conducted public hearings in order to map the opinions of various stakeholders. At the same time, companies and institutions, as well as academic associations and non-governmental organizations, have proposed the creation of mechanisms to measure, monitor and evaluate nanotechnology development to assist policy design. Currently, there exists much lobbying reflecting the interests of the different actors engaged in this discussion. In this paper, we draw attention to the key role that organized workers have played in regulating nanotechnologies, a subject mostly ignored by the mainstream.

11.2 The Context Surrounding the Regulation of Nanotechnologies

Scientific research on ultra-small particles and nanomaterials became a strategic area for development at the beginning of the 1980s. The development of the scanning tunneling microscope allowed for the visualization and rearrangement of atoms, thus the known chemical and physical processes were subjected to more rigorous measuring, analysis, and evaluation. The possibility of manipulating the different properties of resistance, magnetism, reactivity, conductibility, flexibility, and other attributes of materials at the nanometric scale became a major target in science. These innovative properties made possible the discovery of new applications for products already in the market; the creation of products with novel features; and the transformation of manufacturing systems. Of course, they are currently changing the market. The early claim that "needs are the force behind the research and development (R&D), which creates products or devices to fulfill them" turned into its opposite; today, manufacturers are constantly looking for ways to make use of the new properties that the materials present at the nanoscale.

Many countries have coordinated efforts to use public funding to encourage the development of the so-called "new technological revolution". This revolution is full of promises such as treating polluted water through simple and cost-effective techniques; administering personalized drugs to treat cancer; creating smart textiles; manufacturing micro- and nano-sensors and incorporating them into more sophisticated systems or paints, tools and everyday utensils; adapting nano-mechanisms to their surrounding environment; and many other applications. As a result of these potential uses, the United States launched its National Nanotechnology Initiative (NNI) in 2000. Many other countries followed with, of course, many fewer resources.

The accumulated technological know-how of the last two decades of the twenty-first century captured the interest of industry to transform the new attributes of nanomaterials into sources of profit. According to Lux Research, as of 2005, private investments in nanotechnology R&D exceeded public investments (Lux Research 2006). Since then, private investment has been under the pressure of profit maximization, and the rapid transition from the laboratory to the market is paramount. Since the beginning of the first decade of the twenty-first century, the number of manufactured nanocomponents in the market has grown steadily, as well as their presence in almost all industrial sectors (WWICS 2006).

The introduction of nanoproducts into the market did not come hand in hand with regulatory measures, even though many voices advocated for them. These groups warned about the potential risks to health and the environment that the new products would present. There were other issues raised to be taken into consideration before the commercialization of nanoproducts (such as the lack of labeling standards) to help in their trading, patenting, and legal control. The critics also forewarned that some of these products entailed potential ethical implications—particularly those pertaining to the enhancement of the human being. These issues have been neglected, even though nanotechnology is already becoming the platform for the new industrial revolution worldwide. It is indeed a disruptive technology, with the potential to alter the use of raw materials in industry; the international division of labor; trade relations among countries; the way economic progress is seen; and many other areas.

An analysis of the press communiqués of the last 10 years shows that some of the products obtained from nanotechnology or using nanotechnology in their functioning imply risks to the health and the environment. Some groups have endorsed the democratization of nanotechnology R&D, and many countries have implemented mechanisms to encourage the participation of citizens. The need for establishing standards and classification protocols for nanoproducts has been recognized as well, particularly if industry wants to cope with legal, commercial, and environmental aspects and issues related to health and safety. All of these issues fall within the promise that nanotechnology could become an important tool to alleviate problems associated with inequity and poverty in the world.

In contrast however, there is another analysis that demands that attention be paid to different groups and actors (and the interests they represent) in endorsing the development of nanotechnologies. This analytical view shows a panorama that contrasts with the first analysis because it reflects the level of risks to which different actors are subjected and the benefits that each one of them can eventually obtain from the development of nanotechnologies.

11.3 The Main Actors in the Regulation of Nanotechnologies

The public institutions in charge of regulating toxic compounds, processes, and products, are not the only actors involved in the regulation of nanotechnologies. This issue is of interest to many other sectors and groups, such as chambers of foreign trade, investment agencies, chambers of industry and commerce, environmental organizations, academic associations, international institutions, local governments, and many other sectors.

For instance, a declaration of any association of industries on the necessity of labeling (or not) nanotechnological products is part of the discussion on the best way to regulate these technologies. A non-governmental organization (NGO) warning about the potential impact of nanotechnologies in developing countries is also an important issue for the discussion table. The exclusion of certain nanoproducts from coverage under certain insurance policies is a private action, but affects the public discussion about nanotechnology regulation. These and many other examples illustrate the diverse array of arguments that are currently being negotiated.

Mapping the different actors and their proposals is not a simple task because of the variety of arguments they use. In this paper, without trying to be exhaustive, we group the different actors that have an influence over regulation according to their role in production and how they obtain their share of social wealth. Accordingly, we aggregate them into four groups: governments and supranational institutions like the European Union and the Organization for Economic Co-operation and Development (OECD); companies and industrial organizations from different sectors (e.g., financial, manufacturing, commercial); trade unions; and NGOs. It is clear however, that this classification is far from being complete. Representing the interests of a developed nation or a developing country is quite different; NGOs can have different scopes and proposals from those described earlier. Nonetheless, this classification sheds some light on the differences and interests behind the actions of each actor taking part in the discussion.

11.3.1 Governments

Governments make up the first group. For them, the development of nanotechnologies is a way to increase international competitiveness; in other words, to be better positioned in the international market in relation to other countries. In the National Nanotechnology Initiative of the U.S., this goal is explicit, but it can also be read between the lines in the declarations of several governments, including Thailand, Mexico, Brazil, Argentina, the European Union, and others. The main objective then, is to achieve significant progress in nanotechnology before other countries do. Even though there are several partnerships and agreements between actors at the state level, the issue of competition, the struggle to obtain benefits to the disadvantage of the competitor, is an immediate justification for allocating public and private funding in the development of this technology.

In the case of nanotechnology R&D, the interest of most governments goes hand in hand with the interests of the industrial sector. One can argue that the national nanotechnology initiatives are essentially commercial programs, and the rationale behind the trickle down of benefits for the population is taken for granted: more and better science leads to more products and increased productivity, eventually to an increase in competitiveness, and thus more profits for the country. In the end, these steps are assumed to lead to an improvement in the living conditions of the population. Following this logic, and during the global economic crisis, many spokesmen have pointed out the importance of the use of nanotechnologies as a strategy for recovery (Gupta Ray 2009; Prensa Latina 2008; Peiris 2008).

However, while an increase in competitiveness is an ongoing goal, the real risks for governments include: (1) not being able to jump on the bandwagon of nanotechnologies and (2) that eventually these technologies may cause environmental or human harm. While making the leap into nanotechnology can be seen as a triumph or failure for the government in question (depending on the outcome), any harm can be characterized as an unforeseeable accident. The repercussions are thus borne by the world population, if no regulatory measures are taken.

11.3.2 Capitalists

The second group is companies. Here the advantages are more clear and direct. No company is willing to incorporate new technologies into its manufacturing process without knowing for a fact that they will generate greater profits than the old ones. From the view of companies, the potential risks to health and the environment depend on regulation; if there is no regulatory framework, accidents are not punishable beyond the standard legal framework associated with any manufacturing process. Therefore, what is of most interest for these actors is not the real risk, but the risks perceived by society, which are, in the end, directly related to commercial acceptance and profits.

But the capitalist class is not homogeneous. It is segmented into factions according to the place a particular firm holds in the productive structure. And, according to this position, the source of profits and regulatory interests change as well. Hence, it is not surprising that spokesmen from the financial world, the center of speculative capital, like *Forbes*, advocate no regulation of nanotechnologies (Wolfe 2005), and use as an example of success, the case of the Internet, which extended itself rapidly thanks to a deregulated market. Financial capital benefits from the ups and downs of the stock market because these changes are the origins of profits, and they are larger when there is no regulation.

This position against regulation within the capitalist class appears only in the financial group. The companies manufacturing raw nanomaterials for industry such as single-walled nanotubes (SWNTs), multi-walled nanotubes (MWNTs), buckyballs, iron nanoparticles, silver nanoparticles, titanium dioxide nanoparticles, zinc oxide nanoparticles, nanoceramics, and many more (Royal Commission on Environmental Pollution 2008), are concerned with the quality of their nanoproducts. Since these products are difficult to manufacture on a very large and homogenous scale, buyer companies are worried about obtaining basic nanomaterials that do not perform consistently as expected and might create complaints from their customers. These companies support the certification of their products (Zyvex 2004); they want some sort of classification and certification system. As a result, the International Organization for Standardization (ISO) developed the classification *TC 229* to categorize nanotechnologies (ISO 2005). In addition, in

2006 the International Society for Testing and Materials (ATSM) created the ASTM E2456–06 classification for nanotechnologies.

Other companies have made public a series of Voluntary Codes of Conduct (VCC). These are declarations of good will and responsibility to persuade governments and consumers to accept the introduction of nanoproducts into the market. For example, Rusnano, the Russian corporation of nanotechnologies, has created a *nanocertificate* for its associated companies (Rusnano 2008). There are others that have followed the strategy of adopting VCCs such as Bayer (2007), BASF (Badische Anilin und Soda Fabrik 2008), the European Union (Commission of the European Communities 2008), and many others.

Nonetheless, the fact that other factions of the capitalist class do not accept these VCCs presents serious doubts about their reliability and integrity. As a consequence, the commercial capital that sells directly to the consumer is still concerned. For instance, the insurance company Swiss Re filed a complaint in 2004 about the lack of labeling in nanoproducts. The insurance company Swiss Re (2004). When companies avoid labeling, they do not disclose the kind of nanoparticles used and their potential risks to health and the environment. Therefore, the insurance company Swiss Re (2004) is refusing to insure these companies for anything that happens to the customer. By the same token, the Investor Environmental Health Network (IEHN) (Lewis et al. 2008), a European institution in charge of incorporating environmental conditions into corporate policy, published a briefing paper in 2008 with a suggestive title: *Toxic Stock Syndrome: How Corporate Financial Reports Fail to Apprise Investors of the Risks of Product Recalls and Toxic Liabilities.*

Another faction of financial capital, one linked to the insurance sector, needs clear governmental regulation in order to function; that is, they need clear rules about the regulation of nanotechnologies to set up processes and activities. For example, Continental Western refuses to insure companies processing or manufacturing carbon nanotubes (Continental Western Insurance Group 2008). The insurance company Lloyd's warned in 2009 about the urgent necessity of regulating nanotechnologies to avoid unnecessary legal actions (Lloyd's 2009).

11.3.3 Workers

Workers, unions, and union councils form the third group in the classification of actors in favor of regulating nanotechnologies. These groups have completely different interests and concerns behind their endorsement of nanotechnologies regulation. First, research shows that the first people to suffer the effects of risks to health caused by nanoparticles and nanotechnologies are the workforce employed in laboratories. Second, basic nanomaterials are expected to be in use in sectors where the workers are constantly exposed, such as the automobile, electronic, textile, building, etc. (and some are, in fact, already in use). Third, consumers are also exposed to the risks, and workers are, at the same time, consumers (RS&RAE 2004; Mantovani et al. 2009).

The advantages of nanotechnologies for workers remain ambiguous. The marketing of nanoproducts claims that they are more efficient than regular products; that they will be more precise in delivering treatments for certain diseases; that they will facilitate the treatment of polluted water; that they will produce more efficient energy; and many more potential benefits. One can argue however, that these benefits are more general and are not limited to workers per se. *Prima facie* there is little, if any, advantage of the new technologies for the workers, but there are certainly many risks. The history of capitalism is also the history of constant development of new technologies that have not delivered concrete benefits for the working class. It has been the workers' struggle that managed to win, after many decades, reductions in working hours, an increase in real wages, and social benefits such as pensions, retirement, health insurance, and others.

Nevertheless, many of these achievements have been lost in favor of the interests of the capitalist class. We can, in fact, observe how some of the steps backwards are being applied today as a result of the economic crisis. Those measures include termination of labor agreements; reduction of wages; addition of working hours to the regular schedule; obligations to work on holidays; reduction of vacation days; reduction of social security and health benefits; cutbacks in education and housing support programs; and many other examples.

11.3.4 Non-governmental Organizations

NGOs constitute the fourth group. For them, the analysis is more complex. They differ from the other groups in many aspects, but there is one that stands out from the rest: the way they obtain their income. The government obtains its income from taxes and from negotiations and political struggles; companies get their returns from profits, and workers obtain their income from wages and salaries. In contrast, NGOs obtain their income from donations. Thus they experience the risk of losing donations and the pressure to increase them. The survival of NGOs is related to their ability to achieve their self-imposed objectives and to advance the interests of their sponsors.

Many NGOs have an environmental orientation, so their focus of concern is the potential risks to both health and the environment of nanotechnologies. However, their attention changes according to their associations and alliances and the constraints imposed by industrialists, governments, and sponsors. An example of two active NGOs with opposing positions on commercialization of nanotechnologies illustrates. On one hand, we find the stance of the ETC Group (2002), which is in favor of a moratorium on the commercialization of nanotechnologies. And on the other hand, we find the Environmental Defense Found, another recognized NGO, which signed a partnership with DuPont in 2005 to create a framework for the responsible development, production, use, and end-of-life disposal or recycling of engineered nanoscale materials (DuPont & Environmental Defense Fund 2007).

11.3.5 Other Actors

Academics, scientists, and intellectuals have also participated in the discussion about regulation of nanotechnologies. Their participation has been more active since several developed countries adopted technology assessment exercises. However, in both policy design and technology assessment programs, intellectuals have been considered the most important actors in the discussions among the following: public institutions in charge of toxicity management, standards and classifications; commercial and business chambers; associations and institutions of scientific character; and NGOs with an environmental character. Paradoxically, unions have been ignored in these exercises. Furthermore, on more than one occasion they have been regarded as NGOs, neglecting their distinctive class character and their role in history. As a consequence, their weight in negotiations is undermined. The participation of the public as consumers has been incorporated into the discussion indirectly through the mechanisms of public inquiry; nevertheless, these proceedings do not have political relevance and have very little impact on the regulation process.

11.4 The Role of Workers in the Regulatory Agenda of Nanotechnologies

Worker organizations, in association with some NGOs, have defended the position of the working class. In fact, their activism has been critical in the implementation of the regulatory agenda for nanotechnologies. Paradoxically, they have not been invited to take part in the negotiations on the matter.

When conducting a chronological analysis of the different proposals about the regulation of nanotechnologies, we find an eloquent symbiosis between those proposals and the interests of each actor. Words of warning about the potential and proven toxic effects of nanoproducts and nanoparticles did not come from corporations, even though they are the ones manufacturing, selling, and distributing these products. Companies have publicized only their products' positive features and the advantages they offer for the customer. The pattern is similar in the case of certain governments in countries where nanotechnology R&D is more advanced, such as in the U.S., Germany, Japan, Russia, China, South Korea and others.

11.4.1 Early Warnings

The words of caution about the potential threats to workers and consumers, and about the potential damage to the environment by nanoproducts, came from environmental NGOs. They were the ones that started the discussion on this issue. Their concern pertains to the idea that market-driven R&D, in this case for nanotechnologies, is advancing faster than safety measures are being implemented in products and manufacturing. This gap represents a risk for the general population, but especially for workers. In 2001, the ETC Group, headquartered in Ottawa, Canada, launched a call at the Sustainable Development Summit in South Africa for a moratorium on the commercialization of nanotechnologies until appropriate research was conducted (ETC group 2002). This NGO has been active in the form of publications looking at the impacts and implications of nanotechnologies for the environment, social and economic relations, risks to health, and policy design and implementation in several countries. Ironically, very few academic associations or intellectuals echoed the claims of the ETC Group. Exceptions included the Royal Society and the Royal Academy of Engineering (RS&RAE), and Prince Charles, who endorsed the idea (Highfield 2003; RS&RAE 2004). In contrast, many other organizations tried to discredit the ETC Group by arguing that its proposal was just a reflection of activism and Luddism.

Another organization, the Salleh, since 2005 has been demanding more research and information about the occupational and environmental risks of nanotechnologies (ACTU 2005). Since 2004, the Trade Union Congress (TUC) of the United Kingdom has been informing its affiliates about the potential risks of nanotechnologies (TUC 2004).

11.4.2 Perceived vs. Real Risk

In contrast, the concern by government and companies about these issues has been linked to the *perceived* risks rather than the *real* risks represented by nanotechnologies. This has been the status quo at least for the first part of the 21st century. Some governments and most companies made efforts to create, among the public, a positive or neutral image of nanoproducts. As a result, these actors left the responsibility of regulating nanotech R&D to the market. Accordingly, there have been two types of proposals endorsed by governments and companies in the last 5-year period.

On one hand are proposals linked to self-regulatory systems and VCC, that is, schemes created to reduce the negative perception of nanotechnologies among consumers by trusting the word of companies. There is however, nothing related to occupational health in these VCC. For example, the *Nanomark* of Taiwan, adopted in 2004, is an official certificate for nanotechnological products that only shows their origins but not their toxicity (Industrial Technology Research Institute 2004). Other examples include the adoption of VCCs by several corporations like BASF in 2004 (BASF 2008), and other eye-catching proposals like the N*ano Risk Framework Methodology* to evaluate manufacturing processes implemented in 2005 by a partnership between DuPont and the Environmental Defense Fund (DuPont & Environmental Defense Fund 2007). In 2006 and 2007, governments asked companies voluntarily to declare the production, use, or commercialization of nanotechnology to prevent the emergence of potential risks to health and environment. The first request of this kind was the U.K.'s *Voluntary Reporting Scheme (VRS) for Manufactured Nanomaterials* in 2006 (DEFRA 2008), followed shortly afterwards by a call from the U.S. E.P.A. (EPA-USA 2007) and some other countries.

On the other hand, we find proposals associated with the classification and the standardization, as well as ones related to quality evaluation of basic nanomaterials. These proposals are designed to regulate trade between companies and to facilitate patenting. In fact, the patent mechanism in the case of nanomaterials was consolidated in 2007 with the creation of the *UP Class 9977 Nanotechnology* of the U.S. Patent and Trademark Office. Other examples are the creation of the *ISO TC 229 Nanotechnologies* classification in 2005; the certification for nanotechnologies of the American Society for Testing and Materials (ATSM 2006); and the effort of the Zyvex company to certify carbon nanotubes. In the case of Zyvex, the initiative had a clear objective to facilitate commercialization of its product. The heterogeneous nature of nanoproducts has complicated operations between producers and companies that used them as raw materials.

The only sector that has demonstrated a concern about risks from the very beginning of the discussion was, naturally, the insurance sector. In 2004, insurance company Swiss Re warned that the lack of regulation was a big impediment to providing insurance coverage for processes and products associated with nanotechnologies (Swiss Re 2004). This position had repercussions for several companies, since the lack of insurance coverage meant potential losses. It is clear that the interest of insurance companies is not correlated to the risks per se, but to the need for a legal framework to be able to proceed with business.

Academic associations took several divergent paths. The most significant were the ones set in 2004 by the Royal Academy and the Royal Academy of Engineering, which warned about the potential risks of nanotechnologies. The International Council on Nanotechnology (ICON), which is a partnership between Rice University, several corporations, and some NGOs, has been conducting research on the risks to health of some nanotechnologies since 2004 (ICON 2008). However, in contrast with the positions of other NGOs and unions on the full recognition of nanotechnology risks, most academic commentary, although critical, did not question the distribution of nano-products in the market before scientific risk assessment was complete.

11.4.3 Starting to Take Action

While governments and companies were concerned about advancing standardization and reducing the risks perceived by the general population, NGOs and unions were more concerned with potential risks to health and the environment of the nanotechnological revolution.

Between 2006 and 2008 the calls to impose moratoriums on the use and commercialization of nanotechnologies attracted the attention of several sectors. A complete report about the potential effects of nanoparticles in cosmetics, published in 2006 by the environmentalist group Friends of the Earth Australia (FoE 2006), and in 2008 about the risks of nanotechnologies in the agriculture and food sector (Miller and Rye 2008), were added to the voice of the ETC Group. These pronouncements led several other organizations to engage with the issue and stimulated other declarations.

For instance, the Latin American Regional Secretariat of The International Union of Food, Agricultural, Hotel, Restaurant, Catering, Tobacco and Allied Workers' Associations (IUF) met in October 2006 in Santo Domingo, for its thirteenth regional conference. With the presence of 39 workers' organizations from fourteen countries and 95 delegates, a resolution was passed on nanotechnologies. In general terms, the declaration called for public debate, warning that products containing nanocomponents were being launched into the market before civil society and social movements had a chance to assess their possible implications in economic, environmental, and social terms and their effect on human health. Furthermore, the declaration warned of the need to make sure that the debate of a matter that will lead to deep social changes should not be left to the "experts." This is possibly the first declaration issued at a continental level by a federation of trade unions. Months later, in March 2007, the Twenty-fifth Congress of the IUF was held in Geneva. Rel-UITA (The Latin American Regional Secretariat of IUF) introduced the Santo Domingo resolution into the talks, and it was approved, thereby extending its impact to all 122 countries and over twelve million workers (Rel-UITA 2007).

Immediately afterwards, in June 2007, the Principles for the Oversight of Nanotechnologies and Nanomaterials were published by the International Center for Technology Assessment (ICTA). These principles have been endorsed by more than forty organizations such as international environmental NGOs, unions, organic consumer organizations, academic networks, and other groups (Kimbrell 2007). This document represents an important step: it is based on eight principles developed to guarantee the defense of the consumer and the protection of workers against the potential risks and legal implications of nanotechnologies.

The ICTA declaration starts with the precautionary principle, calling for the creation of a legal framework to protect consumers and workers when potential harm is not yet known. The protection goes into effect even when the cause-effect relation has not been completely proven. This principle, widely used within several legal frameworks, seems functional for the case of nanomaterials, since the most important feature in them is precisely the development of unknown properties, different from the ones presented by the same materials at the bulk scale. The latter point leads to the second principle: the need for specific regulations for materials and structures that present novel properties and unknown risks. The three subsequent principles—health and safety of the public and workers, environmental protection and transparency—are part, as well, of most VCCs adopted by corporations. However, in the case of the VCC there is no guarantee that companies will follow them; what society has in hand is just the "good will" and the "word" of the manufacturers. In contrast, the principles endorsed by the ICTA are intended to be incorporated into a legal framework and supported by judicial power.

The other three principles are related to encouraging public participation, with other potential impacts that nanotechnologies could have, and with the manufacturer's liability if there are eventual harmful effects. Overall, these principles are embedded in a simple but straightforward declaration that protects the interests of the general public and workers. Nonetheless, it represents more than what governments and companies can handle. For instance, the principle in favor of more transparency that endorses labeling of nanoproducts has activated a heated debate backstage in the European Parliament and the European Council, mainly in the context of nanofoods.

A year after the publication of the principles, in mid-2008, the European Trade Union Confederation issued a declaration of warning and asked for clearer information regarding the implications of nanotechnologies (Chapter 12). In addition, the Salleh since 2005 has been requesting information from its congress about the impacts of nanotechnologies, and developed a more hostile position by demanding the speedy implementation of regulatory frameworks for nanoproducts.

Between 2008 and 2009, some governments began to make small steps towards implementing regulatory measures adapted from the VCCs and from the reports asking for mandatory measures. There are some examples, including the following: the Afsset of France (l'Agence française de sécurité sanitaire de l'environnement et du travail) (Afsset 2008); the FSAI (Food Safety Authority of Ireland) (FSAI 2008); the promise of implementing similar measures by Canada (RSC 2009); and the United States E.P.A. (Goodman 2009). There has also been action at the municipal level in the U.S., for example by the municipal government of Berkeley in 2006 (City of Berkeley 2006), followed by Cambridge in 2008 (Cambridge Public Health Department 2008), and the State of California, this last one in relation to carbon nanotubes in 2009 (DTSC 2009).

11.4.4 Summary

The preceding chronological analysis showed, first, that the public debate about the risks of nanotechnologies did not originate in either the industrial sector or government, but from NGOs and unions looking after the interests of consumers and workers. In addition, the analysis showed that both companies and governments have been pushing forward the scheme of VCCs over the mandatory regulations. We also observed that unions and NGOs have argued actively in favor of stopping the advance of nanotechnologies in the market until the other actors develop, or at least become willing to create schemes to provide guarantees to protect consumers and workers. Thus far however, commercialization has moved forward rather than democratic dialogue.

11.5 What is at Play Within the Discussion on the Regulation of Nanotechnologies?

The two positions that we have analyzed in this paper are a mirror of contrasting class positions that at the same time endorse opposing economic and social interests.

The industrial sector justifies its position with the argument that technology development is beneficial per se. It is clear that there are many examples of products that are either manufactured from nanotechnology or that use nanotechnology in their functioning that would fall under this parameter. It is not true, however, that the entire population would automatically enjoy all the benefits of new technological developments. New products enter the market because they represent an increase in profits for companies; otherwise they are never introduced. Consumers have access to them through only the market system. This means that people who are not able to afford to pay for those benefits—who ironically are the ones more in need—never get access to them. The most eloquent case of the latter is what happens with the so-called "neglected diseases" or "diseases of the poor." Not only do the poor lack access to the potential benefits of any given technology, but there is no research being conducted in areas where the market is too small or not able to return enough profits.

For their part, unions and environmental NGOs argue that the need for government to regulate nanotechnologies is based in historical experience. This is the case, for instance, of the declaration by the IUF, which asks for a moratorium on the use and commercialization of nanotechnologies. Most of the six points in this declaration refer to the need for research on possible risks to health and the environment, as well as on social and economic implications, along with the role for government regulation and regulation by international organizations before nanotechnology enters the market. There is, however, a background story that justifies this position, which from a political, ethical, and methodological point of view is important to note. This is the historical experience of the IUF in relation to agro-corporations that are currently patenting, researching, and commercializing nanotechnology-based products. The historical experience of unions with transnational agribusiness companies demonstrates that the primary motivation for these companies is profit making. But what is more important to realize is that these companies downplay any kind of health and environmental risk in the pursuit of that primary objective. It is natural then, that when corporations develop a new technology, as is presently the case with nanotechnology, the workers may assume with some historical precedent that the new technologies are only meant to attain economic benefit, and that potential risks will be taken by ignoring safety concerns for them and for consumers.

Another example of using the historical experience to evaluate the motivation behind corporations and their priorities is what happened with the Silicon Valley Toxics Coalition (SVTC), which in the beginning of 2008, developed a public campaign for regulation of research in nanotechnology and of the consumption of its products. The argument is simple; Silicon Valley had previously experienced problems with semiconductor companies like IBM and Fairchild Camera and Instrument that caused contamination in the 1980s of drinking water for more than eighty thousand residents of Santa Clara. There was a series of health impacts attributed to the water, primarily in terms of birth defects. Even today, Santa Clara has 29 contaminated sites designated for cleaning. The semiconductor industry had been advertised as a clean industry, but it was not so for Silicon Valley.

Now it is nanotechnology's turn. Although nanotechnology is advertised as being very clean, it could potentially trigger similar environmental damage. Without a doubt, there are many concerns about its possible impacts on human health and on the environment. The Silicon Valley Toxics Coalition is calling for public regulation and monitoring, on the basis of its historical experience with semiconductors. Moreover, they argue that without the pressure of social organizations, public institutions have difficulty in opposing the experiments of companies. The SVTC report compared the deficiencies that were part of the semiconductor industry with the deficiencies in current legislation. Suggestively, the subtitle of the report is "*Lessons Learned from 1981 Chemical Spills in the Electronics Industry and Implications for Regulating Nanotechnology*" (SVTC 2008).

These two important positions on regulating nanotechnologies and nanotechnology R&D—those of private industry and community and worker organizations—are reflected in various government actions. Over the last decade of rapid development of nanotechnologies, governments have in general supported the positions of the industrial sector. As a result, nanoproducts are now available on the shelves of several stores. The activism of workers and consumers is critical in order to endorse the creation of appropriate legislation to protect their interests.

11.6 Conclusion

The distribution of nanotechnology in the market, which has been growing in the last few years, has triggered an international discussion about its regulation. Several social sectors have created proposals that reflect their interests. Scientists studying human health agree that the workers in laboratories and those working in the industrial sectors that use nanocomponents are the first groups at risk from nanotechnologies, followed by consumers. However, unions, workers, NGOs, and consumers are not there at the negotiating tables or in the panels discussing the regulation of nanotechnologies. This means that those who will be most affected by nanotechnologies have the least say in determining the path that these technologies should follow. At the same time, governments of the countries where nanotechnology R&D is more advanced have endorsed the industrial position based on the adoption of voluntary codes.

In this paper we have illustrated that neither the governments nor the industrialists initially elaborated on the need to create regulatory mechanisms to prevent the potential risks associated with nanotechnology. This claim began with NGOs and trade unions. The chronological analysis also showed that after NGOs and unions took up the issue, some governments in Europe and in the U.S. are starting to acknowledge the need to regulate nanotechnologies. However, unless the rapid introduction of nanoproducts into the market is stopped, it will be difficult to start a negotiation. Currently the discussion is taking place after the fact, and only after the effects become apparent.

Notes

1. *The International Union of Food, Agricultural, Hotel, Restaurant, Catering, Tobacco and Allied Workers' Associations* (IUF) is an international federation of trade unions of workers in agriculture and crops, the preparation and processing of food and drinks, hotels, restaurants

and catering services, and all phases of the production and processing of tobacco. It is a huge federation with a long history, stretching back to 1920. Today its membership is made up of 365 unions from 122 countries, representing a total of twelve million workers (Rel-UITA n/d).
2. The European Parliament suggested in 2009 under the framework of the principle "No data, no market" that nano-foods would have to go under a risk assessment evaluation and labeling process. In the meantime, they should be removed from the market. But a few months later the European Council decided not to impose a mandatory labeling for nano-foods and not to take products off the market, although it stressed the need of implementing risk assessment protocols.

References

Afsset (l'Agence française de sécurité sanitaire de l'environnement et du travail). 2008. Les nanomatériaux. Sécurité au travail. http://www.afsset.fr/upload/bibliotheque/258113599692706655310496991596/afsset-nanomateriaux-2-avis-rapport-annexes-vdef.pdf. (accessed November 6, 2008).
ATSM International (American Society for Testing and Materials). 2006. ASTM E2456-06 Standard terminology relating to nanotechnology. http://www.astm.org/Standards/E2456.htm. (accessed July 9, 2008).
Basf (Badische Anilin und Soda Fabrik). 2008. Code of conduct nanotechnology. http://www.basf.com/group/corporate/en/sustainability/dialogue/in-dialogue-with-politics/nanotechnology/code-of-conduct. (accessed March 12, 2009).
Bayer. 2007. Bayer position on nanotechnology. http://www.baycareonline.com/nano_stewardship.asp. (accessed July 31, 2010).
Cambridge Public Health Department. 2008. Recommendations for a municipal health & safety policy for nanomaterials: A report to the Cambridge City Manager. http://www.loe.org/images/080801/NanoRecommendations.pdf. (accessed August 11, 2008).
Commission of the European Communities. 2008. On a code of conduct for responsible nanosciences and nanotechnologies research. Commission Recommendation of 07/02/2008. C(2008) 424 final. Brussels: Commission of the European Communities.
Continental Western Insurance Group. 2008. Nanotubes and nanotechnology exclusion. Policy CW 33 69 06 08.
City of Berkeley. 2006. Manufactured nanoparticle health and safety disclosure. http://www.ci.berkeley.ca.us/citycouncil/2006citycouncil/packet/120506/2006-12-05%20Item%2013%20Manufactured%20Nanoparticle%20Health%20and%20Safety%20Disclosure.pdf. (accessed March 15, 2008).
DEFRA (Department of Environment, Food and Rural Affairs), United Kingdom. 2008. The UK voluntary reporting scheme for engineered nanoscale materials. http://www.defra.gov.uk/environment/quality/nanotech/documents/vrs-nanoscale.pdf. (accessed July 31, 2010).
DTSC (California Department of Toxic Substances Control). 2009. Chemical information call-in: Carbon nanotubes. http://www.dtsc.ca.gov/TechnologyDevelopment/Nanotechnology/index.cfm#Chemical_Information_Call-in:_Carbon_Nanotubes. (accessed March 3, 2009).
DuPont & Environmental Defense Fund. 2007. Nano risk framework executive summary. DuPont & Environmental Defense Fund. http://www.nanoriskframework.com/content.cfm?contentID=6498. (accessed March 22, 2008).
EPA-USA. 2007. Nanoscale program approach for comment. http://www.epa.gov/oppt/nano/nmspfr.htm (accessed March 15, 2008).
ETC group (Action Group on Erosion, Technology and Concentration). 2002. Patenting elements of nature. http://www.etcgroup.org/upload/publication/220/01/nanopatentsgeno.rtf.pdf. (accessed July 31, 2010).
FoE—A. (Friends of Earth—Australia). 2006. Nanomaterials, sunscreens and cosmetics: small ingredients big risks. http://nano.foe.org.au/nanomaterials-sunscreens-and-cosmetics-small-ingredients-big-risks. (accessed July 31, 2010).

FSAI (Food Safety Authority of Ireland). 2008. The relevance for food safety of applications of nanotechnology in the food and feed industries. Dublin. http://www.fsai.ie/WorkArea/DownloadAsset.aspx?id=7858. (accessed July 31, 2010).
Goodman, Sara. 2009. Nanotech: EPA issues first nanomaterial rule. *E&ENewsPM*, May 24. http://www.eenews.net/eenewspm/rss/2009/06/24/3. (accessed June 26, 2009). (subscription required).
Gupta Ray, Shashwat. 2009. Nanotech, a big way to beat recession. *Sakaal Times*, March 11.
Highfield, Roger. 2003. Prince asks scientists to look into 'grey goo'. *The Telegraph*, June 5. http://www.telegraph.co.uk/news/uknews/1431995/Prince-asks-scientists-to-look-into-grey-goo.html. (accessed July 31, 2010).
ICON (International Council on Nanotechnology). 2008. International council on nanotechnology. http://www.icon.rice.edu/. (accessed July 31, 2010).
Industrial Technology Research Institute. 2004. NanoMark. Industrial Development Bureau. Ministry of Economic Affairs. http://proj3.moeaidb.gov.tw/nanomark/Eng/. (accessed July 31, 2010).
ISO (International Organizations for Standardization). 2005. TC 229 Nanotechnologies. http://www.iso.org/iso/iso_technical_committee?commid=381983. (accessed July 9, 2008).
Kimbrell, George. 2007. Broad international coalition issues principles for strong oversight of nanotechnology. http://www.icta.org/press/release.cfm?news_id=26. (accessed March 20, 2008).
Lewis, Sanford, Richard Liroff, Margaret Byrne, Mary S. Booth, and Bill Baue. 2008. Toxic stock syndrome: How corporate financial reports fail to apprise investors of the risks of product recalls and toxic liabilities. *The Investor Environmental Health Network (IEHN)*. http://www.iehn.org/publications.reports.toxicstock.php. (accessed July 31, 2010).
Lloyd's (2009). Nanotechnology: Balancing risk and opportunity. http://www.lloyds.com/News-and-Insight/News-and-Features/360-News/Emerging-Risk-360/Nanotechnology_balancing_risk_and_opportunity. (accessed July 31, 2010).
Lux Research, Inc. 2006. The Nanotech Report. 4th Edition. New York: Lux Research, Inc.
Mantovani, Elvio, Andrea Porcari, Christoph Meili, and Markus Widmer. 2009. Framing nano project: A multistakeholder dialogue platform framing the responsible development of nanosciences & nanotechnologies. Mapping Study on Regulation and Governance of Nanotechnologies. www.framingnano.eu (accessed March 10, 2009).
Miller, Georgia, Rye Senjen. (2008). Out of the Laboratory and Onto our Plates http://nano.foe.org.au/sites/default/files/Nanotechnology%20in%20food%20and%20agriculture%20-%20text%20only%20version_0.pdf. (accessed July 31, 2010).
Peiris, Manjari. 2008. State sector has to play an important role in riding present financial crisis urges Professor Tissa Vitarana. *Asiantribune.com*, November 21. http://www.merid.org/NDN/more.php?articleID=1590&search=%2FNDN%2Farchive.php%3FdoSearch%3D1%26implications%255B%255D%3DGovernance%26page%3D15%26items%3D15&scorePrecent=44. (accessed January 10, 2009).
Prensa Latina. 2008. Russia: Nanotechnology against world crisis. http://www.cns.ucsb.edu/clips/russia-nanotechnology-against-world-crisis-prensa-latina-12-7-08-3/. (accessed July 8, 2009).
Rel-UITA. 2007. Resolución sobre nanotecnología. Presented at 25 Congreso de la IUF Ginebra, March 19–22, 2007. http://www.rel-uita.org/sindicatos/congreso-uita-2007/resoluciones/resolucion-nano.htm. (accessed April 24, 2007).
Royal Commission on Environmental Pollution. 2008. *Novel Materials in the Environment: The case of nanotechnology*. London: TSO (The Stationery Office). http://www.rcep.org.uk/reports/27-novel%20materials/documents/NovelMaterialsreport_rcep.pdf. (accessed July 31, 2010).
RS&RAE (The Royal Society & The Royal Academy of Engineering). 2004. Nanoscience and nanotechnologies: opportunities and uncertainties. http://www.nanotec.org.uk/finalReport.htm. (accessed March 1, 2009).

RSC (Ministerial notice requiring mandatory reporting of information relating to nanoscale materials). 2009. Nano-regulation creeps closer http://www.rsc.org/chemistryworld/News/2009/February/25020901.asp. (accessed March 1, 2009).
Rusnano (Russian Corporation of Nanotechnology). 2008. Certification. http://www.rusnano.com/Post.aspx/Show/17980. (accessed July 31, 2010).
Salleh, A. 2005. Unions say nano-loopholes may hurt workers. *ABC Science Online.* http://www.abc.net.au/cgi-bin/common/printfriendly.pl?/science/news/stories/s1451929.htm 5 September. (accessed July 31, 2010).
SVTC (Silicon Valley Toxics Coalition). 2008. Regulation emerging technologies in Silicon Valley and beyond. http://www.etoxics.org/messages/SVTC_Nanotech_Report(April-2008).doc. (accessed April 24, 2008).
Swiss Re (Swiss Reinsurance Company). 2004. Nanotechnology: Small matter, many unknowns. http://media.swissre.com/documents/nanotechnology_small_matter_many_unknowns_en.pdf. (accessed July 31, 2010).
The Swiss Retailer's Organisation & Innovation Society. 2008. Code of conduct.
Trade Union Congress (TUC). 2004. Nanotechnology.
Wolfe, Josh. 2005. Nanotech vs. the green gang. *Forbes.com,* April 6. http://www.forbes.com/2005/04/06/cz_jw_0406soapbox_inl_print.html. (accessed May 7, 2007).
WWICS (Woodrow Wilson International Center for Scholars). 2006. A nanotechnology consumers product inventory. Project on Emerging Nanotechnologies. http://www.nanotechproject.org/inventories/consumer/. (accessed July 31, 2010).

Draft Resolution on nanotechnology

Considering

That we are in a world in which science is advancing faster than society, a world driven by business profit where nanotechnology (NT) is launching products on the market before society has the opportunity to analyse their effects.

That civil society and social movements must embark on a broad debate on NT and its economic, environmental, social and health implications. We must not fall into the error of accepting that discussions of NT should be left in the hands of "experts".

The 25th IUF Congress meeting in Geneva, March 19–22, 2007

Resolves:

1. To mobilize our affiliated organizations and urge them to discuss with the rest of society and governments the possible consequences of NT.
2. To demand that governments and the international organizations concerned apply the **Principle of Precaution**, prohibiting the sale of food, beverages and fodder, and all agricultural inputs which contain nanotechnology, until it is shown that they are safe and to approve an international system of regulations specifically designed to analyse these products.
3. To demand that the World Trade Organization (**WTO**) suspend the grant of patents related to nanotechnology in the food industry and agriculture, until the countries affected and social movements can carry out an evaluation of their impact.
4. To demand that the World Health Organization (**WHO**) and the United Nations Food and Agriculture Organization (**FAO**) update the *Codex Alimentarius*, taking into account the use of nanotechnology in food and agriculture.
5. To request the **WHO** to initiate short and long-term studies into the potential effects of nanotechnology—especially nanoparticles—on the health of the technicians and workers that produce them, users and consumers.
6. To request the International Labour Organization (**ILO**) to carry out an urgent study into the possible impact of nanotechnology on conditions of work and employment in agriculture and in the food industry. Following completion of the study, a Tripartite Conference on the subject must be convened as soon as possible.

Submitted by the 13th Conference of Rel-UITA

Chapter 12
ETUC Resolution on Nanotechnologies and Nanomaterials

European Trade Union Confederation

The interests of, and concerns about, labor outlined in the previous chapter are not simply an academic exercise. Since the idea that nanotechnology can create a revolution in manufacturing was first discussed, labor organizations have raised concerns and sought a voice. In this chapter we reproduce the resolution issued by the European Trade Union Council (ETUC) that Foladori and Zayago Lau refer to at the end of the previous chapter. It is a political statement that argues that trade unions need and deserve a seat at the table when the future of nanotechnology is discussed. The ETUC contends that when certain groups are likely to receive a disproportional amount of the risk generated by nanotechnologies, they should have a greater voice in how those technologies are handled. Thus the ETUC does not simply want an equitable distribution of the risks of nanotechnology. Perhaps even more importantly, it wants a fair distribution of the power to make decisions about how nanotechnology will be implemented. The ETUC resolution was adopted by its Executive Committee on 25 June 2008.—eds.

Originally published on the ETUC website: http://www.etuc.org/a/5163/. Republished with permission from the European Trade Union Confederation.

European Trade Union Confederation (ETUC)
Confédération européenne des syndicats (CES)

ETUC RESOLUTION ON NANOTECHNOLOGIES AND NANOMATERIALS

Resolution adopted by the ETUC Executive Committee in their meeting held in Brussels on 24-25 June 2008

Introduction

Nanotechnologies are emerging, trans-disciplinary technologies that enable structures or objects to be designed, manipulated and manufactured on a nanometer scale,[1] i.e., the size of a handful of atoms or molecules. At this scale, the physico-chemical properties of matter can differ significantly from those obtained at larger scales. What all these technologies have in common, therefore, is to produce objects, called nanomaterials, that have new properties and behaviors that cannot be obtained easily or at all with conventional technologies.

Described as the "engine of the next industrial revolution", nanotechnologies have a far-reaching development and application potential, especially in the fields of biotechnologies and medicine (diagnostic, treatment and prevention tools), information and communication technologies (miniaturization and increased storage capacities); energy (more efficient energy storage, conversion and production), agriculture and the environment (soil, water and air cleanup), Etc.

Industry and Governments have taken this firmly on board. Public funding for nanotechnologies in the United States and Europe alike has risen steadily year on year. The European Union, for example, has decided to put 3.5 billion euros into nanotechnology research between 2007 and 2013 on top of private sector investment and national research budgets. The most frequently cited estimate is that the world market in nanotechnologies will amount to 1 000 billion dollars by 2015.[2]

In terms of employment, it is claimed that nanotechnology development is likely to require an additional two to ten million workers across the world by 2014. Many of these jobs are likely to be created in Europe, mainly in start-up companies and in SMEs[3].

---

[1] Usually somewhere between 1 nm and 100 nm. A nanometer (nm) is equal to one billionth of a metre.

[2] The economic development of nanotechnology, European Commission, 2006 http://cordis.europa.eu/nanotechnology.

[3] *Ibid.*

Hundreds of consumer and manufactured products containing engineered nanomaterials or made with the use of nanomaterials are already on the market[4], for example in the areas of cosmetics, sporting goods, textiles, food, paints, constructions and electronic equipment.

Products are been made today and placed on the market without knowing whether nanomaterials are released from them and what their potential impacts on human health and the environment may be. Workers all along the production chain from laboratories through to manufacturing, transport, shop shelves, cleaning, maintenance and waste management are exposed to these new materials. Nevertheless it is unknown whether the safety procedures implemented are adequate or the protection measures applied are sufficient. Workers and consumers are being exposed to products[5] that contain nanomaterials unbeknown and uninformed about the potential risks. Nanomaterials are discharged and disseminated out into the open without knowing what the consequences may be and without effective ways of detecting and measuring them.

There is a growing body of scientific evidence to suggest that some manufactured nanomaterials harbour new and unusual dangers[6, 7]. Because smaller particles have a greater (re)active surface area per unit mass than larger particles, their toxicity may also increase.

While nanotechnologies may bring major benefits to our society, they also raise many concerns about their potential risks to our health and the environment.

In 2005, the European Commission adopted an action plan on nanotechnologies and nanosciences for 2005-2009, which called for an assessment of risk to human health, the environment, consumers and workers at all stages of the life cycle of the technology (conception, manufacture, distribution, use, and recycling).

Most research programmes, however, are still in the very early stages and it will be a long way down the road before comprehensive information is available to give a clear picture of what risks the different manufactured nanoparticles may pose.

The European Trade Union Confederation (ETUC), its member federations and confederations, wish to do a first contribution to this important societal debate by pointing out those elements of the European policy that they see as essential to the responsible development of these emerging technologies.

---

[4] www.nanotechproject.org/consumerproducts

[5] In our understanding "product" encompasses a substance, a preparation or an article.

[6] SCENIHR (Scientific Committee on Emerging and Newly-Identified Health Risks), The appropriateness of the risk assessment methodology in accordance with the Technical Guidance Documents for new and existing substances for assessing the risks of nanomaterials, 21-22 June 2007.

[7] IARC (International Agency for Research on Cancer): http://monographs.iarc.fr/ENG/Meetings/93-carbonblack.pdf; http://monographs.iarc.fr/ENG/Meetings/93-titaniumdioxide.pdf

The ETUC position

Nanosciences and nanotechnologies are new approaches to research and development (R&D) that aim to control the fundamental structure and behaviour of matter at the level of atoms and molecules. These fields open up the possibility of understanding new phenomena and producing new properties of matter that can be utilised in virtually all technological sectors.

The ETUC is convinced that nanotechnologies and manufactured nanomaterials might have considerable development and application potential. These technological advances and the new jobs they might bring may address peoples' needs, help make European industry more competitive and contribute to the achievement of the sustainable development goals set out in the Lisbon Strategy.

However, the ETUC notes that significant uncertainties revolve around both the benefits of nanotechnologies to our society and the harmful effects of manufactured nanomaterials on human health and the environment. The development of these emerging technologies and the products from them also poses huge challenges to our society in terms of regulatory and ethical frameworks.

The ETUC considers that if the past mistakes with putatively "miracle" technologies and materials are not to be repeated, preventive action must be taken where uncertainty prevails. This means the precautionary principle must be applied. This is the essential prerequisite for the responsible development of nanotechnologies and for helping ensure society's acceptance of nanomaterials.

The ETUC welcomes the European Commission's action plan 2005-2009 on nanosciences and nanotechnologies which is based on the safe, integrated and responsible strategy put forward in its 2004 Communication. Nevertheless, our analysis of the first Commission Report on its implementation over the period 2005-2007 reveals large gaps and deficiencies which ought to be eliminated without delay.

Where investment in R&D is concerned, we see and note a gross imbalance between budgets for the development of commercial applications of nanotechnology and those for research into their potential impacts on human health and the environment. The ETUC calls for at least 15% of national and European public research budgets for nanotechnology and the nanosciences to be earmarked for health and environmental aspects and to require all research projects to include health and safety aspects as a compulsary part of their reporting.

The ETUC considers that a standardised terminology for nanomaterials is urgently needed to prepare meaningful regulatory programmes. In particular, ETUC calls on the Commission to adopt a definition of nanomaterials which is not restricted to objects below 100 nanometers in one or more dimensions. This is important to avoid many nanomaterials already on the market to be left out of the scope of future legislations.

The ETUC is concerned at the holdup in the Commission departments' examination of the current legislative framework and its identification of the regulatory changes needed to address workers' and consumers' concerns about the health and environmental implications of nanomaterials.

After the asbestos scandal which cost the lives of hundreds of thousands of workers, and when the EU has recently introduced new legislation on chemicals that puts the onus of proof onto manufacturers, the ETUC finds it unacceptable that products should now be manufactured without their potential effects on human health and the environment being known unless a precautionary approach has been applied and made transparent to the workers.

In particular, ETUC considers that manufacturers of nano-based products should be obliged to determine whether insoluble or biopersistent nanomaterials can be released from them at all stages of their life cycle. In the absence of sufficient data to prove that those released nanomaterials are harmless to human health and the environment, marketing should not be permitted.

The ETUC therefore demands full compliance with REACH's "no data, no market" principle. It calls on the European Chemicals Agency (ECHA) to refuse to register chemicals for which manufacturers fail to supply the data required to ensure the manufacture, marketing and use of their nanometer forms that has no harmful effects for human health and the environment at all stages of their life cycle.

Strict application of this principle must be used to encourage industry to fill the gaps in the scientific knowledge about the safety of engineered nanomaterials, especially the fate and persistence of nanoparticles in human beings and the environment.

The ETUC calls on the Commission to amend the REACH regulation so as to give better and wider coverage to all potentially manufacturable nanomaterials. Nanomaterials may indeed evade the REACH registration requirements because they are manufactured or imported below the threshold of 1 tonne per year. The ETUC demands that different thresholds and/or units (e.g. surface area per volume) are used for registration of nanomaterials under REACH.

The ETUC considers that the obligation to produce a chemical safety report for production volumes only above 10 tpa is another loophole that will allow many manufacturers or importers to avoid doing a risk assessment before putting nanomaterials on the market. The ETUC wants a chemical safety report to be required for all substances registered under the REACH regulation for which a nanometer scale use has been identified.

The ETUC also demands Annexes IV and V of REACH (exemptions from registration) currently under revision not to permit manufactured nanomaterials to evade the REACH requirements.

Workers engaged in research, development, manufacture, packaging, handling, transport, use and elimination of nanomaterials and nanotechnology products will be most exposed, and therefore most at risk of any harmful effects. The ETUC therefore demands that health and safety at work must have priority in any nanomaterials surveillance system. There is a great need for training, education and research in order to allow health and safety specialists (e.g. labour inspectors, preventive services, occupational hygienists, company physicians) preventing known and potential exposures to nanomaterials.

The ETUC calls on the Commission to amend Chemical Agents Directive 98/24/EC which it believes does not afford adequate protection to workers exposed to substances for which there are gaps in our knowledge about their toxicological properties. Employers must be required to implement appropriate risk reduction measures, not only when known dangerous substances are present in the workplace, but also when the dangers of substances used are still unknown. This would enable all manufactured nanomaterials to be covered, along with many other substances that carry unknown health risks to which workers are exposed.

Workers and their representatives (e.g. safety reps) must be fully involved in risk assessment and the selection of risk management measures without fear of retaliation or discrimination. Moreover, they must be informed of the nature of the products present on their work places. The ETUC therefore considers that safety data sheets must clearly state whether nanomaterials are present. If toxicological or ecotoxicological data are missing, that must also be indicated in safety data sheets. The ETUC considers that significant efforts must be made without delay to prevent occupational exposures to already known manufactured nanomaterials. That will involve, in particular, exposure monitoring, health surveillance for workers and appropriate training.

The ETUC believes that consumers also have the right to know what is in a product. In many cases, manufacturers have published no information on tests done on nanotechnology products and their health hazards, or have not labelled consumer products as containing nanomaterials. Not being fully informed prevents the public from making informed decisions about the purchase and use of such products.

The ETUC wants all consumer products containing manufactured nanoparticles which could be released under reasonable and foreseeable conditions of use or disposal to be labeled. In addition, as part of the precautionary approach, ETUC calls on Member state authorities to set up a national register on the production, import and use of nanomaterials and nano-based products. Those measures would make it easier to monitor any human or environmental contamination and to identify where responsibility lay for any harmful effects.

The ETUC believes that Industry Voluntary Initiatives and Responsible Codes of Practices may serve a useful purpose pending implementation of the necessary changes to the current legislative framework and/or the introduction if need be of specific new European legislation to support responsible nanotechnology development.

However, the ETUC is prepared to endorse such initiatives only if the signatories undertake to involve workers' representatives in their design and monitoring, if there is an independent and transparent system for assessing compliance (e.g. by involving labour inspectorates) and if sanctions are foreseen in case of non-compliance. In addition, the ETUC demands that companies which adopt such systems disclose information on the hazards and risks associated with their products and commit themselves to be fully accountable for liabilities incurred from their products.

Finally, since nanotechnologies have the ability to profoundly alter the social, economic and political landscape of our societies, it is essential that all interested parties have a full say in the discussions and decisions that affect them. The ETUC therefore calls on the European Commission and Member State governments to commit sufficient funds to ensure real civic participation in the current debate on these new technologies.

Part III
Equalizing Processes

Chapter 13
Materializing Nano Equity: Lessons from Design

Dean Nieusma

Many engineers and designers got into their professions in large part because they wanted to create products that help people. Translating this desire into material objects is not a straightforward process. Contexts and complexities often make it difficult for such visions to be realized. In this chapter Dean Nieusma offers advice for designers who want to assist the world's poor and disadvantaged. Unfortunately most of the existing structures and institutions that shape or direct the practice of designers are geared towards the wealthy and powerful. And, as Nieusma points out, understanding both the needs and the context of the "have nots" can be a significant challenge for the "haves." One way to address these challenges, Nieusma contends, is through participatory design—having the potential users (and perhaps even "affected non-users") of technology engage in the design process to help limit the biases and misconceptions of the designer. Nieusma stresses the importance of such actions because equity is not just an issue of distributing technologies, it can be built into the design of objects—what he calls "materialized inequity." If new nanotechnologies are designed that can only be used by certain people or fit directly into existing systems of inequity, they will further those inequities for a long time to come. Nieusma implores designers to be proactive in their design to help ensure that their work does not accidentally make life more difficult for specific groups.—eds.

D. Nieusma (✉)
Department of Science and Technology Studies, Rensselaer Polytechnic Institute, Troy, NY, USA
e-mail: nieusma@rpi.edu

This chapter was peer reviewed. Originally presented at the Workshop on Nanotechnology, Equity, and Equality at Arizona State University on November 21, 2008.

13.1 Introduction

Design scholarship has much to offer in thinking through the making of nanotechnologies and the equity challenges that result. This paper reviews several threads in design scholarship that are especially useful in analyzing the design, production, and commercialization of nanotechnology-enhanced or enabled products, with an emphasis on understanding *and responding to* inequities resulting from those activities. My focus on products and the processes used to create them—as opposed to basic laboratory research, innovation policy, or nano potentialities more generally—directs attention toward the material embodiment and dissemination of nanotechnologies. This attention, in turn, enables consideration of a particular dimension of nano inequity, what I will call *materialized inequity*.

The approach taken in this paper is congruent with calls for more careful, more democratic deliberation of nanotechnology futures. Rather than merely "raising the flag" regarding potential areas of concern, the paper attempts to chart out productive ways forward. This includes considering the technology's potential to reshape our economies, improve the environment, and, especially in the context of this volume, respond to economic and political power imbalances through more robust models of stakeholder participation. Calls for deliberation on nanotechnology are especially urgent now, while flexibility remains in the technology's development trajectories—before the momentum of the current trajectories becomes overwhelming and path dependencies set in. By considering how various of the potential applications of nanotechnology are materialized—and hence how the technology's benefits and burdens are distributed—the paper contributes an important perspective from which to view inequity, how it is maintained or extended, and how it might be reduced.

Design scholarship is an appropriate lens for viewing nanotechnology for three important reasons. First, design scholars understand the subtleties of materiality—the objects and spaces that make up all human environments and shape every person's experience—and the interplay between material and social forces. Second, several threads of design scholarship attend explicitly to the intersection of materiality and inequity, showing how different material configurations impact various stakeholder groups in positive and negative ways. I have elsewhere identified and analyzed several domains of "alternative design" scholarship, defined as that which: "seeks to understand how unequal power relations are embodied in, and result from, mainstream design practice and products" (Nieusma 2004, 13). This paper borrows from that framework and extends it to consider its implications for nanotechnology in particular.

A third reason for drawing on design scholarship is that it addresses designer "agency," the latitude designers have in shaping products and outcomes. While all social actors—nanotechnology researchers and product designers included—face substantial structural constraints in their work limiting which paths they may reasonably pursue, those constraints are not absolute. By considering how designers might take better advantage of the latitude that remains, those seeking to redirect nanotechnologies toward enhancing social equity might plot out strategies for change congruent both with designers' individual intentions and with broader social aspirations and ideals.

Leveraging insights gained from alternative design scholarship, this paper makes two arguments. First, it argues that a more complete understanding of nano-inequity must attend to the *subtle and pervasive material mechanisms* through which inequity is produced or exacerbated as nanotechnologies are disseminated throughout economies, ecologies, communities, and cultures. Through such mechanisms, political and economic power imbalances become entrenched. Second, the paper argues that policy dialogue aimed at addressing these imbalances must go further than simply redirecting the same basic nanotechnology products and processes to new user groups. Instead, *a new relationship* between nanotechnology developers and a wide range of stakeholder groups, including diverse user communities, is necessary. Nanotechnology developers and policy makers are each in a unique position to advocate and enable this new relationship.

To set the stage for this two-part argument, the next section discusses the market structure that contextualizes nanotechnology product development. Then, in the following sections, several threads of alternative design scholarship are reviewed in the context of broad challenges facing new technology product design and production, with special attention given to nanotechnology applications. Finally, the paper concludes by reviewing the challenge of materialized inequity and sketching out an alternative grounded in broad-based participation that could be achieved in both product-development and policy-making processes.

13.2 Background: Nano Products and the Market

With over 1,000 nano-enhanced or enabled consumer products on the market as of late 2009 and the number steadily increasing,[1] now is a critical time in the technology's developmental trajectory (Project on Emerging Nanotechnologies 2009). Decisions made today could have wide-ranging, long-term influence over what is done—and what is possible to do—in the future (Davies 2008). While the concepts of path dependencies from economics (David 2007) and technological momentum from the history of technology (Hughes 2000, 1983) are not new contributions to understanding how new technology development trajectories quickly narrow and stabilize in response to social, political, and financial contingencies, deliberately altering development paths and reducing or slowing the building of momentum present themselves as considerable policy and design challenges.

In the highly-globalized contemporary economic context, perhaps the most significant component of this challenge is moderating the influence of the market in shaping new technologies' development trajectories. The market creates numerous opportunities for advancing nanotechnology. One way is by incentivizing new technology applications with first-to-market advantages and intellectual property protection. Another way is by providing a rapid signaling mechanism between those who commercialize new products and their customer base, especially in terms of consumer preferences. However, the market also creates challenges to advancing nanotechnologies, especially in directions that are consistent with broad social goals but are incongruent with existing market incentive structures.

Since market feedback mechanisms operate primarily between buyers and sellers, the priorities of those without buying power are left unaddressed: market feedback is overwhelmingly skewed toward the interests of relatively powerful actors. As a result, economically marginalized groups go underserved in two respects. They typically cannot afford to take advantage of the benefits offered by new technologies, and even when they can afford it, the new technologies have not been carefully tuned to their particular needs and interests (Dyer 1999). Market feedback mechanisms are also limited by the fact that consumers can only choose among *available* product options, not among all options that possibly could have been pursued (though consumer-interests research may offset this limitation partially). While market signals may facilitate efficient distribution of resources among economically potent actors, they clearly do not direct nanotechnology in many of the directions that are both possible and broadly desirable. In particular, the market is poorly equipped to address the needs of those social groups with relatively less economic and social/political power.

A quick review of the Project on Emerging Nanotechnologies' consumer product inventory underscores what many market critics might already assume: the majority of nano products currently available are non-essential items targeted to relatively well-off consumers. Prevalent on the list are cosmetics, dietary supplements, and high-end athletic and active wear. Design scholars are by no means ignorant of the market's influence in shaping new product trajectories, especially as they recognize that the vast majority of product designers work in corporate contexts to create successful consumer products. For example, Victor Papanek's seminal work, *Design for the Real World* (1984), offers a blazing critique of mainstream design, particularly industrial design, in the service of consumer capitalism. It also offers a series of corrective design interventions that include designing for the world's poor, for the disabled, and for other areas missed by existing market incentives.[2] Nigel Whitely extends and elaborates Papanek's analysis in *Design For Society* (1993). He first systematically critiques the dominant influence of market incentives in steering design activity, particularly by debunking the notion that "consumer needs" direct design activity. He then discusses a range of alternative design practices that serve as correctives to market-led missteps.

My own (2004) categorization scheme for alternative design practice, which guides this paper's analysis, is derived largely from Whiteley's work and particularly his quest to identify alternatives to mainstream market incentives in steering design activity. Given the dominance of the market in directing product development trajectories, we should ask: What major problems arise, for which stakeholder groups, and how might those inequities be responded to? To begin answering these questions, the following sections review several threads of design scholarship—namely, appropriate technology design; accessibility, feminist, and anti-racist approaches to design; and participatory design—drawing connections to various facets of nanotechnology at each stage.

13.3 Appropriate Technology Design

As with most other areas of new technology development, nanotechnology arrived on the public scene with its share of hyperbole: Drexler's (in)famous "universal assemblers" would "let us build almost anything that the laws of nature allow to exist" (1986, chapter 2). Some more current articulations of the "promise of nanotechnology" are just as ambitious: As Rober Freitas of the Institute for Molecular Manufacturing claimed, "not only will nanotech provide us with a lot of cool stuff and eliminate global poverty; it will also help us live a really long time" (cited in Pelletier 2008). Similar claims persist to the present, rooted in widely shared cultural narratives of the near-miraculous power of technology (Nye 1996) or in more routine practices of rationalizing investment in new or existing research initiatives (see, e.g., the National Nanotechnology Initiative, undated). Regardless of their origins, such claims almost certainly have been offered prematurely—at present, realistically anticipating the potential of nanotechnology to, say, eliminate hunger is unlikely. Such claims also ignore the complexity of entrenched social problems, problems that have long persisted in the face of wave after wave of "technological fix" even where the ability of a given technology to address specific, narrow facets of the problem is proven (Invernizzi and Foladori 2005).[3]

Granting the potential of new nanotechnologies to contribute to solving complex social problems like global hunger is an important step, but it is only a first step. An essential second step is to consider the likelihood that such potential will be realized and the mechanisms that would enable it. According to trickle-down theory, existing market incentives will eventually (and automatically) direct nanotechnologies toward those social problems where it could have a positive impact, even if those areas are not deliberately targeted in early phases of development. Similarly, the logic of spill-over benefits would have basic or applied research in nanotechnology's major domains of inquiry—advanced materials, pharmaceutical delivery, etc.—eventually flow toward non-targeted problems, though with less predictability than trickle-down dissemination.

There is good reason to question the efficacy of either such mechanism for distributing the benefits of technology: (1) Trickle-down benefits of new technology rarely keep pace with the more-direct imposition of far-reaching burdens (e.g., environmental injustices surrounding the citing of locally unwanted land uses such as industrial production or waste facilities) and (2) even as spillover benefits sometimes do follow investment in new technology (e.g., the oft-cited microwave oven and Tang examples), public benefits usually could have been achieved more reliably if they had been sought after directly (Sarewitz 1996, chapter 2). The alternative to merely asserting or assuming that social problems will be solved by nanotechnologies via business as usual is to deliberately redirect the technology's development trajectory toward important social goals. In addition to Real Time Technology Assessment (Guston and Sarewitz 2002; Chapter 26), another useful precedent for such redirection can be found in the appropriate technology movement.

The appropriate technology (AT) movement squarely attends to the question of how the benefits of technology might be more widely distributed given global economic inequities and market incentives that direct technology solutions away from the needs of the world's poorest peoples. Like other approaches to alternative design, AT has roots in the late-1960s counter-culture movements and, in particular, E. F. Schumacher's 1973 classic, *Small is Beautiful*.[4] The basic thrust of AT is to adapt technologies developed in economically well-off settings to the problems and circumstance confronting peoples in poorer settings. Since poor national, regional, or local economies often cannot afford investment in the infrastructure necessary for basic research and development of contextually suitable technologies and expertise, those resources must be imported if they are to be had (Smillie 2000). Rather than waiting for a particular technology's benefits to trickle-down through ordinary market mechanisms, appropriate technology designers deliberately and systematically redirect technologies to meet the needs of the poor.[5]

The historical evolution of AT, as concept and social movement, is relevant in many respects to understanding how nanotechnologies might be directed along trajectories most relevant to the world's poor,[6] but here I limit consideration to the central characteristic defining "appropriateness," namely, fit to context. Certainly, considerable differences typically exist between a given technology's *context of research and development* (R&D) in the wealthy developed economies of the global North and a potential *context of use* in developing countries in the South (Smillie 2000).[7] The major lesson of 40 years of AT scholarship and practice is that adapting technologies to fit diverse contexts of use is easier said than done, and a host of questions arise: Does the technological solution address the problem as understood by the people experiencing it? Does the technology align with social and cultural assumptions and expectations—and the financial and institutional capacities—that exist where it will be deployed? Does the technology's implementation draw on and extend existing skills and capacities of local people? Satisfactorily answering such questions requires conceptual nuance as well as considerable context-specific knowledge, not the least of which is being able to identify and then develop strategies to accommodate differences among contexts when transferring technology or expertise from a group of "haves" to a group of "have-nots" (Chambers 1997).[8]

However admirable the desire to distribute the benefits of nanotechnology to the world's poor, appropriate technology designers have shown that the strategy of simply "dropping" technologies conceptualized and developed in economies of the North into those of the South is based on misguided assumptions about how technologies "work" with respect to their contexts (Pytlik et al. 2001). So what direction should nanotechnology developers take to respond to contextual differences? From the perspective of appropriate technology design, the first step is to identify the context of application/use with some level of specificity: (1) who is targeted to benefit and how, (2) what market and non-market financial incentives are already in place and what challenges they create, and (3) what variety of understandings of nanotechnology circulate among local populations and where fault lines exist. Another early step would be to develop a clear picture of the relationships between nanotechnology proponents in the context of R&D and intended beneficiary populations[9] as well as between allied proponents in the context of use and the intended direct beneficiary population: What parallel or ulterior motives might exist among proponents in addition to their desire to help?

More challengingly, AT designers would advocate careful analysis of any cultural differences in how the specific "problem" being addressed is understood and experienced as well as more general assumptions about the role of technologies or objects in community life. Lack of grid electricity, for example, may be understood as a problem *in itself* for people accustomed to it, but for those living without grid electrification, the problem is usually more specific: fire hazards associated with kerosene or the expense of batteries for cassette players. At best, misunderstandings of these dimensions of cultural difference result in the wrong approach to solving a perceived problem. At worst, new solutions to ill-defined problems create much larger disruptions to previously functional social, political, or economic systems.[10] Only after a good understanding of the "context of use" has evolved through on-going interactions with intended beneficiaries is it sensible to seriously consider particular technological means. And only after problem and solution have been aligned conceptually does it make sense to begin the methodical work of responding to pragmatic concerns such as distribution logistics, service or maintenance requirements, end-of-life disposal, etc.

The appropriate technology movement proceeds under the assumption that new technologies can, in fact, be directed to meet the needs of the poor, but that such applications will not "automatically" occur through some inherent logic of technology diffusion. If nanotechnologies in particular are to address global problems such as hunger, then they will need to be deliberately and systematically steered in that direction. A critical component of that effort is transforming the technologies, sometimes radically, from how they are conceptualized and instantiated in their contexts of R&D so as to fit "appropriately" in their proposed contexts of use.

Appropriate technology designers have identified a wide range of factors relevant for assessing and achieving this fit to context, and they have produced handbooks, manuals, and more general analyses to help designers interact better with local communities and account better for contextual variables in the technology design process (Wicklein 2001, Smillie 2000, Hazeltine and Bull 1999, Darrow and Saxenian 1993). Since the point is to respond to contextual *differences*, however, no single set of criteria will apply in every situation. The general thrust, however, is coherent and clear: Because technologies are imbued with the assumptions circulating within their contexts of development, applications to different contexts require careful attention to the resource base available within contexts of use as well as integration of intended users' knowledge and values. Generalized assertions of the potential for nanotechnologies to address poverty and hoped for spillover benefits are no substitute for deliberate steering of the technology toward those ends.

13.4 Gender, Race, and Physical Ability in Design

Free-market technology steering mechanisms are limited in their ability to respond to the interests of a variety of social groups besides the global poor, in part because economic marginalization tends to align with social or political marginalization. When groups marginalized (on any grounds) are poorly represented within the institutions responsible for directing technology development, it may be unsurprising that technologies and products poorly respond to the marginalized groups' needs or expectations. Stated plainly, groups with relatively less power in society generally tend to have less influence on technology development trajectories, and correspondingly are less well served by those technologies.[11]

This claim is as true within a given society or nation as it is among societies or nations. Whereas above I considered the redistribution of technology benefits from populations in rich countries to those in poor countries, that distinction fails to account adequately for inequitable distribution of benefits within a given country, rich or poor.[12] This is especially true when the distribution of benefits is not exclusively determined by "ability to pay." While the lessons extracted from the appropriate technology movement focus on understanding the context of use vis-à-vis the context of development at a national or regional level, I did not address the matter of diversity—and potential divisions—*within* a given context.

This section shows that however important it may be to understand the main assumptions and values within a given context, it is equally important not to assume such assumptions and values are universally held within that context or that accommodating all such assumptions and values is necessarily conducive to enhancing equity across the board. In fact, many of the dominant assumptions and values within a given community lie at the very heart of inequity—with racism, sexism, and "ableism" central among the mix.[13] Addressing inequities arising from these phenomena, and their material manifestations in technologies and products, requires moving beyond "development assistance." It requires squarely facing the structural inequities that operate in all social contexts, including where the contexts of R&D and context of use considerably overlap.

Several threads of design scholarship have addressed the intersection of technology development and social structural inequality (as distinguished from solely-economic inequity), including feminist design (e.g., Rothschild 1999; Weisman 1992), race and design (e.g., Wright 2000; Walton 1999), and accessibility design (e.g., Imrie and Hall 2001; Covington and Hannah 1997), among others. Scholars and practitioners representing each of these approaches share a commitment to addressing social structural inequality through design. Because the nature of inequity is somewhat different for each marginalized group, however, the core insights offered by each approach are distinct. Since there is insufficient space to treat each of these approaches with the depth it deserves, I will briefly highlight one important contribution each approach offers to thinking about materialized inequity and provide an example from the literature.[14]

The accessibility design community targets its efforts primarily to characteristics of the built environment—buildings, signage, roadways, transit systems. Despite this shared topic of inquiry, the community is diverse, including proponents of people with a wide range of physical and mental disabilities, the elderly, children, mothers with children, and others. Because of the diversity of targeted "users" of the built environment, many different types of exclusion are addressed and many strategies for accommodating diverse user groups are offered.[15]

For people with physical disabilities, for example, accessibility design scholars highlight the many ways built environments physically preclude certain types of access to certain facilities and resources: Stairs (without accompanying ramps or lifts), for example, prevent wheelchair users access to the space beyond.[16] But accessibility design scholarship goes further to note that physical exclusion has both physical and psychosocial dimensions, the paradigmatic example of which is backdoor wheelchair ramps. While such ramps enable wheelchair users (and others) physical access to the services available within the building, they do so in a way that implies the users are second-class citizens and, therefore, upholds their marginal status. Simply providing physical access, therefore, does not necessarily eliminate inequities arising from systematic discrimination. By recognizing the confluence of material and psychosocial exclusionary mechanisms, accessibility designers remind us that even the groups targeted to benefit from a technology application may experience, at least in part, that benefit as a burden.

Feminist design scholarship makes a similar contribution in recognizing how multiple, overlapping modes of material exclusion and/or marginalization work together to reinforce inequities. In terms of steering technology development trajectories, feminist design scholarship highlights both opportunities for promoting more equitable distribution of technology, but also limitations in using material change or technology innovation, by itself, to create equity, since no single solution is likely to be sufficient in countering marginalization. As with accessibility design scholars, feminist design scholars identify a variety of ways inequity is materialized by design.

One example of "marginalization by design" includes physical barriers to women's participation, a simple and direct example of which is the history of designing military aircraft cockpits according to the physiological characteristics of men (Weber 1997). Irrespective of prevailing legal and cultural norms circulating at the time male-oriented cockpits were designed through the 1980s, the decisions that were made then physically precluded most women from piloting the corresponding aircraft for years into the future. As long as these aircraft were in operation, women were excluded, and this was true regardless of changes in the social norms concerning women's roles in the military, in the airline industry, or society at large.[17]

In a more nuanced take on materialized gender inequity, women's interests and needs may be attended to in the design of certain technologies, but that attention may well *reinforce* stereotypical assumptions about women and their social roles, thereby further entrenching inequities. The 1993 redesign of the Ford Probe, for example, was widely cited as a successful "design for women": the car's redesign was led by a woman, Mimi Vandermolen, and was targeted to professional women customers, who industry analysts recognized were increasingly making automobile purchasing decisions (Hess 1995). Despite these "successes," however, the project was not without its critics. While the Probe's project manager was a woman, she nevertheless reported to a predominantly-male group of directors within an almost exclusively-male organizational hierarchy. And while the car was explicitly designed for women customers, this was achieved through features such as accommodation for high heels in the accelerator and brake pedals design and accommodation for long fingernails in the design of the controls. Surely, high heels and long fingernails are worn frequently by the professional women whom Ford targeted, but using these as central characteristics of a "car for women" reinforces problematic assumptions of how women are socially defined. Finally, the irony of the car's name, the Probe, and its associated symbolism, was not lost on feminist critics. None of this is to say that the Ford Probe redesign was anti-woman, or even anti-feminist (though there are grounds for making the latter argument). Instead, the point is that what constitutes empowerment, especially in contexts with deep-seated inequity, is far from straightforward; attention must therefore be directed to the multiple layers of marginalization and the role of new technologies in entrenching some even as it counters others.

Race and design scholarship is a third approach to understanding the intersection of technology development and inequity. It shares with feminist design attention to the subtle ways deep-seated social inequality manifests in material objects and how designed objects, in turn, reinforce those inequities. Extending the lines of analysis above, Richard Dyer's (1999) work on the historical development of photographic equipment shows that racism need not be intentionally enacted on the part of individuals for it to be real. Without elaborating the details of his analysis, Dyer convincingly shows how, in the context of a variety of challenges getting color tones to look "natural," photography developers invested considerable effort to achieve pleasing "flesh tones"—that is to say, white people's skin tones. At the same time, through no malicious intent by film developers, the parallel problem of light reflectivity and effective representation of black people's skin tones was ignored.

Thus, in critical early stages of development, the technology was systematically refined so as to achieve acceptable outcomes for light-colored skin tones without similar attention to darker tones, ultimately resulting in the familiar black blobs in the place of identifiable faces in high-school yearbooks and newspapers through the 1980s and beyond. At no point were there insurmountable technological barriers to addressing the problems of attractively representing darker skin tones; sufficient effort simply was not invested.

Dyer does not ignore the role of the market economy in his analysis; despite the existence of a non-trivial market for photographic portraits among wealthy black persons, their combined influence was not sufficient to redirect attention toward their needs. But neither does Dyer reduce the phenomenon to a simple, temporal market-economy shortcoming; as both racial segregation and market conditions changed over time, photographic equipment continued to "carry forward the assumptions that had gone into [making] them" (135). In the case of the development of photography equipment, lack of racist intent by technology innovators did not preclude a racially inequitable outcome—photography equipment that was ill-suited to represent dark skin tones. As later innovations built on early innovations, the problem of representing dark skin tones was increasingly assigned to the skin tones themselves (and those people with them) instead of being seen as a deficiency of the technology innovation process that systematically ignored them. In Dyer's words, "It may be—certainly was—true that photo and film apparatuses have seemed to work better with light-skinned peoples, but that is because they were made that way, not because they could be no other way" (135).

A similar problem has surfaced in public health and medical care technology development, where marginalized groups like blacks and Latinos have historically been excluded from some testing protocols for potential benefits or, worse, systematically targeted in high-risk testing as in the Tuskegee experiments (Fisher 2007). Manifestation of this situation in the context of nanotechnology development is addressed by Slade (Chapter 4), which explores the ethical issues surrounding unequal access to clinical trials testing the efficacy and safety of certain cancer therapies. Perhaps with the exception of the Tuskegee experiments, each of the cases of racial inequity illustrates a phenomenon where the intent of technology innovators is mostly independent of the racially inequitable outcome. The structural racism that permeates the institutions in which technology innovation takes place finds its way into the material configurations of technologies that result. The distinction between individual racism and structural racism is crucial for making sense of this phenomenon, and hence is something that must be kept in mind when assessing the developmental trajectories of any new technology.

The processes by which ableism, sexism, and racism are materialized in technologies and products need not be intended for them to be real. Indeed, the very opposite intent may inadvertently reinforce marginalization by unreflectively relying on stereotypical assumptions about the target user group during the design process. Although impossible to verify, it seems reasonable to assume that most of the individual designers whose decisions cumulate into materialized inequity do not start with malicious intent. Yet in the context of market-based incentives and the fragmented characteristic of alternative approaches to innovation, institutionalized structures of inequity combine with ignorance of the underlying value systems that prop those structures up and culminate in unintended technological development outcomes.

Rather than diminishing the significance of the various "isms"—ableism, sexism, racism, and others—in technology innovation, the *social* structures of inequality make materialized inequity even more noteworthy. Because we cannot always immediately see it and because we cannot always assign responsibility for it to individual actors, materialized inequity is as intractable as the assumptions, values, and social stereotypes that are bound up with it. Left with objects and spaces that contradict even their inventors' intentions, one may become frustrated by the seeming impossibility of technology development that addresses social equity. A more optimistic lesson, however, is that by recognizing that social and material inequities overlap one another—or better, that they are mutually constituted—variously situated social actors will be better prepared to address inequities by attending to both domains simultaneously.

As with any process that contributes to shaping the built world, participants in nanotechnology innovation cannot be absolved from responsibility for extending or reducing social inequity. To embrace such responsibility, we need to move beyond simply extending new technology applications into a variety of new contexts of use and toward applications that contribute productively to reducing inequities within a given context. As with other domains of coordinated social activity, technology innovation shares responsibility for countering discrimination in all its manifestations.

13.5 Participatory Design

In critically analyzing the technologies and products that make up the built world, it would be mistaken to limit our consideration to consumer products and other objects that are selected by their users. A wide range of objects in many different environments are not chosen by those who directly use them or otherwise come into contact with them. In many settings all over the world, people are required to use—on a regular basis—technologies chosen by others, including some with characteristics that run contrary to the interests of their proximate users. Most notably, the objects that occupy workplace environments—office equipment and supplies, production equipment and materials, and an increasing diversity of information systems—are rarely selected by their worker-users and yet directly impact quality of work-life in fundamental ways. Consideration of workplace technologies has special significance in anticipating nanotechnology's potential impacts.

Tensions over labor-saving and deskilling applications of workplace technologies, in particular, have a long and well documented history, dating at least back to the Luddites of the early nineteenth century (Bailey 1998). Far from being resolved in modern times, these tensions continue today with the added dimension of globalization and the rapid movement of capital, products, technologies, and even industrial production facilities (Lommerud and Straude 2007). Tensions also exist over worker health and safety standards, especially in contexts where new workplace technologies, such as nanotechnologies, increase risks to workers, including by introducing whole new categories of risk.[18] Given that Americans have spent increasing time at work over the past few decades (Schor 1992), ignoring technology choices in the workplace—who makes decisions, on whose behalf, and through what processes—would miss a critical dimension of the distribution of technologies' benefits and burdens, and hence of equity.

Over the past several decades in the North American and northern European contexts, workers and their unions have, for the most part, treated technology adoption as a black or white issue. They either accepted or resisted implementation of new technologies as proposed by management, but have not played an active role in determining the contours of potential new technologies (Cherkasky and Scannell 1999). Participatory design scholars, on the other hand, have rejected the "take it or leave it" position with respect to new workplace technologies, especially information and communication technologies. Instead, these scholars have advocated a middle path, one that embraces new workplace technologies but does so without accepting the burdens of deskilling, health risks, and low quality work-life. Participatory design invites technology users to the drafting board to participate in devising workplace technologies that increase productively and quality of work-life simultaneously (Schuler and Namioka 1993). More broadly defined, participatory design is a process whereby the "people destined to use the system play a critical role in designing it" (Schuler and Namioka 1993, xi).

The roots of participatory design reach into a variety of fields, including labor studies, feminist theory, and user-centered design as well as mid-1970s Scandinavian "co-determination" laws that mandated worker participation in workplace decision making (Greenbaum and Kyng 1991). According to its advocates, worker participation will lead to both smarter and fairer decision making: Smarter because engaging experienced workers in the design process will lead to improved production outcomes and fairer because workers will exercise their rights to participate in the making of decisions that affect their work lives (Nieusma 2004). In sharing workplace decision-making authority with workers, the movement promised to reverse the decades-long trend of deskilling work in industrial settings and to ensure workers' interests more generally were accounted for in shaping new workplace technologies (Ehn 1993).

As noted above, observers of nanotechnology have not failed to appreciate the technology's potential for radically altering workplace environments and the health, safety, and quality of worklife (Foladori and Invernizzi 2008). Workers' unions, in particular, have grappled with the transformative potential of nanotechnologies in the workplace, and have called for better oversight in the technology's application and dissemination (IUF 2009). For example, in coalition with environmental protection and public health organizations, workers' unions issued a declaration entitled, "Principles for the Oversight of Nanotechnologies and Nanomaterials" in 2007:

> The people that research, develop, manufacture, package, handle, transport, use and dispose of nanomaterials will be those most exposed and therefore most likely to suffer any potential human health harms. As such, worker protection should be paramount within any nanomaterial oversight regime.... Any regulatory regime designed to protect workers from the health effects of nanomaterials requires written comprehensive safety and health programs addressing workplace nanotechnology issues.... Workers and their representatives should be involved in all aspects of workplace nanotechnology safety and health issues without fear of retaliation or discrimination. (Principles 2007, 3)

This declaration directs explicit attention to risks to workers and the need for worker participation in the regulation of those risks.

With strong financial incentives to externalize production costs to workers and the environment, especially when risks are highly uncertain, the call for oversight directs attention to enduring tensions in the workplace. Most notably, oversight is needed to manage the tensions that exist between technology decision makers (usually management but also sometimes designers) on one hand and workers (and those who would regulate on their behalf) on the other. Participatory design scholars attend to this category of tension explicitly:

> The design process is a political one and includes conflicts at almost every step of the way. Managers who order the system may be at odds with workers who use it. Different groups of users will need different things from the system, and system designers often represent their own interests. Conflicts are inherent in the process. (Greenbaum and Kyng 1991, 2)

Participatory design processes are posited to mediate such tensions while moving forward with new workplace technology implementation. If implementing new workplace technologies is to be reconciled with workers' short and long-term interests, participatory design advocates aver, workers must be empowered to participate directly in the design of those technologies: workers must have some ability to "select" among alternative possibilities.

In advocating democratic workplace decision making, participatory design directs attention to the distribution of both burdens and benefits of technology implementation, recognizing that a given technology can manifest either for any given audience and will almost certainly manifest both in the case of multiple audiences. And it especially seeks to avoid the disproportionate leveling of burdens upon workers. Other categories of users have also been attended to by participatory design advocates, including intended beneficiaries of international development assistance (e.g., Haggar et al. 2001), citizens impacted by public architecture and urban planning (e.g., Dubbeling et al. 2009), and users of healthcare services (e.g., Sjöberg and Timpka 1998). All of these user groups must engage—often directly and intimately—with technologies and objects they do not explicitly choose to use. Were their perspectives accommodated in the design process, the resulting objects would be more likely to meet their needs.[19]

Besides those users who do not choose the objects they must use, one additional, broad category of stakeholders deserves identification: *affected non-users*. Affected non-users are those people impacted by the use of an object by others. Because they are impacted, they arguably have a right to some level of decision making regarding that object's use. For example, individualized transportation choices, like buying a car, impact not only the car's buyer but also everyone else in the community—through car-oriented urban development, increased air pollution, and contributions to climate change. One can choose not to buy a car, but one cannot choose to abstain from car culture.[20] In many areas of technology innovation, ownership, and practice, individual choices impact others, and an increasing number of product/object choices impact, in modest but cumulative ways, everyone on the planet.

The cumulative and distributed impact of aggregated individual choices is obviously relevant to the case of nanotechnologies; those who would choose to forgo the potential benefits of nanotechnology will still face some of the potential burdens as the technology and its byproducts are increasingly widely disseminated. To ensure nanotechnology development is steered toward enhanced equity, the distribution of potential burdens deserves equal attention as the distribution of benefits. And this is especially true for impacts disproportionately faced by certain social groups, including workers especially but also extending beyond workplace environments.

13.6 Materialized Inequity and Design Alternatives

As exemplified across each of the approaches discussed above, alternative design scholars are interested in understanding how the benefits and burdens of technologies and products are distributed, including in ways that entrench power inequality or, conversely, in designing technologies so as to counter inequity. In each case, limitations in market mechanisms for fairly distributing or directing technologies are evident, as are the implications of technology choice going far beyond the distribution of "technological goods." As stated in the introduction to this paper and shown in each of the following sections, material objects impact how people experience the world, including their social worlds, in subtle yet pervasive and differential ways. These impacts are not only felt by those who explicitly choose to buy or use a given technology; they are felt by many people within the technology's field of influence. As nanotechnologies disseminate throughout markets, communities, and ecological systems, many people will be touched by them, users and non-users alike. But just because the impacts of nanotechnologies will be pervasive, that does not mean the various positive and negative impacts will be distributed evenly. This section investigates more systematically the implications of *materialized inequity*.

Alternative design scholarship emphasizes various facets of the intersection of artifacts and social power, from simple market shortcomings to deeply embedded stereotypes. One common thread spanning these is the reciprocal relationship between the role of artifacts in *reflecting* social priorities and their role in *reinforcing* or stabilizing those priorities. How we construct the material world has clear implications for the distribution of social power, even as there are real limits in the degree to which material innovation, by itself, can create social change. While much in the future of nanotechnologies remains to be determined, we are currently witnessing those futures "in the making." Decisions made about nanotechnology will necessarily embody social values, in both direct and not-so-direct ways.

In a direct way, nanotechnology policy making will shape how the technology is materialized in the present, and as a result how it will continue to evolve for many years to come. Technology funding agencies, regulators, and researchers make decisions at many levels that influence research agenda and, hence, determine which applications of the technology will be pursuable in the near and medium terms. In policy making, broad funding priorities or regulatory frameworks are set based on a complex mix of imagined technological possibilities (both positive and negative) and the diverse interests of participants in the process. Including groups with a vested interest in countering inequities is essential if that goal is to be structurally embedded in policies rather than merely tacked on to provide justificatory rhetoric for activities likely to result in business-as-usual outcomes.

Technology implementers also play an important role in shaping nanotechnology's future in direct and obvious ways. As basic research is translated into specific material objects, whether microelectronics or paint or pharmaceuticals, precedents are put in place, and experience is gained, around where nanotechnology "fits" in the world—socially, politically, and most important economically. As with other areas of design—for example, making buildings accessible to wheelchairs—the challenge is not always what is *possible* to do with technology but instead is what is *actually* done with it. That building designers can make accessible buildings is a different matter than systematically making them such. That nanotechnology can be used to alleviate poverty is distinct from how it is actually, perhaps aggressively, being implemented in the present.[21] The role of "external" stakeholders in the application phase of nanotechnologies is less clear than with policy making, but even providing a broader perspective to participants within the system can be beneficial.

But the point of my analysis is not necessarily to advocate specific instances of nanotechnology that will enhance equity, however welcome such applications would be. This paper is concerned with a different task: identifying the indirect, not-so-obvious ways technologies come to embody social values and materialize social power relations. Market incentives aligned with already powerful "vested interests" are unlikely to steer nanotechnology development, as a whole, in directions that empower marginalized groups and equalize social power. While renegade nanotechnology researchers may work as individuals or small groups outside the dominant institutional structures incentivizing commercializable (or military) applications, they will be poorly positioned to make systemic change to the overall developmental trajectory. Even granting the eventuality of trickle-down benefits to non-targeted future users, such benefits will likely be offset by significant advantages provided to those users who are explicitly targeted: the cosmetics industry, military institutions, advanced medicine, etc.

In order to offset the dominance of market incentives for shaping nanotechnology development in ways that benefit already-powerful social groups, orchestrated efforts are needed to identify those groups whose needs ought to be targeted. Nanotechnology designers of all stripes, from those working on basic research to specific applications, need to be central players in such efforts if they are to be successful. To achieve a more equitable distribution of benefits and burdens surrounding nanotechnology, attention is needed, first, to the various ways inequity is materialized given current decision-making structures and, second, to the strategies for changing those structures so as to materialize new patterns of benefits and burdens. Input from nanotechnology designers is important to both of these. But a guiding strategy, common to each of the design approaches reviewed above, is to integrate along with designers intended user groups and other affected stakeholders in technology decision making, including the design process.

If the goal is to move beyond addressing specific communities' unmet needs and to democratize the whole of nanotechnology development, then the strategy can be generalized to integrate a wide range of stakeholder groups, including a range of representatives from groups marginalized both in the market and in society more generally. This involves systematically reconsidering who participates in nanotechnology decision making and on what terms. "Strong democratic" guidance to technology developers is needed for a wider range of interests and perspectives to inform nanotechnology's overarching development trajectory.[22] As shown by the alternative design scholarship above, single-instance equity-enhancing technologies are inadequate in addressing structural inequities.

But even as I advocate democratizing technology development through radically expanded participation and mechanisms that distribute technology decision making authority, it would be a mistake to suggest that designers and other technology experts do not share a central role in redirecting technology innovation. Nanotechnology designers—those most proximate to the shaping of potential nanotechnology products or processes—need not be construed as barriers to redirecting trajectories toward more democratic ends. To the contrary, technology innovators (including those working within dominant institutions) are often strong proponents of alternatives to raw market incentive structures, and some even seek to contribute to more equitable futures for those marginalized. At the same time as I advocate inclusion of more diverse stakeholder groups in the design process, I also look to identify technology experts capable of and willing to facilitate a highly participatory technology design process.

Redirecting nanotechnology development to address enduring social inequities, then, will almost certainly require a combined effort by nanotechnology experts, representatives of the social groups subjected to materialized inequities, and policy makers committed to experimenting with new decision-making protocols. For this collaboration to be effective, already-enfranchised decision makers not only need to allow the participation of currently disenfranchised groups, but they also need to reframe the problems they are attempting to solve with nanotechnology. Understanding how structures of inequity are both reflected in and reinforced by new technologies is a critical component of that reframing exercise.

13.7 Conclusion: Lessons for Nanotechnology Development

Leaving the market to its own devices will likely result in nanotechnology applications in upgraded luxury consumer items and advanced medical treatments for those already overwhelmingly served by existing technologies. Achieving a different outcome, namely to serve the needs of marginalized groups across sectors of the global society, will require deliberate, systematic efforts to do so. Rather than waiting to see how inequality is materialized in nanotechnologies of the near future, decision makers can deliberate now over what sort of alternative futures nanotechnology might work to achieve. Such deliberation could identify strategies for redirecting the technology's development trajectory and then monitor nanotechnology performance as decisions are made, as the technology is materialized and disseminated, and as stakeholders are impacted.

Current nanotechnology decision makers, and those involved in funding nanotechnology research in particular, are in a unique position to change the distribution of benefits and burdens into the future. They can create incentives for technology developers to target new user communities and their distinct needs and circumstances, to include more stakeholders in the technology's design processes, and generally to take seriously the extent to which seemingly innocuous design decisions can materialize inequity—that is, cumulate into technologies, products, and environments embodying economic and political power imbalances.

This paper has reviewed a variety of mechanisms through which technologies and products disempower members of marginalized groups across a wide range of categories of marginalization. It has also shown how social and cultural assumptions and values become embodied in technologies, not of necessity but as a matter of practice, as people and their social and cultural institutions direct (and fail to redirect) new technology development trajectories. Once their assumptions and values are materialized in the built world, they are reinforced in the social world. And as inequity is materialized in what is done now, momentum and path dependencies limit alternative trajectories and, hence, what is doable—practically if not absolutely—in the future. By investing now in trajectories that explicitly and directly counter materialized inequities, before the momentum of status-quo strategies becomes unyielding, new paths of nanotechnology development can be established to achieve more equitable distribution of benefits and burdens.

Acknowledgements I would like to acknowledge the assistance of the editors of this volume and anonymous reviewers for their help in framing this paper and articulating its arguments. In particular, I would like to thank Jameson Wetmore for his close reading of and detailed comments on an early draft. I would also like to thank the participants of the Workshop on Nanotechnology, Equity, and Equality (20 November 2008, Tempe, Arizona) for feedback on an early iteration of ideas presented here.

Notes

1. The Project on Emerging Nanotechnologies' consumer product inventory shows sustained increases in the number of identified products from March of 2006 to December 2009, with an average growth rate of over 200 documented new products each year. While this data may under-represent the number of products offered at any given interval, the trend matches most observers' expectations of increased market penetration of the technology.
2. Papanek's array of areas under-addressed by design ranges from the mundane (as listed above), to what have since become major areas of design activity (e.g., medical devices and experimental research), to what remains marginal but has no direct connection to underserved social groups (e.g., design for life in marginal conditions, such as underwater).
3. As we are frequently reminded, the problem of global hunger is not only one of technological capacity (i.e., insufficient food production), but a social and political problem as well—ineffective distribution mechanisms, barriers raised by war and political strife, lack of motivation by the rich to act on behalf of the poor, etc.
4. Schumacher's approach was influenced by his exposure to a variety of technology practices across Asia, in particular, that of Gandhism. Gandhi is sometimes cited as the grandfather of the appropriate technology movement (Akubue 2000).
5. A relatively new entrepreneurial trend in appropriate technology practice promotes dissemination of appropriate technologies through normal market mechanisms in poor countries (e.g., Polak 2008), lending some credence to the trickle-down model. It should be noted, however, that technology development is often heavily subsidized in these models—a hybrid extra-market/market approach is used rather than a purely market-based approach.
6. See Willoughby (1990) for an excellent review of the appropriate technology movement through the mid-1980s.
7. Feminist scholars also have a history of "attending to context" in the development of technology (e.g., Bush 1997).
8. Although this insight may seem self-evident to contemporary technology developers, it was hard-won through many years of experimentation and experience with "technology transfer" efforts resulting in broken-down, idled, abandoned, and utterly reappropriated technological systems that remain peppered throughout developing countries today. In some respects, the

appropriate technology movement was the counter-cultural response to failures of the post-WWII technology transfer model, which comprised of literally airlifting entire technological systems from developed countries, installing them in developing countries, and expecting them to function equivalently (Willoughby 1990).
9. The term "intended beneficiary" is borrowed from Dudley and emphasizes the fact that development assistance "does not guarantee that the intended targets will benefit. It only expresses an intention" (1993, 7).
10. Richard Sclove opens his monograph, *Democracy and Technology* (1995), with the early-1970s case of distributed water wells introduced in Ibieca, Spain. Despite meeting all technical specifications and presumably increasing "convenience" among users, the wells inadvertently but drastically undermined social cohesion by removing the community's primary congregation spot for women—the common well.
11. As pointed out by one of the editors (Wetmore), many people within the most marginalized groups have become so disaffected with dominant technology development trajectories that they actually have no "expectations" to speak of.
12. To be fair, many participants in the appropriate technology movement have attended to questions of distribution within a given social group or nation, by directing their efforts to the poorer populations within rich countries or poor countries.
13. Borrowed from Wolbring (2008; Chapter 5), ableism is highlighted by quotation marks because it is a less-well-known "ism" label. It refers here to inequity based in physical or mental ability, or perceptions thereof, especially among the physically and cognitively "disabled," the elderly, and children.
14. I do not intend to imply that each contribution is entirely unique; there is considerable overlap in approach taken and insights offered among the various alternative design communities. This is true both because there are common mechanisms of marginalization at play across groups and because insights made in one area are picked up and applied to other areas.
15. Various labels are attached to design aimed at improving accessibility, including the common but problematic label, "universal design." With other scholars, I take issue with the term "universal," because it contributes to the problem it attempts to solve, albeit inadvertently. If someone with a very rare disability is not accommodated by a "universal design," for instance, then that person's exclusion can be understood to be total: excluded both materially and conceptually by the purported "universal" solution.
16. A considerable body of accessibility design scholarship has arisen, and much of America's built environment has been transformed, in direct response to the Americans with Disabilities Act of 1990.
17. As Weber notes, a 1993 directive sought specifically to address this problem, adjusting design specifications to "accommodate at least 80% of eligible women" (1997, 242).
18. The small size and resulting high reactivity of nanoparticles pose "a challenge to the traditional methods of evaluating occupational health and risk assessment and measurement instruments currently in use" (Foladori and Invernizzi 2008, 32).
19. Increasingly, the logic of user participation in design is applied even to the development of ordinary consumer products—objects that are, in fact, directly chosen by their users (e.g. Bonner and Porter 2002).
20. That is, except if one chooses to extract oneself from his or her ordinary social life. For the very tiny fraction of people raised in extremely remote settings where cars never visit, non-participation in car culture is not really a choice.
21 This is not to say that there are not nanotechnology policy makers and researchers actively pursuing equity goals, as this volume clearly indicates, but that such efforts are not readily evident when reviewing the current state of nanotechnology dissemination, for example by reviewing the Project on Emerging Nanotechnologies' consumer product inventory.
22. The term "strong democracy" comes from Barber (1984). Winner (1986) and Sclove (1995) offer applications of a similar approach to the context of technology policymaking.

References

Akubue, Anthony. 2000. Appropriate technology for socioeconomic development in third world countries. *The Electronic Journal of Technology Studies* 26: 33–43.

Bailey, Brian J. 1998. *The luddite rebellion*. New York, NY: New York University Press.

Barber, Benjamin. 1984. *Strong democracy: Participatory politics for a new age*. Berkeley, CA: University of California Press.

Bonner, John V.H., and J. Mark Porter. 2002. Envisioning future needs: From pragmatics to pleasure. In *Pleasure with products: Beyond usability*, ed. William S. Green and Patrick W. Jordan, 151–158. London: Taylor and Francis.

Bush, Corlann Gee. 1997. Women and the assessment of technology: To think, to be; to unthink, to free. In *Machina ex dea: Feminist perspectives on technology*, ed. Joan Rothschild, 151–170. New York, NY: Teachers College Press.

Chambers, Robert. 1997. *Whose reality counts? Putting the first last*. London: Intermediate Technology Publications.

Cherkasky, Todd D., and Ray Scannell. 1999. Making technology work for workers. *WorkingUSA* 2(6): 28–46.

Covington, George A., and Bruce Hannah. 1997. *Access by design*. New York, NY: Van Nostrand Reinhold.

Darrow, Ken, and Mike Saxenian. 1993. *Appropriate technology sourcebook: A guide to practical books for village and small community technology*. Stanford, CA: Appropriate Technology Project, Volunteers in Asia.

David, Paul A. 2007. Path dependence, its critics and the quest for "historical economics." In *The evolution of economic institutions: A critical reader*, ed. Geoffrey M. Hodgson. Cheltenham, England: Edward Elgar Publishing, 120–142. Originally published in Pierre Garrouste and Stavros Ioannides eds. 2001. *Evolution and path dependence in economic ideas: Past and present*. Cheltenham, England: Edward Elgar Publishing.

Davies, J. Clarence. 2008. *Nanotechnology oversight: An agenda for the new administration*. Project on Emerging Technologies. Washington, DC: Woodrow Wilson International Center for Scholars.

Drexler, Eric. 1986. *Engines of creation: The coming era of nanotechnology*. New York, NY: Anchor Books.

Dubbeling, Marielle, Laura Bracalenti, and Laura Lagorio. 2009. Participatory design of public spaces for urban agriculture, Rosario, Argentina. *Open House International* 34(2): 36–49.

Dudley, Eric. 1993. *The critical villager: Beyond community participation*. London: Routledge.

Dyer, Richard. 1999. Making 'white' people white. In *The social shaping of technology*, 2nd ed. ed. Donald MacKenzie, and Judy Wajcman. Cambridge: MIT. Originally published in Richard Dyer, 1997. *White*. New York, NY: Routledge.

Ehn, Pelle. 1993. Scandinavian design: On participation and skill. In *Design at work: Cooperative design of computer systems*, ed. Joan Greenbaum, and Morten Kyng, 41–78. Hillsdale, NJ: Lawrence Erlbaum Associates.

Fisher, Jill A. 2007. Governing human subjects research in the USA: Individualized ethics and structural inequalities. *Science and Public Policy* 34(2): 117–126.

Foladori, Guillermo, and Noela Invernizzi. 2008. The workers' push to democratize nanotechnology. In *The yearbook of nanotechnology in society, Vol.1: Presenting futures*, ed. Erik Fisher, Cynthia Selin, and Jameson M. Wetmore, 23–36. New York, NY: Springer.

Greenbaum, Joan, and Morten Kyng. 1991. Introduction: Situated design. In *Design at work: Cooperative design of computer systems*, ed. Joan Greenbaum, and Morten Kyng, 1–24. Hillsdale, NJ: Lawrence Erlbaum Associates.

Guston, David, and Daniel Sarewitz. 2002. Real-time technology assessment. *Technology in Society* 24(1): 93–109.

Haggar, Jeremy, Alejandro Ayala, Blanca Díaz, and Carlos Uc Reyes. 2001. Participatory design of agroforestry systems: Developing farmer participatory research methods in Mexico. *Development in Practice* 11(4): 417–424.

Hazeltine, Barrett, and Christopher Bull. 1999. *Appropriate technology: Tolls, choices, and implications*. San Diego, CA: Academic.

Hess, David J. 1995. *Science and technology in a multicultural world: The cultural politics of facts and artifacts*. New York, NY: Columbia University Press.

Hughes, Thomas P. 2000. Technological momentum. In *Technology and the Future*. ed. Albert H. Teich, 26–35. Boston, MA: Bedford/St. Martin's.

Hughes, Thomas P. 1983. *Networks of power: Electrification in western society, 1880–1930*. Baltimore, MD: The John Hopkins University Press.

Imrie, Rob, and Peter Hall. 2001. *Inclusive design: Designing and developing inclusive environments*. London: Spon Press.

International Union of Food, Agricultural, Hotel, Restaurant, Catering, Tobacco and Allied Workers' Associations (IUF). 2009. http://www.iufdocuments.org/cgi-bin/show_about.cgi?l=en (accessed August 7, 2009).

Invernizzi, Noela, and Guillermo Foladori. 2005. Nanotechnology and the developing world: Will nanotechnology overcome poverty or widen sisparities? *Nanotechnology Law and Business* 2(3): 294–303.

Lommerud, Kjell Erik, and Odd Rune Straume. 2007. Technology resistance and globalisation with trade unions: The Choice between employment protection and flexicurity. Working Paper No. 25/2007. Núcleo de Investigação em Políticas Económicas, Escola de Economia e Gestão, Universidade do Minho, Braga, Portugal.

National Nanotechnology Initiative. Undated. 2009. Applications and products: Putting technology to use. http://www.nano.gov/html/facts/nanoapplicationsandproducts.html. (accessed September 14, 2009).

Nieusma, Dean. 2004. Alternative design scholarship: Toward appropriate design. *Design Issues* 20(3): 13–24.

Nye, David E. 1996. *American technological sublime*. Cambridge: MIT.

Papanek, Victor. 1984. *Design for the real world: Human ecology and social change*, 2nd ed. Chicago, IL: Academy Chicago Publishers.

Pelletier, Dick. 2008 (September). Nanotech wonders hyped at Wash. DC Conference. Future Blogger.: http://www.memebox.com/futureblogger/show/841-nanotech-wonders-hyped-at-wash-dc-conference. (accessed September 14, 2009).

Polak, Paul. 2008. *Out of poverty: What works when traditional approaches fail*. San Francisco, CA: Berrett-Koehler.

Principles for the Oversight of Nanotechnologies and Nanomaterials. 2007. http://www.iufdocuments.org/www/documents/Principles%20for%20the%20Oversight%20of%20Nanotechnologies%20and%20Nanomaterials.pdf (accessed August 7, 2009).

Project on Emerging Nanotechnologies. Consumer products inventory. 2009. http://www.nanotechproject.org/inventories/consumer/analysis_draft/. (accessed December 17, 2009).

Pytlik, Edward C., Ernest Frank, III, and Anthony Akubue. 2001. Cultural and gender issues in appropriate technology. In *Appropriate technology for sustainable living*, ed. Robert C. Wicklein, 113–132. Peoria, IL: Glencoe/McGraw-Hill.

Rothschild, Joan, ed. 1999. *Design and feminism: Re-visioning space, places, and everyday things*. New Brunswick, NJ: Rutgers University Press.

Sarewitz, Daniel. 1996. *Frontiers of illusion: Science, technology, and the politics of progress*. Philadelphia, PA: Temple University Press.

Schor, Juliet. 1992. *The overworked American: The unexpected decline in leisure*. New York, NY: Basic Books.

Schuler, Douglas, and Aki Namioka. 1993. Preface. In *Participatory design: Principles and practices*. ed. Douglas Schuler, and Aki Namioka, xi–xiii. Hillsdale, NJ: Laurence Erlbaum Associates.

Schumacher, E.F. 1973. *Small is beautiful*. New York, NY: Harper & Row.

Sclove, Richard. 1995. *Democracy and technology*. New York, NY: Guildford.

Sjöberg, Cecilia, and Toomas Timpka. 1998. Participatory design of information systems in health care. *Journal of the American Medical Informatics Association* 5(2): 177–183.
Smillie, Ian. 2000. *Mastering the machine: poverty, aid and technology*. London: ITDG.
Walton, Anthony. 1999. Technology versus African Americans. *The Atlantic Monthly* 283: 14–18.
Weber, Rachel N. 1997. Manufacturing gender in commercial and military cockpit design. *Science, Technology and Human Values* 22(2): 235–253.
Weisman, Leslie Kanes. 1992. *Discrimination by design: A feminist critique of the man-made environment*. Urbana, IL: University of Illinois Press.
Whiteley, Nigel. 1993. *Design for society*. London: Reaktion Books.
Wicklein, Robert C. ed. 2001. *Appropriate technology for sustainable living*. Peoria, IL: Glencoe/McGraw-Hill.
Willoughby, Kelvin. 1990. *Technology choice: A critique of the appropriate technology movement*. Boulder, CO: Westview.
Winner, Langdon. 1986. *The whale and the reactor: A search for limits in an age of high technology*. Chicago, IL: University of Chicago Press.
Wolbring, Gregor. 2008. Commentary: Nanoscale sciences and technology and the framework of ableism. In *Nanotechnology: Ethics and Society,* ed. Deb Bennet-Woods, 212–215. Boca Raton, FL, Taylor & Francis Group.
Wright, Michelle M. 2000. Racism and technology. *Switch*: 6(2).

Chapter 14
Public Perceptions of Fairness in NBIC Technologies

Ravtosh Bal

In this chapter Ravtosh Bal explores the National Citizens Technology Forum (NCTF), a month long U.S.-wide public engagement focused on applications of nanotechnology for human enhancement (see also Wolbring). Through the process of deliberation, the NCTF participants expressed concerns about a wide array of equity issues. They worried that access to therapeutic applications might be limited by income, gender, and race and argued that systems of governance need to be set up to ensure that society as a whole benefits from these technologies, not just specific individuals or groups. Bal's work first demonstrates that equity is not simply an issue that academics are debating; the public is quite concerned as well. Policymakers looking for justification to include equity in their decision-making can find a great deal of validation in the NCTF reports. Second, Bal's chapter gives further evidence that citizen technology forums may be one route through which more equal futures can be envisioned and advocated.—eds.

R. Bal (✉)
School of Public Policy, Georgia Institute of Technology, Atlanta, GA, USA; Andrew Young School of Policy Studies, Georgia State University, Atlanta, GA, USA
e-mail: rbal3@mail.gatech.edu

This chapter was peer reviewed. It was originally presented at the Workshop on Nanotechnology, Equity, and Equality at Arizona State University on November 22, 2008.

14.1 Introduction

In recent years there has been an increase in the involvement of the public in policy making, often through the design of new institutional arrangements such as citizen juries and consensus conferences. Public participation in policy making has been proposed as a means to a more democratic policy making and to make policy more attuned to the needs of the people. Policy formulation that results from a democratic discourse among various stakeholders can reflect the values and preferences of citizens. As a result, participatory methods are seen as leading not only to better decisions but also to more legitimate decisions. In this paper, I examine the deliberations in the National Citizens Technology Forum (NCTF) which was designed to gather public input into nanotechnology policy making.

Organized as a consensus conference, the NCTF looked at NBIC (nanotechnology, biotechnology, information technology, and cognitive sciences) technologies for human enhancement. Nanotechnology has been heralded as the next big revolution that will bring immense benefits to the people (National Science and Technology Council 2000). Proponents of nanotechnology believe that it has the potential to transform not only society but human life itself. On the other end of the spectrum, critics warn that nanotechnology will lead to great social upheavals (ETC Group 2004). Learning from the controversies around biotechnology and genetic engineering, policy-makers have carved a larger role for social and ethical issues in the debates around nanotechnology development. The twenty-first century Nanotechnology Research and Development Act that established the National Nanotechnology Program requires that all nanotechnology science and engineering research centers include research on societal, ethical, and environmental concerns and integrate public input into the program activities. The integration of ethical, legal, and social implications (ELSI) studies in nanotechnology development and policy is guided by the idea that technology should be shaped by the values and concerns of the public (Bennett and Sarewitz 2006; Fisher and Mahajan 2006). The NCTF was a nation-wide public deliberative exercise to elicit informed public views about the development of nanotechnology for human enhancement.

Recent scholarly work has examined various aspects of the NCTF. Powell and Kleinman (2008) have examined the effect of these deliberations on citizens' perceptions of their efficacy and knowledge; Philbrick and Barandiaran (2009) evaluated the NCTF as a "proof-of-concept" for incorporating consensus conferences in the U.S. policy process; Kleinman et al. (2009) examined the incentives that motivate the public's participation in the debates around emerging technologies; while Delborne et al. (2009) evaluated the online deliberations.

This paper looks at the equity issues that arose during the deliberations utilizing the transcripts of the internet deliberations of the participants, the final reports, and the survey data. I examine whether public participation methods lead to a consideration of equity issues. How important was the issue of equity to the participants? Do the participants articulate equity as a goal? What are the recommendations pertaining to equity? I examine how equity as a value was defined by the participants and how their views of equity shaped their policy recommendations. I also use the survey data to see whether race, gender, income, and education have an impact on perceptions of fairness. An examination of how equity was dealt with in the deliberations of the NCTF can help clarify the ways in which citizens combine values with their views and concerns about nanotechnology. These deliberations can also justify the creation of policies regarding emerging technologies that take equity into consideration as a goal.

14.2 Citizen Participation and Equity

Two academic fields that think about science policy are the Policy Sciences and Science and Technology Studies. Both fields have argued for citizen participation in science policy making. The positivist bent in most policy formulation has been criticized by many policy scientists who believe that successful policy outcomes require integrating norms and values into the policymaking process (deLeon 1988; Dryzek 1990; Stone 1997). The case for participatory policy analysis is made on the grounds that citizens can best articulate their needs and concerns, and their participation can lead not only to more informed decision making but also to the development of a more involved public (deLeon 1988). Participatory policy analysis has been heavily influenced by the concept of deliberative democracy that considers public dialogue and discourse as an alternative to representative democracy. Open, public dialogue permits a broader understanding of different interests leading to learning and a consensus based decision making that is more legitimate (Cohen 1997). Fischer (1993) has argued that this collaborative, participatory approach works best when the problem at hand is a mix of technical and social issues, and adequate time as well as resources are available to solve these problems that more likely than not have long-term impacts.

Most policy issues connected with science and technology represent this mix of social and technical issues. Science and Technology Studies has provided an understanding that science and society are intertwined in complex ways. The "co-production of science and society" and the social embeddedness of science have provided newer insights into the way in which the relationship between citizens, experts, and science is structured and the way public participation in science and technology (S&T) policy is viewed (Jasanoff 2003; Wynne 1996). Non-experts do not bring just a different perspective, but both expertise and lay knowledge are produced and shaped by the interaction between experts and lay citizens (Bucchi and Neresini 2008).

Newer models of technology assessment, such as Constructive Technology Assessment (CTA) (Schot and Rip 1997) and Real Time Technology Assessment (RTTA) (Guston and Sarewitz 2002), emphasize public participation. CTA repudiates the delineation of invention and regulation. The process of technology assessment is one of social learning, and all aspects of a technology are open to deliberation. RTTA is seen as the more reflexive of the two in nature. It:

> ...encourages more effective communication among potential stakeholders, elicits more knowledge of evolving stakeholder capabilities, preferences and values, and allows modulation of innovation paths and outcomes in response to ongoing analysis and discourse (Guston and Sarewitz 2002, 100)

Both these approaches place emphasis on the "co-evolution of science, technology and society" and a broadening of the design criteria to include societal aspects of innovation (Schot and Rip 1997; Guston and Sarewitz 2002). Public participation when a technology is taking shape allows values and norms to be incorporated early in the development of a technology and make it possible for alternative pathways to be explored.

An important value that is associated with technology assessment is equity. In the U.S. the public funding of science is legitimized on the promise that developments in science hold for the betterment of humanity. The argument made is that in addition to solving the problems of health, poverty, and hunger, the economic growth made possible through scientific development will also allow a better quality of life for all people. If universal good is truly the goal of science then equity becomes an important evaluative criterion for publicly funded research and development.

Understanding how S&T policy can lead to greater equity requires an elaboration of the definition of equity as the policy design is determined by the definition. Equity is most commonly defined as fairness or the lack of differences that are unjust. Caplan, Light, and Daniels define fairness as:

> ...equalizing people's opportunities to participate in and enjoy life, given their circumstances and capacities.... a commitment to fair equality of opportunity thus recognizes that we should not allow people's prospects in life to be governed by correctable, morally arbitrary, or irrelevant differences between them, including those that result from disease or disability (1999, 856)

Fairness may be based on the merit principle, a needs rule, an equality rule, or a combination of these factors. Fair procedure and fair distribution of rewards are equally important (Leventhal 1977). Amartya Sen (1995, 2009) has proposed an approach based on "capabilities to function." These capabilities determine a person's well-being and "the person's freedom to lead one type of life or another" (1995, 40). People differ in terms of the freedom they enjoy to transform their resources into the type of functioning they value. Policy should focus on enhancing freedom so that people can choose to use the resources to lead the life they value. The role of reasoning connects the idea of justice to the practice of democracy as "government by discussion." Public reasoning allows divergent views and perspectives to enter the debate and enables a comparative assessment of ways of reducing injustice in the world (Sen 2009).

The relationship between policy and equity in the arena of science and technology however, has not been explored by many scholars in the field (Woodhouse and Sarewitz 2007). Those that have done so emphasize that equity considerations should be central to deliberations regarding science and technology policy. For instance, Cozzens et al. (2002, 102) contend that their "goal is to develop ways to design and evaluate S&T policies and programs so that they reduce inequalities rather than increase them [and] spread benefits widely rather than concentrating them on a few people." A way to do so is by clearly formulating equity goals for S&T policy. They can be framed such that research and development (R&D) addresses the problems of the poor and focuses on the creation of public goods. Policy can even be designed to slow down the pace of development of those technologies that can create new inequities or magnify existing ones so that alternatives can be explored (Woodhouse and Sarewitz 2007). Another way to incorporate equity goals and considerations in policy is by increasing public participation in decision making around science and technology issues. This will allow for a greater diversity of perspectives and values (Crow and Sarewitz 2001; Woodhouse and Sarewitz 2007). Public participation can contribute to a more equitable policy process as well as lead to more equitable outcomes if the views of the participants are taken into consideration in policy formulation. Public participation "can create mutual understanding among scientists and the public, constructively influence the conduct of science in response to evolving ethical norms, and modify the direction of science so that it can better address societal goals and priorities" (Sarewitz 1997, 31). Policy designed in such a fashion can contribute to greater equity in society.

14.3 The NCTF Process

The NCTF focused on converging (or NBIC) technologies for human enhancement. Converging technologies refers to the combination of nanotechnology, biotechnology, information technology, and the cognitive sciences. Researchers in the field believe that this convergence of technologies will transform society as it will lead to "…a tremendous improvement in human abilities, societal outcomes, the nation's productivity, and the quality of life" (Roco and Bainbridge 2002). Human enhancement, or the improvement in human abilities, is often defined by juxtaposing it to the concepts of remediation and therapy. The distinction between therapy and enhancement is problematic as it depends on definitions of normality, and on the concept of health and disease as well as the definition of ability (Wolbring 2008). The transformative power of these technologies to improve human abilities or enhance the human body naturally creates ethical dilemmas in the mind of the public and policy makers. In addition to the desirability of such changes, the cost and access to these technologies are important ethical concerns since unequal access can exacerbate existing divisions in society. An important consideration in these discussions on human enhancement is whether these technologies will contribute to greater social justice. As pointed out by Lin and Alhoff (2008, 259), "…it is important to note the following: advantages gained by enhanced persons also imply a relative disadvantage for the unenhanced, whether in sports, employment opportunities, academic performance, or any other area. That is to say, fairness is another value to consider in the debate."

The National Citizens Technology Forum was organized on six sites across the United States in March 2008. The six sites were Arizona State University, University of California at Berkeley, Colorado School of Mines, Georgia Institute of Technology, University of New Hampshire, and University of Wisconsin-Madison. I was part of the research and facilitation team at Georgia Institute of Technology. Each of the Citizens Technology Forum sites operated independently though they had access to the same experts, read the same background materials, and interacted with each other during the internet sessions. The process combined face-to-face sessions with interactive internet sessions that were a series of 2-h, synchronous, on-line sessions. A panel of content experts was also available during these sessions to answer questions (Hamlett et al. 2008).

Consensus conferences do not seek to have a statistically representative sample of the population as participants but seek broader participation so that a variety of views are brought to the deliberations. The NCTF participants were not strictly a representative sample but were selected to ensure a demographically diverse group in terms of race, gender, income and education levels, political ideology, and party affiliation. There were 72 participants with a median age of 39, and a majority of them had a college degree or graduate degree. Their answers to a series of questions on the survey reveal that the issue of human enhancement was important to the majority; 90% felt that the issue was at least somewhat important to them. The survey also had a number of statements regarding the existing equality in society. These figures provide an indication that for a majority of the participants, a more equitable distribution of wealth and equal rights was desirable and discrimination of minorities was a concern.

14.4 Survey Data

The NCTF participants were administered a pre- and post-survey, the results of which reveal the participant's concerns about equity and corroborate the viewpoints expressed during the deliberations. The pre-survey provides an insight into the participants' perceptions about nanotechnology, government, and science. The cost of the technology was an important concern. The pre-test survey data reveals that 56% of the participants felt that human enhancement technologies would be too costly for the average American and 14% felt that it would be available to only the wealthiest Americans. After deliberation, 56% still felt that human enhancement technologies would be too expensive for the average American, while the number who thought that it would be available to only the wealthiest Americans increased to 25%. This concern about the cost of the technology is also reflected in the participants' views about whether the government should guarantee access to these technologies if these technologies are very expensive. Before deliberation 59% of the participants felt that the government should guarantee everyone equal access to human enhancements. The percentage increased to 64% after deliberation. Although cost was a concern, the majority did not feel that medical insurers should cover enhancements and this percentage decreased post-deliberation. The analysis of the transcripts reveals that participants viewed enhancement technologies as opposed to remediation technologies as elective in nature which can explain why most participants felt that medical insurers need not cover enhancements.

The participants also believed that public input in the use of nanotechnology for human enhancement was important. Pre-deliberation 75% of the participants agreed or strongly agreed that they could contribute to science and technology policy decisions, and an equal percent felt that it is very important for citizens to have a say in how nanotechnology is used for human enhancement. Post deliberation, these numbers increased to 81 and 96%. Pre-deliberation, 72% of the participants felt that the primary goal of government supported research was to find cures to diseases and infections. After a month of deliberation, this number increased to 86%.

Critics of deliberation have highlighted the role that inequalities based on ascriptive characteristics such as race and gender can play in derailing the deliberative process (Mansbridge 1983; Sanders 1997). As pointed out by Sanders (1997, 352), deliberation requires not just equal resources and an equal opportunity to participate, but it also requires equality of what she terms "epistemological authority" or the equal regard of one's arguments. The major problem in deliberation is "how more of the people who routinely speak less-who, through various mechanisms or accidents of birth and fortune, are least expressive in and most alienated from conventional American politics might take part and be heard and how those who typically dominate might be made to attend to the views of others". Research on environmental risk perceptions has also demonstrated that there are race and gender differences in perceived risk (Finucane et al. 2000). These differences in risk perceptions can be explained by the subjective experience of vulnerability of certain groups of people (Satterfield et al. 2004). Since equity is often conceived in group terms, I have tried to see whether group affiliation determines the participants' perception of equity. Do gender, race and income impact perspectives about access and affordability in different ways? Do those who "speak less" have different perceptions of fairness than those who are used to being heard? Was equity a concern for all?

Table 14.1 shows the change in views of different groups of participants regarding access and affordability after deliberations. As can be seen in the table, the deliberations changed the opinions of women and men regarding whether government should guarantee access and whether medical insurance should pay for enhancements but in different directions. Women, in general, were more worried than men about the affordability of the technologies. Affordability was a concern for more white participants than minorities (non white), while a larger percentage of minorities felt that the government should ensure access and that medical insurers should pay for enhancements. A larger percentage of lower income participants felt that the government should guarantee access. In the case of medical insurance paying for enhancements, the percentage of lower income participants who agree with the statements decreased after deliberations, while in the case of higher income participants this percentage increased. Surprisingly, a larger percentage of higher income participants had concerns about affordability as compared with lower income participants.

Table 14.1 Differences in group perceptions pre- and post-deliberation

		Government should guarantee access %	Medical insurance should pay %	At least a little worried about average person's affordability %	At least a little worried about your and your family's affordability %
Males	I	54	43	71	83
	II	69	31	60	74
Females	I	62	38	78	95
	II	59	43	65	78
White	I	48	27	84	91
	II	57	36	70	82
Nonwhite	I	75	61	61	86
	II	75	38	50	68
Low income	I	68	49	68	84
	II	73	35	54	70
High income	I	49	31	83	94
	II	54	40	71	83

I= Pre-deliberation; II= Post-deliberation

All the percentages in Table 14.1 give a raw probability that is not causal. Given their membership in a certain group, how much of the participants' probability of giving a certain answer is explained by those characteristics alone? In order to see whether socioeconomic characteristics have an impact on views regarding access, affordability, and public involvement, I ran a logistic regression with and without robust standard errors on a number of simple models to arrive at some preliminary conclusions that can guide further research. Table 14.2 reports the odd ratios for the models with robust standard errors. However, there were no differences between the models with or without standard errors in terms of the size and significance of the coefficients between the two sets.

My hypothesis is that socio-economic characteristics would have an impact on how the participants view access, affordability, and the importance of citizens' opinions in how nanotechnology develops. I wanted to also see the change, if any, after deliberation. The results show that only a few of the socio-economic variables were statistically significant, but they were consistently so across the different models. Race and income were the statistically significant variables. It should be expected that low income individuals and nonwhites would be more concerned that without safeguards policies would tend to favor the majority or the economically strong. Therefore it is important to build such safeguards into policymaking and to make sure they function in the way intended. This also has implications for deliberative exercises. Group dynamics require that careful attention is paid to the design and facilitation of deliberations. Mendelberg and Karpowitz (2006) have shown how decision rules, such as rule of the majority or consensus and the gender composition of the participants, shapes deliberation and has a bearing on the outcomes of

Table 14.2 Effects of Socio-economic characteristics on equity perceptions

Independent variables	Model 1 Government should guarantee access	Model 2 Medical insurers pay	Model 3 Affordability for average person	Model 4 Affordability for self and family	Model 5 Citizen involvement
Predeliberation					
Gender	1.35	0.71	1.35	2.7	0.63
Education	1.00	0.45	1.41	3.92	2.10
Party affiliation	1.08	1.77	1.70	2.29	1.90
Political ideology	1.31	0.96	0.41	0.53	0.9
Race	3.47**	4.32***	0.25**	0.54	1.28
Income	2.40*	2.38	0.44	0.40	0.88
Postdeliberation					
Gender	0.67	2.32	1.12	1.05	0.67
Education	∧	0.07**	2.05	1.66	3.18
Party affiliation	2.15	0.80	1.40	1.97	0.74
Political ideology	1.29	1.87	0.79	0.65	0.42
Race	2.37	1.13	0.37*	0.41	6.53***
Income	2.36	0.61	0.49	0.53	1.34

Notes: Cell entries are odd ratios with robust standard errors
Levels of significance: * = 10%, ** = 5%, *** = 1%
∧ Education was dropped from this regression due to perfect correlation

deliberation. In the NCTF deliberations, the preliminary statistical analysis reveals that not gender but race and income had an impact on perceptions of equity. The analysis of the internet transcripts and the final reports gives a better sense of these interactions.

14.5 Equity and Internet Sessions

The survey data gives an indication of the participants' views about affordability and access, but the transcripts of the internet deliberations provide a more nuanced understanding of the participants' concerns about the effect of NBIC technologies on the existing inequalities in society as well as the ways in which these can be mitigated. The transcripts[1] of the internet sessions reveal that equity was an important concern that was common to all the sites during the face-to-face deliberations. The first few internet sessions were spent voicing the concerns and initial reactions of all the teams so as to uncover the commonalities across all six sites during the face-to- face deliberations. Equity was a common theme for all the sites. The transcripts reveal that there were marked similarities at all six locations while discussing issues related to equity. The analysis of the deliberations' internet sessions has interesting policy implications. Participatory policy making does bring ethical issues to the forefront, and citizens actively seek to shape a more equitable world through their recommendations.

Equity was defined by the participants in terms of access to the technology. During the internet discussions the participants grappled with the question of what accessibility of the technology means and discussed whether it is even practical to talk about accessibility for all. As one of the participants put it during their discussions: "The discussion is futile as inequality is inevitable," Another stressed the benefits but also the difficulties: "I agree that equal access is the better, but I don't think it's realistic in today's society." There was a recognition that the existing inequalities in society make access for all a difficult principle to implement. But the high cost of technology combined with its transformative potential was seen as a factor that would expand the existing gap between the "haves" and "have-nots" in society.

At one point in the deliberations, one of the participants asked the moderator to define accessibility who described it as "the ability of all people to have equal access to technology." The participants used this definition as a starting point to unravel the concept further. The criteria used to decide the availability of the technology to people was also a point of debate. How to ensure equity of access? The impact of the technology on the quality of life was an important decision factor. As one participant remarked, "How do you measure what is a change for the better? Equal access is better." While there was no decisive definition of equity that was agreed upon, most of the participants recognized that existing inequalities in society were a problem that has to be addressed. As one of them argued, "we need to make the inequality gap smaller not larger and nanotechnology will be a big determinant." Another participant added to the discussion:

> While it may be inevitable (it already exists), we can try to mitigate its effects. We can slap down the insurance industry now so they don't block entire segments of the population from getting coverage. We can try to predict scenarios where new treatments would be available and offer ways to expand those treatments.

The participants recognized that these steps would not be easy.

The notions of accessibility and equity were seen as closely intertwined with the relations of power. Who decides who gets what? Questions of funding and the organization of the research enterprise itself were an integral part of the arguments about equity. Even when discussing regulation, an important concern was who will control funding of the regulators and will they have an agenda? There was an awareness of the differential distribution of power and choice in society. Who decides what the betterment of society is? Who decides who can obtain the advances? Who are the regulators? Who decides who gets what? Lewenstein has argued that ethical issues:

> all involve questions of fairness, equity, justice, and especially power in social relationships. That is what makes them 'ethical' issues. In each case, not only are legitimate questions possible about how nanotechnology research and application should develop, but even more fundamental questions exist about how to make decisions and who should control those

decisions. These fundamental questions are asking about the source of power in societies with unequal social distributions of power (2005, 14).

The transcripts show that power relations, the source of power of the regulators, and the allocation of funding were seen as linked to the issue of accessibility, affordability, and equity. Fair process was felt to be as important as fair outcomes. Even if the power to regulate is unequally distributed, fair procedures can lead to more equitable outcomes. One of the participants emphasized this point by making the following statement: "It does not matter who is regulating it as long as the basis of how they regulate it is fair and accurate."

The funding of technology development was seen as an important factor in access to the technology. Does the largest contributor of funding for the development of technology get an equivalent share in the decision-making, or can the public have a say in the development of the technology? The discussion on funding and equity rested on an assumption that an increased participation by the private sector in the development of these technologies would lead to greater inequity as money would become the driving force in access to the new technology. One of the participants mentioned that one of the major concerns during the face-to-face deliberations was the question, "will money be the driving force in access to new technologies?" In the participants' view, development of these technologies by the public sector ensured public access to the technologies, while a dominant role for the private sector led to "personal access" to the technologies by those who could afford the high prices. The involvement of monopolies and big businesses in the development of these technologies was seen as resulting in the "haves" getting preference. One of the participants saw the capitalist system as a problem because profit rather than need becomes the driver of research:

> The capitalist system itself poses a concern. For potential products in the private sector, which would include health care, the research funds allocation tends to be very strongly influenced by profit potential. This would direct funds to areas where the biggest profit potential is, rather than the actual needs of society. The direction of research activity would also be heavily weighted toward military applications and other matters considered politically desirable.

Public involvement and public input on funding priorities were seen as important factors that could counteract the inequitable results of NBIC technology. One of the participants stated, "I'd like to know if 'We the People' can dictate to science what problems we want them to solve: i.e. Not a dime for Super Soldiers until you fix Downs [Syndrome] or some such." Another agreed with the statement and said, "wouldn't we also want to know if a process is in place (beside what we are doing [NCTF]) so that the public can give input". Participants argued for greater decision making power for the public regarding NBIC research and development. Many participants argued for more public involvement in the process of allocating public funds and for ensuring that all ethnic groups and races, including those who are under-represented such as Native Americans, have a say.

> As a member of a sovereign government that does not have adequate representation in Congress or at any governmental level for that matter but yet holds over 1/4 of the land mass

of Arizona and only 1.5% of the U.S. population and believes that nanotech is harmful to the planet . . . how do we get heard?

Public involvement in decision making requires not only, equal opportunities for all races and ethnic groups to voice their views, but also to be heard. An equal consideration of different viewpoints, marginal and mainstream, is essential for democratic decision making.

The discussions also revealed that the participants envisioned a larger role for government in the formulation of regulation to protect a new category of individuals comprising of those who were "unenhanced" from discrimination. Human enhancement technologies could create "classism" or reinforce the already existing underclass (see Chapter 5, this volume). If we want to prevent discrimination, we would have to create new laws to protect the unenhanced or the unaltered person, as well as those with "undesired" genetic traits. One of the participants stated that we may "have to create laws to protect them like we currently have hiring handicapped or minorities, because at some point an unaltered person will be a minority." Part of the discussions that dealt with concerns regarding cost and access also focused on the role of the health insurance sector. Participants were of the view that the government may need to put into place laws so that insurance companies do not deny coverage to those who have been detected with incurable diseases, using the new technologies.

The discussions also reveal that many of the participants felt strongly about the need to preserve diversity in society. One of the participants felt that by "nurturing the stronger gene we may inadvertently be harming the balance in nature." Another questioned, "What is the level of genetic diversity that is necessary for the health of a population? Is eliminating undesirable genetic effects inadvertently disadvantageous? What would, or who would, decide what is desirable?" There was a concern that testing on humans may lead to a possible exploitation of those disadvantaged by their circumstances. It may lead to selective participation and/or secret testing on specific ethnic groups. It was also deemed important to include all ethnic groups and minorities in testing and clinical trials so that the effects on all groups could be studied.

The lack of a specific time span within which most of the envisioned technologies could come to market was a source of frustration for many participants. One of the participants talked about the fact that experience with earlier technologies like information technology has shown that costs go down the longer a technology has been in the market making it more accessible in terms of affordability. But in the case of nanotechnology there is no knowledge of how long will it take for these technologies to become more accessible. There was also a belief that more cost effective and ethical alternatives to the technologies should be explored.

A participant noted that there seems to be little concern with the applications of these technologies for broader social purposes rather than individual enhancements. Another said that "I'd like to see more overall emphasis in nanotechnology on societal issues, as opposed to individual ones." The transcripts reveal that many participants emphasized the need for equal consideration of the social consequences of the technology. Accessibility of enhancement was viewed as having more social impacts while remedial accessibility was described as more of a health care issue. Another concern was access to technological information that was not only balanced, but also understandable to the lay person so that the public could be better informed. As a participant stated, "One of our group's concerns was balanced information. Do we know that there is currently full disclosure of available technology?"

This broad range of concerns covering access, equity, regulation, and public involvement was present in the questions that the participants formulated for the experts. The panel of experts consisted to five members, one of whom was an ethicist while another dealt with ELSI issues. Since the participants were aware of the backgrounds, it is natural that the questions that were put to these two experts dealt with ethical concerns such as equity, autonomy, and choice. But analysis of these questions reveals how varied the ethical concerns were and how centrally equity featured in them. The questions that pertained directly with equity were:

> What is the public policy process to ensure access to NBIC technologies for underserved populations in the US, and more broadly outside the US?
> Is the goal of NBIC solely monetary or for the betterment of humanity?
> How do our differences actually make us much more valuable than would otherwise?
> Can priorities for funding focus on real needs and involve the public and how can people of all races be heard?
> Do we as a society have an ethical obligation to focus limited resources on technology that affect broader groups of people and how do we ensure the same opportunities for those with less?
> Will increasing the diversity in scientists and engineers lead to a greater thinking of ethics?
> The best way to allocate resources is the fairest way and allocation of tax payer money should be done in an open environment?
> Is there monitoring of research using lower economic individuals in foreign countries?
> How do you include underrepresented groups in research and testing?
> What is the best way to distribute new developing technology that can have elevated prices that make access difficult but could potentially aid individuals in a multitude of ways?
> With the decision making process dominated by interest groups and lobbyists who is going to protect the general public?

The variety and the number of questions that the participants had regarding equity and its interaction with regulation, funding decisions, research agendas, and the organization of the research enterprise, and composition of the scientific workforce, show that the participants were very concerned about the implications of NBIC technologies for equity in society. It was an issue which engaged them right through the month long deliberations till the formulation of the final reports.

14.6 Final Reports

The final reports were written over the last weekend during the face-to-face session. Participants at each site wrote their own report, so there are six reports that represent the consensus view of participants at their respective sites. A review of these reports shows how the concerns expressed during the deliberations were transformed into policy recommendations. The reports are not restricted to the frame of the background materials for the recommendations of the groups cover environment risks and regulation too (Philbrick and Barandiaran 2009, Hamlett et al. 2008). All six reports talked about the importance of equitable access to the technology, and five reports made specific recommendations regarding equitable access to these technologies. Participants at Arizona believed that "all people should have equitable access to the benefits of these technologies." The California report also stated that equitable access to the benefits of the technology should be ensured, while the public's exposure to the risks is minimized. The Colorado report stated that, "everyone, regardless of socioeconomic or cultural status, deserves equal access." The Georgia group was concerned that the high cost of the technology would limit access to the technology, which in turn would amplify existing inequalities in society. The New Hampshire report recommended that access to these technologies must be guaranteed to those who need them most with need defined in terms of the impact the therapy will have on the quality of life rather than an individual's desire for enhancement or availability of financial resources. The Madison report stated that nanotechnology development "should strive to benefit the greatest common good and address basic societal needs." The group recommended that policies need to be formulated to address the possible widening of existing social and economic divisions on account of the cost of the technology, both nationally and globally.

Another factor common to all reports was the recommendation that NBIC technologies should be developed for the collective good rather than individual benefit. The survey results also show that a majority of the participants believed that funding should be directed to remediation purposes rather than for enhancement technologies. The funding of nanotechnology research as well as health insurance reform were other common issues. The recommendations included an allocation of funds such that products which are "widely distributed and easily affordable" are promoted; funding should remain both public and private to avoid monopolies; and a system to prioritize the allocation of funds and increase the publics' ability to have a say in how funds for non-military research are allocated should be established. Public involvement in funding decisions was seen as a way to ensure that funds are utilized for addressing basic social needs. The California report recommended that the "government introduce methods for increasing stakeholders' ability to have a say in how funds for non-military research are allocated. By stakeholders we refer to the public, NGOs, and others that represent the public interest." In addition to a say in funding allocations, many participants had talked about the need to ensure greater public involvement in the development of this technology as well as ensuring public access to balanced information regarding the technology and its development. This is reflected in one of the recommendations made in the Colorado report:

We believe that an informed public can alter the course of this technology, so as to avoid the possible disastrous outcomes of a technology which runs rampant without proper regulation, and to ensure that nanotech is used for the greatest good for the greatest number.

The concerns expressed during the internet deliberations regarding testing, preservation of human diversity, ensuring minority groups have a say in decision making, and enacting new laws to protect the unenhanced from discrimination, all found an expression in the final reports. The California report also talked about the development of alternative low risk and equitable technologies before allocating funds to enhancement technologies.

Woodhouse and Sarewitz (2002) present six categories of science policy that may be effective in reducing inequity: research focused on the problems of the poor and marginal; broadening participation in policy making to include a diversity of perspectives; public funding of research to focus on the creation of public goods; reducing the cost of goods and services already available in the market; greater transparency about the societal impacts of scientific inquiry; and slowing the rate of technological development to allow for developing policy to deal with the issue of equity. Despite the fact that this article was not a part of the background materials provided to the participants, and no one in the group had studied nanotechnology and equity, examples from all of these categories are found in the policy recommendations of the NCTF participants demonstrating that the informed public can develop innovative policy solutions to deal with equity concerns. The participants were aware that the benefits of science are not equally available to all; new technologies can benefit some groups more than others; and all groups do not have the same access to decision making. Their deliberations and recommendations were broad ranging in their treatment of equity as a goal in science policy and were not constrained by the frame provided by the background materials.

A common criticism of deliberative exercises is that they are consciously manipulated by organizers to produce the desired results. The background materials can be the source of such undue influence on the deliberations. Background materials and information are important in consensus conferences, for the process depends on the participation of not lay persons, but an informed public. The background materials do influence the deliberations as they provide a frame for participants to think about and organize the issues. The important point to consider is whether these background materials unduly influence the deliberations. The impact of framing on public perceptions of nanotechnology has been studied by researchers with differing results. Cobb's (2005) study shows that framing nanotechnology as associated with specific risks resulted in a higher proportion of respondents who thought that risks outweighed benefits but when nanotechnology was framed negatively without mentioning specific risks he found no effect of framing. Macoubrie's (2006, 237) analysis shows that the concerns of individuals "are not simply attributable to the information with which they were provided".

The background materials for the NCTF were prepared keeping in mind that they had to be "accurate, balanced, and accessible to ordinary people" (Hamlett et al. 2008). The background materials were balanced not only in terms of the content but also in that no one viewpoint was espoused. A section of the background materials was devoted to ethical and policy issues that highlighted disputed themes such as "remediation versus enhancement", the impact of NBIC technologies on the concepts of human nature, and unpredictable consequences of the technology. A section dwelled on the issue of fairness which was defined in terms of cost and access leading to widening of the gap between the "haves" and "have-nots". Philbrick and Barandiaran's (2009) analysis shows that the background materials did not unduly influence the NCTF participants. While the background materials were balanced in terms of the coverage of risks or concerns and benefits; the final reports focused on health, environment issues, and ethical issues rather than benefits. The participants did not frame their deliberations only in terms of the background materials. Though the issue being deliberated in consensus conferences is framed by organizers and experts, the participants have control over the agenda and the final reports (Cronberg 1995). The time spent in discussions of equity as well as the recommendations regarding equitable access to these technologies demonstrates that the participants were concerned about the issue. Equity was framed by the participants, not only in terms of access to the technologies, but also in terms of the ability to use the technologies; it was seen as not only dependent on fair distribution of resources but also of power. Participants were aware that existing social stratification and social structures interact with new technologies and reinforce existing inequalities.

14.7 Conclusions

Preliminary statistical analysis shows that socio-economic factors do influence equity concerns, and there was a clear differential influence across groups of participants. The hypothesis that socio-economic characteristics would have an impact on how the participants view access and affordability was examined pre and post deliberation. The results show that race and income were consistently significant in their impact on equity considerations. Even though education is closely associated with race and income in the general population, it did not have a similar effect. The other surprising finding was that gender did not have significant explanatory power. Whether the lack of explanatory power of some of these variables was due to unobserved characteristics of the sample or the manner in which the categories were drawn is a matter of further study. Also, the participants form only a very small sample and it may not be correct to extrapolate these findings to the whole population. Despite the problem of generalizing from such a small sample, these tentative findings do have implications for the design of deliberative exercises. For instance, Karpowitz et al. (2009) suggest that marginalized or disenfranchised groups should be provided a space for enclave deliberation within a participatory forum so that the interests of the marginalized group are specifically articulated.

The analysis of the internet transcripts and final reports reveals that equity was an important concern of the participants' at all six sites. The emphasis on consensus in the design of public participation has been criticized as it may lead to results that are an expression of the majority interests. In the case of NCTF, 81% of the participants said that they personally endorse almost every major point in their group's final report, while a mere 2.9% personally objected to many of the major points in the final report of their site (Hamlett et al. 2008) demonstrating that consensus was not achieved at the cost of silencing the marginal voices. The consensus view expressed in the final reports shows that equity was a concern shared by all participants. The comparison of pre- and post deliberation survey data demonstrates that public perceptions of equity can be shaped through reasoned discussion amongst participants with diverse viewpoints. Differences in perceptions exist, and the opportunity to initiate dialogue can help narrow these differences as well as allow divergent views to enrich the debate. The diversity of perceptions was not bounded by the background materials or the agenda set by the organizers. The participants revealed a nuanced notion of equity and a deep understanding of the complicated relationship between science, society, and inequality. They were aware of the opportunities and threats to equity posed by nanotechnology as well as the uncertainty involved.

Their recommendations can be an invaluable tool for policymakers as they present a number of ways of formulating policy that seeks to achieve greater equity in society. These policies can be oriented to increasing diversity in the science and technology workforce; formulating a system of prioritization of funding allocations based on social needs; guaranteeing access to these technologies to the most needy with the latter defined as those whose quality of life will be the most impacted; ensuring a balance of public and private funding; development of less costly technologies; and greater involvement of a more informed public.

This paper tries to bring into focus the importance of equity as a value to the participants. It was a value expressed by them during the deliberations and was reflected in the recommendations formulated by the participants. Equity was an expressed preference of the participants and a pressing concern for action. These recommendations should be taken as an input into the decision making process by the policy makers. Opening up science and technology decision making to the public promotes the expression of what Cozzens (1996, 184) terms "quality of life" goals for science that are of two types, "what-goals" such as education and health, and "how-goals" such as equity and democracy. "The objective has shifted from simply developing technology to doing so responsibly, and the goal of wealth has been transformed to that of prosperity." The participants were not as concerned with the economic returns that nanotechnology promises but saw these quality of life goals as important objectives that science should deliver. Of these, equity was an important concern for the public and has to be taken into consideration by policy makers in decision making for the latter to be responsive to and reflective of the concerns of the public.

Acknowledgments This paper reports findings from the National Citizens Technology Forum (NCTF) held in March 2008. The research was funded by the National Science Foundation (NSF) and the Center for Nanotechnology in Society at Arizona State University (CNS-ASU) (grant # 0531194). I would also like to thank Dr. Michael Cobb, North Carolina State University, and Dr. Clark Miller, Arizona State University, for the survey data.

Note

1. There were nine internet or K2K (keyboard to keyboard sessions), lasting two hours each, interspersed between the two face to face sessions. During the internet sessions participants from all six sites were grouped into different teams, with each team having members from all the sites. I coded these transcripts for discussions about equity.

References

Bennett, Ira and Daniel Sarewitz. 2006. Too little, too late? Research policies on the societal implications of nanotechnology in the United States. *Science as Culture* 15(4): 309–325.
Bucchi, Massimiano, and Federico Neresini. 2008. Science and public participation. In *The handbook of science and technology studies*, 3rd ed., ed. Edward J. Hackett, Olga Amsterdamska, Michael Lynch, and Judy Wajcman, 449–472. Cambridge, MA: MIT Press.
Caplan, Ronald L., Donald W. Light, and Norman Daniels. 1999. Benchmarks of fairness: A moral framework for assessing equity. *International Journal of Health Services* 29(4): 853–869.
Cobb, Michael D. 2005. Framing effects on public opinion about nanotechnology. *Science Communication* 27(2): 221–239.
Cohen, Joshua. 1997. Deliberation and democratic legitimacy. In *Deliberative democracy: Essays on reason and politics,* ed. James Bohaman, and William Rehg, 67–92. Cambridge, MA: MIT Press.
Cozzens, Susan. 1996. Quality of life returns from basic research. In *Technology, r&d and the economy,* ed. Bruce L.R. Smith, and Claude E. Barfield, 184–209. Washington, DC: The Brookings Institution and the American Enterprise Institute.
Cozzens, Susan E, Kamau Bobb, and Isabel Bortagaray. 2002. Evaluating the distributive consequences of science and technology policies and programs. *Research Evaluation* 11(20): 101–107.
Cronberg, Tarja. 1995. Do marginal voices shape technology? In *Public participation in science: The role of consensus conferences in Europe,* ed. Simon Joss, and John Durant, 125–133. London: Science Museum.
Crow, Michael, and Daniel Sarewitz. 2001. Nanotechnology and societal transformation. In *Societal implications of nanoscience and nanotechnology,* ed. Mihail C. Roco, and William S. Bainbridge, 55–67. Dordrecht: Kluwer.
Delborne, Jason A., Ashley A. Anderson, Daniel Lee Kleinman, Mathilde Colin, and Maria Powell. 2009. Virtual deliberation? Prospects and challenges for integrating the Internet in consensus conferences. *Public Understanding of Science.*
DeLeon, Peter. 1988. *Advice and consent.* New York: Russell Sage Foundation.
Dryzek, John S. 1990. *Discursive democracy.* New York: Cambridge University Press.
ETC. 2004. *The little big down: A small introduction to nano-scale technologies.* Winnipeg, Canada: Action Group on Erosion, Technology, and Control. http://www.etcgroup.org. (accessed August 6, 2010).
Finucane, Melissa L., Paul Slovic, Mertz C.K., James Flynn, and Theresa A. Satterfield. 2000. Gender, race and perceived risk: The "white male" effect. *Health, Risk, and Society* 2(2): 159–172.
Fischer, Frank. 1993. Citizen participation and the democratization of policy expertise: From theoretical inquiry to practical case. *Policy Sciences* 26(3): 165–187.
Fisher, Erik, and Roop Mahajan. 2006. Nanotechnology assessment: Contradictory intent? U.S. federal legislation on integrating societal concerns into nanotechnology research and development. *Science and Public Policy* 33(1): 5–16.
Guston, David H. and Daniel Sarewitz. 2002. Real-time technology assessment. *Technology in Society* 24(1–2): 93–109.

Hamlett, Patrick W., Michael D. Cobb, and David H. Guston. 2008. National citizens' technology forum: Nanotechnologies and human enhancement. Tempe, AZ: The Center for Nanotechnology in Society: Arizona State University.

Jasanoff, Sheila. 2003. Technologies of humility: Citizen participation in governing science. *Minerva* 41(3): 223–244.

Karpowitz, Christopher F., Chad Raphael, and Allen S. Hammond IV. 2009. Deliberative democracy and inequality: Two cheers for enclave deliberation among the disempowered. *Politics and Society* 37(4): 576–615.

Kleinman, Daniel Lee, Jason A. Delborne, and Ashley A. Anderson. 2009. Engaging citizens: The high cost of citizen participation in high technology. *Public Understanding of Science.*

Leventhal, Gerald S. 1977. What should be done with equity theory? New approaches to the study of fairness in social relationships. http://www.eric.ed.gov/PDFS/ED142463.pdf. (accessed August 6, 2010).

Lewenstein, Bruce. 2005. What counts as social and ethical issues in nanotechnology? HYLE *International Journal for Philosophy of Chemistry* 11(1): 5–18.

Lin, Patrick, and Fritz Alhoff. 2008. Untangling the debate: The ethics of human enhancement. *Nanoethics* 2(3): 251–264.

Macoubrie, Jane. 2006. Nanotechnology: Public concerns, reasoning and trust in government. *Public Understanding of Science* 15(2): 221–241.

Mansbridge, Jane. 1983. *Beyond adversary democracy.* Chicago: University of Chicago Press.

Mendelberg, Tali, and Christopher F. Karpowitz. 2006. How people deliberate about justice: Groups, gender, and decision rules. http://www.princeton.edu/~talim/RosenbergFeb82006.pdf. (accessed November 19, 2009).

National Science and Technology Council, Committee on Technology, and Interagency Working Group on Nanoscience, Engineering and Technology. 2000. National nanotechnology initiative: Leading to the next industrial revolution, supplement to President's FY 2001 budget. Washington, DC: NSTC.

Philbrick, Mark, and Javiera Barandiaran. 2009. National citizens' technology forum: Lessons for the future. *Science and Public Policy* 36(5): 335–347.

Powell, Maria, and Daniel Lee Kleinman. 2008. Building citizen capacities for participation in nanotechnology decision-making: The democratic virtues of the consensus conference model. *Public Understanding of Science* 17: 329–348.

Roco, Mihail C., and William Sims Bainbridge eds. 2002. *Converging technologies for improving human performance: Nanotechnology, biotechnology, information technology and cognitive science.* NSF/DOC-sponsored research. Arlington, VA: National Science Foundation.

Sanders, Lynn M. 1997. Against deliberation. *Political Theory* 25: 347–376.

Sarewitz, Daniel. 1997. Social change and science policy. *Issues in Science and Technology* 13(4): 29–32.

Satterfield, Terre A., Mertz C.K., and Paul Slovic. 2004. Discrimination, vulnerability, and justice in the face of risk. *Risk Analysis* 24(1): 115–129.

Schot, J, and A Rip. 1997. The past and future of constructive technology assessment. *Technological Forecasting and Social Change* 54(2–3): 251–268.

Sen, Amartya. 1995. *Inequality re-examined.* Cambridge, MA: Harvard University Press.

Sen, Amartya. 2009. *The idea of justice.* London: Allen Lane.

Stone, D. 1997. *Policy paradox: The art of political decision making.* New York: Norton, W.W. & Co. Inc.

Woodhouse, Edward, and Daniel Sarewitz. 2007. Science policies for reducing societal inequities. *Science and Public Policy* 34(2): 139–150.

Wolbring, Gregor. 2008. Why NBIC? Why human performance enhancement? *Innovation: The European Journal of Social Sciences* 21(1): 25–40.

Wynne, Bryan. 1996. Misunderstood misunderstandings: Social identities and public uptake of science. In *Misunderstanding science? The public reconstruction of science and technology,* ed. Alan Irwin, and Bryan Wynne, 19–46. Cambridge: Cambridge University Press.

Chapter 15
Equity and Participation in Decisions: What Can Nanotechnology Learn from Biotechnology in Kenya?

Matthew Harsh

Discussions of equity and technology often focus on the distribution of risks and benefit. But key to addressing these issues is how power to make decisions that shape those issues is distributed. In the last chapter, Bal reported that members of the American public expressed a desire for equity once they were given a voice in nanotechnology decision making. In this chapter, Matthew Harsh looks at how biotechnology regulation was developed in Kenya to help think about how nanotechnology might be regulated in a developing country. A variety of horizontal inequalities criss-cross his account, but they are different from the categories presented in the first part of this book: Europeans versus Kenyans; government versus non-government; and experts versus lay people. The agora is not open; existing power constellations are built into access to meetings and where decisions are made. Harsh shows that the ultimate distribution of risk is certainly not influenced most by those who are affected most, namely, those who live in rural areas. The lesson is a sobering one with regard to both the complexity and obdurateness of politics. Harsh argues that if equity concerning the effects of nanotechnology is to be achieved, equity of access to decision making about nanotechnology will have to come first.—eds.

M. Harsh (✉)
Center for Nanotechnology in Society, Arizona State University, Tempe, AZ, USA
e-mail: mharsh@asu.edu

This chapter was peer reviewed. It was originally presented at the Workshop on Nanotechnology, Equity, and Equality at Arizona State University on November 22, 2008.

15.1 Introduction

When examining news about the latest developments in nanotechnology, it appears that nanotechnology offers ever-increasing promise for developing countries. For example, nanotechnology is being used to develop a coating for sandy soil so that the soil will retain more water and be more fertile (Baldwin 2009). Nanotechnology could provide better water filtration (Dickson 2009), enable easier HIV testing (Gill 2009), and produce more options for generation of energy (Bourzac 2009). These technologies, it is said, will benefit (poor) people living in developing countries. However, as this volume shows, even if nanotechnologies such as these become technically feasible, it is unlikely that the benefits of new nanotechnologies will be distributed evenly. Distributional consequences of new nanotechnologies raise questions about equity and equality: Who benefits more from these technologies and who less? Who is susceptible to more risk or less risk? Who *should* benefit more?

While distributional outcomes of nanotechnologies are undoubtedly important, the intent of this essay is to examine equity and nanotechnology in a converse way (see Fig. 15.1). I focus on unequal inputs to the nanotechnology development process. By inputs, I am not referring to the material or labor inputs used in technology creation, although there are certainly equity issues surrounding these kinds of inputs (see chapters in Part I of this volume). Rather, I am referring to *decision inputs* that guide and shape the nanotechnology development process (see the top half of Fig. 15.1). Conceived broadly, these are decisions about which technologies to develop and how. As opposed to questions about who benefits and who is at risk, the equity and equality questions here are: Who decides? Who should decide? For instance in the case of the nano-coating for soils mentioned above, relevant questions include: How much have farmers who might use the coating in their arid fields taken part in decisions about the development of the coating? Will they take part in decisions about how it might be deployed in their region or regulated by their national government? How much should they be involved in all these decisions? How does that compare with how much scientists, environmental organizations, or manufacturers take part in these same decisions?

The goal of this essay is not to give direct answers to these types of questions. Rather, this essay looks laterally across questions about participation in decision making from a perspective of equity. The essay asks: How does focusing on equity help us understand these important questions about "who decides"? Or more explicitly: How is equity relevant to analysis of participation in decision making about nanotechnology in developing countries or regions?[1]

I must state a few caveats about context and comparison before going forward. Context is utterly important—the answers to questions about who decides and who should decide will be different depending on the specific nanotechnology and the specific country or region in which it is to be developed. Context is especially important for this essay because I use a comparative method. I draw out lessons for nanotechnology by reflecting on recent developments in agricultural biotechnology in Kenya. The nanotechnology-biotechnology comparison has been made many times before (see for instance Mehta 2004; David and Thompson 2008)

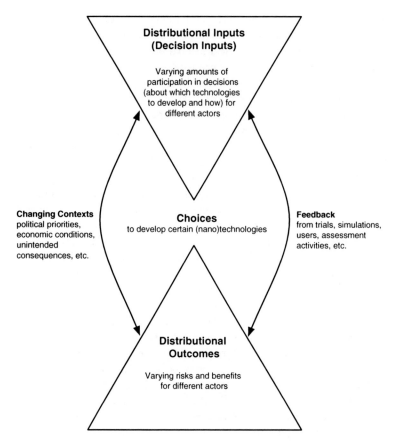

Fig. 15.1 A distributional view of technology development

and has its shortfalls (Mody 2007). Comparisons usually focus on how nanotechnology and biotechnology are technically similar and different and how probable impacts of the technologies vary in terms of risks and benefits. The overall goal of such comparisons is often to try to ensure that public reactions to nanotechnology are less adverse than previous reactions to biotechnology, particularly to biotechnologies involving genetic-modification (GM). Making these comparisons between nanotechnology and biotechnology is fraught with difficulty, not least because of the broadness and still emergent nature of both nanotechnology and biotechnology. Furthermore, previous nanotechnology-biotechnology comparisons have also not focused on developing countries or regions. In light of these criticisms, this essay avoids technical comparisons and comparisons of claimed impacts. It instead focuses on *process* and decision making that leads to technology development. The choice to focus on decision making is partly due to the fact that it is still too early to judge many of the impacts of agricultural biotechnologies in Kenya, particularly GM technologies, because no GM crops have been commercialized in Kenya at the time of writing.

The essay presents two arguments about how equity is relevant to analysis of participation in decisions about nanotechnology in developing regions. The first is that equity is relevant because there are clear inequalities surrounding access to decisions about emerging technologies. Experience from biotechnology in Kenya shows that who can and cannot access decisions depends on the vertical dimension of inequality (inequalities between the rich and poor) and the horizontal dimension (inequalities between groups distinguished by culturally-defined differences such as race, gender, and religion) (Cozzens and Wetmore 2008). Horizontally, the Kenya case shows how science itself becomes a basis for culturally-defined difference that shapes the distribution of access to decisions. Furthermore, focusing on equity helps address criticisms of other approaches to analyzing participation in decision making, which can lose sight of differential agency resulting from varying access to resources or from social and cultural hierarchies. The lesson for nanotechnology is that, like all new technologies, nanotechnologies will be developed among existing inequalities that affect access to decisions. However, especially to the extent that nanotechnology utilizes more complex and advanced science, it could create a new basis for unequal access to decisions based on unequal access to new scientific knowledge.

While uneven access to any given decision is clearly important, the possibilities for how, and if, certain decisions can even occur are also strongly influenced by institutional constraints. My second argument is that focusing on equity can help us understand how institutional mechanisms[2] structure decisions and thus influence access and participation. Different institutional mechanisms lead to different distributions of when and where decisions occur. Compared to the linear institutional mechanisms of the Green Revolution model of agricultural research and development, new partnership models at work in Kenya have offered some actors more access to decisions. However, this increased access could be considered to be just or unjust depending on the theory of distributive justice adopted. The lesson here for nanotechnology is that it is important to look beyond the more headline-catching technological innovations and focus on the perhaps more mundane realm of changing institutional structures behind nanotechnology developments, because these will shape participation in decisions. Any new institutional mechanisms must in turn be examined in historical context. Previous institutional trends have led to uneven development of biotechnology in Africa, and the same could be true for nanotechnology.

In the conclusion I reflect on how focusing on unequal inputs helps provide a more complete picture of equity. It is also a step towards understanding how governance can be more anticipatory (Barben et al. 2008) and technology assessment might be more real-time (Guston and Sarewitz 2002) in development contexts.

15.2 Distribution of Access

Ideas about *access* are the key to my first argument about how equity is relevant to analyses of participation in decisions about emerging technologies in developing regions. Discussions of equity generally begin by examining something that people value (for example, something tangible such as food or intangible such as education) and a population or society among which this valued or desired thing is unevenly distributed (Cozzens et al. 2008). Access to decisions about emerging technologies is such a thing. Experience from agricultural biotechnology in Kenya shows how access to decisions is both highly valued and highly unevenly distributed.[3]

The development of agricultural biotechnology involving GM in Kenya over the last two decades[4] embodies the trajectory of an emerging technology, especially as related to uncertainty. In the early emergence period of GM in Kenya (in the 1990s), high levels of technical uncertainty existed about whether certain technologies could actually be developed. For instance in 1991, Kenya began a project to create a GM virus-resistant sweet potato. The project involved a partnership between the Kenya Agricultural Research Institute (KARI), the United States Agency for International Development, the private company Monsanto, and later a non-governmental organization (NGO) called the International Service for the Acquisition and Application of Agricultural Biotechnologies (ISAAA) (ABSP 2002).[5] Much has been written about this project in the academic literature and the international media, as it was the first attempt to develop and cultivate a GM crop in East Africa, and in all of sub-Saharan Africa except South Africa (New Scientist 2004; Odame et al. 2003a, b; Qaim 1999). The initial research and development was performed in the United States, and in 2000 GM plants were transferred to Kenya for confined testing. By 2003 tests showed that the sweet potato plants failed to be resistant to Kenyan viruses because they were engineered to be resistant to American strains of the viruses (Cohen and Paarlberg 2004; Gathura 2004 Interview with former KARI scientist, 2005). As late as 2003 then, there was still uncertainty about the technical viability of the GM sweet potato.

Technical uncertainty surrounding GM in Kenya has decreased since 2004, yet other intimately connected non-technical uncertainties have not decreased as quickly. The sweet potato project has been regrouping its technical efforts, attempting to create a new cultivar that hopefully fares better against Kenyan viruses (Interview with manager of NGO, 2004). Several other GM crop development projects are underway in Kenya including insect resistant cotton and maize and virus-resistant cassava (See Harsh 2005). These projects have been bolstered by commercialization of similar crops in other countries (including some African countries).[6] Despite these technical efforts, there is still a great deal of uncertainty about how GM crops will reshape Kenyan farming systems and broader Kenyan society and political economy. Projection analyses of GM crops have been performed but these mainly use quantitative socioeconomic methods to justify the development of GM technologies, such as that of Qaim for the sweet potato project (1999). Missing in these analyses are discussion of cultural and social uncertainties that are harder to quantify and will likely decrease at a much slower pace than technical uncertainties.[7]

It is the possible pervasive and far reaching effects of GM technologies, coupled with uncertainty, that make decisions about GM so important and make access to those decisions such a highly sought after commodity. This can be seen in examining some of the primary and most visible decision points about the development of GM in Kenya. For the most part, these decision points are related to the establishment of Kenya's national biosafety system.

Biosafety involves regulation to protect human and natural environments from hazards associated with the release of a new species. Biosafety is one of the five "policy venues" that, according to Robert Paarlberg, constitute the broader decision making or governance of biotechnology. The other four are: "intellectual property rights (IPR) policy; trade policy; food safety and consumer choice policy; and public research policy" (2000, 4).

National governance of biotechnology in Kenya essentially defaulted to decisions about biosafety mainly because capacity to govern GM, like the development of GM technologies themselves, was mostly donor-driven, and biosafety was the specific focus of many donor efforts. One of the most significant donor-sponsored biosafety projects was coordinated by the United Nations Environment Program-Global Environmental Facility (UNEP-GEF) (UNEP-GEF 2003).[8] Eventually biosafety became operationalized in the form of Kenya's Biosafety Bill and joint Biotechnology and Biosafety Policy.[9] A key issue for governance of biotechnology in Kenya thus becomes examining how the Biosafety Bill and joint Policy were created. Specifically for my argument about the distribution of access, it is important to ask: Who was and was not involved in making decisions about the Bill and Policy?

The Biosafety Bill and Policy have had long and bumpy histories[10] that have provided a hodgepodge of opportunities for participation of different kinds of actors. Workshops were a main mechanism for participation, and thus main decision inputs. Each donor initiative sponsored several workshops that shaped decisions about the Bill and Policy. The workshops were often not solely organized by the donor or the coordinating government body, the National Biosafety Committee (NBC).[11] They were often co-organized or solely organized by NGOs, such as ISAAA (the NGO that led the GM sweet potato project mentioned above), the Agricultural Biotechnology Stakeholders Forum (ABSF), the Kenya Agricultural Biotechnology Platform, and later Africa Harvest Biotech Foundation International (AHarvest). As organizers, these NGOs had privileged access to biotechnology decisions, and also the ability to effectively exclude other organizations or individuals by not informing them about workshops.[12] These NGOs had further access in that they were the force keeping the development of biosafety moving forward between donor initiatives. This meant sometimes using their own funding to sponsor their own workshops (interview with senior policy maker in Ministry of Education, Science and Technology, 2005). Eventually, government-mandated comment periods provided the most formal mechanism for participation of broader actors in the shaping of the Biosafety Bill. Finally, less formal ways of participating in shaping the Biosafety Bill and joint Policy chosen by some actors included carrying out protests, distributing press releases, and writing editorials and letters to the editor in local newspapers.

Mid 2004–2005 was a particularly active and turbulent time for biotechnology and biosafety in Kenya, and one that highlights disparity of access to decisions about biotechnology. In early 2004, a coalition of NGOs formed whose goal was to stop the development of GM crops in Kenya (hereafter referred to as the anti-GM coalition). The coalition included at least five NGOs. Two were national branches of large well-known international NGOs (the headquarters of one was in Europe and the other in Africa) that work on poverty alleviation and a number of general issues related to rural poor livelihoods. Another NGO was a smaller European-based organization which worked only on GM issues. The coalition also included an NGO that was a regional (i.e., sub-national) network of community-based organizations that worked with small-scale farmers on land management. One of these community-based organizations was a small-scale farmers association and was itself an active member of the anti-GM coalition. All of these NGOs came to work on GM through their advocacy work related to the environment, small-scale farming, or ecological land management. The groups engaged in biosafety work in addition to their other projects. Thus, they were not well-funded when it came to work on biotechnology or biosafety. Of all the NGOs, only one person was employed to work full-time on biotechnology issues. The staff members from these groups that participated in the anti-GM coalition were generally Kenyan (only one was European) but were not scientists.

The coalition never gained much access to decisions about biotechnology and biosafety, a situation that—in the eyes of the coalition—led to bad policy choices, and certainly led to a non-constructive policy environment. When the coalition first formed, it began to investigate how the biosafety process had occurred to date, and tried to participate in the process. The coalition wrote letters to the NBC and requested meetings. However, they did not receive responses to their letters. They were denied entry to meetings and denied copies of the draft Bill and Policy. None of the NGOs ever heard about any workshops being held (interviews with coalition members, 2004). A general lack of communication between the coalition and the NBC further fueled the coalition's use of ideological and conspiratorial language when discussing the actions of the NBC and KARI in press releases and public meetings. This led to antagonistic relationships between the NBC and the coalition. It also led to the adoption of highly polarized viewpoints which prevented opportunities for any real debate about how biotechnology should be developed in Kenya (Harsh 2005, 2008). Overall, the coalition charged that they were left out of decisions about biotechnology and biosafety, and that because of this, the interests of small-scale farmers and the environment were left out of the NBC and of decisions about biotechnology (See Harsh 2009). This coalition clearly desired access to decisions, but had very little success in attaining it.

On the other hand, in this same period ISAAA and ABSF had a relative abundance of access to decisions about the development of biotechnology and biosafety. As opposed to the NGOs that make up the anti-GM coalition, ISAAA and ABSF are very well-funded, and working on biotechnology and biosafety development constitutes their core missions. Senior staff members of ISAAA and ABSF are generally Kenyan and hold graduate degrees in the sciences. During 2004–2005, key members of these organizations were active members of the UNEP task forces whose job was to draft the Policy and Bill. Also during this period and part of the UNEP program, ISAAA and ABSF organized several workshops. The crucial role that ABSF played in developing biosafety was noted by a top official in the National Council for Science and Technology (Interview, 2005). One NGO employee claimed that ABSF effectively wrote the Biosafety Bill and had the ear and personal mobile phone numbers of key policy-makers (Interview, 2008).

Both vertical and horizontal dimensions of inequality (Cozzens and Wetmore 2008) are clearly relevant to this story about unequal access to decisions about biotechnology in Kenya. With respect to the vertical dimension (inequalities between the rich and poor), the Kenyan case shows that access to resources is clearly an important factor that influences access to decisions. However, it is not just that well-funded NGOs and actors often have more access to decisions than less well-funded actors. The coordinating body in the Kenyan Government, the NBC, also claimed not to have enough resources to coordinate with all the NGOs interested in biosafety and biotechnology decisions. Moreover, perhaps the poorest group of actors in the story, small-scale farmers, had very little access to decisions. Generally, only the very small percentage of farmers that worked with NGOs on either side of the biotechnology issue knew anything about the development of biotechnology or biosafety in Kenya. And even of this small subset, only a smaller subset could come to Nairobi for workshops and meetings, depending on whether NGOs could pay for their travel and expenses.

The horizontal dimension of inequality (inequalities between groups distinguished by culturally-defined differences such as race, gender and religion) is also clearly important. In the case of decisions about agricultural biotechnology in Kenya, race and gender are significant, but possibly not the most significant factors. Perhaps paradoxically, the main actors within key NGOs who have had the most access to decisions (such as ISAAA, ABSF, and AHarvest) were women and African (native Kenyans).[13] In the period of late 2004–2005 the only NGO staff member who was European was a member of the anti-GM coalition—the side with significantly *less* access.

The most significant culturally-defined difference that led to inequalities in the Kenyan case was science itself. The staff members of NGOs in the anti-GM coalition were not scientists. Senior staff members of ISAAA, ABSF and AHarvest were generally PhD scientists. Members of the anti-GM coalition repeatedly stated that it was primarily this difference that kept them out of decisions about biosafety and biotechnology. Because they were not scientists and lacked scientific expertise, they were not able to participate in decisions about biotechnology (Interviews, 2004 and 2005). As in Europe and North America, decisions about GM in Kenya were framed in a scientific manner (Levidow 2007; Rayner 2003); the focus was on risks to human health or environment that could be quantified by scientific-based risks assessment. If actors were not fluent in scientific parlance, or could not frame their concerns in a risk-based scientific manner, they had a very difficult time accessing and participating in decisions (Harsh 2009).

Examining this case from Kenya shows that equity is relevant to participation in decisions about emerging technologies in developing regions because access is a highly desired and unevenly distributed commodity. Equity is also relevant though, because focusing on equity addresses criticisms of other frameworks that analyze participation in decisions. An equity analysis that examines horizontal and vertical inequalities, as in the Kenyan case study, brings differential agency of actors to the forefront.

Other approaches to participation in decision making do not focus so explicitly on how horizontal and vertical inequalities keep certain actors from decisions. For instance, participation in policy decisions is a major topic of discussions in political science and international relations discussions about governance (Hajer and Wagenaar 2003; Rhodes 1997).[14] Discussion of governance in this context focuses on collectivity in decision making, but does not necessarily problematize how varying access to resources or social ties can prevent certain actors from getting a seat at the table (Levidow 2007; Pestre 2008; Harsh and Smith 2007). In development studies since the 1970s there has been a focus on participation of the poor in their own development (e.g. Chambers 1983) that has led to many participatory policies, methodologies, and practices in development agencies. However, activities done in the name of participation can maintain or reinforce social hierarchies, such as those within a village (Cornwall and Pratt 2003). Participation and governance are discourses that can serve to cover up the politics of development initiatives; they depoliticize development (Ferguson 1994). An equity approach to participation can perhaps bring politics to the surface with its focus on inequalities. Some evidence for this comes from literature in Science and Technology Studies that has been focusing on the politics of participation involving science for some time (see for example Petersen 1984, and the other chapters in this section of this volume). Equity and justice are issues that are explicitly discussed (Nelkin 1984; Krimsky 1984).

There are clear lessons for nanotechnology from this first argument about access. The most obvious lesson is that no matter what nanotechnologies are being developed, or in which developing region they are being developed, they will be developed among existing inequalities that will affect access to decisions. These inequality concerns will not be as straightforward as who has access to resources and who does not, but access to resources will always be one key issue. Importantly, nanotechnologies could create the basis for new horizontal inequalities based on science in the same way that biotechnology has done in Kenya. Especially to the extent that nanotechnology will draw on more advanced or complex science, the distribution of who has access to this scientific knowledge will shape the distribution of who has access to decisions about how technologies should be developed, regulated, and used.

15.3 Institutional Structuring

My second argument is that equity is relevant to understanding participation in decisions about emerging technologies in developing regions because focusing on inequality issues can help show how different institutional mechanisms structure decision points, and thus who can access decisions and participate in them. By institutional mechanisms I refer to the configuration of public and private organizations (at international, national, and sub-national scales) that fund and carry out research and development of emerging technologies specifically for use in developing regions. Different institutional mechanisms can lead to different spatial and temporal distributions of decision points for different actors. In terms of *space* one can ask: In what geographical area do decisions occur, in what cities and countries? In what settings do they occur? Do they occur in laboratories, board rooms, or government offices? In terms of *time* one can ask: When in the technology development process does a given actor have opportunities to participate in decisions? Is it in upstream activities such as deciding which research projects to fund? Or is it in more downstream activities such as decisions about regulation, distribution, or use? I return to agricultural biotechnology in Kenya to give examples of how institutional configurations structure distributions of decisions.

The Green Revolution has had a significant impact on institutional mechanisms for developing agricultural technologies in Kenya, and throughout other developing regions, since the 1960s. The Green Revolution set out to increase food production by providing high-yielding crop varieties to farmers in Asia, Africa, and South America (Pearse 1980). Based on the model of land grant universities in the United States, scientific knowledge and new crop varieties, along with new fertilizers, pesticides, and irrigation technologies, were produced in international research laboratories as technological packages. These were then passed to national agricultural research centers (such as KARI in Kenya) and then to farmers via state-run extension services (Clark et al. 2003). Figure 15.2 illustrates the model, often called the "pipeline model" (Clark et al. 2003, 1847). The whole process was largely donor supported (some of the most prominent donors were the Ford and Rockefeller Foundations).

This model is still influential in agricultural technology development largely because it has created entrenched international and national institutional structures. The international research centers established by the Green Revolution

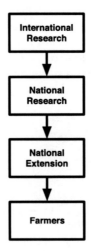

Fig. 15.2 Institutional model of the Green Revolution

are collectively known as the Consultative Group for International Agricultural Research system (CGIAR 2009). There are now 15 CGIAR centers world-wide, two of which have their headquarters in Kenya. Nationally, Kenya's agricultural research system is one of the most developed in all of Africa. Over twenty-five institutes conduct agricultural research in Kenya and 833 full-time equivalent employees were employed in the late 1990s (Beintema et al. 2003). Kenya spends over three times as much as the average sub-Saharan African country on agricultural research (Beintema et al. 2003).

The Green Revolution model significantly structures the distribution of decision points about what technologies should be developed. Spatially, there are two important issues. Initial decisions about what crops to develop happen in laboratories and test fields at research institutes. They also often happen far from the country and region of eventual use. Temporally, under the Green Revolution institutional model, major decisions that lock in the specific technologies to be developed occur upstream, early on in the development process. As knowledge moves linearly down the pipe (see Fig. 15.2) and technologies are developed, there are fewer options to control technology development; there are fewer decision points open to actors. Generally, the only decision in which farmers (end users) can participate is whether or not to adopt the new technologies and techniques. For the initial high yielding varieties of the Green Revolution, biosafety was not an issue as it is for GM, so few discussions or debates took place like those described above related to regulating GM in Kenya.

Since at least the 1990s, several new institutional mechanisms to develop agricultural technologies have emerged in Kenya.[15] These models essentially involve reconfiguring institutions; they are attempts to form new kinds of connections or new partnerships between institutions other than the linear connections associated with the Green Revolution model. These new mechanisms also involve new actors.

For instance, a now common model is the development of new technologies via Public-Private Partnerships (PPPs). The motivations for moving to these types of partnerships forge together technical, legal, economic, and moral concerns. The logic is that for new advanced biotechnologies, it is the private sector, and not the public sector, which possesses the relevant scientific expertise and owns the relevant intellectual property necessary to develop technologies. The private sector thus partners with a public sector research institute (a national research institute such as KARI or one of the international CGIAR research institutes) by sharing knowledge with the public partner and by licensing or donating intellectual property. The public partner gains access to new technologies, and the private partner benefits from a more philanthropic public image as well as the development of possible new markets (Horsch and Montgomery 2004). Due to the convergence of multifaceted motivations, donors are widely supporting these kinds of partnerships. Some of these are the same donors that funded the Green Revolution (such as the Rockefeller Foundation), but some major new donors have also emerged, such as the Bill and Melinda Gates Foundation. The GM sweet potato project has been developed via this kind of PPP mechanism, as are many of the other current GM projects in Kenya.

In addition to the private sector, NGOs are another important category of new actors prevalent in new institutional mechanisms. NGOs are often co-ordinating actors in PPPs, liaising between public and private actors, as well as with local farmers, policy-makers, and local university scientists. ISAAA is playing this role for the sweet potato project. NGOs can also play more specialized roles, such as the African Agricultural Technology Foundation (AATF, also based in Nairobi) whose role is specifically to negotiate intellectual property rights in PPPs.

Beyond PPPs, other new types of partnership arrangements are also emerging in Kenya. One such mechanism is direct collaboration between national research institutes (or other similar institutions) in two or more countries (so-called South-South partnerships or North-South partnerships depending on the countries involved). Another mechanism is direct collaboration between a national and an international agricultural research center to co-develop a specific product in a specific context, such as animal vaccines (Smith 2005).[16] Lastly, an important new model is the development of Centers of Excellence which bring public sector expertise from several countries together, focused on one area of science, such as the Bioscience for Eastern and Central Africa initiative which was founded in 2004 and is based in Nairobi (ILRI 2003).

Compared to the Green Revolution model, these new partnership mechanisms seem inherently more systemic. They implicitly or explicitly encourage more interconnections between different kinds of actors. However, in attempting to forge new connections between actors, how have partnership models changed when and where decision points occur and thus who can be involved in decisions?

The evidence from Kenya suggests that the distribution of decisions has not changed as much as one might expect given the greater focus in PPPs on partnership and forming new linkages, and on broader views of technological change. Yet, the changes in distribution of decisions that have occurred have enabled some new actors to have more access to decisions compared to the Green Revolution model.

For instance in the GM sweet potato project, ISAAA argues that despite the failed trials the project has actually been a success (Interviews, 2004). According to ISAAA, if one adopts a broad view of technological change, then even though a technically viable product has not emerged, the project was a success because it helped develop capacity to regulate GM technologies in Kenya that did not exist before the project began.[17] However this more systemic view did not significantly alter where and when decisions occur. The story of the genesis of the project involves a Kenyan scientist doing her doctoral training in the United States (Odame et al. 2003a). Similar to the Green Revolution model then, initial decisions that locked in the specific technological choice were still mainly made outside Kenya and occurred upstream. One scientist explicitly stated that farmers did not need to be involved in the sweet potato project until a technical product was fully created (Interview, 2005).

There is further evidence that other PPPs involving GM in Kenya have operated with similar distributions of decisions as that of the sweet potato project. When asked about how initial decisions were made to begin GM projects in

Kenya, one actor[18] told the story of traveling to Australia for a scientific conference with a KARI scientist. After learning about the current state of GM science, the two Kenyan scientists were inspired to try to create GM products in Kenya (Interview, 2004). Regarding the specific decision to use GM to develop insect resistant maize in Kenya, a university scientist/NGO manager told the story of sitting around a board room table at the International Maize and Wheat Improvement Centre (a CGIAR center). A gentleman from the private company Novartis happened to be at this meeting and it was decided that work should begin on GM maize (Interview, 2004). That work evolved into a PPP now called the Insect Resistant Maize for Africa initiative. Spatially, these vignettes show how decisions were made in settings such as board rooms at research facilities and at scientific conferences. Some key decisions were clearly made geographically far from Kenya. Temporally, the decisions again locked in technological choices very upstream.

Nevertheless, new partnership mechanisms have changed the distribution of decisions in some significant ways. One main way is that NGOs, and to some extent private sector actors, now have much more access to decisions and more control of technology developments because of the move to PPPs. ISAAA is just one of many NGOs that play a managing role in PPPs in Kenya and throughout Africa.[19] In 2005, AHarvest received US$16.9 million from the Gates Foundation to coordinate a partnership to develop bio-fortified sorghum (AHarvest 2005). AATF received US$47 million from the Gates Foundation and the Warren G. Buffet Foundation to develop maize that is more water efficient. For many NGO staff members, working at NGOs gives them more access to decisions about biotechnology than they might have had in previous and alternate careers as university researchers. To a much lesser extent, the shift to PPP mechanisms also has increased the amount of access to decisions for some farmers. NGOs, such as ISAAA and AHarvest, generally have networks of farmers with which they work. Oftentimes these farmers are geographically close to NGO offices. Essentially, the situation is that some rural poor farmers (i.e., one group considered to be disadvantaged in an equity analysis) are gaining some additional access to decisions through these new institutional mechanisms.[20] But that increase pales in comparison to the increased access that NGO staff members (i.e., one group considered to be advantaged) are gaining because of new partnership mechanisms.

Given these changes, does the shift from the Green Revolution model to more partnership mechanisms represent a just intervention? This depends on the theory of distributive justice adopted.[21] From a libertarian perspective, the answer is most likely yes. Justice in this perspective focuses on preservation of property rights, and at the heart of PPP mechanisms are licensing agreements and contracts to protect intellectual property. From a utilitarian perspective that focuses on growth and not distributional issues (such as the distribution of access to decisions) the answer is also probably yes. The motivation for GM in African policy discourses is often linked to economic growth (Harsh 2009). From a contractarian perspective, the answer is again, yes. The disadvantaged actors are slightly better off because of PPP mechanisms. Some rural farmers are gaining more access to decisions. Finally, from a communitarian perspective where just interventions decrease the gap between advantaged and disadvantaged actors the answer is surely no. The gap in access to decisions between the advantaged and disadvantaged actors is not decreasing, and probably even increasing. Scientists working for NGOs are experiencing a much larger increase in access than poor farmers.

There are evident lessons for nanotechnology from this discussion about institutional mechanisms and biotechnology. Given that institutional constraints strongly structure the distribution of decisions, it is important to pay as much attention to changes in institutional structures supporting the development of nanotechnology in developing regions as to changes in technical developments. Institutional developments rarely make headlines the way technical stories can. However, studying emerging institutional developments will reveal if the PPP model will be the dominant model for nanotechnology development, or if new models will emerge. Some new developments are already occurring. For example, the Gates Foundation recently announced a partnership with the United States National Science Foundation (NSF). The Gates Foundation will give US$24 million to the NSF (NSF will match this with an additional US$24 million) for research relevant to developing regions (including nanotechnology) and allow the NSF to vet projects (Doughton 2009).

A final point is that any new institutional mechanisms that do emerge to develop nanotechnology need to be historically contextualized. A lesson from the Green Revolution is that institutional strategies lead to infrastructural entrenchment that can create geographic disparities in development of new technologies. Partly because the Green Revolution led to significant infrastructural investments in Kenya, later biotechnology investments were also made in Kenya (favorable biosafety policies also played a role). The same cycle may be repeating itself again. PPP mechanisms are creating new infrastructure in terms of new networks of NGOs, many of which are based in Kenya. This is an important lesson for nanotechnology, especially to the extent that nanotechnology development will be infrastructure-dependent.

15.4 Conclusion

Through comparison with agricultural biotechnology in Kenya, this essay has argued that equity is relevant to analyses of participation in decisions about nanotechnology in developing regions. A focus on equity helps us understand that uneven distribution of access to decisions is linked to vertical inequalities related to access to resources, as well as to horizontal inequalities related to scientific identity and access to scientific knowledge. Framing analyses in terms of equity brings differential agency and politics to the forefront. Furthermore, distributional analyses can help us see how institutional constraints structure where, when, and even if, decisions take place.

There are perhaps some larger benefits of focusing on equity in analysis of participation in decision making. Such a focus provides a more complete and complementary approach to the study of equity and nanotechnology (see Fig. 15.1). This is not merely about providing analytical symmetry, balancing a focus on distributional consequences or outputs with a focus on unequal inputs, but about providing a means to think about how the two are connected. Inputs and consequences are not connected in a linear way. Deliberate feedback mechanisms are often put in place to provide information about possible outcomes based on technical trials, simulations, or modeling (see Fig. 15.1). Changing contexts, such as shifting political priorities, changing economic conditions, or unintended consequences, can all also have bearing on technological trajectories and their outcomes. In the end however, decision making usually involves speculation about distributional consequences and performing some sort or cost/benefit or risk/benefit calculation. This not only involves weighing the benefits that some will receive against the risks that others will face—a difficult political calculus made even more difficult because of uncertainty and unforeseen consequences—but also involves determining what constitutes a risk and a benefit. These calculations and constitutions could both drastically change depending on who is taking part in a decision. The distribution of access to decisions is thus importantly linked to distributional consequences.

In making my arguments to this point, I have avoided normative arguments about which different actors should have access to which decisions. Likewise, I have not argued in favor of one institutional mechanism to develop a nanotechnology over another. However, decision makers regularly need to make these commitments.

The second larger benefit of focusing on equity and participation in decisions, then, is that it helps equip decision makers with the ability to make normative commitments. For instance, one way that has been suggested to deal with uncertainty and unintended consequences of technology, is to make governance more anticipatory (Barben et al. 2008) and assessment of technologies more real-time (Guston and Sarewitz 2002). The conceptual framework of anticipatory governance involves creating anticipatory knowledge about how technologies might develop and how these developments are related to societal outcomes (Barben et al. 2008). The methodology of real-time technology assessment (RTTA) has tried (in developed countries) to implement anticipation by engaging with publics and end users of technologies and feeding their views into anticipatory knowledge. Anticipatory governance and RTTA thus generally try to widen the distribution of decision inputs and analyze and develop new feedback mechanisms between decisions and outcomes. In the context of institutional mechanisms to develop emerging technologies in developing regions discussed in this essay, implementation of an RTTA approach is most consistent with a communitarian perspective of distributive justice. From this perspective, one would argue for more access for farmers to decisions earlier on in the technology development process, and for institutional mechanisms that can do a better job of applying this so-called bottom-up approach than PPP mechanisms have been able to do thus far for GM crop projects in Kenya. Just what these institutional mechanisms might look like, and indeed how governance might generally be made to be more anticipatory in developing regions, requires much more research. However, a starting point might be exploring how donors and other architects of emerging technological interventions in developing regions think about societal outcomes in their current institutional models.

Notes

1. I use the term developing regions instead of developing countries to account for variations within countries.
2. By institutional mechanisms I mean the configurations of institutions (various public and private organizations operating at international, national, or sub-national scales) that fund and carry out research and development of technologies specifically for developing regions.
3. Information in this section and the next comes from ethnographic fieldwork carried out in Kenya from 2004 to 2008. Facts coming from individual interviews are cited as such.
4. Kenya has been engaging with other non-GM biotechnologies such as biopesticides and tissue culture since the 1960s, see the timeline in Harsh (2005).
5. Other partners that have more recently joined the project include the Roodeplaat Vegetable and Ornamental Plant Institute of South Africa, another public sector institute, and the Danforth Plant Science Center in the USA (Horsch and Montgomery 2004).
6. As mentioned above, currently there are no GM crops commercialized in Kenya. However, three other countries in Africa have commercialized GM crops. These are South Africa, Egypt and Burkina Faso.
7. For instance, in another (non-GM) agricultural biotechnology project involving tissue culture bananas developed by ISAAA (the same NGO that is a partner in the GM sweet potato project), farmers were concerned about how tissue culture bananas might lead to the disappearance of a banana variety important for betrothal ceremonies (Harsh and Smith 2007).
8. Bilateral support to develop biosafety in Kenya also came from the Netherlands Directorate-General for International Co-operation and the United States Agency for International Development. To be sure, some donor initiatives did lead to the development of the other policy venues that make up the governance of biotechnology, but these were far outshone by efforts to develop biosafety. See Harsh (2005 and 2008) for more detail.
9. The Biosafety Bill essentially provides the legal authority to regulate biotechnology. The joint Biotechnology and Biosafety Policy is a statement of intent to develop biotechnology in Kenya.
10. The first draft of the Biosafety Bill was produced in 2003 and the Bill was finally passed into law in 2008. Some of the factors leading to the prolonged history include: the signing and ratification of the Cartagena protocol in 2000 and 2002 respectively; and the decision in 2002 to create a new law to provide legal authority for regulating biotechnology instead of using existing laws, which was the strategy up to 2002. Drafting of the Biotechnology and Biosafety Policy started in 2002–2003 and the Policy was accepted by the Cabinet in 2006. (Policies in Kenya do not need Parliamentary approval.) See Harsh (2005 and 2008) for more details.
11. At the time the NBC was within the National Council of Science and Technology, a division of the Ministry of Education Science and Technology. With the passing of the Biosafety Bill into law in 2009, the NBC became the National Biosafety Authority. The hosting ministry was also split after the 2007 elections and is now the Ministry of Science and Technology.

12. For some workshops invitees were limited, for instance to only Members of Parliament. However, most workshops were nominally open to the public, but were not advertised broadly. Thus if one was not invited, one would generally not know about the workshop and would not be able to attend.
13. I have previously argued that these actors represent what Tvedt (2006) calls a new elite class of actors development system (Harsh 2009).
14. Here, governance is used to describe a change in the nature of decision making of states from one where the state operated in a command-and-control manner to one where the state is but one actor negotiating amongst other actors to make policy decisions, such as NGOs and private sector organizations (Pierre and Peters 2000).
15. These new institutional mechanisms have not entirely displaced pipeline mechanisms. Experience varies between different research centers and different developing regions.
16. This is different than the Green Revolution model in the that the international and national centers are jointly researching and developing the same product. This involves two-way knowledge flow and partnership. For instance, the international center must focus much more on specific national and even sub-national contexts.
17. Importantly, this was a revision from ISAAA's prior statements about the motivation for the project that focused more narrowly on the technological product—on the economic impact that the GM potato would have.
18. This actor was highly involved in the Nairobi biotechnology community having worked as a university scientist, a manager of an NGO, a coordinator for a donor project on biosafety, and as the head of a new research partnership at various overlapping points in her career.
19. Note that although many of the PPPs are based in Kenya, their remit is to develop GM technologies for all of Africa.
20. For example, one such group of farmers working with ISAAA is the Murang'a Banana Marketing Collective, a network created by ISAAA located in Murang'a district of Central Province approximately 70 km from Nairobi. Through their interactions with ISAAA they have learned about tissue culture technology. However, their access to decision making related to the technology is mostly limited to a decision to adopt the technology or not. In general, ISAAA does not release figures about how many farmers they worked with on the tissue culture project. See Harsh and Smith (2007) and Smith (2007).
21. The four theories of distributive justice used here in very general terms. For a much more detailed treatment, see Cozzens (2007).

References

ABSP. 2002. *Biotechnology research and policy activities of ABSP in Kenya: 1991–2002.* Michigan State University: Agricultural Biotechnology Support Project.
AHarvest. 2005. Africa harvest helping Africa fight hunger and malnutrition with biofortified sorghum. http://africaharvest.org/files/ABS.pdf. Accessed 17 November 2008.
Baldwin, Derek. 2009. Green dream: A world in a grain of sand. http://www.xpress4me.com/news/uae/abudhabi/20011769.html. Accessed 6 May 2009.
Barben, D., E. Fisher, C. Selin, and D. Guston. 2008. *Anticipatory governance of nanotechnology: Foresight, engagement, and integration.* In *The handbook of science and technology studies*, ed. Edward J. Hackett, Olga Amsterdamska, Michael Lynch, and Judy Wajcman, 979–1000. Cambridge, MA: MIT Press.
Beintema, N., F. Murithi, and P. Mwangi. 2003. *Agricultural science and technology indicators: Kenya.* The Hague: International Service for National Agricultural Research.
Bourzac, Katherine. 2009. A hybrid nano-energy harvester: The device harnesses both sunlight and mechanical energy. http://www.technologyreview.com/energy/22410/. Accessed 6 May 2009.
CGIAR. 2009. The consultative group on international agricultural research. http://www.cgiar.org/. Accessed 6 May 2009.

Chambers, R. 1983. *Rural development: Putting the last first*. London: Longman.
Clark, N., A. Hall, R. Sulaiman, and G. Naik. 2003. Research capacity building: The case of an NGO facilitated post-harvest innovation system for the Himalayan Hills. *World Development* 31: 1845–63.
Cohen, Joel I., and Robert Paarlberg. 2004. Unlocking crop biotechnology in developing countries: A report from the field. *World Development* 32: 1563–77.
Cornwall, A., and G. Pratt. 2003. The trouble with participatory rural appraisal: Reflections on dilemmas of quality. *Participatory learning and action notes* 47: 1–7.
Cozzens, Susan E. 2007. Distributive justice in science and technology policy. *Science and Public Policy* 34: 85–94.
Cozzens, Susan E., Isabel Bortagaray, Sonia Gatchair, and Dhanaraj Thakur. 2008. Emerging technologies and social cohesion: Policy options from a comparative study. Paper presented at the PRIME Latin America Conference, September 2008. http://prime_mexico2008.xoc.uam.mx/papers/Susan_Cozzens_Emerging_Technologies_a_social_Cohesion.pdf. Accessed 3 August 2010.
Cozzens, Susan E., and Jameson Wetmore. 2008. *Introduction to the workshop on nanotechnology, equity, and equality*. Tempe, AZ: Arizona State University, Center for Nanotechnology in Society.
David, Kenneth, and Paul B. Thompson, eds. 2008. *What can nanotechnology learn from biotechnology?* Burlington, MA: Academic.
Dickson, David. 2009. How nanotech can meet the poor's water needs. http://www.scidev.net/en/new-technologies/nanotechnology-for-clean-water/editorials/how-nanotech-can-meet-the-poor-s-water-needs.html. Accessed 6 May 2009).
Doughton, Sandi. 2009. Gates foundation takes on a partner in new venture. http://seattletimes.nwsource.com/html/localnews/2008952016_gatesag31m.html. Accessed 6 May 2009.
Ferguson, James. 1994. *The anti-politics machine: Development, depoliticization, and bureaucratic power in lesotho*. Minneapolis, MN: University of Minnesota Press.
Gathura, G. 2004. GM technology fails local potatoes. *The Daily Nation* http://allafrica.com/stories/200401290011.html. Accessed 14 January 2004.
Gill, Victoria. 2009. Mobile technology battles HIV. http://news.bbc.co.uk/1/hi/sci/tech/7989856.stm. Accessed 6 May 2009.
Guston, David H., and Daniel Sarewitz. 2002. Real-time technology assessment. *Technology in society* 24: 93–109.
Hajer, Maarten A., and Hendrik Wagenaar, eds. 2003. *Deliberative policy analysis: Understanding governance in the network society*. Theories of Institutional Design. Cambridge: Cambridge University Press.
Harsh, Matthew. 2005. Formal and informal governance of agricultural biotechnology in Kenya: Participation and accountability in controversy surrounding the draft biosafety bill. *Journal of International Development* 17: 661–77.
Harsh, Matthew. 2008. "Living technology and development: Agricultural biotechnology and civil society in Kenya." Science and Technology Studies, PhD thesis, University of Edinburgh, Edinburgh.
Harsh, Matthew. 2009. Non-governmental limits: Governing biotechnology from Europe to Africa. In *The limits to governance*, ed. Catherine Lyall, Theo Papaioannou, and James Smith. London: Ashgate.
Harsh, M., and J. Smith. 2007. Technology, governance and place: Situating biotechnology in Kenya. *Science and Public Policy* 34: 251–60.
Horsch, R., and J. Montgomery. 2004. Why we partner: Collaborations between the private and public sectors for food security and poverty alleviation through agricultural biotechnology. *The Journal of Agrobiotechnology, Management, & Economics* 7: 80–83.
ILRI. 2003. *New biosciences facility for east and central Africa*. Nairobi: International Livestock Research Institute.

Krimsky, S. 1984. *Beyond technocracy: New routes for citizen involvement in social risk assessment.* In *Citizen participation in science policy*, ed. James C. Petersen. Amherst, MA: University of Massachusetts Press.

Levidow, Les. 2007. European public participation as risk governance: Enhancing democratic accountability for agbiotech policy? *East Asian Science, Technology and Society* 1: 19–51.

Mehta, MD. 2004. From biotechnology to nanotechnology: What can we learn from earlier technologies? *Bulletin of science, technology & society* 24: 34.

Mody, Cyrus C.M. 2007. Nanotechnology: Where did it come from? What is it for? http://scienceblogs.com/worldsfair/2007/06/nanotechnology_from_where_did.php. Accessed 6 May 2009.

Nelkin, D. 1984. *Science and Technology Policy and the Democratic Process.* In *Citizen Participation in Science Policy*, ed. James C. Petersen. Amherst, MA: University of Massachusetts Press.

New Scientist. 2004. Monsanto's showcase project in Africa fails. *New Scientist* 2433: 7.

Odame, H., P. Kameri-Mbote, and D. Wafula. 2003a. *Innovation and policy process: The case of transgenic sweet potato in Kenya.* Nairobi: African Centre for Technology Studies.

Odame, H., P. Kameri-Mbote, and D. Wafula. 2003b. *Globalisation and the international governance of modern biotechnology implications for food security in Kenya.* Nairobi: African Centre for Technology Studies.

Paarlberg, R.L. 2000. Governing the GM crop revolution: Policy choices for developing countries. *Food, Agriculture, and the Environment Discussion Paper* 33: 1–44.

Pearse, A. 1980. *Seeds of plenty, seeds of want: Social and economic implications of the green revolution.* Oxford: Clarendon Press.

Pestre, Dominique. 2008. Challenges for the democratic management of technoscience: governance, participation and the political today. *Science as Culture* 17: 101–20.

Petersen, James C, ed. 1984. *Citizen participation in science policy.* Amherst: University of Massachusetts Press.

Pierre, Jon, and B. Guy Peters. 2000. *Governance, politics and the state.* Basingstoke: Palgrave Macmillan.

Qaim, Matin. 1999. The economic effects of genetically modified orphan commodities: Projections for sweet potato in {Kenya}. *International Service for the Acquisition and Application of Agricultural Biotechnology Brief* 13: 1–47.

Rayner, Steve. 2003. Democracy in the age of assessment: Reflections on the roles of expertise and democracy in public-sector decision making. *Science and Public Policy* 30: 163–70.

Rhodes, R. 1997. *Understanding governance: Policy networks, governance, reflexivity and accountability.* Buckingham: Open University Press.

Smith, J. 2005. Context-bound knowledge production, Capacity building and new product networks. *Journal of International Development* 17: 647–59.

Smith, James. 2007. Culturing development: Bananas, petri dishes and 'mad science'. *Journal of Eastern African Studies* 1: 212–33.

Tvedt, T. 2006. The international aid system and the non-governmental organisations: A new research agenda. *Journal of International Development* 18: 677–90.

UNEP-GEF. 2003. *UNEP-GEF Project on the implementation of the national biosafety framework of Kenya: State of developments.* Nairobi: United Nations Environment Program.

Chapter 16
Nanotechnology: How Prepared Is Uganda?

Kikonyogo Ngatya

This chapter is a newspaper article written by Kikonyogo Ngatya that was originally published in the Uganda Sunday Monitor *on June 28, 2009. Ngatya reports on the controversy surrounding a small drinking vessel being sold in Uganda for approximately US$250 to US$500 that merchants claim offers health benefits from nanotechnology. The point of republishing this article is not to demonstrate that Ugandans don't understand nanotechnology. After all the claims made for the glass aren't entirely dissimilar from the promises of nanotechnology products being developed in the North. Rather it emphasizes the difficulties inherent in ensuring that nanotechnologies will be safe and provide tangible benefits. Europe and the United States have begun the process of developing regulations, but have a long way to go before any robust and enforceable national policies are put into place—and these countries have the benefit of substantial resources in terms of money, infrastructure, and scientists. Developing countries like Uganda lack many of these assets. As can be seen in this article there are a number of concerned officials, scientists, and even reporters who are taking steps to do what they can to make sure that developing countries don't experience adverse effects from nanotechnologies, but it will likely be an uphill struggle.—eds.*

K. Ngatya (✉)
Freelance reporter based in Uganda

Originally published in the *Uganda Sunday Monitor* on June 28, 2009. Reprinted with permission from the *Uganda Sunday Monitor*.

SUNDAY MONITOR

Business & Technology | June 28, 2009

Nanotechnology: How Prepared is Uganda?

KIKONYOGO NGATYA

Kampala

There is frenzy in Kampala, especially among the middle class, of a new type of small glass, with near magical powers, claimed to enhance body mood and replenish water and other beverages with lost essential minerals. The glass is believed to have been developed at high altitude.

It costs between Shs500,000-1,000,000. The glass, whose brand name is withheld, claims to make sick people get nutrients from its use. One pours water and drinks. It is also claimed that carrying it in one's pocket makes them healthier.

It is one of the numerous products imported into the country based on a new era of advanced research based on nanotechnology, a science that manipulates matter at the scale of atoms and molecules. The atoms can be used in a wide range of applications like food, medicine, cosmetics, pesticides and crop sector development.

But a senior official working with the Uganda National Council for Science and Technology (UNCST) said there is a likely public health danger to consuming products of nanotech, which have not undergone a bio-safety check.

The officer, speaking on condition of anonymity because the council has no policy in place yet to monitor or evaluate the safety and effectiveness of nanotech products, said some of the products are counterfeits. The officer believes that Uganda is still lagging behind in nanotechnology generation.

"Like this glass in town. It is kiwaani (fake). But we cannot tell if it is right or wrong, because there is no capacity to taste it," the officer said. The officer's comments followed numerous complaints from the public that the glass was not working.

But Ms Ruth Mbabazi, the Secretary to the National Bio-safety Committee acknowledges that Uganda had no policy or tools to evaluate products of nanotechnology, before pronouncing them to be safe and sold on the market.

She however said some products were already on the market. The committee comprising of experts from various research and policy institutions, also looks at UNCST approved technology research in Uganda has a big mandate in monitoring products of genetic medication like those from biotechnology.

So far, the committee monitors confined genetically modified banana research in Kawanda, and cassava at Namulonge agriculture research institutes. But biotechnology, involving the transfer or modification of a specific gene with desired traits is quite different from nanotechnology, yet Uganda is also still struggling with developing a regulatory framework.

Ms Mbabazi said it was important that the environmental, health and safe use of genetically modified organism (GMOs) is understood. "Nanotechnology should fall within our mandate. Uganda needs to be aggressive with developing laws to monitor its use," she said.

But Dr. Terry Kahuma, the Executive Director of Uganda National Bureau of Standards (UNBS), said there is a need to develop standards to ascertain and quantify the effectiveness of new products of nanotech. He said UNBS had come up with a mechanism of developing standards for health products, if they have a "perceived danger".

"But buying a product is voluntary. Use your eyes and ask questions," Dr Kahuma said.

Dr Charles Mugoya, an expert on biotechnology, the science that makes GMOs, said the use of both nanotechnology and biotechnology together is good but should be regulated.

Dr Mugoya, working with the Eastern Africa's research body, ASARECA, said many developments in science were good for food production, but may also be misused by criminals, by developing undesirable products and not adhering to ethical standards.

Developed GMOs are helping to produce disease and drought tolerant food varieties in the wake of warming global temperatures. Climate change is fuelling new diseases and occurrences of previously suppressed ones. There has been a rise of Uganda's temperature from 0.2-0.3 degrees in the last fifty years.

The implications are that it's no longer sustainable with the current technologies to either develop medicine or grow enough food. The insects and pests, as they adjust to the changes in temperature, are gaining further strength to attack. According to Equinet, a southern African regional body, monitoring the access and fairness in access to health by the population, governments in the region need to develop policies to monitor new technologies and their impacts on food production.

Part IV
Nanotechnology and the World System

Chapter 17
Nanotechnology and the Poor: Opportunities and Risks for Developing Countries

Todd F. Barker, Leili Fatehi, Michael T. Lesnick, Timothy J. Mealey, and Rex R. Raimond

Among the high hopes surrounding nanotechnology is the hope that it will solve a number of pressing problems in developing countries. The counterpoint to that hope is the fear of a nano-divide—benefits accruing only in the global North. This volume has already included several references to possible disparities, in Foladori and Zayago Lau's discussion of workers, Harsh's portrayal of decision processes in Kenya, and the reprinted article on regulation of nanotechnology in Uganda that closed the last section. This chapter charts the issues involved in a more comprehensive way. The chapter consists of excerpts from a report on a series of meetings called the "Global Dialogue on Nanotechnology and the Poor" that was facilitated by the Meridian Institute, a non-governmental organization that "helps people solve problems and make informed decisions involving multiple points of view." The dialogues developed an overview of the pluses and minuses of applying nanotechnologies to several prominent challenges facing developing countries. Admirably, the chapter reflects needs expressed by participants from the global South as well as a solid base of technical information on feasibility.—eds.

This chapter was originally published in Frtiz Allhoff and Patrick Lin (eds.), *Nanotechnology & Society: Current and Emerging Ethical Issues*. Dordrecht: Springer, 2008, pp. 243-263. Reprinted with permission from Springer.

17.1 Introduction

Millions of people worldwide continue to lack access to safe water, reliable sources of energy, healthcare, education, and other basic human development needs. Since 2000, the United Nations Millennium Development Goals (MDGs) have set targets for meeting these needs. In recent years, an increasing number of government, scientific, and institutional reports have concluded that nanotechnology could make significant contributions to alleviating poverty and achieving the MDGs. Concurrently, these and other reports have also identified potential risks of nanotechnology for developing countries (UN Millennium Project 2005).

Perceived by many as the next "transformative technology"—like electricity or the Internet—nanotechnology encompasses a broad range of tools, techniques, and applications that manipulate or incorporate materials at the nanoscale in order to yield novel properties that do not exist at larger scales. These novel properties may enable new or improved solutions to problems that have been challenging to solve with conventional technology. For developing countries, these solutions may include more efficient, effective, and inexpensive water purification devices, energy sources, medical diagnostic tests and drug delivery systems, durable building materials, and other products. Additionally, nanotechnology may significantly increase developing countries' production capacities by enabling manufacturing processes that create less pollution and have modest capital, land, labor, energy, and material requirements.

Both the public and private sectors in developed and developing countries are investing heavily in nanotechnology research and development. More than 20 countries, including developing countries such as China, South Africa, Brazil, and India, have national nanotechnology programs, and many more are developing or expanding nanotechnology research and development capacity. The collective public and private sector investment in 2005 was approximately US$10 billion, up 20% from 2004 (Salamanca-Buentello et al. 2005). In addition, the number of patents on nanotechnology-related inventions (including those from developing country researchers) (Singh 2007), scientific literature citations (now up to 12,000 publications per year) (Colvin 2002), and nanotechnology-based products reaching the market are skyrocketing globally.

The rise in nanotechnology investments and proliferation of applications has contributed to growing international dialogue about implications of the rapid evolution of nanotechnology, including potential near- and long-term social and economic disruptions, human health and environmental risks, and ethical, legal, and other impacts. Governments, companies, NGOs, universities, international institutions, standardization bodies, and other stakeholders have initiated a number of efforts to discuss, develop, and implement risk assessment, governance, standardization, and public involvement strategies to address these potential implications.

Despite these efforts, there are few processes to collectively engage multiple stakeholders in addressing the opportunities and risks of nanotechnology for developing countries. Moreover, where processes do exist to identify linkages between nanotechnology and development, these activities remain disengaged from the predominant risk assessment, governance, standardization, and other key initiatives.

These gaps are a significant concern, as current decisions in both developed and developing countries may result in policies, practices, and systems that have long-term impacts on whether nanotechnology will help or hinder the effort to address specific human development needs.

To address this need, Meridian Institute, a non-profit organization that specializes in helping people solve problems and make informed decisions about complex and controversial societal issues,[1] has convened the Global Dialogue on Nanotechnology and the Poor: Opportunities and Risks (GDNP) to close these gaps through a variety of strategies that raise awareness about the implications of nanotechnology for developing countries, catalyze actions that address specific opportunities and risks, and identify ways that science and technology can play an appropriate role in the development process.

As part of the GDNP process, Meridian is convening a series of sector-specific activities, beginning with the International Workshop on Nanotechnology, Water, and Development, held October 2006 in Chennai, India. This workshop brought together participants from developed and developing countries and with a broad range of perspectives and expertise to discuss the range of challenges people in developing countries may face when developing and implementing strategies for improving access to clean water and opportunities for using nanotechnology to address water supply challenges, as well as risks, and other issues that need to be addressed in relation to specific nanotechnology applications. Meridian plans to convene other sector-focused workshops in the areas of commodities (agricultural, mineral, and non-fuel commodities), energy, and health care.

In the following pages, we describe, as illustrative examples, possible opportunities and risks presented by nanotechnology for developing countries. Drawing extensively from the International Workshop on Nanotechnology, Water, and Development, we focus, in particular, on opportunities, risks and other issues in the context of nanotechnology applications for water, but also provide examples of applications for energy, health, agriculture, and food.

17.2 Opportunities – Nanotechnology and Development

17.2.1 Water

Nanotechnology for water purification has been identified as a high priority area because water treatment devices that incorporate nanoscale materials are already available and human development needs for clean water are pressing. Poverty and water are closely linked and access to water resources has become widely equated with ensuring that basic human needs are met. It is predominantly the poor of the world who depend directly on water and other natural resources for their livelihoods.

In 2002, 1.1 billion people lacked access to safe drinking water, and 2.6 billion people lacked access to adequate sanitation (UN'Millennium Project 2005). The consequences of lack of access to clean water and adequate sanitation are overwhelming: waterborne diseases and water-related illnesses kill more than five million people a year worldwide, 85% of these being children, according to the World Health Organization (UN Millennium Project 2005). Additionally, many children miss school because neither their homes nor schools have adequate drinking water or sanitation facilities and hundreds of millions of African, Asian, and Latin American families lose vital income from the lack of access to reliable drinking water and sanitation services (meeting the MDG target for clean water and sanitation is estimated to yield economic benefits close to US$12 billion a year).

Given the importance of clean water to people in developed and developing countries, numerous organizations are considering the potential application of nanoscience to solve technical challenges associated with the removal of water contaminants. Technology developers and others claim that these technologies offer more effective, efficient, durable, and affordable approaches to removing specific types of pollutants from water. A range of water treatment devices that incorporate nanotechnology are already on the market and others are in development. These nanotechnology-based products include nanofiltration membranes; nano-ceramic, clay, and polymer filters; zeolites; nanocatalysts; magnetic nanoparticles; and nanosensors.

Nanofiltration membrane technology is already widely applied for removal of dissolved salts from salty or brackish water, removal of micro pollutants, water softening, and wastewater treatment. Nanofiltration membranes selectively reject substances, which enables the removal of harmful pollutants and retention of nutrients present in water that are required for the normal functioning of the body. It is expected that nanotechnology will contribute to improvements in membrane technology that will drive down the costs of desalination, which is currently a significant impediment to wider adoption of desalination technology.

Zeolites, clays, and nanoporous polymers are also materials used for nanofilters. While these materials have been used for many years to purify water, recent improvements in scientists' ability to manipulate on the nanoscale allow for greater precision in designing these materials, for instance, allowing much greater control over pore size of membranes (Cientifica 2003).

Zeolites are microporous crystalline solids with well-defined structures. Generally they contain silicon, aluminium, and oxygen in their framework and cations, water, and/or other molecules within their pores. Many occur naturally as minerals and are extensively mined in many parts of the world. Others are synthetic and are made commercially for specific uses or produced by research scientists trying to understand more about their chemistry. Zeolites can be used to separate harmful organics from water and to remove heavy metal ions from water.

Nanocatalysts include enzymes, metals, and other materials with enhanced catalytic capabilities that derive from either their nanoscale dimensions or from nanoscale structural modifications. These substances promote the chemical reaction of other materials without becoming permanently involved in the reaction. Controlling a material's size and/or structure at the nanoscale can produce catalysts that are more reactive, more selective, and longer lasting. Consequently, smaller quantities of catalysts are needed, reducing raw materials consumption, byproducts and waste, and, potentially, the overall cost of catalysis.

Nanocatalysts such as titanium dioxide (TiO_2) and iron nanoparticles can be used to degrade organic pollutants and remove salts and heavy metals from liquids. People expect that nanoelectrocatalysts will enable the use of heavily polluted and heavily salinated water for drinking, sanitation, and irrigation (UN Millennium Project 2004). Using catalytic particles either dispersed homogeneously in solution or deposited onto membrane structures could chemically degrade pollutants instead of simply moving them somewhere else. Catalytic treatment of polluted water could be specifically targeted to degradation of chemicals for which existing technologies are inefficient or cost prohibitive.

Magnetic nanoparticles are being investigated for a variety of chemical separation applications including water treatment because they have high surface areas and can bind with chemicals without the use of auxiliary adsorbent materials. Additionally, the application of surface coatings can functionalize the chemical reactivity of magnetic nanoparticles, making them suitable as nanocatalysts for the chemical decomposition of chemicals. Once adsorption or catalysis has occurred, the magnetic nanoparticles can be removed from the water using a magnet or a magnetic field and reused.

Nanosensors for the detection of contaminants and pathogens can improve health, maintain a safe food and water supply, and allow for the use of otherwise unusable water sources. Nanosensors can detect single cells or even atoms, making them far more sensitive than counterparts with larger components. Conventional water quality studies rely on a combination of on-site and laboratory analysis, which requires trained staff to take water samples and access to a nearby laboratory to conduct chemical and biological analysis. New sensor technology combined with micro- and nanofabrication technology is expected to lead to small, portable, and highly accurate sensors to detect chemical and biochemical parameters.

17.2.2 Energy

Access to electricity is not specifically among the MDGs, but it could help with most of them: pumping water for human use and for agriculture, powering rural clinics and refrigerating medicines, lighting schools, and helping people earn sustainable livings in their own businesses. "Access to basic, clean energy services is essential for sustainable development and poverty eradication, and provides major benefits in the areas of health, literacy, and equity. However, over two billion people today have no access to modern energy services," according to Practical Action (Intermediate Technology Development Group).

Some 2.4 billion people use traditional biomass energy—wood, crop residues, and dung—for cooking and heating, a number that is increasing rather than decreasing. This is inefficient for most purposes; it can cause burns and respiratory problems due to indoor pollution and, depending on the source of the biomass, can degrade environmental systems and resource bases.

Cheap solar-powered electricity has long been an aspiration for tropical countries, but glass and silicon photovoltaic panels remain too expensive and delicate. Nanotechnology may allow for the production of cheap photovoltaic films that can be unrolled across the roofs of buildings. It may even be possible to paint solar power films onto surfaces.

17.2.3 Health

Nanotechnology offers a range of possibilities for healthcare and medicinal breakthroughs, including targeted drug delivery systems, extended-release vaccines, enhanced diagnostic and imaging technologies, and antimicrobial coatings. ...

Nanotechnology could also enable simple, accurate, small, and stable diagnostic tests and devices. ...

Nanoporous membranes may help with disease treatment in the developing world. They are a new way of slowly releasing a drug, important for people far from hospitals. Making the nanopores only slightly larger then the molecules of drugs can control the rate of diffusion of the molecules, keeping it constant regardless of the amount of drug remaining inside a capsule. ...

Nanomaterials such as silver nanoparticles are being used in a variety of products such as textiles, paints, and coatings to provide antibacterial and antimicrobial protection.

17.2.4 Food and Agriculture

Several studies suggest that nanotechnology will have major, long-term effects on agriculture and the production of food, but it remains unclear how these changes will affect developing countries. Many of the promised advances for agriculture are similar to some promised advances in drug delivery in human medicine: time-controlled release; remotely regulated, pre-programmed, or self-regulated delivery of nutrients or disease treatments; transplanted cells protected by membranes; bio-separation; and rapid sampling and diagnosis of plant or animal health (Roco 2003; see also Scott and Chen). Nanotechnology may also help make food products cheaper and production more efficient and more sustainable through using less water and chemicals. ...

17.2.5 Risks and Cross-Cutting Issues

Nanotechnology products are entering the global marketplace at an increasing pace and strong competitive and economic drivers will likely accelerate this trend, leading some observers to argue that the sheer momentum of efforts to develop nanotechnology could be overwhelming the need to examine and manage associated risks such as near- and long-term socioeconomic disruptions, human health and environmental impacts, and ethical, legal, and trade implications.

17.2.6 Environmental, Human Health, and Safety Risks

There is currently slow but growing availability of studies on the environmental, human health, and safety (EHS) impacts of engineered nanomaterials, including data on the toxicity, fate, and transport of nanoparticles. Only a limited number of studies have been published on the potential toxicity of specific nanoscale materials, and the incongruity of results of initial research results demonstrates the complexity of assessing EHS risks.

Several fundamental aspects of nanotechnology cause concern that the risks associated with nanomaterials may not be the same as the risks associated with the bulk versions of the same materials. For instance, as a particle decreases in size, a larger proportion of atoms is found at the surface as compared to the inside. Thus, nanoparticles have a much larger surface area per unit mass compared with larger particles. Also, as the size of matter is reduced to tens of nanometers or less, quantum effects can begin to play a role, and these can change optical, magnetic, and electrical properties of materials. Since growth and catalytic chemical reactions occur at surfaces, a given mass of nanomaterials will be more reactive than the same mass of materials made up of larger particles. These properties might have negative health and environmental impacts and may result in greater toxicity of nanomaterials (Bruske-Hohlfeld et al. 2005).

The study of nanoparticles' toxicity is complicated by the fact that they are highly heterogeneous. Not only are they exclusively engineered to specification but in many cases nanoscale materials will alter in physical size upon interaction with aqueous systems. Furthermore, the surface coating of nanoparticles can be altered to completely change the material's toxicity. For example, changing the surface features of the material can change a hydrophobic particle into a hydrophilic one (Goldman and Coussens 2005).

Nanotechnology handlers, consumers, and other people could be exposed to nanoparticles through inhalation, ingestion, skin uptake, and injection of nanoscale materials. Nanoparticles could also interact with ecosystems, animals, plants, and microorganisms.. Furthermore, use of nanomaterials in the environment may result in novel by-products or degrades that also may pose risks. To date, very few ecotoxicity studies with nanomaterials have been conducted. Studies have been conducted on a limited number of nanoscale materials and in a limited number of aquatic species. There have been no chronic or full lifecycle studies reported.

17.2.7 Socioeconomic Issues

Given the potential rapid and radical technology innovations that may be enabled by nanotechnology, some people and publications have expressed concerns that nanotechnology applications could have adverse socioeconomic effects on developing countries (ETC Group 2004). Other people, however, have said nanotechnology applications – as described extensively above – may yield significant economic and social benefits for developing countries.

Product research & development	Systematic activities to increase knowledge and apply it to the (further) development of new applications. In the context of the workshop, participants focused on assessing the maturity of specific nanotechnology applications and the steps that would be necessary for further development.
Environmental, human health, & safety risks	Potential harm that may arise from a material, combined with probability of an event (e.g., exposure). In the context of this document, the focus is on potential risks to the environment, human health or worker safety.
Socio-economic issues	Impacts on individuals, institutions, or society resulting from a policy or project (e.g., the introduction of a product, of a market intervention) such as price changes, welfare changes, and employment changes.
Ethics	A branch of philosophy concerned with evaluating human action, in particular what is considered right or wrong based on reason. In the context of nanotechnology, ethical questions have focused, for instance, on applications related to human enhancement and performance, privacy questions resulting from research into nanotechnology monitoring systems, and questions about possible malevolent or military uses of nanotechnologies.
Intellectual property rights & access	Intellectual property rights (IPRs) are legal protections for intellectual property claimed by individuals or institutions. Copyrights, patents and trademarks are common mechanisms for protecting intellectual property. IPRs are intended to spur innovation and commercialization, but may limit the ability of individuals and institutions to access technology.
Public participation & engagement	Processes that affect whether and how individuals participate in societal discourse, including public information, public education, and public discussion and dialogue regarding nanotechnology.
Governance	Processes, conventions, and institutions that determine how power is exercised to manage resources and societal interests, how important decisions are made and conflicts resolved, how interactions among and between the key actors in society are organized and structured, and how resources, skills and capabilities are developed and mobilized for reaching desired outcomes. This includes risk governance (i.e., comprehensive assessment and management strategies to cope with risk) and governance for innovation (i.e., programs targeting nanotechnology R&D for public objectives). Using this definition, governments, governmental and intergovernmental institutions, as well as public and private corporations, non-governmental organizations, and informal associations are examples of institutions involved in governance.
Capacity building	Assistance provided to develop a certain skill or competence, including policy and legal assistance, institutional development, human resources development, and strengthening of managerial systems.
International collaboration & cooperation	Collaborative partnerships between individuals, and institutions from developed and developing countries at a local, national, regional level on any aspect of nanotechnology.
Scalability, delivery, & sustainability	The ability to scale-up production and distribution of products so they reach large numbers of people (i.e., success not limited to pilot projects) and the sustainability of products, which relate to numerous factors including, for example, costs, ease of use, and durability.

Fig. 17.1 Cross-cutting nanotechnology issues

Cross-Cutting Issues	Nanoparticle Filter (Indian Institute of Technology and Eureka Forbes)
Product research & development	No specific comments on this issue for this technology.
Environmental, human health, & safety risks	1. Laboratory studies have determined that the filter is effective for removing the contaminants of concern, in particular pesticides.
	2. Laboratory studies by certified third party labs demonstrated that no nanoparticles were found in the filtered water at the limits of detection of existing testing systems and current standards.
	3. Participants asked how the spent cartridges will be disposed.
Socio-economic issues	Participants asked whether this technology should be applied up-stream and not just at the point of use of drinking water, for example, to prevent people bathing in pesticide contaminated water.
Ethics	Some participants asked whether, given the potential benefits that would accrue to human health due to the successful removal of pesticides from contaminated groundwater, applications such as this be should pursued or whether regulatory frameworks should be set up first.
Intellectual property rights & access	The technology was patented by IIT and licensed to Eureka Forbes.
Public participation & engagement	• Questions were raised about whether the communities and households that will receive these filters were provided an opportunity to learn about the devices and to choose whether they wish to make use of them above and beyond whatever "social marketing" was undertaken.
Governance	• No specific comments on this issue for this technology; see cross-cutting comments in section above and comments directly above regarding EHS risks and Ethics.
Capacity buliding	• No specific comments on this issue for this technology; see cross-cutting comments in section above.
International collaboration & cooperation	• See Next Steps section for details related to collaboration and cooperation.
Scalability, delivery, & sustainability	• A factory is currently under development that will produce 40,000 filters per month. • These filters have undergone accelerated testing and have been determined to produce enough drinking water for a household for 1 year, approximately 5,000 l. • The company producing these filters is developing a video to explain use of the technology. • The filter is gravity driven and does not require power. • The filters will cost US$2.90 and the nanomaterial costs US$0.67. • The filter cartridge needs to be replaced once a year. The company producing the filters will replace the filter cartridge as part of its contract with users. • Participants discussed this technology as an example of the potential of public-private partnerships to develop and deploy such technologies. In conjunction, they discussed the need for an NGO partner to distribute and facilitate the use of the technology and the need for social marketing.

Fig. 17.2 Example of application of cross-cutting nanotechnology issues matrix

In particular, some people have raised questions about the socioeconomic effects of nanotechnology applications that could impact global demand for agricultural, mineral, and other non-fuel commodities (South Centre 2005). The term "commodities" usually refers to "undifferentiated, widely traded raw materials and agricultural products that are traded principally on the basis of price." Impacts on commodity markets are important, because 95 of the 141 developing countries derive at least 50% of their export earnings from commodities. In 2003, fifty-four of those countries depended on non-fuel commodities for more than half of their export earnings (e.g., copper and zinc account for 61% of Zambia's export earnings; cotton makes up 72.7% of Mali's earnings). UNCTAD estimates that a total of two billion people—a third of the global population—are employed in commodity production, half of those in agriculture (United Nations Conference on Trade and Development 2005).

Although the exploitation of natural resources may contribute to economic development and enhanced public welfare, many developing countries that are highly dependent on commodity export as a primary source of revenue appear low on the United Nations Development Programme's Human Development Index. Dependence on revenue from a narrow range of commodities is risky for countries and producers, because they depend on international markets that have shown long-term price declines and sharp short-term price fluctuations. These countries and producers often find themselves with limited access to credit for production inputs, capital for investments, and know how.

Nanotechnology applications are being developed that could impact global demand for agricultural, mineral, and other non-fuel commodities. Some applications of nanotechnology could increase global demand, while others could lead to a decrease in demand for specific commodities. Applications that result in reductions or increases in the demand for commodities could have potentially far reaching socio-economic and other effects in developed and developing countries. The dependence of many developing countries on one or two commodities is likely to accentuate the socio-economic effects resulting from changes in commodity markets in comparison to countries with more diversified economic bases.

17.2.8 Ethics

Some people have identified an ethical tension in balancing the urgency of alleviating the critical needs of developing countries with nanotechnology and ensuring that those populations are not exposed to the unknown, but potentially significant, risks posed by those technologies. This ethical dilemma includes questions about whom (e.g., developed or developing country governments, international organizations, civil society, poor people) should make such risk-benefit decisions and what should be their motivation (e.g., economic competitiveness, humanitarianism, modernization).

A number of reports have discussed the ethical issues related to nanotechnology and equity. Some have said that past science and technology advances such as vaccines have enabled mass application of solutions to human development needs.

But, empirical studies have also shown that introduction of new technologies have resulted in further marginalization of the poor, for instance because the underlying commercial and distribution infrastructure remains in the control of developed countries. Some groups have also argued that the current approach to nanotechnology research and development is overly top-down and does not take into account existing solutions including traditional, alternative, or complementary practices, such as herbal medicine and traditional pest management, despite the cultural significance and an increase in use of these practices in many developing countries.[7]

The U.S. and European governments and researchers in other developed and developing countries are working, to different extents, on nanotechnology research related to human enhancement and performance. Some groups are now asking about the ethical implications of nanotechnology applications, such as materials that enable bone, tissue, and nerve regeneration, that would benefit disabled people in developed countries, but possibly be unaffordable for those in developing countries.

There are also a number of ethical considerations regard nanotechnology's implications on privacy. Some say that nanoscale information gathering systems such as transmitters could be both ubiquitous and invisible because of their small size. Some say that nanoscale health monitoring devices that could be temporarily or permanently implanted in the body could also have negative privacy implications.

17.2.9 Intellectual Property Rights and Access

Several publications have expressed concerns related to IPRs and the impact of patents and technology management strategies on the ability of developing countries to access new technologies. For example, the potential for broad nanotechnology patents on conventional and natural materials at the nanoscale raises the possibility that such patents could give patent owners excessive control over the use of nanoscale materials.

Some groups are urging that the effects of patents, conditions in technology licenses, and impacts of government and corporate policies on people's ability to use nanotechnology for meeting human development be considered now, even though some of the potential benefits of nanotechnology may be years away. Without this discussion, they argue, the technology will be controlled by developed countries and multinational corporations, primarily benefit consumers in developed countries, and lead to a deepened divide between developed and developing countries.

17.2.10 Public Participation and Engagement

With nanotechnology investments continuing to rise and applications proliferating, awareness and understanding regarding the implications of nanotechnology for developing countries is increasing. However, this awareness is still generally limited – few people involved with nanotechnology are considering development issues; few people involved in the development community are considering the potential role of nanotechnology in addressing humanitarian needs. These gaps continue to be a significant concern, as current and future decisions in both developed and developing countries may result in policies, practices and systems that can have long-term impacts on whether nanotechnology can help address specific human development needs and, if so, how quickly. More robust mechanisms are needed to engage the public in dialogue about the responsible innovation and governance of nanotechnology.

17.2.11 Governance

The growing realization that nanotechnology applications are already being developed and used in both developed and developing countries is contributing to heightened interest in discussions about the governance of nanotechnology. A small, but growing number of national and international initiatives are addressing components of the issue, yet, there is still significant confusion about what is meant by "governance of nanotechnology". There are currently a range of opinions on the adequacy of existing systems for governance of nanotechnology versus the need for new nanospecific governance frameworks. Additionally, existing initiatives that are addressing governance are largely taking place in developed countries (national initiatives) or dominated by developed countries (international initiatives).

Many commentators feel that North-South communications about nanotechnology risks are weak among scientists and policy makers, as well as among national-level ministries and international institutions. Addressing these challenges will be critical to the development of a governance framework that addresses both opportunities and risks, is broadly supported by all sectors of society, responds to the needs of developing countries, and can inform decision making at national, regional, and international levels.

17.3 Conclusion

While application of nanotechnology is still regarded to be a relatively young scientific discipline, there are already hundreds of products available (with hundreds more in development) to consumers that utilize the novel characteristics of nanotechnology. With strong competitive and economic drivers likely to accelerate this trend, the perception that nanotechnology products were "years away," which seemed to be the prevailing view only a few years ago, is now being displaced by a growing interest in catalyzing specific actions that support the emergence of creative and appropriate approaches to nanotechnology innovation and governance.

Our analysis of the current dynamics and changing landscape of nanotechnology, informed by the GDNP and other processes, leads us to conclude that there is a pressing need for innovative approaches that: enhance the role of developing countries in responsible nanotechnology development and governance; encourage the development of appropriate products targeted to help meet critical human development needs; and include methods for addressing the safety, appropriateness, accessibility and sustain-ability of nanotechnology to meet the needs of developing countries.

Notes

1. Meridian Institute is a non-profit organization whose mission is to help people solve problems and make informed decisions about complex and controversial societal problems. Meridian's mission is accomplished through facilitation, mediation, information, and consultation services. Meridian's work focuses on a wide range of issues related to environment and natural resources, climate and energy, agriculture and food security, international development, science and technology, health and safety, and security. Meridian Institute works at the local, national and international levels. For more information, please visit http://www.merid.org/.
2. For more information on the International Workshop on Nanotechnology, Water, and Development, please visit http://www.merid.org/nano/waterworkshop/.
3. To inform these discussions, Meridian provided the following background papers: "Nanotechnology, Water and Development"; "Overview and Comparison of Conventional and Nano-Based Water Treatment Technologies"; "Examples of Enabling Technologies for the Development of Nano-Based Water Treatment Technologies"; and "Water and Development News Compilation."
4. Information obtained by Meridian Institute after the workshop indicates that: (1) the filter cartridges will be in the field for a year and replaced, as part of the sales agreement, by Eureka Forbes' local service units; (2) plastics and the useful metals from the used cartridges will be recycled; (3) the remaining balance of material will be incinerated; material remaining from incineration will be landfilled or used as filler (e.g., in brick manufacturing).

References

Bruske-Hohlfeld, I et al. 2005. Do nanoparticles interfere with human health? *GAIA* 14.1: 21–23.
Cientifica. 2003. Nanoporous Materials. http://www.cientifica.com/. Cited 1 June 2007.
Colvin, V. 2002. Responsible Nanotechnology: Looking Beyond the Good News. EurekAlert. http://www.eurekalert.org/context.php?context = nano&show = essays& essaydate = 1102. Cited 1 June 2007.
ETC Group. 2004. Down on the Farm: The Impact of Nano-Scale Technologies on Food and Agriculture (23 November 2004): http://www.etcgroup.org/en/materials/publications.html?id = 80. Cited 1 June 2007.
Goldman, L. and C. Coussens, eds. 2005. *Implications of Nanotechnology for Environmental Health Research*. National Academy of Sciences Roundtable on Environmental Health Sciences, Research and Medicine, Environmental Health Research. http://www.nap.edu/catalog/11248.html. Cited 1 June 2007.
Intermediate Technology Development Group. Power to the People. http://www.itdg.org/html/advocacy/power_to_the_people_paper.htm. Cited 28 November 2004.
Roco, M.C. 2003. Nanotechnology: Convergence with modern biology and medicine. *Current Opinion in Biotechnology* 14: 337–346.

Salamanca-Buentello, F et al. 2005. Nanotechnology and the Developing World. *PLoS Medicine*. doi: 10.1371/journal.pmed.0020097.

Scott, N and H. Chen. *Nanoscale Science and Engineering for Agriculture and Food Systems*. Cooperative State Research, Education and Extension Service, US Department of Agriculture.

Singh, K.A. 12 January 2007. Intellectual Property in the Nanotechnology Economy. *Nanoforum*. http://www.nanoforum.org/. Cited 1 June 2007.

South Centre. 2005. The Potential Impacts of Nano-Scale Technologies on Commodity Markets: The Implications for Commodity Dependent Developing Countries. Geneva, Switzerland.

UN Conference on Trade and Development. 2005. Trends in World Commodity Trade, Enhancing Africa's Competitiveness and Generating Development Gains. Report by the UNCTAD secretariat for the 2nd Extraordinary Session of the Conference of African Union Ministers of Trade, 21–24 November, 2005, Arusha, Tanzania.

UN Millennium Project. 2004. Forging Ahead: Technological Innovation and the Millennium Development Goals. Task Force on Science, Technology, and Innovation. 8 November 2004. http://www.cid.harvard.edu/cidtech/TF10Edit11-8.pdf. Cited 1 June 2007.

UN Millennium Project. 2005. Innovation: Applying Knowledge in Development. Task Force on Science, Technology, and Innovation. http://www.unmillenniumproject.org/reports/reports2.htm. Cited 1 June 2007.

UN Millennium Project. 2005. Health, Dignity, and Development: What Will It Take? Task Force 7 on Water and Sanitation. http://www.unmillenniumproject.org/documents/tf7interim.pdf. Cited 1 June 2007.

Chapter 18
Science Policy and Social Inclusion: Advances and Limits of Brazilian Nanotechnology Policy

Noela Invernizzi

In this chapter Noela Invernizzi explores equity in the Brazilian nanotechnology initiative. From the viewpoint of increasing equality, the initiative has several strong points, according to her analysis. For instance, national capacity is a good thing for a developing country since it helps decrease dependence on foreign sources, and Brazil has moved quickly to establish such capacity. Nevertheless, the initiative holds potential for increasing inequalities. Despite the priority that the Brazilian government places on "social inclusion" and "poverty reduction," the initiative takes no concrete steps in these directions. National competitiveness is the dominant goal, input from social actors has been narrowly drawn, and there is little attention to assessment of risks. The Brazilian example is, unfortunately, not an outlier. National governments frequently justify their science and technology programs by claiming they will benefit all citizens, but rarely are mechanisms put in place that help to distribute the goods generated widely. Invernizzi offers some suggestions for remedying this problem as does Cozzens at the end of this volume.—eds.

N. Invernizzi (✉)
Universidade Federal do Paraná, Curitiba, Brazil
e-mail: noela@ufpr.br

This chapter was peer reviewed. It was originally presented at the Workshop on Nanotechnology, Equity, and Equality at Arizona State University on November 22, 2008.

18.1 Introduction

When the Workers' Party (*Partido dos Trabalhadores*)[1] came into office in Brazil in January 2003, one of its three main goals was to achieve social inclusion and reduce social inequality. The 2004–2007 Pluri Annual Plan, prepared during President Lula's first term, pinpointed concentration of income and wealth, social exclusion, the low rate of job creation, and the fact that increased productivity did not result in higher income for the average working family as the main problems to be faced (Brasil 2004, 5). The new government considered science and technology strategic tools for developing its social agenda. This resulted in social inclusion becoming one of the three main axes of the Strategic Plan of the Science and Technology Ministry (MCT 2004a)[2] and the establishment of the Secretariat of Science and Technology for Social Inclusion.

In the early twenty-first century, nanoscience and nanotechnology (N&N) began to occupy a strategic spot in investment in science and technology (S&T) in the more industrialized countries. Brazil soon followed suit, taking the first steps toward establishing a policy for the sector in late 2000. This policy has been consolidated during President Lula's two terms (the first started in January 2003 and the second term is set to end in December 2010) and has produced rapid results in terms of strengthening Brazilian capabilities in the field of N&N.

The purpose of this chapter is to evaluate to what extent nanotechnology policy design and actions are in line with the government's goals for social inclusion and greater equality. It is my understanding that Brazilian N&N policy is creating some capabilities that are essential to foster an equality-driven development of nanotechnology. I highlight, in Section 18.2, the effective construction of a national policy for N&N, the sustained investment in the upgrading of research infrastructure and qualification of human resources, and the efforts to stimulate better distribution of resources throughout the country.

However, as I will later argue, in Section 18.3, the design and the implementation of the Brazilian nanotechnology policy have a number of limits that may hinder the purpose of the government to use S&T as a tool for promoting social inclusion and reducing inequality. I will point out three limitations: a) social goals are not clearly embedded in the nanotechnology policy rationale; b) there has been little discussion of nanotechnology policy among different social actors; and c) little attention is paid to the assessment of the social implications and risks of nanotechnologies.

I argue throughout this second section that, in order to enlarge the scope of the nanotechnology policy to include social goals more resolutely and directly, this policy needs to supersede, on the one hand, the still prevalent linear model of innovation that connects S&T to social welfare almost automatically. On the other hand, it needs to integrate a more complex perspective of social construction of technology, paying more attention to the context in which nanotechnology is emerging and its social implications, as well as allowing participation of a broader set of social groups to orient nanotechnology development. I will reinforce these points in a brief final comment in which I place the preceding arguments within the current discussion on nanotechnology, poverty, and development.

18.2 Brazilian N&N Policy: Building National Capabilities

Brazil has the strongest capabilities in N&N in Latin America. The country stands out in the region in terms of its research infrastructure, number of researchers, number of publications, and budget allocated to N&N research (OICTI 2009, Kay and Shapira 2009, Chapter 19, this volume). These capabilities are the result of the decisive stimulus given by the Brazilian Ministry of Science and Technology (MCT) to nanotechnology over the current decade and, also, of the conditions created by the historical construction of a system of science and technology for many decades.

In the context of the global emergence of nanotechnology policies in the early 2000s, the Brazilian Ministry of Science and Technology and its National Council for Scientific and Technological Development (CNPq) moved quickly to establish the basis for a national N&N policy. In late 2000, the MCT and the CNPq organized a workshop called "Tendencies of Nanosciences and Nanotechnologies." At this meeting, 32 researchers from different fields reached an agreement about the necessity of a national program of N&N. A working committee was created at the meeting with the purpose of identifying the expertise of Brazil in the field and developing research policy recommendations (CNPq Notícias 2000).

In response to the advice of these groups, in 2001 the CNPq called for inter- and multidisciplinary research projects to run the Cooperative Networks of Basic and Applied Research on Nanosciences and Nanotechnologies with the purpose of creating and consolidating national expertise in this field (Knobel 2002). Three million *reais* (US$1 million according to the exchange rate at the time) were allocated for the creation of four research networks, and another 5 million *reais* were added in 2003. The networks involved 300 researchers, 600 graduate students, 77 universities and research centers and 13 companies from different parts of the country (Toma 2005).[3] Other scientists gathered at four Millennium Institutes oriented to research in different areas of nanotechnology, which were funded with 22.5 million *reais* from 2001 to 2003.[4]

In 2003, during the first year of the Lula government, a group at the MCT began working on a Program for the Development of Nanoscience and Nanotechnology.[5] The Program was submitted for public consultation on the MCT website,[6] and it was later incorporated into the MCT's Multi-year Plan for 2004–2007. The objective of the program was to develop new products and processes from nanotechnology in order to increase the competitiveness of domestic industry. It recommended actions to build and support laboratories and research networks, as well as the implementation of institutional projects focusing on research and development (R&D). The estimated budget for these actions was 78 million *reais* (approximately US$28 million) (MCT 2003).

This first program was strengthened in 2005, giving way to a more comprehensive National Nanotechnology Program launched by President Lula da Silva and the Minister of Science and Technology with a budget of 71 million *reais* (US$31 million) for 2005–2006. The National Nanotechnology Program is better aligned with the goals of another strategic piece of policy, the Industrial, Technological and Foreign Trade Policy established in 2004. Nanotechnology, described in the latter as "a gateway to the future," is considered a strategic area for the enhancement of national competitiveness and the increase of the international share of Brazilian industry.

In the context of this approximation of research policy and industrial policy, and following a positive evaluation of the outcomes of the first four research networks, ten new cooperative research networks, connecting around one thousand researchers, were funded in 2005. These networks were supported with 27.2 million *reais* (US$12 million) for 4 years, under the Brazil Nano Program.[7] Their research profile reflects an orientation towards industrial application, involving cooperation with the productive sector. At the same time, several projects for starting the incubation of nanotechnology companies and R&D projects involving companies and research institutions were funded by FINEP (Financer of Studies and Projects), an agency of the MCT.

The budget earmarked for nanotechnology by the MCT between 2004, when the first nanotechnology program was launched, and 2008 totaled 191.1 million *reais* (MCT 2008).[8] Although this investment is small in the context of global nanotechnology R&D figures, it certainly is significant at the Brazilian and Latin American level. Figures can be seen in Table 18.1 below.

The Brazilian Nanotechnology policy gives priority to a set of fields in which the country would be better prepared for N&N development. The S&T Ministry's Plan of Action for 2007–2010 states that the food, biotechnology, electro-electronics, aerospace, textiles, metal-mechanics, and energy sectors should be given priority for the development of the National Nanotechnology Program (MCT 2007a, 144). Considerably comprehensive, these areas were selected according to their strategic importance for innovation in the national economy. Particular attention is directed to sectors that need to resolve technological bottlenecks (as in the case of the textile sector); sectors in which Brazil has competitive advantages or opportunities, such as tropical agriculture, food, and mineral resources; and sectors with potential for the development of new products and nationalization of technologies (MCT 2007a, 146).

All these actions in tackling research into nanotechnology have benefited from considerable advantages resulting from previous investment in S&T. For several decades, the country has carried out a systematic policy for the qualification of

Table 18.1 Nanotechnology Budget, Brazil (in Brazilian *Reais*)

	2004	2005	2006	2007	2008	Total
Allocated budget	17,500,000	60,300,000	28,400,000	57,700,000	27,200,000	191,100,000
Executed budget	98.32%	91.10%	96.6%	99.8%	n.a.	–

Source: Ministry of Science and Technology. General Coordination of Micro and Nano Technologies

human resources, both at home and abroad, and has constructed a relatively wide-ranging research infrastructure at a number of universities and public research centers. Nanotechnology policy also benefits from the historical learning that has been amassed by several public laboratories and companies in the development of S&T with a view to domestic problems. Of particular importance are the Fundação Osvaldo Cruz, a research center of the Ministry of Health, which is the most important health research institution in Latin America, with considerable experience in tropical diseases; *Embrapa* (Brazilian Agricultural Research Corporation) oriented to research, technology generation and technology transfer for the agribusiness; and *Petrobrás* (Brazilian Oil Company), internationally renowned for its achievements in technological innovation for deep and ultra-deep water oil exploration and production, to name but a few examples. Today, these companies and research centers and a number of other research institutions, such as the Brazilian Synchrotron Light Laboratory (LNLS), the National Institute for Space Research (INPE), the Brazilian Center for Physics Research (CBPF) and the Renato Archer Research Center (CENPRA, dedicated to information technology, microelectronics and automation), are involved in nanotechnology research.

According to the MCT, there are at least 48 universities or other research institutions with laboratories operating in nanotechnology in Brazil (Baibich 2008). The Nanotechnology Program has invested in improving infrastructure for research. Several of the existing laboratories have been upgraded, and new ones were set up in 2006 and 2007, such as the Inmetro Nanometrology Center (the National Institute of Metrology, Normalization and Industrial Quality), the Multi-user Nanotechnology Laboratory of the CETENE (Center for Strategic Technology of the Northeast), the National Nanotechnology Laboratory for Agribusiness at Embrapa, and the Cesar Lattes Center of Nanoscience and Nanotechnology at the Brazilian Synchrotron Light Laboratory.

In terms of human resources, the policy to support research in networks, starting out with four and rising to ten, in addition to sub-networks, has quickly connected researchers from universities and research centers from practically all over the country, thereby strengthening the potential of invested resources and the sharing of infrastructure. Support from the CNPq, specifically aimed at young doctors, has been annually renewed since 2005, and includes a growing number of research projects to ensure that there will be more qualified personnel for the coming years.[9] The MCT estimates that Brazil today has around 3000 nanotechnology researchers, including professors and students (Agência Brasil 2007).

At the end of 2008, the CNPq proposed and financed a new institutional structure, the National Institutes of Science and Technology, intended to occupy a strategic position within the National System of Science, Technology and Innovation.[10] The institutes are characterized by a research network format involving several institutions from different regions of the country led by an excellence-level institution. Ten National Institutes of Science and Technology (NIST), out of 123, are involved in areas of research in nanotechnology.[11]

Another important action when it comes to training personnel, gathering research groups, and sharing infrastructure has been international cooperation within the region, mainly with Argentina. The Bi-national Center of Nanotechnology (CBAN),[12] founded in 2005 by the Argentinean and Brazilian governments, has organized working meetings to join Brazilian and Argentinean researchers, and several editions of the Brazilian-Argentinean School of Nanotechnology, oriented to graduate students of both countries. In 2008, cooperative research projects began to be funded by the CBAN. A similar form of cooperation with Mexico is under study. There is also cooperation with France, through joint research projects carried out by Brazilian and French researchers, and with India and South Africa, under the IBSA Nanotechnology Initiative. The MCT organized two Fact Finding Missions in Switzerland and Japan in order to assess possibilities for collaborations with those countries. In addition, there are several informal international cooperation networks established directly by researchers.

A noteworthy directive of the Science and Technology Ministry under Lula's government has been the policy to channel 30% of the budget for a number of financed projects to the North and North-eastern areas of the country, those which are less developed economically and socially and also have poorer S&T resources. Several calls for research projects and infrastructure upgrading in nanotechnology have obeyed this principle. Although it is difficult to reverse the historical concentration of S&T facilities and human resources in the South-East region (mainly in São Paulo state), this equalizing policy, already in place for several years, may contribute to creating N&N capabilities in less developed regions. It is interesting to note that the North-Eastern region has been dynamically integrated into the development of N&N by a core group of highly qualified researchers at the Federal University of Pernambuco. Table 18.2 shows the distribution of funding through public calls for projects among diverse regions of the country. Table 18.3 shows the distribution of the research networks by region. Brazilian regions can be identified in Fig. 18.1.

Table 18.2 Nanotechnology budget allocated through public calls for projects by regions (in Brazilian *Reais*)

Region	2004	2005	2006	2007
Center-West	–	3,074,070	516,100	7,363,720
North	–	15,000	50,000	136,650
North-East	402,500	8,948,960	1,420,520	7,298,230
South	641,240	6,553,960	5,814,390	6,893,200
South-East	4,244,080	18,279,920	18,277,200	32,237,530

Source: Ministry of Science and Technology. General Coordination of Micro and Nano Technologies

Table 18.3 Research networks by region

Research network	Activity period	Head institution	State and region of head institution	Other Regions involved
Nanostructured materials	2001–2004	Universidade Federal do Rio Grande do Sul, UFRGS	Rio Grande do Sul/South	North-East South-East South
Molecular nanotechnology and interphases	2001–2004	Universidade Federal de Pernambuco, UFPE/ Universidade de São Paulo, USP	Pernambuco/North East; São Paulo/ South-East	South-East North-East South
Nanobiotechnology	2001–2004	Universidade Estadual de Campinas, Unicamp	São Paulo/South-East	South South-East North-East Center-West
Semiconductor, nanoinstruments and nanostructured materials	2001–2004	Universidade Federal de Pernambuco, UFPE	Pernambuco/ North-East	North-East South-East South Center-West
Nanophotonics	2005–2008	Universidade Federal de Pernambuco, UFPE	Pernambuco/ North-East	North-East South-East
Nanobiotechnology and biostructured materials,	2005–2008	Universidade Federal do Rio Grande do Norte, UFRN	Rio Grande do Norte/North-East	North-East Center-West
Molecular nanotechnology and interphases—phase II	2005–2008	Universidade Federal de Pernambuco, UFPE	Pernambuco/ North-East	North-East South-East South
Nanobiomagnetism	2005–2008	Universidade de Brasília, UnB	Brasília, Distrito Federal/Center-West	Center-West South-East
Nanostructured coatings	2005–2008	Pontifícia Universidade Católica do Rio de Janeiro, PUC-RJ	Rio de Janeiro/ South-East	South-East South
Scanning Probe Microscopy. Software and hardware	2005–2008	Laboratório Nacional de Luz Sincrotron, LNLS	São Paulo/ South-East	South South-East
Carbon Nanotubes: Science and applications	2005–2008	Universidade Federal de Minas Gerais, UFMG	Minas Gerais/ South-East	North-East South-East South

Table 18.3 (continued)

Research network	Activity period	Head institution	State and region of head institution	Other Regions involved
Simulation of nanostructures	2005–2008	Universidade de São Paulo, USP	São Paulo/South East	South-East South
Glyco-nano-biotechnology	2005–2008	Universidade Federal do Paraná, UFPR	Paraná/South	South North-East South-East
Nanocosmetics: Conceptualization and applications	2005–2008	Universidade Federal do Rio Grande do Sul, UFRGS	Rio Grande do Sul/South	South South-East

Source: Ministry of Science and Technology. General Coordination of Micro and Nano Technologies

The preparation of a national policy for N&N was a fundamental step towards the rapid build-up of national capabilities, which, furthermore, tend to be better distributed regionally. The systematic investments in the field favor the consolidation of research groups who have good working conditions in the country, preventing

Fig. 18.1 Brazilian regions

brain-drain. All these conditions are essential for including nanotechnology research in a development program for a country that aims to reduce social inequality. Nevertheless: are these conditions sufficient?

18.3 Will Nanotechnology Help to Reduce Social Inequalities? Policy Limits and Possibilities

The elaboration of a national policy that guides nanotechnology development coherently and the construction of national capabilities throughout the country are essential points of departure for directing nanotechnology to social goals. In this respect, Brazilian MCT has been very coherent and successful. However, as I will argue in this section, Brazilian development of nanotechnology faces the risk of not contributing in a significant manner to the social inclusion and inequality reduction goals that are guidelines of the MCT's strategic plan. The risk lies on three limitations of nanotechnology policy and actions. First, social goals are not clearly embedded in the nanotechnology policy rationale; they are just expected as an output of competitiveness enhancement. Second, there has been little discussion of the nanotechnology policy among different social actors, so they can be informed and advance their interests. Third, up to now nanotechnology policy and actions have not focused enough attention on the assessment of social implications and risks of nanotechnologies, a factor that may reinforce existing inequalities and create new ones.

18.3.1 The Need to Embed Social Goals

Concerning the first limitation, it is useful to start by asking: What does the MCT's establishment of social inclusion and inequality reduction as one of the three main guiding axes for its plan for science and technology mean? I understand that this statement has at least two very important connotations. It means, on the one hand, that people traditionally excluded from the benefits of science and technology should be included. On the other hand, it means that the science and technology research agenda should address specific problems to improve the living conditions of socially disadvantaged groups. To what extent does nanotechnology policy respond to these two issues? The former is expected to be reached as a trickle-down effect of a more competitive economy. The latter is absent. Let us discuss this further.

Brazilian nanotechnology policy places almost total emphasis on advancing this emerging technology to increase national competitiveness, as clearly stated in the objective of the MCT Plan for Nanotechnology (2004–2007): "To encourage research activities, development of new products and processes and the transfer of technology between academia and companies with a view to technological innovation in order to encourage competitiveness in domestic industry" (MCT

2004a, 143).[13] As affirmed in another MCT document, by increasing competitiveness and the added value of production through nanotechnology, a set of positive results for society are expected, such as reaching a national mass market, creating higher-quality jobs, and reducing dependence on imports (MCT 2003, 9).

In terms of policy rationale, this way of understanding the connection between science, technology, economic growth, and social benefits, is very much like the so-called linear model of innovation. Certainly, enhancing competitiveness is important and necessary for the country, and economic growth may have some of the expected positive trickle down effects that help reduce inequality, such as employment opportunities for some labor force sectors (while having adverse effects for others) and access to some less expensive goods by a wider population. However, it is doubtful that a nanotechnology policy oriented to increased competitiveness *only* will be sufficient to achieve benefits for disadvantaged people, helping to reduce inequality. The insufficiency is twofold. First, it is very uncertain that the expected broad social benefits from the linear chain of innovation will in fact happen in a context characterized by notorious social inequality, a high informal labor market, and a great deal of people almost excluded from market production and consumption. The second problem with this linear rationale is that stimulus for innovation and competitiveness does not ensure that the resulting technological developments will satisfy the needs or solve the problems of those who are excluded. Guided by market pressures, the nanotechnology trajectories and products may be mostly oriented to satisfy the needs of affluent consumers, helping to intensify social inequalities. The well known 90/10 gap in health research, or the fact that only a mere 10% of all health research funding is being used to address 90% of the world's burden of disease, suffered primarily in developing countries, is a paradigmatic example of how innovation may not correspond to the needs of large poor populations (Global Forum for Health Research 2000).

Thus, a limitation of the National Nanotechnology Program, when it comes to its capacity to contribute toward reducing inequalities, is the lack of objectives that are *directly* tackling problems affecting the less privileged sectors of the population. In this respect, it would be necessary to enlarge the nanotechnology policy horizons establishing some priority research areas oriented to search for solutions to the problems resulting from persistent social inequality such as nutrition, housing, diseases, environmental degradation, etc. Another important focus is to explore the potential of nanotechnologies for low scale production processes that are the source of living for a large low income population. Current programs on social technologies carried out by the MCT Secretariat of Science and Technology for Social Inclusion may stimulate, in this way, the development of some emerging nanotechnologies as social technologies.[14]

18.3.2 Including More Groups in the Discussion

I turn now to another aspect that reveals a weak insertion of social goals in nanotechnology policy design and actions. Steering nanotechnology towards greater social equality requires a wide-ranging discussion of the objectives and aims that include perspectives, interests, and values from a number of social groups and organizations. Nevertheless, the National Nanotechnology Program has been developed with limited public discussion. Scientists have been the most relevant actors in conceiving and managing it. For the most part, policy makers involved are also researchers in the field that occupy positions in the government (Invernizzi 2008). Besides scientists and the government, the only representatives from society have been businessmen.

When the first proposal of a national policy for nanotechnology was opened for public consultation on the MCT website in late 2003, the main actors that sent their opinions were researchers in the field and, occasionally, corporate leaders. One social scientist and one science journalist were the most "diverse" participants (MCT 2004b). Scarce culture of public participation, a short consultation period, and the fact that the subject was little known to the public at that time were the main factors for the lack of participation.

All of the representatives of the policy management committees, such as the Director Council of the Brazilian Nano Network (a program that manages the ten cooperative research networks), are scientists, government officials, and people from the business community. The Science and Technology Ministry's Consulting Committee for Nanotechnology, which advises the Ministry on the goals to be attained and the use of funding for research, is entirely made up of researchers and government officials.[15] Discussions concerning regulations are being held by the Brazilian Technical Regulations Association. Among the issues under consideration are process and product safety. But they are being discussed only by specialists in regulations, nanotechnology researchers, and business representatives.[16] Consumer associations and workers are not represented, even though they are interested parties. Additionally, all participation by scientists is exclusively that of researchers from the fields of physical and natural sciences and engineering; the presence of social scientists is almost non-existent.

Scientists, with their expertise, are obviously key actors in discussing and constructing nanotechnology development. However, they are not representatives of the interests of different social groups in society. Moreover, they have their own specific interests in research and funding issues. Businessmen, in turn, are clearly representative of their corporate interests. But there are many other groups with interests concerning nanotechnologies, such as workers and unions that will see production processes and employment conditions change and eventually face new risks in the workplace; consumer organizations that demand regulations, safety, and transparent public information on new products; environmental groups that are interested in potential nanotechnology solutions to environmental problems and are also worried about potential new problems; and solidarity economy associations that are looking for technologies adapted to their particular forms of production, etc. If nanotechnology is such an important technology, so relevant for the development of the country in the coming years, and has a potential to contribute to ameliorate the living conditions of the people, as nanotechnology policy states then, more voices should be heard when it comes to discussing the goals and directions of this development. It is time for the country to advance the democratization of science and technology discussion.

18.3.3 Assessing the Social Implications of Technological Change

The third point I wish to highlight is the scarce practical importance given to the evaluation of the social implications and potential risks of nanotechnology. All major technological changes have broad social implications, and these implications are not evenly distributed within society. On the contrary, there are winners and losers (Crow and Sarewitz, 2001). Poorer and powerless social groups are likely to be more affected by implications such as employment shifts and obsolescence of skills. This is also the case of environmental and health risks related to technology. Even if some technological risks are likely to affect the whole society, they may hit harder those groups that have the worst living conditions and are weakly organized. Social implications and risks, even if unintended, may generate new forms of inequality. For this reason, assessing them in real time, accompanying the development of nanotechnologies, and taking preventative measures, should be a central aspect in a nanotechnology policy with a view to social equality.

The main documents that deal with Brazilian nanotechnology policy include brief references to its economic, social, or ethical implications. For instance, the 2003 document that served as a basis for the Nanoscience and Nanotechnology Development Program of the 2004–2007 Pluri Annual Plan stated that one of its goals was to "inform society of the impacts of nanotechnology on the life of the average citizen, new opportunities and risks of obsolescence that the technology may have in store for current products and processes" (MCT 2003, 9); however, this subject was not included among the topics that were considered as a priority. The lack of a social and environmental component in the program was raised during the public consultation of the aforementioned document. The Science, Technology and Innovation Program for Nanotechnology for the S&T Ministry's 2007–2010 Plan touches on the need to "set policies on ethical matters and the social impact of nanotechnology based products" (MCT 2007a, 144), but does not go into detail about what these policies should be.

Therefore, the very subtle presence of social, economic, and ethical implications of N&N in documents has been translated into marginal importance being given to these issues in concrete actions such as funding research and qualification of human resources. Between 2001 and 2008, a period during which the number of nanotechnology research calls and the number of projects funded by the CNPq increased significantly, only one call for research had the economic, social, and ethical implications of nanotechnology as a specific goal, and only four projects were funded.[17] Other calls included the subject, but projects were not funded. For instance, the call for projects for the formation of cooperative research networks in 2005 included a proposal for studies into the social, ethical, and environmental impacts of nanotechnology and information and education concerning nanoscience and nanotechnology in addition to basic research and the development of innovative products and processes. However, none of the ten networks funded had these subjects as a research focus. Likewise, in addition to basic experimental or theoretical research and the development of new nanotechnology-based products, the calls aimed at young doctors in 2005, 2006, and 2007 were intended to support research into their social, ethical, and environmental impacts. In this case too, no project has been funded concerning the latter issues. In the most recent call for young doctors, this subject has been excluded. None of the calls for projects I analyzed stressed the importance of putting together multidisciplinary research teams made up of social and natural scientists. In a nutshell, the support for nanotechnology has not been accompanied by a more active concern about increasing knowledge regarding its economic, social, and ethical implications.[18]

The potential risks of nanotechnology play an even more marginal role on the research agenda. The policy documents and the calls for research do not clearly include any directive concerning the assessment of potential risks. A call for research for the assessment of the potential risks of products containing nanotechnology, which was supposed to be made in 2007 by the CNPq, was reworked a number of times and ended up not being issued for unexplained reasons (MCT 2007b). There are at least two government agencies involved in nanotechnology risks assessment. One is the Brazilian Technical Regulations Association, which is currently analyzing risks as one of the topics included in international debates over nanotechnology regulations. The other is Fundacentro, an agency of the Labor and Employment Ministry for Health and Safety in the Workplace, which is carrying out studies on the risks nanoparticles pose to workers. Fundacentro studies were begun because of the interest of researchers and not because they were requested by the government.

To sum up, the National Nanotechnology Program lacks the direct guidelines for the social goals proposed in the MCT plans and the government program in general. In spite of its potential, nanotechnology could hardly be expected to make a contribution towards improving the living conditions of the less privileged groups only through increased competitiveness and economic growth. More specific social goals, defined by distinct social actors, should be on the research agenda for an equality-driven development of nanotechnology. On the other hand, new social problems will arise from the development of nanotechnology. It is necessary to make more space on the agenda for research into the economic, social, and ethical implications of nanotechnology and its potential risks by multidisciplinary teams of natural and social scientists. More knowledge of these issues is essential when it comes to preparing preventative and compensatory policies aimed at reducing the risk of creating new types of inequality and social exclusion as a result of the technological and economic transformations we have begun to face.

18.4 Final Remarks

In recent years, the implications of nanotechnology for developing countries and poor populations have been the focus of a heated debate (Leach and Scoones 2006;

Salamanca-Buentello et al. 2005; Brahic and Dickson 2005; Meridian Institute 2005; Juma and Yee-Cheong 2005; Hassan 2005; RS and RAE 2004). In this debate, I and other authors have pointed out the limits of some positions that we have called instrumentalist interpretations, those which highlight the technical potential of nanotechnology in order to overcome some poverty problems such as poor access to drinking water, energy, food, cures for diseases, etc. Focusing on the technology itself, this approach usually overlooks the strength of the economic and social trends at the heart of which nanotechnology is being developed and shaped. We have argued that, in a context in which there are considerable social and economic inequalities, it would be difficult for the benefits of nanotechnology to be widely distributed (Invernizzi, Foladori and Maclurcan 2007; Invernizzi and Foladori 2005).

From a contextual viewpoint that stresses the social context in which technology is produced, used, or adapted, technologies materialize social relationships, interests, political power, and values. Discussing the potential of nanotechnology for development with greater social equality implies recognizing that factors such as profit-driven innovation, intensified global competition, increasingly stringent and restricted intellectual property rights, the concentration of innovation in more industrialized countries with their affluent markets, and the elevated levels of poverty and global inequality are significant barriers. For this reason, the goals of competitiveness which guide Brazilian nanotechnology policy are not enough to ensure the social distribution of its benefits.

In order to guide nanotechnology to make a significant contribution to social inclusion and reduced inequality, improving the living conditions of a large part of the population of the country, a new rationale and new lines of action need to be integrated into the nanotechnology policy. In the first place, problems that lie at the heart of social inequality should be incorporated as an input to design the research agenda in order to spur on research priority areas that orient nanotechnology outcomes to their solution. A second decisive point is to assure public discussion and the widespread participation of concerned social groups in discussing the development of nanotechnology, guaranteeing that it will benefit the society as a whole. Finally, it is urgent to encourage research into the social implications and risks of nanotechnology, and to take measures to avoid or compensate potential new social inequalities that may arise from it. In doing so, the Brazilian National Nanotechnology Policy will make a more decisive commitment to the social goals of the current government.

Notes

1. President Luis Inácio Lula da Silva was elected to office with 61% of the popular vote, with strong support from those in lower income brackets. His Workers' party (*Partido dos Trabalhadores*) has deep roots in the trade unions and workers' movements led by Lula, himself a steel worker in the late seventies and early eighties. The union movement and the party itself played an important role in the country's return to democracy in 1985, following a long period of military dictatorship that had led to heavy economic growth (the so-called "economic miracle") and extremely high income concentration. Lula took office after a decade of

neo-liberal reforms with adverse social effects such as unemployment, increased gray market labor and lower salaries in real terms, which had helped maintain the status quo of inequality and poor distribution of wealth.
2. The MCT Strategic Plan for 2004–2007 established three guiding axes: (1) activities to support the Industrial, Technology and Foreign Trade policy; (2) National Strategic Goals, including the space program, the nuclear program, the Amazonian program, and international cooperation; and (3) Social Inclusion, comprising several programs such as: popularization of science, technological education, local productive arrangements, social technologies, popular cooperatives, among others (MCT 2004a).
3. The four networks were: Nanostructured Materials; Molecular Nanotechnology and Interphases; Nanobiotechnology; and the Network of Semiconductor, Nanoinstruments and Nanostructured Materials.
4. The Millennium Institutes program was supported by the World Bank in several Latin-American countries. The goal of the program was to integrate research groups in networks, boost the use of the national research infrastructure and connect national scientists with international research centers in order to promote excellence level research in strategic areas, including nanotechnology. This program has recently been substituted by the more extensive National Institutes of Science and Technology Program.
5. Also during Cardoso's government the first proposal for a National Program of Nanotechnology was designed. However, with the change of government, this project was interrupted.
6. This document, prepared by scientists from the field of nanotechnology and policy makers (with these latter often being researchers in the field themselves), was posted on the Science and Technology Ministry's website for public consultation. However, there were very few opinions from people other than researchers in the field and, eventually, businessmen (MCT 2004b). I will return to this consultation later.
7. The ten networks research in the areas of nanophotonics, nanobiotechnology and biostructured materials, molecular nanotechnology and interphases, nanobiomagnetism, nanostructured coatings, microscopy, carbon nanotubes, simulation of nanostructures, glyco-nanobiotechnology, and nanocosmetics.
8. This figure corresponds to US$95,600,000 with the exchange rate of 1 dollar = 2 *reais* (an average of the rates in October 2008). However, this rate has changed through the period analyzed.
9. The support of young doctors has been secured through MCT/CNPq Bulletins 028/2005, 42/2006, 09/2007 and 62/2008.
10. CNPq Bulletin Number 15/2008.
11. National Institute of Science and Technology of Molecular and Nano-structured Systems, National Institute of Science and Technology of Micro and Nanoelectronics Systems, National Institute of Science and Technology of Carbon Nanomaterials, National Institute of Science and Technology for Integrated Markers, National Institute of Science and Technology of Semiconductor Nano Devices, National Institute of Science and Technology of Nanobiotechnology and National Institute of Science and Technology of Nano Bio Pharmaceuticals. (http://www.cnpq.br/resultados/2008/015.htm). (accessed August 5, 2010).
12. http://www.mct.gov.br/index.php/content/view/24251.html. (accessed August 5, 2010).
13. This and the following quotes from MCT documents are translations made by the author from the original in Portuguese.
14. Social Technologies are defined as technologies that help to make viable social inclusion and self-managed undertakings (Dagnino 2004)
15. Cf. MCT Regulation Number 614, of 01 December, 2004, which establishes the BrasilNano Network as one of the elements of the Nanoscience and Nanotechnology Development Program within Industrial, Technological and Foreign Trade Policy and MCT Regulation Number 322, of May 28, 2005, which establishes the Consultant Committee for the field of nanotechnology (http://www.mct.gov.br/index.php/content/view/72255.html). (accessed August 5, 2010).

16. http://www.fisica.ufc.br/redenano/abnt2007.pdf. (accessed August 5, 2010).
17. MCT/CNPq Bulletin Number 013/2004
18. Nevertheless, a network for the study of Nanotechnology, Society and the Environment (Renanosoma) was founded in 2004, uniting thirty-five researchers from twenty-eight institutions (see www.nanotecnologia.incubadora.fapesp.br). (accessed August 5, 2010). Renanosoma has organized an international seminar every year since then, promoting interaction between natural and social scientists, and is currently very active in the public communication of nanotechnology matters. This network is not funded by government-sponsored actions for nanotechnology. However, members of this network have at times been granted funding from the CNPq or other agencies' calls for research in social sciences or public communication of science to develop research or outreach activities in nanotechnology and society issues.

References

Agencia Brasil. 2007. Brasil investiu cerca de R$150 milhões em nanotecnologia em cinco anos. http://www.agenciabrasil.gov.br/noticias/2007/05/11/materia.2007-05-11.7164383133/view . (accessed January 24, 2010).

Baibich, Mario. 2008. Fortalecimiento del uso de nuevas tecnologías. El Programa Nacional de Nanotecnología en Brasil. Seminario Internacional sobre Políticas de Ciencia, Tecnología e Innovación, Colciencias, Bogotá, http://www.slidefinder.net/a/articles_157811_archivo_ppt7/952976. (accessed August 5, 2010).

Brahic, Catherine, and David Dickson. 2005. Helping the poor: The real challenge of nanotech. *SciDev.Net* February 21. http://www.scidev.net/content/editorials/eng/helping-the-poor-the-real-challenge-of-nanotech.cfm. (accessed August 5, 2010).

Brasil Governo Federal. 2004. Plano Pluri Anual 2004–2007. Orientação estratégica de Governo Um Brasil para todos: Crescimento sustentável, emprego e inclusão social. www.defesanet.com.br/docs/ppa_2004_2007.pdf. (accessed August 5, 2010).

CNPq Notícias. 2000. Reunião de Trabalho "Tendências em Nanociências e Nanotecnologias". Uma iniciativa da Secretaria de Políticas e Programas do MCT e do CNPq. Brasília, 22 de Novembro de 2000. http://www.memoria.cnpq.br/noticias/noticia05_040401.htm. (accessed August 5, 2010).

Crow, Michael, and Daniel Sarewitz. 2001. Nanotechnology and societal transformation. Paper presented at the National Science and Technology Council Workshop on Societal Implications of Nanoscience and Nanotechnology, Sept. 28–29, in Washington, DC, USA.

Dagnino, Renato. 2004. A tecnologia social e seus desafios. In *Tecnologia Social, uma estratégia para o desenvolvimento*. ed. Antonio De Paulo. São Paulo: Fundação Banco do Brasil, 187–210. http://www.ige.unicamp.br/site/publicacoes/138/A%20tecnologia%20social%20e%20seus%20desafios.pdf. (accessed August 5, 2010).

Global Forum for Health Research. 2000. The 10/90 report on health research 2000. *Geneva: WHO*. http://www.globalforumhealth.org/Media-Publications/Publications/10-90-Report-on-Health-Research-2000. (accessed August 5, 2010).

Hassan, Mohamed. 2005. Nanotechnology: Small things and big changes in the developing world. *Science* 309(5731): 65–66.

Invernizzi, Noela. 2008. Visions of Brazilian scientists on nanosciences and nanotechnologies. *NanoEthics. Ethics for Technologies that Converge at the Nanoscale* 2: 133–148.

Invernizzi, Noela, and Guillermo Foladori. 2005. Nanotechnology and the developing world: Will nanotechnology overcome poverty or widen disparities? *Nanotechnology, Law and Business* 2: 2–11.

Invernizzi, Noela, Guillermo Foladori, and Donald Maclurcan. 2007. Nanotechnologys' controversial role for the South. *Science, Technology and Society* 13(1): 123–148.

Juma, Calestous, and Yee-Cheong Lee. 2005. Innovation: Applying knowledge in development. London, Sterling, VA: Earthscan, Millennium Project. www.unmillenniumproject.org/documents/Science-complete.pdf. (accessed August 5, 2010).
Kay, Luciano., and Philip Shapira. 2009. Developing nanotechnology in Latin America. *Journal of Nanoparticle Research* 11: 259–278.
Knobel, Marcelo. 2002. Nanoredes. *Com Ciência Revista Eletrônica de Jornalismo Científico* 37. http://www.comciencia.br/reportagens/nanotecnologia/nano11.htm. (accessed August 5, 2010).
Leach, Melissa, and Ian Scoones. 2006. The slow race: Making technology work for the poor. London: Demos. http://www.demos.co.uk/files/The%20Slow%20Race.pdf. (accessed August 5, 2010).
MCT (Ministério da Ciência e da Tecnologia). 2003. Desenvolvimento da nanociência e da nanotecnologia. Proposta do Grupo de Trabalho criado pela Portaria MCT n° 252 como subsídio ao Programa de Desenvolvimento da Nanociência e da Nanotecnologia do PPA 2004–2007. http://www.mct.gov.br/upd_blob/0002/2361.pdf. (accessed August 5, 2010).
MCT (Ministério da Ciência e da Tecnologia). 2004a. Plano estratégico do MCT 2004-2007. http://ftp.mct.gov.br/sobre/pdf/plano_estrategico.pdf (accessed January 25, 2010).
MCT (Ministério da Ciência e da Tecnologia). 2004b Relatório sobre a Consulta Pública ao Documento elaborado pelo GT de nanotecnologia. http://nanotecnologia.iv.org.br/portal/referencias/documentos/RELATORIO%20SOBRE%20A%20CONSULTA%20PUBLICA%20AO%20DOCUMENTO%20ELABORADO%20PELO%20GT%20DE%20NANOTECNOLOGIA. pdf/view (accessed August 5, 2010).
MCT (Ministério da Ciência e da Tecnologia). 2007a. Ciência, Tecnologia e Inovação para o Desenvolvimento Nacional. Plano de Ação 2007–2010. http://www.mct.gov.br/upd_blob/0021/21439.pdf (accessed August 5, 2010).
MCT (Ministério da Ciência e da Tecnologia). 2007b. Relatório de gestão Exercício 2007. Secretaria de Desenvolvimento Tecnológico e Inovação Coordenação Geral de Micro e Nanotecnologias. http://www.mct.gov.br/upd_blob/0025/25096.pdf (accessed August 5, 2010).
MCT (Ministério da Ciência e da Tecnologia). 2008. Relatório analítico Programa de C,T&I para Nanotecnologia. Coordenação Geral de Micro e Nanotecnologias. http://www.mct.gov.br/upd_blob/0028/28213.pdf. (accessed August 5, 2010).
Meridian Institute. 2005. Nanotechnology and the poor: Opportunities and risks. Closing the gaps within and between sectors of society. http://www.meridian-nano.org/gdnp/NanoandPoor.pdf. (accessed January 24, 2010).
OICTI. 2009. La nanotecnología en Iberoamérica. Situación actual y tendencias. Observatorio Iberoamericano de Ciencia, Tecnología e Innovación del Centro de Altos Estudios Universitarios de la Organización de Estados Iberoamericanos. http://www.oei.es/observatoriocts/index.php?option=com_content&view=article&id=12&Itemid=3. (accessed August 5, 2010).
RS and RAE (Royal Society and Royal Academy of Engineering). 2004. Nanoscience and nanotechnologies: Opportunities and uncertainties. *Policy* document 20/04. London: The Royal Society and The Royal Academy of Engineering. http://www.nanotec.org.uk/finalReport.htm. (accessed August 5, 2010).
Salamanca-Buentello, Fabio, Deepa L Persad, Erin B. Court, Douglas K. Martin, Abdallah. S. Daar, and Meter A. Singer. 2005. Nanotechnology and the developing world. *PLoS Medicine* 2(5) http://medicine.plosjournals.org/perlserv/?request=get-document&doi=10.1371/journal.pmed.0020097. (accessed August 5, 2010).
Toma, Henrique E. 2005. Interfaces e organização da pesquisa no Brasil: Da química à nanotecnologia. *Química Nova* 28(suppl.): S48–S51. http://www.scielo.br/pdf/qn/v28s0/26775.pdf. (accessed August 5, 2010).

Chapter 19
The Potential of Nanotechnology for Equitable Economic Development: The Case of Brazil

Luciano Kay and Philip Shapira

Luciano Kay and Philip Shapira take a more detailed look at the consequences of Brazilian nanotechnology, using publication and patent data. Brazil is the powerhouse of Latin America in terms of research output in nanotechnology, and national policy tries hard to link that output to innovation outcomes. However, as with most Brazilian science, the regional distribution is highly uneven, and nanotechnology shows no signs of equalizing it. Kay and Shapira examine Brazil's performance on four criteria of equitable economic development: agenda setting; R&D investment; R&D outcomes; and risk awareness and allocation. Echoing some of Invernizzi's observations, they note that broader participation in agenda setting is low; little attention is being paid to social or environmental issues in R&D investments; commercialization may be weak since corporate engagement is low; and little attention has been given to risk. They conclude that the potential for nanotechnology to contribute to equity in Brazil is not yet being fulfilled.—eds.

L. Kay (✉)
School of Public Policy, Georgia Institute of Technology, Atlanta, GA, USA
e-mail: Luciano.kay@gatech.edu

This chapter was peer reviewed. It was originally presented at the Workshop on Nanotechnology, Equity, and Equality at Arizona State University on November 22, 2008.

19.1 Introduction

The links between scientific research and economic development have long been a topic of interest for science and technology (S&T) policy, particularly with regard to how research in science stimulates, transfers to, and interacts with the development of new knowledge-intensive industries and enterprises. In general, the literature finds that scientific research can be an important factor in economic and business development, although the mechanisms through which science contributes to development are many and often diffuse (Fleming and Sorenson 2004). These mechanisms may operate at different levels within and across national systems of innovation. For instance, from a macro-level perspective, it has been suggested that scientific research generates capabilities that are the base for the creation of knowledge and the development of new technologies. In turn, such capabilities may strengthen the technological performance of national systems of innovation. In other words, a positive relationship is implied between scientific capabilities and national technological performance (Van Looy et al. 2006).

Universities and public research organizations are important contributors to the development of scientific and technological capabilities in different forms across countries and economic sectors (Mazzoleni and Nelson 2007). Of course, other processes and factors (such as access to finance or the inflow of talented people) also contribute to the fostering of technological innovation. Nonetheless, universities and research institutions are key actors in the process of increasing indigenous capabilities, particularly when their research programs occur in application-oriented sciences and engineering, and are directed towards problem-solving and the advancement of technologies of interest to demanding user communities. Furthermore, science may have strong effects on fostering convergence in inventive outcomes (i.e., clarifying the solution space) leading inventors directly to useful applications, while providing researchers with motivation to continue their inventive search even when facing failure (Fleming and Sorenson 2004).

However, the linkages between scientific research and economic development are not sequential or linear. For instance, although there is evidence of a strong correlation between scientific publication and a country's gross domestic product (GDP,) a direct relation between new knowledge production and economic development is more difficult to prove. A general implication of this for policy-making is that investment in science does not necessarily result directly in marketable products or enterprise development. This implication is particularly relevant for developing countries with limited resources and where open-ended increases in spending for science are likely to be neither feasible nor effective (Vinkler 2008). For most developing economies, increased spending in science needs to be balanced against investments in education or infrastructure where more direct development effects are probable.

Nanotechnology is among the most recent of scientific domains to raise issues about the relationship between science and economic development. Nanotechnology involves "the understanding and control of matter at the scale of approximately 1–100 nanometers where unique phenomena enable the design and production of materials, devices and systems which have novel applications" (NTSC 2007). The application of nanotechnology is anticipated to foster a new wave of technology-led business growth and innovation over the next decade (Lux 2007). Many developed economies, including the U.S., Japan, and member states of the European Union, have greatly increased public research investment in nanotechnology since 2000. A fresh twist has been the rise of China—still a developing economy, albeit a fast-growing one—as the second largest producer of research publications in nanotechnology (Youtie et al. 2008a). This has motivated other medium-level developing countries, including some in Latin America, to increase their research investments in nanotechnology—or risk falling further behind.

An important feature of nanotechnology lies in the breadth of its potential impacts. Similar to information and communication technology (ICT), nanotechnology has been characterized as an emerging general purpose technology with pervasive effects spread across all sectors of the economy and having profound consequences for growth and productivity as well as individual firms (Shea 2005; Youtie et al. 2008b). Nanotechnology may simultaneously disrupt current business practices, create needs for new technological skills in the workforce, and foster new opportunities for start-up and existing enterprises. The extent of nanotechnology's potential reach—not only in high-technology sectors but also in agriculture, energy, medicine, and traditional industries such as pulp and paper—bolsters arguments that developing countries must also expand capabilities across nanotechnology's multiple disciplines in order to retain, if not advance, industrial competitiveness.

Additionally, developing countries need to develop capabilities in nanotechnology to avoid potential problems associated with some applications of the technology. It has been pointed out that diverse risks are associated with aspects of this emerging technology, including health and environmental risks (Glänzel et al. 2003; Roco 2003; Maynard 2006). Expertise across the domain of nanotechnology and related technologies will be necessary to pinpoint and minimize risks and to regulate applications, including in developing countries (Invernizzi and Foladori 2005).

Given this mix of opportunities and constraints, can the development of nanotechnology contribute to the reduction of gaps between rich and poor in developing countries? This is immediately a problematic question, because we are still uncertain about how scientific research influences economic development and about what will be the balance of effects that nanotechnology may have on agriculture, industry, and overall economic growth. Moreover, we note that the process of scientific research in developing countries (as perhaps elsewhere) is itself not evenly distributed in terms of social participation, geographic location, or institutional participation. So, absent other changes, investing more resources in science in developing countries may reinforce, rather than reduce, existing patterns of inequality. Nonetheless, the goal of this chapter is to probe the potential roles of nanotechnologies in influencing social inequalities, drawing on the case of a developing country. To reduce social inequalities, we will argue that S&T policy should include nanotechnology research components that are able to build capabilities in lagging regions, solve environmental problems in poor communities, or empower disadvantaged communities to enable knowledge absorption and innovation spawning, as already suggested in other literature (Cozzens 2007; Youtie et al. 2008b). These are not the only ways to reduce inequalities, and there is much more to reducing inequality than nanotechnology development can be expected to address. However, if nanotechnology research is targeted along these lines, it could make a valuable contribution. Whether nanotechnology research is already doing so is the empirical question that we explore in the chapter.

Particularly for developing economies, it may be the case that scientific research in subject areas aligned with natural endowments, existing industrial assets, or urban problems have useful direct effects. Those areas include research that may assist primary economic activities such as agriculture that poorer communities depend on for their sustenance. However, if indigenous scientific research is focused instead on the needs of the largest or most capital-intensive farmers, it may not help poor agricultural small-holders or laborers and may even lead to further displacement in developing countries. In this regard, discussions about equity and fairness need to link scientific research and economic development with broader concerns such as the distribution of property rights, strategies of rural development, or enabling the poor to achieve greater economic security, access, and control (Cozzens 2007). This may require new policies and procedures to redefine the role of S&T and associated planning and institutions and to ensure that natural resource and economic developments reduce poverty and inequality (Anderson et al. 2006; Woodhouse and Sarewitz 2007).

In the rest of this chapter we explore the linkages between scientific research and economic development by looking at the trajectory and potential contribution of nanotechnology to equality and equity. *Equality* is broadly defined as a similar distribution of resources and opportunities, including of wealth, income, health, or power (Cozzens et al. 2007). *Equity* is a normative concept that embodies fairness (although not necessarily similarity) in the distribution of resources and opportunities; it can be understood as the absence of potentially problematic inequalities resulting from discussions of "who should get what" (Woodhouse and Sarewitz 2007). In many developing countries, concerns about equality and equity are prominent since there are typically wide disparities in income distribution between rich and poor, as well as concentrations of decision-making power held by privileged elites. In this chapter, we focus particularly on the development and implications of nanotechnology in Brazil, which has been identified as one of the world's most unequal countries in terms of income distribution (World Bank 2005). However, we also seek to draw broader insights for the relationships between science, technology, and equitable economic development.

19.2 Nanotechnology Research, Equality and Equity

The potential impacts of nanotechnology research on economic development depend on multiple factors. Impacts from and to science are mediated not only by the features of a country's innovation system, but also by institutional arrangements and path-dependent existing distributions of power, income, human capital, and access to decision-making. Moreover, other non-S&T policies and factors, including those related to procurement, multi-national investment, industry structure, antitrust, and market regulation, may influence the commercialization and application of nanotechnologies. We have developed a concept map to represent (in simplified form) this complex array of factors, and our expectations about relationships between them (Fig. 19.1).[1] The concepts are arranged to represent potential causal propositions and linkages rather than the typical hierarchy of more general to less general concepts often used in such maps (Novak and Cañas 2008). We acknowledge that, as depicted, our model abbreviates the complex processes and forces that influence the relationships between science and economic development. There are multiple feedback links, many of which are rather subtle, that are not fully depicted in our representation. In other words, the resulting linearity of the graphic as a consequence of simplification should not be understood as linearity in the overall process. Nonetheless, we suggest that this model is a helpful starting point.

In our model, nanotechnology research is centrally located, since this is at the core of our inquiry. However, the model also emphasizes the importance of existing institutional arrangements and political processes, which are interrelated and ultimately frame the characteristics and outcomes of any S&T policy. Institutional arrangements include the structure and operation of the various sectors (public, private, and non-profit) in the economy; the functioning of system elements (such as education or health care); the characteristics of regulatory systems; the nature of interrelationships (including university-industry linkages); and institutional flexibility, capability, and ability to change. Political processes include governance procedures, methods of stakeholder engagement, means of deliberation, anticipation, and decision-making. Together, these arrangements and processes determine the access that different groups have to resources and opportunities, and enable, to a greater or lesser extent, the participation and consideration of those groups in policy development and implementation. These arrangements and processes also have a bearing on how new knowledge is generated and used in the economy and society and on capabilities for, and trajectories of, innovation. The process of science and the development of new technologies can contribute to inequality and inequity (Woodhouse and Sarewitz 2007). Such outcomes are more likely where institutional arrangements and political processes reinforce them.

However, it is also possible that science and technology may lead to the creation of new knowledge, enhanced technological capabilities, and innovations in ways that reduce inequalities by re-distributing the benefits of new technologies. For example, disadvantaged communities and social groups may be assisted by the creation of technologies and knowledge that add value to (and do not displace) the economic activities they are engaged in. The development of new materials to improve quality and performance in traditional manufacturing industries or the introduction of more efficient agricultural implements and new methods to produce outputs with fewer demands on natural resources or energy in small-scale enterprises could be examples of such technologies. Disadvantaged communities and groups may also be empowered if they (or their representatives) are able to participate in decision-making about the development and implementation of

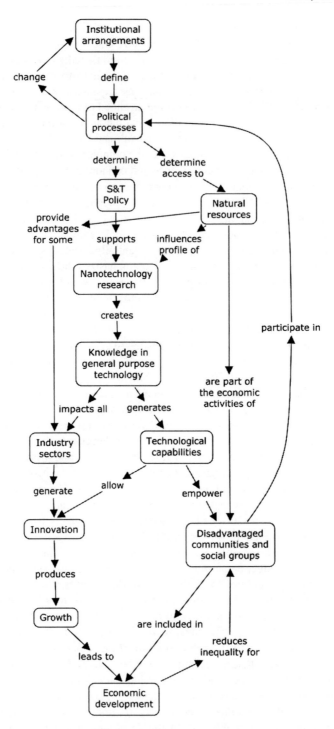

Fig. 19.1 Nanotechnology research, institutions, policy, capability, and economic development

new technologies. From an equity perspective, the consideration and inclusion of communities and social groups in deliberations about science, technology, and innovation would signify progress towards more equitable development particularly if they can participate actively in shaping the agenda for research, development, and use of those new technologies that lead to innovation and economic growth. Creating and managing means to enable participation of all groups (and not only more affluent and better educated citizens) is admittedly a challenging task; it is not easily achieved in advanced countries with relatively well-developed and comprehensive educational systems and traditions of citizen engagement, let alone in developing countries where educational systems are not comprehensive, literacy is less widespread, and problems of day-to-day survival are foremost.

With this contextual discussion in mind, if a science policy in an emerging domain—in our case, nanotechnology, which is still in its early phases of development—is to reduce inequality in the subsequent distribution of its benefits

Table 19.1 Prospective linkages of science and technology policy to developmental equality and equity

Science and technology policy component	Links to equality and equity	Examples of potential contribution of policies
Agenda setting	Extent of participation of communities and social groups to provide inputs and define priorities	Deliberative democratic processes of agenda setting could enrich and legitimate policies, and reduce instances of "unfairness" in policy goals and, subsequently, in the impact of R&D outcomes
Public R&D investment	Criteria for funding allocation (e.g., lagging regions or industrial areas, smaller research institutions or leading research centers) and evaluation of alternative programs (e.g., corporate priorities, triple bottom line, or other benchmarks)	Funding of lagging regions or disadvantaged communities and triple bottom line evaluation would contribute to more even production of knowledge and development of technologies
R&D outcomes	Access to knowledge and technologies (e.g., patents vs. public domain, incentives to develop and use certain technologies, pricing)	Promoting research and development of useful and affordable technologies that target needs of disadvantaged groups would increase access and empower communities
Risk awareness and allocation	Concern with environment and health protection for population groups when undertaking research and developing new technologies	Provision of information to general public and dialogue with relevant stakeholders would legitimate policies if the risks are known and more evenly distributed

and foster inclusion and equity, what might we look for in terms of its characteristics and relationships? There is a broad range of possibilities here, from measurable considerations of the distribution of R&D activity to more intangible and normative dimensions related to the interests and perspectives of different communities and social groups. However, we suggest that we can address our question through probes in four key areas: (i) agenda setting (the participation of communities and social groups to provide inputs and define priorities); (ii) R&D investment (the criteria and mechanisms for the distribution of public R&D funding, the evaluation of alternative programs, and the incentives influencing private R&D investment); (iii) R&D outcomes (types of knowledge and technology development, mechanisms of access, incentives to develop certain technologies, and relationships to economic and societal needs); and (iv) risk awareness and allocation (concern with environment and health protection for all population groups when undertaking research and developing new technologies). For each of these four probes, it is possible to operationalize indicators to assess activity and progress (See Table 19.1). We acknowledge that the extent to which significant outcomes will be achieved even when factors and processes hypothesized to be equality- or equity-enhancing are present will vary according to circumstances and by country (i.e., similar measures may have varied outcomes in different national innovation systems). In the following two sections, we deploy this framework using the case study example of Brazil.

19.3 Patterns of Nanotechnology Research in Brazil

We focus our empirical investigation on nanotechnology research in Brazil for two interrelated reasons. First, Brazil is highly unequal in terms of income distribution. Among 22 Latin American and Caribbean countries, Brazil has the greatest level of income inequality (as indicated by Gini coefficient measures of income distribution), according to World Bank (2005). Since the 1960s, the income share of the richest 20% of Brazil's population has been about 32 times that of the poorest 20% (Fields 2001). Brazil thus offers an appropriately challenging testbed to explore the contribution of scientific and technological research to reducing inequality. Second, Brazil has been at the forefront of developing countries in implementing policies to promote the development of nanotechnologies, with the first programs implemented in 2001. In short, Brazil offers an important starting point to understand how scientific research in general and nanotechnology in particular may lead to or help the process of development with (or without) equality and equity. We anticipate that our findings may have potential implications for other Latin American countries as well.

To begin our assessment, it is helpful to situate Brazil in continental and global contexts in terms of nanotechnology scientific output. Latin America (including the Caribbean) comprises about 8.3% of the world's population and produces about 8.9% of global gross domestic product, but the continent contains only 2.5% of all scientists in the world and publishes about 2.6% of global scientific articles

(UNESCO 2005). In 2007, Latin America contributed about 3% of the world's nanotechnology publications (Kay and Shapira 2009).[2] The continent's low investment in scientific personnel and low publication output relative to its population and economic scale has been considered as evidence of weaknesses in the development of nanotechnology research there (Roco 2005; Besley et al. 2008).

On the other hand, within the domain of science, Latin American scientists have been at least as quick as elsewhere in pursuing research at the nano-scale, as illustrated by the continent's global share of nanotechnology publications relative to all scientific publications. Moreover, Latin America is a diverse continent. Brazil, although perhaps a third-tier country in nanotechnology output at the worldwide level, is clearly the leader in terms of nanotechnology publications in the region. Brazil contains about one-third of all scientists in Latin America (UNESCO 2005) and contributes more than 50% of the continent's nanotechnology research output (Kay and Shapira 2009). Only three other Latin American countries—Mexico, Argentina, and Chile—show any significant level of performance in total nanotechnology publications (Fig. 19.2). Brazil and these three countries invest more in R&D as a percentage of GDP than most other Latin American countries. Brazil is the leader on this score, spending more than 1% of its GDP in R&D, compared with an average of 0.6% for all of Latin America (UNESCO 2005). At the other extreme, there are multiple Latin American countries (such as Peru, Ecuador, and Bolivia) with very low overall R&D investment, scarce research activity in nanotechnology, and non-existent nanotechnology patenting activity. In general, Latin American countries that are not among the four leaders have seen decreasing shares in the continent's nanotechnology publication output.

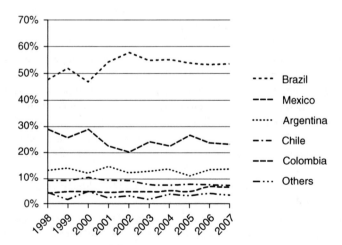

Fig. 19.2 Nanotechnology research publication output, Latin America, by share of leading countries, 1998–2007 *Note*: the chart shows percentages of the overall Latin America's research publications output in nanotechnology *Source*: Georgia Tech global databases of nanotechnology publications

The overall picture of nanotechnology research for Latin America suggests that only Brazil and probably few of its neighbors can expect a significant role in the development of nanotechnology research and in influencing the impact of nanotechnologies in their economies. But, we do not anticipate any easily observable direct relationships between nanotechnology research and economic development as measured by macro-economic aggregates. For example, in Brazil, real per capita income (in constant 2000 US$) grew by about 10% between 1998 and 2006 (analysis of data from World Bank 2007), while there was a slight diminishment in the Gini coefficient for income distribution (where 1 represents perfect equality and 0 represents perfect inequality) from 0.60 to under 0.57 over the same period (CGEE 2007). During this period, Brazil's annual nanotechnology publication output doubled. Nonetheless, we do *not* attribute any relationship whatsoever but rather emphasize and illustrate how difficult is to disentangle the complex set of relationships to which nanotechnology research may contribute. According to official reports, the drivers of such economic growth and the small reduction of inequity have been primarily the increased demand for formal employment and the expansion of social programs, both in a context of reduction of inflation (CGEE 2007).

However, we do anticipate that there should be an effect from Brazil's nanotechnology research efforts on the country's technological capacity and ability to develop innovations that relate to nanotechnology. Previous research has found that countries that develop strong technological and innovation performance in emerging research fields also produce relatively large numbers of publications and have multiple research producing institutions (Van Looy et al. 2006). Brazil has performed well on such measures in recent years. Nanotechnology publications increased by 17% annually between 1998 and 2007, to reach more than 1,070 publications in the latter year. The number of research institutions involved in nanotechnology also increased by the same rate for this period, to 185 research institutions involved in 2007.

Brazil's science and technology policymakers have attempted to ensure that the country's nanotechnology research leads to innovation outcomes (Rediguieri 2009). Promoting the creation and consolidation of integrated cooperative networks of basic and applied research, including participation by industry, has been a goal of implemented programs (Knobel 2002). A program of Cooperative Networks for Basic and Applied Research in nanotechnology was announced by the national research funding agency (Conselho Nacional de Desenvolvimento Científico e Tecnológico) in 2001, leading to the establishment of four major nanotechnology networks involving about forty research institutes, also with corporate involvement, in the following year (de Almeida 2003).

Around this time, the Brazilian government identified nanotechnology as a priority area and established a National Program of Nanotechnology in 2002 (OSEC 2005). This program includes such goals as creating innovative enterprises, increasing the value-added of products, and capturing an "outstanding position in the global scenario of nanotechnology" (MCT 2003). Moreover, three institutes were established in the nanotechnology field under the Millennium Institutes program (a partnership between the Brazilian Ministry of Science and Technology and the World Bank to establish fifteen research centers). A series of other complementary research initiatives have been established to provide instrumentation and support scholarships and research awards in nanotechnology (de Almeida 2003). In 2003, nanotechnology was identified by the federal government as one of three sectors for the future development of industry and trade in Brazil (OSEC 2005). There has also been a federal component aimed at generating more equal development by targeting lagging regions with specific funding. In addition, several Brazilian states, especially São Paulo, have provided additional support to nanotechnology. Overall, Brazil added US$30 million to its nanotechnology R&D spending in the 2005–2006 period, although on a purchasing power parity basis Brazil's nanotechnology investment was under fifty cents per capita in 2006, compared with more than US$4 per capita in the U.S., much of Europe, and Japan (Lux 2007).

Whether these policy initiatives can spur innovation and economic development throughout the whole country is less clear. At first sight, this seems unlikely when looking at how nanotechnology research activity is distributed at the geographic and institutional levels. As in other research fields, in Brazil there are disparities in the distribution of nanotechnology research. Most of the country's nanotechnology research resources remain focused on supporting projects at top research institutions, mostly elite universities (Invernizzi 2007). This pattern suggests that new distributional and access goals have yet to outweigh traditional public research priorities and allocation methods. The majority of the research activity in all sciences is concentrated in the southeastern states, with Sao Paulo as the leading one (Packer and Meneghini 2006). This is the most economically developed state as well, contributing about 30% of the country's GDP and 45% of the total public funding for science and technology (IBGE 2007; Bound 2008). Moreover, almost 30% of all Brazilian nanotechnology research institutions are located in São Paulo, authoring or co-authoring about 65% of the country's nanotechnology publications in the period 1998–2007 (this proportion is higher than the average for all sciences). Rio de Janeiro is the second-ranked state, concentrating almost 13% of the research institutions and authoring or co-authoring about 14% of the nanotechnology publications in that period (Fig. 19.3). Meanwhile, northern states like Amazonas and Para, and the northeastern state of Maranhão, hold about 9% of the country's population and produce 5% of its GDP, but only 2% of the nanotechnology research output (IBGE 2007).

These figures are not encouraging if we expect nanotechnology research to change existing regional patterns of inequality in Brazil. The production of scientific knowledge in the field remains highly concentrated, with the majority of states in Brazil still in a lagging position. Although the integration of multiple research institutions and scientists has been an explicit goal of Brazilian nanotechnology networks programs, the scene remains dominated by few large research institutions. For example, most of the 36 institutions involved in the first nanotechnology program implemented in 2001 were universities that co-authored 87% of the total nanotechnology research output of Brazil in the period 1998–2007 (Kay and Shapira 2009).

We have examined the research profile of nanotechnology in Brazil to understand the alignment of scientific activity with the particular needs and comparative

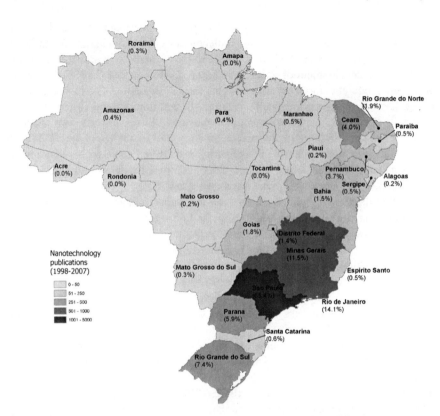

Fig. 19.3 Nanotechnology research publications, Brazilian states, by publication output range and share of national total, 1998–2007 *Source*: Georgia Tech global databases of nanotechnology publications

advantages of the country. We also have explored how Brazil's research profile matches or differs from global patterns. We do have some prior expectations. For example, Brazil has a relatively high natural resource dependency. More than two-fifths of Brazil's exports are comprised of primary products (including coffee, iron ore, and other agricultural, forestry and mining commodities). Although this is much reduced from the 1960s when primary products comprised over 90% of the country's exports, Brazil's economy remains significantly reliant on natural resource production and trade (Barbier 2005). The agricultural sector alone employs nearly 21% of Brazil's workforce and receives considerable public sector support, with government spending accounting for nearly 37% of value-added in Brazilian agriculture in 2004 (World Bank 2007).

We might thus expect the natural resources sector to be among those targeted by nanotechnology research. Indeed, it has been argued that Brazilian science as a whole follows a "bio-environmental" pattern characterized by concentrations of research publications in agriculture, biology, and earth and space sciences (Glänzel et al. 2006; Scarano 2007; Bound 2008). Since this model is aligned with the comparative advantages and natural endowments that the country has, we searched our nanotechnology research databases for patterns of publications consistent with these characteristics. We find that Brazil (as elsewhere in the world) concentrates most of its nanotechnology research in the areas of physics, materials science, and chemistry (Glänzel et al. 2003). The share of engineering research is still relatively important but decreased slightly between 1998 and 2007. Research areas associated with agriculture, biology, and earth-related sciences have a minor although growing participation in our dataset (Fig. 19.4).

For purposes of comparison, we defined a combined category for bio-environmental research, including research areas that might be included in that model suggested for Brazil.[3] This category increased its share from 5% in 1998 to more than 10% of all nanotechnology publications in 2007. Although promising, this share is below the levels observed for the similar category in Brazilian science as a whole. Other authors have found that, when considering total research output in relation with worldwide averages, the emphasis on bio-environmental research area is relatively greater in Brazil. For instance, in Brazil's publications, the relative weight of agricultural science is 3 times higher, and of biology 2.6 times higher, than the average for the world as a whole (Glänzel et al. 2006; Bound 2008). Moreover, although the bio-environmental category of nanotechnology is increasing in Brazil, most research output remains concentrated in physics, materials science, and chemistry. This suggests that there has not been any strongly effective targeting to Brazil's bio-environmental needs and opportunities as a result of national nanotechnology policy.

Patterns of research production are only a part of the story of how science may or may not contribute to development and the reduction of inequality and inequity.

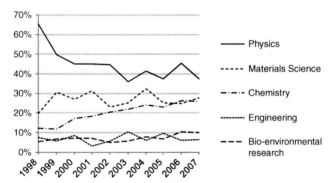

Fig. 19.4 Leading research areas and bio-environmental research, by share of total nanotechnology publication output, Brazil, 1998–2007 *Note*: the bio-environmental research category comprises the following subject categories: biochemistry & molecular biology, energy & fuels, cell biology, biophysics, environmental sciences, biotechnology & applied microbiology, biology, medicine, plant sciences, zoology, food science & technology, agriculture, geosciences, marine & freshwater biology, genetics & heredity, tropical medicine, developmental biology, and water resources *Source*: Georgia Tech global databases of nanotechnology publications

More important is the *relevance* and *use* of research. For example, it is possible for scientific research production to be concentrated, but for the benefits and spillovers from that research to be broadly disseminated, geographically and among communities and social groups. In Brazil, however, we doubt that this is likely to happen in the short or medium term. Based on an analysis of patent data—a proxy frequently used to measure industry involvement and potential commercialization or application of new technologies—we observe low enterprise activity in nanotechnology in Brazil, and therefore fewer chances for research to be transformed into applications and commercial products. Our data indicates only 157 nanotechnology-related patents in the period 1990–2008 for Brazil (Kay et al. 2009).[4] This low level of patenting activity limits our analysis. We did find a few patents that might be linked to bio-environmental domains but no clear trends or clusters in the leading patent categories (by International Patent Classification) which included Medical or Veterinary Science (class A61), Organic Macromolecular Compounds (class C08), and Physical or Chemical Processes or Apparatus In General (class B01).

An analysis of corporate entry into nanotechnology discovers 75 unique Brazilian companies with either a nanotechnology publication or patent during the period 1990–2008 (Kay et al. 2009). Through to 2008, Brazil ranked 30th in the world by the number of corporations that have entered into nanotechnology, but 12th by total number of research organizations involved in nanotechnology publications (almost all of them, universities and public research organizations). This suggests that corporate take-up of nanotechnology activity is somewhat weaker in Brazil than would be indicated by its public research effort. Additionally, we observe that only a handful of companies have participated in the official nanotechnology network programs (according to MCT 2006), which also suggests that channels for commercialization are weakly developed in Brazil. It is fair to note that nanotechnology is still an emerging technology and many promised applications are years (even decades) away. Yet, other countries have many hundreds (in the cases of China or South Korea) and thousands (in the cases of the U.S., Japan, or Germany) of companies that have started to enter into the nanotechnology domain through publications and patents.[5] More than 800 nanotechnology-enabled consumer products are now reported to have entered the world's marketplaces, none of which is identified as originating in Brazil (PEN 2009). Some of these products may already be in the Brazilian market (produced by outside corporations), although none of the products in international markets is yet reported to be made in Brazil, notwithstanding the national emphasis on promoting Brazilian nanotechnology exports. This gap between nanotechnology research (where Brazil is at least maintaining a position) and corporate entry (where relatively few companies have emerged so far) presents a major challenge to ensuring that outcomes from nanotechnology R&D are equality- or equity-enhancing in Brazil. Without an active indigenous corporate base, Brazil will remain dependent on international companies who may well be less likely to be influenced to develop applications specifically customized to Brazilian needs.

19.4 Equality and Equity Assessment

In this section, we revisit our four aspects of how S&T policy can be linked to equality and equity and compare them with our data and analysis for Brazil. In doing this, it is important to emphasize that nanotechnology is only one part of the broader S&T policy. Although nanotechnology is expected to have increasingly pervasive effects, its contribution to reducing inequality and inequity is relative and complementary to other development-specific social and economic policies. At the very least, however, we might hope that nanotechnology would not contribute to further inequalities. At best, if there is effective and well-targeted governance, nanotechnology might contribute to the reduction of inequality and inequity.

So how does nanotechnology development to date in Brazil fare when compared against the framework of how S&T policy might influence aspects linked to equality and equity? We highlighted four aspects: agenda setting, public investment, R&D outcomes, and risk awareness and allocation. In relation to the first aspect, agenda setting, we would expect to find instances of widened participation as a necessary condition for a positive contribution to equality and equity. Our data show that an increasing number of institutions in Brazil are involved in nanotechnology research, yet most of them are universities and only a few of them have a significant share of the total research output. As mentioned before, industry involvement in both research and commercialization is still very low and few other non-governmental institutions participate in nanotechnology research. Moreover, we have only limited evidence of the participation of other organizations in the governance of nanotechnology development. One example is the Fundação Jorge Duprat Figueiredo de Segurança e Medicina do Trabalho—a labor advocacy foundation undertaking research to assess the impact that nanotechnology might have on workers' health and quality of life in Brazil. A handful of other interest groups, mostly also labor advocates, are involved in activities for disseminating nano-related information to the general public.[6] Yet, these interest groups have not been involved in the core formulation of nanotechnology policy in Brazil and, overall, the evidence indicates low participation of different stakeholders in the ongoing process of governance. This raises concerns relating not only to targeting and accountability but also to risk control for nanotechnology (Wilsdon 2004).

On the second aspect, public investment in nanotechnology research in Brazil, we looked at the targeting and allocation of research activity. We examined targeting towards bio-environmental opportunities and needs that might afford valuable comparative advantages for Brazil. Among applications of such research would be those in energy storage, production, and conversion; agricultural productivity enhancement; water treatment and remediation; and disease diagnoses and screening (Scarano 2007). However, we did not observe any special or strongly-weighted targeting of nanotechnology research to bio-environmental domains. The same occurs with patenting areas, although in this case the low levels of commercialization activity impede to draw definitive conclusions on trends and relationships with Brazil comparative advantages.

Our data also show that publicly-sponsored nanotechnology research activity is unevenly distributed in Brazil. There are significant disparities in scientific publication by region and still only a small number of institutions participating in nanotechnology network programs. Since most nanotechnology research is publicly funded, this suggests that research funding is still highly concentrated in a few states and institutions. There has not been an opening for lagging regions to take special advantage of nano-related activities. Indeed, there is debate in this regard after a recent proposal from the Ministry of Science and Technology to build a new national nanotechnology laboratory in São Paulo—the state with the highest share of research—which would further concentrate research activity (Invernizzi 2007). Additionally, our review of Brazilian policy documents does not suggest strong implementation of social and environmental criteria (as well as economic ones) in nanotechnology program development. Although policy documents do not explicitly neglect social or environmental criteria for program evaluation, they do not stress them either (MCT 2003, 2006).

As for the third aspect, R&D outcomes, we again emphasize that nanotechnology is still in an early stage of development in Brazil, as elsewhere. Applications and uses are still emerging. Yet, already, we can see that there is a very low level of corporate engagement and patenting activity in the field (Kay et al. 2009) in Brazil. This indicates that to date there is only a weak effort in indigenous commercialization of nanotechnologies in the country, a finding that indicates that a broad diffusion of R&D outcomes is less likely. Moreover, patterns of concentrated research activity are also emerging, which is related to the patterns of allocation of public R&D spending, as noted above. Investment in nanotechnology research may lead to spillovers of economic activity in regions where research centers are located through the creation of higher wage research-related scientific, managerial, and technical jobs. However, the major part of the benefits of nanotechnologies is expected to emerge from commercialization. This may occur by both manufacturing locally nanotechnology-based products or using new technologies that increase industry productivity, and by bringing more value to other business users and consumers through enhanced materials, innovative devices, and other solutions. The nature of these technologies and products, and the extent to which different groups are able to afford and access them, will influence the distribution of benefits from nanotechnology in the future. While we cannot fully judge at this point whether the distribution of benefits from nanotechnology development is fair or not, the early indications are not greatly encouraging.

The fourth aspect in our framework is awareness and allocation of risk. As suggested by other authors, the novel characteristics of nanotechnology demand careful risk-benefit assessment and risk management, particularly since innovation in the field is running ahead of policy and regulatory actions (Wilsdon 2004; Renn and Roco 2006). There is increasing attention to potential issues related to nanomaterial exposure, the relative hazard nanomaterials pose to humans and the environment, and the deficits in the understanding of risks (Stern and McNeil 2008). Concern about environment and health protection for all population groups when pursuing different applications of nanotechnologies should be an integral part of S&T policy if we hope to minimize potential risks and optimize possible positive contributions to development. However, to date, official presentations and reviews of nanotechnology programs in Brazil do not include any significant broad-based risk assessments or related considerations of safety (MCT 2003, 2006). Furthermore, the lack of involvement of non-governmental organizations suggests that groups who may be exposed to risks may not have proper representation in governance activities. If this is the case, nanotechnology research might contribute to widening existing gaps instead of contributing positively to equality and equity. Indeed, without communicating potential risks to stakeholders, Brazil is unlikely to develop forums where different actors may exchange ideas, concerns, and insights to reach consensus for appropriate regulatory actions, including risk protection of disadvantaged groups (Renn and Roco 2006).

19.5 Concluding Remarks

For Brazil and other developing countries, there is *potential* for developing nanotechnology in ways that contribute to broader processes of development and reductions in inequality and inequity. Based on a review of policy documents and analyses of data of nanotechnology publications and patents, we have developed insights into the current nanotechnology research profile of Brazil. We have compared the country's nanotechnology policies and profile against a framework to probe links with and possible impacts for equality and equity. We find that Brazil's nanotechnology profile may not be appropriately exploiting the comparative advantages that Brazil may have in natural knowledge-economy areas (Bound 2008). The higher concentration of research activity of nanotechnology in favored regions and elite research institutions is another warning regarding the potential contribution of nanotechnology to equitable development. Arrangements to develop indigenous corporate as well as other socially-relevant uses for nanotechnology products and applications have yet to be strongly developed. Participation in governance and policymaking is weak by organizations and groups outside of policymaking and scientific elites. Considerations of risk in nanotechnology development have yet to be fully embraced in policy and practice. Similar considerations may apply to other developing countries in Latin America and elsewhere, particularly those that, while less developed in terms of nanotechnology research, are still endowed with natural resources and other comparative advantages. Those countries may learn from the Brazilian experience in order to better integrate emerging technologies in their processes of development.

We have assessed nanotechnology development in Brazil and discussed its potential contribution to a broader process of development based on four aspects related to the reduction of inequality and inequity: policy agenda setting, public R&D investment, R&D outcomes, and risk awareness and allocation. Overall, there is no clear evidence of a potential contribution of nanotechnology to increasing equality and equity from the public spending point of view. Yet, the benefits of nanotechnology may be more evenly and fairly distributed in the future if policies include economic, social, and environmental benchmarks ("triple bottom line") to define targets. Moreover, the network of institutions linked to nanotechnology development is mostly comprised of academic research institutions, with low involvement of other research institutions and non-governmental organizations that might participate in the process of governance of new technologies. This undermines a potential positive contribution of nanotechnology to reducing inequality and inequity.

From the point of view of risk assessment, the lack of involvement of a full range of stakeholders indicates a pending task in terms of policy and regulatory actions to protect the environment and health for all population groups when pursuing potentially hazardous applications of the new technologies. Finally, as nanotechnology increasingly shifts from the research laboratory to applications, we anticipate that policy will need to focus more intensely on the relevance and use of nanotechnology. This may include working with existing firms to develop nanotechnology-enabled products to target applied research and development of applications and technology transfer that addresses societal needs (including in bio-environmental areas), and anticipating, communicating, and regulating any risks associated with nanotechnology. Although research policy for nanotechnology in Brazil has not made any significant contributions to reducing inequality and inequity thus far, in forthcoming applied development and application phases, there may still yet be opportunities and pathways to address these concerns.

Acknowledgements This study uses data from the large-scale global nanotechnology publication and patent datasets developed by the group on Nanotechnology Research and Innovation Systems at Georgia Institute of Technology – a component of the Center for Nanotechnology in Society (CNS-ASU). Support for the research was provided through CNS-ASU with sponsorship from the National Science Foundation (Award No. 0531194). The findings and observations contained in this paper are those of the authors and do not necessarily reflect the views of the National Science Foundation.

Notes

1. An earlier version of this concept map was presented at the Workshop on Nanotechnology, Equity and Equality, at Tempe, Arizona in November 2008. We appreciated the comments received from colleagues at this workshop and have made several changes which are incorporated into the version in this chapter.
2. The nanotechnology research data cited in this chapter are gathered from the global databases of nanotechnology publications and patents developed at the Georgia Institute of Technology, using the definition of nanotechnology and methods described in Porter et al. (2008). The sources of data are Thomson Reuters ISI-WoS for scientific publications and Patstat for patents. In this chapter, nanotechnology publications for a specific country or region refer to scientific publications authored or co-authored by any authors affiliated with research or other organizations in that country or region. Nanotechnology patents for a specific country are those in which patent inventors report an address in such country or patent assignees are organizations established in such country, as reported by patent authorities.
3. The research areas from our database included in this combined category are: biochemistry & molecular biology, energy & fuels, cell biology, biophysics, environmental sciences, biotechnology & applied microbiology, biology, medicine, plant sciences, zoology, food science

& technology, agriculture, geosciences, marine & freshwater biology, genetics & heredity, tropical medicine, developmental biology, and water resources.
4. This refers to all patents (applications and grants) between 1990 and 2006 in which patent inventors report an address in Brazil or patent assignees are Brazilian organizations, as reported by Patstat databases.
5. Authors' analysis of corporate nanotechnology publications and patents using Georgia Tech global databases of nanotechnology publications and patents (see also Porter et al. 2008).
6. For example, among others, IEP (Intercambio, Informação, Estudos e Pesquisas), Dieese (Departamento Intersindical de Estatística e Estudos Sócio-Econômicos), and DIESAT (Departamento Intersindical de Estudos e Pesquisas de Saúde e dos Ambientes do Trabalho).

References

Anderson, Jon, Charles Benjamin, Bruce Campbell, and Daniel Tiveau. 2006. Forests, poverty and equity in Africa: New perspectives on policy and practice. *International Forestry Review* 8(1): 44–53.
Barbier, Edward B. 2005. *Natural resources and economic development*. New York: Cambridge University Press.
Besley, John C., Victoria L. Kramer and Susanna H. Priest. 2008. Expert opinion on nanotechnology: Risks, benefits, and regulation. *Journal of Nanoparticle Research* 10(4): 549–558.
Bound, Kirsten. 2008. *Brazil. The natural knowledge economy*. London: Demos.
CGEE. 2007. *Analise da pesquisa nacional por amostra de domicilios PNAD 2005. Livro 3: pobraza e desigualdade*. Brasilia, DF: Centro de Gestao e Estudos Estrategicos (CGEE).
Cozzens, Susan. E. 2007. Distributive justice in science and technology policy. *Science and Public Policy* 34(2): 85–94.
Cozzens, Susan. E., Rob Hagendijk, Peter Healey, and Tiago Santos Pereira. 2007. *A framework for analyzing science, technology and inequalities: Preliminary observations*. ResIST: Researching Inequality through Science and Technology. Working Paper 3. Oxford: James Martin Institute.
de Almeida, Alexandra, O. 2003. Responses to questionnaire on nanotechnology: Brazil. Evidence to the Royal Society and Royal Academy of Engineering study on nanoscience and nanotechnologies. São Paulo: British Consulate General.
Fields, Gary. S. 2001. *Distribution and development: A new look at the developing world*. Cambridge, MA: MIT Press.
Fleming, Lee, and Olav Sorenson. 2004. Science as a map in technological search, *Strategic Management Journal* 25(8–9): 909–928.
Glänzel, Wolfgang, Jacqueline Leta, and Bart Thijs. 2006. Science in Brazil. Part 1: A macro-level comparative study. *Scientometrics* 67(1): 67–86.
Glänzel, Wolfgang, Martin Meyer, M. Du Plessis, Bart Thijs, Tom Magerman, and Balazs Schlemmer. 2003. *Nanotechnology, analysis of an emerging domain of scientific and technological endeavor*. Leuven: O&O Statistieken.
IBGE. 2007. IBGE divulga as contas regionais 2002–2005. http://www.ibge.gov.br/home/presidencia/noticias/noticia_visualiza.php?id_noticia=1039&id_pagina=1 (retrieved November 2, 2008).
Invernizzi, Noela. 2007. *Los científicos brasileños legitiman las nanotecnologías*. Red Latinoamericana de Nanotecnología y Sociedad—ReLANS. http://estudiosdeldesarrollo.net/relans/documentos/Noela-Visiones-esp.pdf. (retrieved December 12, 2007).
Invernizzi, Noela, and Guillermo Foladori. 2005. Nanotechnology and the developing world: Will nanotechnology overcome poverty or widen disparities? *Nanotechnology Law & Business Journal* 2: 1–10.
Kay, Luciano, Noela Invernizzi, and Philip Shapira. 2009. The role of Brazilian firms in nanotechnology development. Paper presented at Atlanta Conference on Science and

Technology, October 2009. http://www.cherry.gatech.edu/PUBS/09/Kay-Invernizzi-Shapira-09atlanta%20conf.pdf. (retrieved December 13, 2009).
Kay, Luciano, and Philip Shapira. 2009. Developing nanotechnology in Latin America. *Journal of Nanoparticle Research* 11: 259–278.
Knobel, Marcelo. 2002. Nanoredes. http://www.comciencia.br/reportagens/nanotecnologia/nano11.htm. (retrieved October 20, 2008).
Lux. 2007. *The nanotech report. Investment overview and market research for nanotechnology.* 5th Ed. New York: Lux Research Inc.
Maynard, Andrew. D. 2006. Nanotechnology: assessing the risks. *Nano Today* 1: 22–33.
Mazzoleni, Roberto, and Richard R. Nelson. 2007. Public research institutions and economic catch-up. *Research Policy* 36(10): 1512–1528.
MCT. 2003. Programa de desenvolvimento da nanociência e da nanotecnologia. Proposta do Grupo de Trabalho criado pela Portaria MCT n° 252 como subsídio ao Programa de Desenvolvimento da Nanociência e da Nanotecnologia do PPA 2004–2007. http://www.mct.gov.br/upd_blob/0002/2361.pdf. (retrieved June 18, 2009).
MCT. 2006. *Relatório nanotecnologia investimentos, resultados e demandas.* Secretaria de Desenvolvimento Tecnológico e Inovação (SETEC)—Coordenação-Geral de Micro e Nanotecnologias (CGNT). Brasília: Ministério da Ciência e Tecnologia.
Novak, Joseph D., and Alberto J Cañas. 2008. *The theory underlying concept maps and how to construct and use them.* Pensacola, FL: Florida Institute for Human and Machine Cognition.
NTSC. 2007. *The national nanotechnology initiative: Strategic plan.* Washington, DC: Executive Office of the President, National Science and Technology Council, Nanoscale Science, Engineering, and Technology Subcommittee.
OSEC. 2005. *Brazil: Nanotechnology overview.* São Paulo, Swiss Business Hub Brazil, Business Network Switzerland. http://www.osec.ch. (retrieved January 23, 2010).
Packer, Abel L., and Rogerio Meneghini. 2006. Articles with authors affiliated to Brazilian institutions published from 1994 to 2003 with 100 or more citations: I—The weight of international collaboration and the role of the networks. *Anais Da Academia Brasileira De Ciencias* 78: 841–853.
Porter, Alan L., Jan Youtie, Philip Shapira, and David J. Schoeneck. 2008. Refining search terms for nanotechnology. *Journal of Nanoparticle Research* 10: 715–728.
PEN. 2009. Inventory of nanotechnology-based consumer products currently on the market. Project on emerging nanotechnologies. http://www.nanotechproject.org/inventories/. (retrieved October 23, 2009).
Rediguieri, Carolina F. 2009. Study on the development of nanotechnology in advanced countries and in Brazil. *Brazilian Journal of Pharmaceutical Sciences* 45: 189–200.
Renn, Ortwin, and Mihail C. Roco. 2006. Nanotechnology and the need for risk governance. *Journal of Nanoparticle Research* 8: 153–191.
Roco, Mihail C. 2003. Broader societal issues of nanotechnology. *Journal of Nanoparticle Research* 5: 181–189.
Roco, Mihail C. 2005. International perspective on government nanotechnology funding in 2005. *Journal of Nanoparticle Research* 7: 707–712.
Scarano, Fabio. R. 2007. Perspectives on biodiversity science in Brazil. *Scientia Agricola* 64: 439–447.
Shea, Christine M. 2005. Future management research directions in nanotechnology: A case study. *Journal of Engineering and Technology Management* 22: 185–200.
Stern, Stephan T., and Scott E. McNeil. 2008. Nanotechnology safety concerns revisited. *Toxicological Sciences* 101: 4–21.
UNESCO. 2005. *UNESCO science report 2005.* Paris: United Nations Educational, Scientific and Cultural Organization.
Van Looy, Bart, Koenraad Debackere, Julie Callaert, Robert Tijssen and Thed van Leeuwen. 2006. Scientific capabilities and technological performance of national innovation systems: An exploration of emerging industrial relevant research domains. *Scientometrics* 66: 295–310.

Vinkler, Peter. 2008. Correlation between the structure of scientific research, scientometric indicators and GDP in EU and non-EU countries. *Scientometrics* 74: 237–254.

Wilsdon, James. 2004. The politics of small things: Nanotechnology, risk, and uncertainty. *IEEE Technology and Society Magazine*, Winter 2004.

Woodhouse, Edward, and Daniel Sarewitz. 2007. Science policies for reducing societal inequities. *Science and Public Policy* 34: 139–150.

World Bank. 2005. *World development report 2006: Equity and development.* New York: Oxford University Press.

World Bank. 2007. *World development report 2008: Agriculture for development.* New York: Oxford University Press.

Youtie, Jan, Philip Shapira, and Alan L. Porter. 2008a. National nanotechnology publications and citations. *Journal of Nanoparticle Research* 10: 981–986.

Youtie, Jan, Maurizio Iacopetta, and Stuart J.H. Graham. 2008b. Assessing the nature of nanotechnology: Can we uncover an emerging general purpose technology? *Journal of Technology Transfer* 33: 315–329.

Chapter 20
Open Access Nanotechnology for Developing Countries: Lessons from Open Source Software

Dhanaraj Thakur

Experience has shown that there is no quick fix that enables developing countries to benefit from technological development. Even strategies designed with specific social goals in mind rarely succeed unless there is great attention to the details. For instance some observers have claimed that tight intellectual property (IP) protection prevents developing countries from accessing the benefits of emerging technologies and they should, therefore, develop other strategies. In this chapter, Dhanaraj Thakur explores one such strategy—an "open access" system of IP in developing countries. He uses open source software to illustrate the range of possible property arrangements. He notes, however, that the application of these principles in nanotechnology has thus far been limited and that earlier studies have shown that open source technologies still require significant complementary facilities and skills that are less likely to be available in developing country contexts. While open source might be part of a strategy that can help a developing country capitalize on nanotechnology, Thakur argues that it will not be a panacea.—eds.

D. Thakur (✉)
School of Public Policy, Georgia Institute of Technology, Atlanta, GA, USA
e-mail: dthakur@gatech.edu

This chapter was peer reviewed. It was originally presented at the Workshop on Nanotechnology, Equity, and Equality at Arizona State University on November 22, 2008.

20.1 Introduction

In a variety of cases governments have emphasized science and technology as a way to address socio-economic problems. Policy-makers are therefore concerned with ensuring the right environment to develop, diffuse, and effectively use new technologies. Part of this process then is to understand how to leverage the potential of emerging technologies such as nanotechnologies. However, this potential might be unrealized under aspects of intellectual property rights regimes that limit the research, development, and diffusion of nanotechnologies (Maclurcan 2005). In this regard, there have been calls to adopt open access practices particularly for the benefit of developing nations (Bruns 2004). However, there has been little analysis to date on the efficacy of such an approach. Proponents of open access approaches essentially argue that they can lead to a more equitable distribution of production and consumption and improve innovation related to nanotechnology. This chapter explores the merit of such arguments by looking at the open access experience of software from a developing country point of view.

The term "developing country" represents a broad range of countries which I further specify for purposes of this chapter. First let's imagine the set of all developing countries.[1] Many of these countries would have the capacity to consume (use and absorb) nanotech products. Others would also have the capacity to produce (or engage in the product development of) certain nanotechnologies. In addition another set, including large countries such as India and Brazil, would also have existing nano research and development (R&D) activities and networks. For purposes of this chapter I focus on the production (as in product development) and consumption (the use and absorption of products via the market and otherwise) of nanotechnology. While the issue of R&D is very important and can be particularly interesting in any discussion related to open access, I want to keep this article focused and manageable. I then examine two issues: (1) what is the feasibility of applying open access approaches to nanotech and (2) what are the implications of adopting these approaches to the production and consumption of nanotechnology for developing countries.

To examine the feasibility issue I draw on some lessons in the way open access licenses and patents have been used in the computer industry, in terms of both open source software (OSS) and computer hardware. I show how such licenses and patents can be applied to open access nanotech development by looking at two existing examples and the use of patent pools.

In terms of implications, I also draw on the impact of OSS in developing countries and specifically case-studies from a related research project. This project found that the experience of OSS in developing countries does not necessarily follow the rhetoric of OSS advocates or international development agencies. That is, while benefits do exist, the uptake of OSS is actually highly skewed and concentrated both at the producer and consumer levels. This is a result of a combination of certain skill requirements and the policy environment. I argue that the potential for open access nanotech can similarly be limited by human resource and equipment costs, lack of enforcement of intellectual property rights and the overreach of these rights, lack of open standards and inadequate alternatives to the use of patents. Furthermore, these factors can contribute to a skewed distribution in the development and use of open access nanotech. These findings however do not discount the value of an open access approach particularly in comparison to the status quo. Accordingly if developing countries want to truly embrace this approach then governments need to develop the appropriate human resources, legal environments, and infrastructure to facilitate open access nanotech development.

To discuss this further I begin by looking at some of the suggested benefits of nanotech for developing countries and more specifically at adopting open access principles to maximize these benefits. Also, to better understand the feasibility of this concept I look at the application of open access to software and hardware in the computer industry and how similar approaches are already being applied to nanotech. Finally, I examine related research results from open source software cases in several developing countries and apply these results to the case of nanotech. Based on this I suggest issues that policy-makers, particularly in developing countries, should consider and possible approaches to deal with some of the associated challenges.

20.2 Nanotechnologies and Developing Countries

Nanotechnology and nano-science generally refers to the development of materials at the scale of a nanometer (nm) or 1 billionth of a meter and more specifically at 1–100 nm. The opportunity to create physical structures at such a small scale has led to a host of applications in areas such as medicine, energy, material science, electronics, and national security. Accordingly, nanotech proffers new solutions to old problems such as those in developing countries. Nanotech represents the future as a pervasive technology with relevance to a variety of sectors for developing countries. For example, Invernizzi and Foladori (2006, 114) note that there is a lot of emphasis on nano-medicine as a tool to deal with the many health problems that plague the poor through improved diagnosis, drug delivery, and the development of better implants and prostheses.

Other areas to which nanotechnologies can be applied include developing low-cost renewable sources of energy, potable drinking water, controlling the transmission of HIV/AIDS, creating employment through the development of new industries, and food security (Hassan 2005, 65; Bürgi and Pradeep 2005, 645; Schummer 2007, 291; Bruns 2004). In fact, Salamanca-Buentello et al. (2005) posit several existing nanotechnologies as ways to achieve key Millennium Development Goals. The potential benefit to developing countries should be seen as addressing both the unique problems of a given country as well as finding cheaper or more feasible options to already established solutions.

The extent to which nanotechnologies are actually beneficial to developing countries depends of course on several factors that are unique to each country (Invernizzi, Foladori, and Maclurcan 2008, 123). For example, some developing countries (e.g.

Brazil, China, and India) are much more advanced in terms of nanotechnology for a variety of reasons including large domestic economies. These diverse experiences can potentially lead to a South-South gap in terms of nanotech diffusion and production (Hassan 2005, 65; Maclurcan 2005). In addition, some argue that these differences will only help to widen the gap between the haves and have nots within a country (Schummer 2007, 291).

There are also many similarities in terms of the challenges that developing countries face when trying to realize the benefits of nanotechnology. These include low incomes, small internal markets, limited human resources, training and educational facilities, trade barriers, and current patterns in the application of intellectual property rights (IPRs) (Maclurcan 2005). Given my concentration on open access and without assuming that addressing one challenge alone can improve the efficacy of nanotechnology in developing countries, this chapter focuses on problems related to IPRs.

20.3 Intellectual Property Rights and Nanotechnology

IPR regimes grant certain protections and rights to owners of artistic, scientific, and other intellectual property such as music, formulas, brand names and software. These rights can take various forms with different degrees of exclusivity depending on the motivation of the owner of the property. I purposely use this definition to recognize the fact that IPRs cover not only what one might call traditional property rights (such as copyrights, patents, trademarks, etc.), but also open access licenses which are less restrictive in what users can do with a given intellectual property.

To further specify what is meant by traditional IPRs, I briefly elaborate on two already mentioned as these are more pertinent to this discussion. Copyrights make the holder the only entity that can legally reproduce, distribute or sell an original expression or idea (e.g. a book). This is for a fixed period of time, for example in the United States it is up to 50 years after the author's death. A patent also provides a similar legal monopoly, but it is used for the invention of a process or device. It gives an entity the sole right to develop and sell its invention and therefore prevents others from doing so. Patents also last for a limited amount of time depending on the jurisdiction in question.

Several observers have discussed the issue of IPRs as a critical and preliminary factor in creating the appropriate environment for the development of nanotechnologies. Bastani and Fernandez (2002, 472) suggest that in the case of nano-tech, IPRs can generally be applied at two levels. First there are the computational techniques that are part of nanotech design. This includes software and hardware that can be used to model the design of molecular structures. Second, there are manufacturing methods and processes. This includes, for example, improvements in positional assembly or the use of tiny robot arms or similar techniques to move molecular building blocks into required positions for bonding and ultimately the construction of nano-materials. Most patenting in nanotechnology occurs at these two levels

(Lemley 2005, 601). Another category of the commercially available products that result from the use of these computational techniques and manufacturing processes is also becoming apparent. Looking at all levels, Chen et al. (2008, 123) note that the leading countries in the patenting of nanotech include the United States, members of the European Union, and Japan.

Of note are the patterns in which IPRs are now applied in developed countries and what this means for the South. There is a trend to file patents much earlier and more often in the production process than with previous technologies (Lemley 2005, 601). In addition, the scope of what can be protected through patents has widened substantially. The fact that some nanotechnologies could be applied in multiple and different fields means that firms can hold patents in industries where they are not active. This implies that more research and effort is often required to file such patents (Lounsbury et al. 2008). This in turn can increase the already substantial litigation and other costs associated with filing patents which can make their use prohibitive for smaller firms and those with limited resources.

Einsiedel and Goldenberg (2004, 28) argue that these patterns could reduce access and limit commercial opportunities for nanotechnologies While the costs to acquire sufficient intellectual property protection is one thing, if such practices lead to monopolistic markets it could also limit innovation. Furthermore, examples of the way patents have been employed in areas such as genetically modified maize have led to concerns about the concentration of ownership of technologies with a potentially wide socio-economic impact.

20.4 The Open Access Approach

Taken together these trends imply that the potential for nanotechnology to address the various socio-economic challenges facing developing countries appears difficult to realize. This is particularly relevant in terms of the product development or consumption of such technologies. For example, paying for licenses could create an expensive product for the consumer. Another issue is that in some cases this could create a reliance on a few firms that control the relevant patents. Often these firms will be concentrated in developed countries. If a nanotech patent has a large potential scope for application across different fields this could be detrimental to small firms operating in related fields. Also, apart from costs, the concentration of patents to a few owners can make it difficult for all but a few firms in developing countries to develop products that are suitable to address local problems. Given this reality, it has been argued that there exists an alternative set of IPRs that could address these issues which is called "open access."

I define the open access model in part by referring to the work Benkler (2006).[2] It is a collaborative production process that involves multiple organizations or individuals, usually active outside the context of the market and distinguished from the use of traditional IPR regimes (i.e. copyright, patents, etc.) in three ways. First, this collaborative production is not driven by price signals as much as production under traditional IPRs. Second, it is not guided by a managerial hierarchy typical of firms whose production is based on traditional IPRs. Instead, there are usually networks of collaborators with a loose structure of seniority. Third, the inputs and outputs of production are shared with everyone, with similar constraints in terms of distribution of the final product, placed on everyone. This last condition is typically codified in terms of licenses that are the functional equivalents of copyrights. That is, they specify how others can use the product and how to treat derivative works. The extent to which a given intellectual property falls under a traditional IPR model or an open access model therefore depends on the type of protection being used.

Given that there is widespread and usually inclusive participation in open access projects and that open access licenses place less restrictions on consumers than traditional IPRs, the result is a potentially more equitable distribution of production and consumption. In this sense, some have pushed it as an alternative approach to the development of nanotech products which could better serve both developed and developing countries (Bruns 2004; Vallance 2000). More specifically, it could help to overcome some of the limitations mentioned above by reducing costs to the consumer, the potential for monopolies and the concentration of patents in the North.

Two additional advantages are worth noting. First, when it comes to product development, it can be argued that more contributors can help develop new and innovative uses of a technology. This is one of the main features of any open access production model. Thus new and relevant nanotech products could be developed for specific developing country contexts especially where contributors to an open access nanotech project are from such countries. Second, and related to the first point, is the possibility of alleviating some of the concerns about ethical and safety issues associated with nanotech applications for humans. Through the collaboration a large and potentially diverse set of participants can bring different ethical and safety concerns to bear on a project (Bruns 2001, 198).

One should not, however, assume that developers must choose either traditional property protection or open access as their approach to intellectual property protection for nanotechnology. In practice, firms can combine both approaches and this would be true for nanotech as well. In addition, both approaches can be applied to almost all economic sectors. While this is obvious for patents and copyrights, open access can also have a wide applicability in capital intensive sectors such as biotechnology (Hessel 2006) or networked based informal sectors such as the design and production of boards for windsurfing and snowboarding (Shah 2006, 339).

20.5 The Feasibility of Open Access

Although one can see the potential benefits of open access to nanotech production and consumption in developing countries, if the approach is not feasible it makes little difference. To explore the possibilities I consider the two main components of nanotech development: computational techniques and manufacturing methods.

Each of these components can be associated with existing open access practices in the computer industry such as Open Source Software (OSS) and open hardware initiatives. I first review how these applications of open access have generally worked in practice and then look at two specific examples of open access nanotech initiatives. These examples show how the application of open access licenses to nanotech products can work.

Under OSS, the source code of a software package is typically made available to users without any fees. Software can be copied, modified, expanded, or re-distributed, usually under certain restrictions. These constraints can vary depending on the type of license used. For example, some licenses such as the General Public License (GPL)[3] stipulate that all derivative works be issued under the same (open access) terms and conditions. Alternatively, the Berkeley Software Distribution (BSD)[4] license states that derivative works can be distributed under other terms, even for commercial purposes, as long as credit is given to the original author. In general, most OSS licenses include some form of free distribution and allow modification to the source code which allows for collaborative production.[5]

For a firm or individual, the typical business model under OSS would be to focus on support or consultancy services for a particular software package (as the software itself is available for little or no cost). Large firms such as Red Hat follow this model by providing basic versions of their Linux computer operating system for free while charging for a host of support services such as more advanced Linux systems, software customization, training, and support. Similarly individual software consultants can also provide support and training services for existing open source applications to other companies who lack such in-house skills.

As mentioned earlier one of the components of nanotech product development involves the use of computational techniques including software. Here the use of software can be distributed under open source licenses without any additional effort. That is, licenses such as the GPL (General Public License) or BSD (Berkeley Software Distribution) are already designed for software and include the relevant legal nomenclature for this technology. Some examples of OSS used in the design of nanotech include NanoHive-1[6] software for the modeling of objects at the nanoscale, which was released under a GPL license, and Fungimol,[7] a molecular CAD program released under an LGPL license.[8] A more recent example is the Open Molecular Mechanics project which can augment existing nanotech software by allowing users to carry out complex simulations on desktop machines at faster rates than before (Friedrichs et al. 2009). This software was released under a more liberal license similar to the BSD.

One hypothetical solution that could move elements of nanotech design into the mold of open access is the use of a "Molecular Description Language" (Prisco 2006). Prisco postulates that with such a language interested persons can then create the designs for objects by specifying their underlying molecular structures. The analog of this would be programming languages used in software development such as C++. Once the molecular design language is standardized, production facilities, regardless of their location in the world, can also be standardized to incorporate these designs. Furthermore, these designs can then be shared, modified, improved, etc. on the basis of licenses such as those used by OSS.

The other major component of nanotech production is manufacturing methods including physical equipment. With the inclusion of physical objects, it is ostensibly difficult to apply licenses used in OSS. However in the computer industry open access licenses have been developed for hardware as well as software. Thus in order to understand the feasibility of applying open access approaches to the physical requirements of nanotech development I again refer to examples from the computer industry including open hardware licenses.

The treatment of open hardware is more complex than an information good such as software. Fried and Torrone (2008) suggest that one way to address these complexities is to divide computer hardware into seven different components for which licenses for open access intellectual property protection can be applied. These include mechanical diagrams, circuit and layout diagrams, parts lists, and software (source code that runs a micro-processor chip, drivers or firmware for other devices). By focusing on the information components of these hardware it is possible to use existing licenses that apply to OSS such as the GPL, BSD, etc. An example of this is Sun Microsystems' OpenSPARC project. In this case, Sun Microsystems released the designs and code for the UltraSPARC T2 micro-processor under a GPL license. As with other OSS projects, Sun's intention was to leverage some of the advantages of the open access model mentioned earlier for micro-processors such as an increase potential for innovation through a larger set of contributors and users.

There has also been some recent work to develop a separate set of open hardware licenses to address the specific requirements and nomenclature of computer hardware. These include the TAPR Open Hardware License[9] and the Chumby HDK License,[10] both of which were developed in 2007, but have yet to achieve widespread use. While there is as yet no consensus on the efficacy and utility of these types of new licenses, they mark a first step. In reality, one can find producers of computer hardware that employ an overall mix of traditional and open access licenses depending on the specific component in question, with some taking a more explicit open access approach than others. Similarly, it is possible to see how open access nanotech production could consist of hybrid production models (open and traditional) applied to different components of the same system (Bruns 2001, 198).

The business model that has emerged for open hardware is to make the designs and related software available to customers at little or no cost (as with OSS) while selling the parts and materials for the product for a profit (Singel 2008). The consumer can then assemble the product based on these designs or modify it accordingly. The types of computer hardware products available using this approach include motherboards, MP3 players, complete computer systems, in-house or private telephone exchange PBX systems, etc.[11] One example is the Italian company, Arduino, that released a micro-controller board on a Creative Commons Attribution-Share Alike License.[12] Its business model also consisted of making available the designs and specifications of its product for free while making a profit on the sale of the parts. Arduino has since built a reputation for having an expertise in micro-controller boards and has been able to acquire consultancy work, similar to OSS developers. Consumers have also made improvements on the original design or created designs for specific purposes (Thompson 2008).

While the above discussion highlights the applicability of OSS and open hardware licenses there are still limitations in terms of patents. One can see that OSS licenses can function as the equivalent to copyrights by stipulating how the intellectual property can be distributed. However in the case of an invention of a process or device there is as yet no open access alternative to patents in the way that OSS licenses can be used as an alternative to copyrights. One interesting option, however, is the Open Invention Network (OIN)[13], which acquires software patents relevant to the development of the OSS operating system Linux. Specifically, participants agree not to sue one another for patent infringement while being able to access each other's patents on a royalty-free basis through various cross-licensing agreements. Note that this does not remove the practice of filing for patents as they are currently defined, rather it changes the way patents are used.

This form of cooperation is often referred to as a patent pool and does exist in many sectors other than software. The difference with the OIN initiative is that its purpose is to promote the open access model of production. Such pools could also work in terms of patents relevant to nanotech product development but would require a significant incentive structure for companies to participate. This structure could stem from the perceived benefits of cooperation or could be policy-induced. Lemley (2005, 601) for example, mentions the case of the United States government's intervention to push airline companies into cross-licensing agreements during World War I. This was to enable greater production and innovation of airplanes to support the war effort at the time. Also, a future nanotechnology patent pool could limit the problems associated with a potential patent thicket (Clarkson and Dekorte 2006, 180).

20.6 Open Access Nanotechnology

The discussion of OSS licenses, open hardware licenses, and patent pools has thus far pointed to the feasibility of applying open access to nanotechnologies. However, it is also useful to briefly look at two existing examples of projects that employ an open access approach to nanotechnology production that is relevant for developing countries. In the first example, a group of researchers demonstrated a simple and effective method for the removal of arsenic from contaminated water (Lounsbury et al. 2008). Their method involved synthesizing magnetite nanocrystals. The magnetic properties of these crystals were enhanced and specified at the nano-level. While previous methods exist for this process, they are very expensive because of the materials and equipment required. As a result, the researchers sought to develop an alternative low cost method. They were able to replace many of the expensive material requirements with everyday products such as soap, rust and vinegar while moving the crystallization process out of a lab and into a kitchen. This method is cheap, can have a significant impact in some developing countries, and is released under a Creative Commons Attribution license. Their motivations were similar to other exponents of open access approaches in that they wanted to promote a production model that focuses on collective benefits as opposed to the use of traditional IPRs and private benefits and ownership (Lounsbury et al. 2008, 11). In addition they wanted to show how an open access approach could be successfully applied to nano-technology. They argue their initial work could further spur the use of open access for nano-tech.

Another more nascent example is a project working on the development of aerogel. Aerogel Technologies has made available (primarily via their website: http://www.aerogel.org/) several methods for developing different types of aerogels. Aerogel is a substance derived from gel where its liquid component has been removed. This gives it a very low density and several associated uses. For example it nullifies heat transfer and is therefore a good thermal insulator. Aerogels have an underlying structure of nano sized pores, the size and density of which can be adjusted in the production process. However, unlike the arsenic removal example, one important piece of equipment in this process is a supercritical dryer required to remove the liquid from the gel. The aerogel.org website notes that these are typically found in university labs working in this field and can cost over US$10,000. Thus instructions are also provided for the construction of a supercritical dryer that, given the costs of parts, can total between US$550 and US$1500.[14] The two founders of aerogel.org, Stephen Steiner and Will Walker, describe one of their goals as trying to leverage the benefits of collaborative production. That is, they want to promote greater innovation and discovery by ensuring that their work and the methods they have developed are available through an open access license. All the information on the website is also released under a Creative Commons Attribution license.

In the arsenic removal case above, the researchers were concerned about private benefits and ownership of nano-technologies under traditional IPRs. While in the aerogel case, the authors are concerned about getting more people to further work on developing aerogels. Thus these examples might not show a business rationale but they do show two related but different motivations for using open access licenses and the potential reasons why others might also employ such licenses.

20.7 Implications for Open Access Nanotechnologies in Developing Countries

While open access can feasibly be applied to both the information and hardware components of the development of nanotech products, and we are now observing some nascent projects in this regard, the other issue I want to examine is what implication does this have for developing countries. To begin with I look at open source software in developing countries. The application of OSS to developing countries also parallels the arguments for nanotechnologies in terms of the breadth of applications. Benefits include skills development, supporting the growth of a local software industry that is linked to the rise in the use of OSS (e.g. low barriers to entry), localization of popular software, greater diffusion of associated technologies such as personal computers (through cheaper operating systems such as Linux), and achieving some independence from foreign software companies since governments could rely more on local firms (Wong 2004; Dravis 2003; Kshetri 2004, 74; van Reijswoud and de Jager 2008).

A recently completed cross-national research project (Cozzens et al. 2008) on the distributional consequences of OSS can help to assess these benefits. This will make it possible to go beyond some of the postulated outcomes of open access on nanotechnologies in developing countries mentioned earlier. The research project was based on a framework that assessed the distributional consequences of emerging technologies on various components of socio-economic value (such as employment, health, income, political power, etc.). It hypothesized that the distribution of benefits that derive from OSS are mediated by national conditions and public interventions. Case studies were completed for seven countries (including three developing countries). These were based primarily on interviews with key persons in each country as well as secondary data from government and other sources.

From this analysis, Cozzens et al. found that the scale of the OSS industry varies across countries but was ostensibly larger in developed economies with several global players standing out. In most developing countries no significant OSS commercial activity was observed. In some cases the government was a major customer for the domestic software industry and therefore the government's preference for OSS or software using traditional IPRs (proprietary software) was significant for local firms. In other cases, there was some dissonance between the rhetoric of policy and actual practice. For example, even where public policies called for greater use of OSS (typical in many developing countries), there were instances where government actions indirectly hindered this call. This included public procurement policies of governments that were skewed toward proprietary software, the implicit requirement of proprietary software for accessing government services, and the strengthening of existing IPR regimes through the use of software patents.

Cozzens et al. also found low levels of diffusion across all countries. The use of OSS was even lower in the developing countries in the sample. Of even greater concern were the horizontal (i.e. group or sectoral) and vertical (i.e. income) differences among the small segments within our country sample using OSS. For example, larger firms were more likely to use OSS than were those involved in information and communication technologies (ICT) related businesses. Larger educational institutions were also more likely to use OSS. This could stem from the costs of acquiring the requisite skills to develop and use OSS. The skill requirement is most evident when Cozzens et al. looked at a survey of actual OSS developers which showed that this area was dominated by a small group with very similar characteristics. Much of this group was comprised of highly educated, high-income males from North America and Europe. Other surveys have pointed to increased participation in OSS projects by programmers from India, Brazil and Russia (Ghosh 2006).

Thus even though the benefits of OSS (in terms of economic efficiency and innovation) appear great, the distribution of these benefits where they do exist appeared to be highly skewed across all countries in Cozzens et al.'s sample. This qualifies some of the rhetoric from international development agencies and others that calls for increased OSS use in developing countries without considering the distribution of benefits (Dravis 2003; UNCTAD 2004). I argue that the distribution of benefits and the factors behind it are also relevant to the efficacy of open access nanotech because of its reliance on a combination of OSS and open hardware licenses and particular types of patent pools. However, it does not mean that open access lacks utility for these countries. Rather, in order to realize such benefits on a larger scale, governments need to address several issues.

First, there is the expertise cost which developing countries must address in terms of training in order to leverage the benefits of the open access model. This is important for product development and even consumption. In some cases, this cost proved too prohibitive to enable its use. For example, in the study by Cozzens et al. they note how one government ministry in Mozambique was very keen on using OSS given its benefits. However it eventually found that it was cheaper to go with the proprietary option (Microsoft) as the human resource costs for maintaining an OSS system outweighed the license fees. This is quite pertinent in the nanotech case where high level skills are critical. In developing countries it might be plausible to expect an even smaller cadre of people with the requisite skills to contribute to an open access nano project. Addressing this human resource need will have to be a policy priority for developing countries interested in applying open access approaches to technological development.

There are also capital costs, which in the case of software is computer equipment, electricity, etc. Such costs are also important to the nanotech case as, in many instances, the initial equipment costs are prohibitively high. However unlike the aerogel.org example above, the use of open access information might not always significantly reduce these expenses. This points to the challenge of dealing with the sometimes substantial capital costs required to engage in the development of nanotech products commercial or otherwise. One option is for government collaboration with firms. For example, in a recent development Hoyle and Tolfree (2008, 308) note the establishment of the Manufacturing Engineering Centre at Cardiff University (United Kingdom) which provides access to its micro and nano manufacturing equipment to British firms and organizations. The provision of facilities and equipment through a government-industry-university collaboration is not new or unique of course (for example there is the National Nanotechnology Infrastructure Network[15] in the U.S.) but points to one of the ways that the capital cost burden can be reduced for the individual firm. The United Kingdom example does not specify an open access approach. However, where interests are similar it is foreseeable that this approach could work in the funding of such a facility. In developing countries, particularly countries in the same region, such collaboration could be pursued. Furthermore, where academic institutions are involved the development of collaborative facilities could then be tied to the development of human resource and training.

Another point is the overall legal approach governments take with regard to intellectual property. Governments in developing countries should not inflate existing

IPRs to the point where they constrain the use and development of technologies using open access. For example, the passage of the Digital Millennium Copyright Act (DMCA) in the United States is geared towards the promotion of proprietary software rather than OSS by making reverse engineering and interoperability more difficult; important aspects of the open access development process (Wong 2004). In the same way, the inflation of traditional IPRs could reduce the efficacy of open access nanotech. For example, it might become more difficult to access and therefore improve upon software used in nanotech development that is protected by copyright. Through international agreements that stem from the World Trade Organization and the World Intellectual Property Organization, or through bilateral agreements with their partners in the North, developing countries are being encouraged to follow suit by adopting similar laws (Shadlen, Schrank, and Kurtz 2005, 45).

At the same time the implementation of intellectual property laws require proper enforcement. That is, all open access licenses are simply another form of intellectual property protection and require enforcement as would any copyright protected work. A lack of IPR enforcement in general can be detrimental to the development of open access. For instance, Cozzens et al. found that local OSS developers in one developing country had a difficult time finding employment because it was easy (yet illegal) to get free copies of proprietary software. This made the whole concept of open access (and their expertise) moot. In the same way, existing business models for nanotech would be dependent on IPR enforcement, whether they focus on consultancy and design, the provision of parts and materials, or a combination of both. Thus although seemingly counterintuitive, any strategy for pushing open access nanotech in a developing country must also be accompanied by adequate IPR protection which takes care to not overextend the reach of copyrights and patents.

A related and important consideration in ensuring the efficacy of open access models is the use of standards. Standards refer to the infrastructure on which nanotech designs, methods and software are implemented. If there is no standard there will be no interoperability, a critical requirement for firms developing locally relevant products under an open access approach. This problem can be illustrated by an initial lack of standardization in one aspect of a related field: molecular biology. In molecular biology the use of micro-arrays has enabled advances in genetic research. However, with no standard for reporting the results of micro-array studies, there have been limits on the exchange of data. Thus Brazma et al. (2001, 365) proposed a standard for delivering this data which has helped alleviate the problem. Also, Lausted et al. (2004, R58) note how the release of an early DNA micro-array in 1998 based on an open access model helped develop a standard design around which subsequent work has developed. The implication is that open access researchers (regardless of region) would be better off by establishing and working with common standards from early on. This is a process that could be facilitated by governments.

This last example points to another issue related to standards in open access models. In that case, not only did an early standard help develop subsequent DNA micro-array use, it had the added benefit of being open. The use of open standards has significant advantages over closed or proprietary standards in that they can also benefit from increased scrutiny, security, and innovation. More importantly, without open standards open access models would be severely limited. Another example from the software industry, the file format for word processor documents, is useful here. The main format that most consumers are familiar with is the ".doc" format used in Microsoft Word. Since this is a closed format, OSS developers could not effectively develop alternative word processors using the same format, although some versions did exist. With the advent of the open ".odt" format several OSS developers have since been able to offer improved open source word processors. In a similar way open access nanotech researchers can also contribute to the development and improvement of open standards used in their work. This is particularly important where there can be significant diversity in the types of institutions working in nanotech as well as potential overlap across sub-fields. Thus different stakeholders can have a say in the development of standards. Note that nanotech firms or individuals operating under either a traditional IPR model or an open access model can work with open standards. Thus it is important not just to establish standards early on but that these standards be open. For small developing countries in particular open standards can also provide independence from large foreign firms in the same way that OSS does.

Ultimately, successful open access nano initiatives will require a combination of different licenses using the existing models from OSS and hardware. OIN's patent pool to promote OSS could also be applied to open access nano. However the challenge lies in making sure that the incentives for each patent owner are relatively symmetric to ensure participation. In a field as diverse as nanotech this will be difficult as patents might be applicable to more than one sub-field—a situation that will likely lead to asymmetry. Also, for developing country firms to be participants in such pools they would have to contribute their own patents. Thus ultimately, while open access patent pools offer some promise, to ensure long-term success, open access models might require an equivalent form of intellectual property protection to the patent without having to resort to the public domain.

20.8 Conclusion

Based on calls for the application of open access models to nanotech development in developing countries, this chapter first looked at the feasibility of this approach. I did this by looking at licenses used in open source software and how these had been applied to both computer software and hardware. I argued that open access licenses can be applied to nanotech projects and, to a lesser extent, patent pools were also applicable. By looking at two actual projects I showed how such licenses were used and also discussed the motivation behind the proponents of these projects. These included both moral and scientific motivations. While there was no business rationale in these cases, it is possible that such projects might develop in the future as was the case with OSS and computer hardware.

I then examined the implications that an open access approach would have by looking at the experience of OSS in some developing countries. This experience demonstrated that many of the benefits of OSS have not been realized on a larger scale because of several challenges, all of which could influence the development of open access nanotech in these regions. It is these challenges then that policy makers need to address if open access is to be truly effective.

This includes the consideration of the human resource and equipment needs for any nanotech development, particularly through a long-term national lens. Collaboration among industry, government and academic groups in this regard could be effective and could follow similar initiatives in other sectors and countries. These initiatives could also be established under existing regional cooperation agreements.

Another policy issue is the need for a balanced approach to IPRs that effectively enforces property protection so as to facilitate the growth of open access firms while not overextending the reach of existing copyright and patent laws. Tied to this is the need for open standards. Governments collaborating with firms and universities could also help promote the use of such standards, without which open access would be less effective.

Much of the existing discourse around nanotech for developing countries treats those countries as consumers, not producers. They would be receivers rather than central to the development of their own nanotechnologies. Open access nano can offset this trend by allowing more persons from developing countries to participate in nanotech production and consumption. Also some of the associated capital costs could be lowered particularly where there is collaboration among developing countries. Finally, at a minimum, the value of an open access approach to nanotech development is to offer an improvement to existing trends in IPRs for developing countries. Without appropriate interventions, however, open access nano could lead to a skewed distribution of limited benefits similar to the experience of OSS thus far. Therefore if these countries want to truly embrace this approach then governments need to develop the appropriate human resources, legal environments and infrastructure to facilitate open access nanotech development.

Acknowledgements I would like the thank Jameson Wetmore and an anonymous reviewer for their useful comments on the original version on this chapter.

Notes

1. There are of course innumerous ways to define "developing country." To use a very simple definition, the World Bank suggests that countries with a Gross National Income per capita of less than US$11,905 (i.e. classified as low or middle income) can be referred to as developing. This implies that 144 out of 210 countries classified by the Bank are developing (IBRD 2009).
2. Benkler uses the term commons based peer production model in his discussion of the open access model.
3. http://www.gnu.org/copyleft/gpl.html
4. http://www.opensource.org/licenses/bsd-license.php
5. For a more comprehensive discussion on the diversity and nuances of OSS licenses see Rosen (2005).
6. http://www.nanoengineer-1.com/nh1
7. http://sourceforge.net/projects/fungimol/

8. LGPL stands for "Lesser GPL" which is similar to the GPL but allows for the coexistence of multiple licenses.
9. http://www.tapr.org/ohl.html
10. http://www.chumby.com/developers/agreement
11. For a recent list of available open hardware products see http://blog.makezine.com/archive/2008/11/_draft_open_source_hardwa.html. Accessed 20 Feb 2009.
12. The Creative Commons Attribution license[0] is similar to the GPL and states that derivative works must be distributed under an equivalent license and credit given to the original author. See http://creativecommons.org/licenses/by-sa/3.0/
13. http://www.openinventionnetwork.com
14. See http://www.aerogel.org/. Accessed 28 Feb 2009
15. http://www.nnin.org/

References

Bastani, Behfar, and Dennis Fernandez. 2002. Intellectual property rights in nanotechnology. *Thin Solid Films* 420–421:472–477.
Benkler, Yochai. 2006. *The wealth of networks: How social production transforms markets and freedom*. London: Yale University Press.
Brazma, Alvis, Pascal Hingamp, John Quackenbush, Gavin Sherlock, Paul Spellman, Chris Stoeckert, John Aach, Wilhelm Ansorge, Catherine A. Ball, Helen C. Causton, Terry Gaasterland, Patrick Glenisson, Frank C. P. Holstege, Irene F. Kim, Victor Markowitz, John C. Matese, Helen Parkinson, Alan Robinson, Ugis Sarkans, Steffen Schulze-Kremer, Jason Stewart, Ronald Taylor, Jaak Vilo, and Martin Vingron. 2001. Minimum information about a microarray experiment (MIAME)—toward standards for microarray data. *Nature Genetics* 29(4): 365–371.
Bruns, Bryan. 2001. Open sourcing nanotechnology research and development: issues and opportunities. *Nanotechnology* 12(3): 198–210.
Bruns, Bryan. 2004. Applying nanotechnology to the Challenges of global poverty: Strategies for accessible abundance. In First Conference On Advanced Nanotechnology: Research, Applications, And Policy. October 21–24. Washington, DC.
Bürgi, Birgit R., and T. Pradeep. 2005. Societal implications of nanoscience and nanotechnology in developing countries. *Current Science* 90(5): 645–658.
Chen, Hsinchun, Mihail C. Roco, Li Xin, and Lin Yiling. 2008. Trends in nanotechnology patents. *Nature Nanotechnology* 3(3): 123–125.
Clarkson, Gavin, and David Dekorte. 2006. The problem of patent thickets in convergent technologies. *Annals of the New York Academy of Sciences* 1093(1): 180–200.
Cozzens, Susan E., Isabel Bortagaray, Sonia Gatchair, and Dhanaraj Thakur. 2008. Emerging technologies and social cohesion: Policy options from a comparative study. Paper presented at the PRIME Latin America Conference, September 24–26, 2008. http://prime_mexico2008.xoc.uam.mx/papers/Susan_Cozzens_Emerging_Technologies_a_social_Cohesion.pdf. Accessed 21 Nov 2008.
Dravis, Paul. 2003. Open source software—Perspectives for development. Washington, DC: InfoDev—The World Bank.
Einsiedel, Edna F., and Linda Goldenberg. 2004. Dwarfing the social? Nanotechnology lessons from the biotechnology front. *Bulletin of Science Technology Society* 24(1): 28–33.
Fried, Limor, and Phillip Torrone. 2008. Open source hardware. In Emerging technology conference, San Diego, March 3–6 2008. San Diego.
Friedrichs, Mark S., Peter Eastman, Vishal Vaidyanathan, Mike Houston, Scott Legrand, Adam L. Beberg, Daniel L. Ensign, Christopher M. Bruns, and Vijay S. Pande. 2009. Accelerating molecular dynamic simulation on graphics processing units. *Journal of Computational Chemistry* 30(6): 864–872.

Ghosh, Rishab Aiyer 2006. Economic impact of open source software on innovation and the competitiveness of the Information and Communication Technologies (ICT) sector in the EU. Maastricht, the Netherlands: UNU-MERIT.

Hassan, Mohamed H.A. 2005. Nanotechnology: Small things and big changes in the developing world. *Science* 309(5731): 65–66.

Hessel, Andrew. 2006. Open source biology. In *Open Sources 2.0*. eds. C. DiBona, D. Cooper, and M. Stone. California: O'Reilly Media Inc.

Hoyle, Robert, and David Tolfree. 2008. Bridging the micronanomanufacturing gap. *International Journal of Technology Transfer and Commercialisation* 7(4): 308–327.

IBRD. 2009. World development indicators 2009. Washington, DC: The World Bank.

Invernizzi, Noela, and Guillermo Foladori. 2006. Nanomedicine, poverty and development. *Development* 49: 114–118.

Invernizzi, Noela, Guillermo Foladori, and Donald Maclurcan. 2008. Nanotechnology's controversial role for the South. *Science Technology and Society* 13(1): 123–148.

Kshetri, Nir. 2004. Economics of linux adoption in developing countries. *IEEE Software* Jan/Feb: 74–81.

Lausted, Christopher, Timothy Dahl, Charles Warren, Kimberly King, Kimberly Smith, Michael Johnson, Ramsey Saleem, John Aitchison, Lee Hood, and Stephen Lasky. 2004. POSaM: A fast, flexible, open-source, inkjet oligonucleotide synthesizer and microarrayer. *Genome Biology* 5(8): R58.

Lemley, Mark A. 2005. Patenting nanotechnology. *Stanford Law Review* 58(2): 601–630.

Lounsbury, Michael, Christopher Kelty, Cafer T. Yavuz, and Vicki L. Colvin. 2008. Towards open source nano: Arsenic removal and alternative models of technology transfer. *Technology, Innovation and Institutions Working Paper Series* TII-1-2008.

Maclurcan, Donald C. 2005. Nanotechnology and developing countries. *Journal of Nanotechnology Online* (September), http://www.azonano.com/details.asp?ArticleID=1428.

Prisco, Giulio. 2006. Globalization and open source nano economy *Nanotechnology Perceptions: A Review of Ultraprecision Engineering and Nanotechnology* 2(1a): 35–40.

Rosen, Lawrence. 2005. *Open source licensing—Software freedom and Intellectual Property Law*. New Jersey: Prentice Hall.

Salamanca-Buentello, Fabio, Deepa L. Persad, Erin B. Court, Douglas K. Martin, Abdallah S. Daar, and Peter A. Singer. 2005. Nanotechnology and the developing world. *PLoS Medicine* 2(5): e97.

Schummer, Joachim. 2007. The impact of nanotechnologies on developing countries. In *Nanoethics: The ethical and social implications of nanotechnology*, eds. F. Allhoff, P. Lin, J. Moor and J. Weckert. Hoboken, NJ: Wiley.

Shadlen, Kenneth C., Andrew Schrank, and Marcus J. Kurtz. 2005. The political economy of intellectual property rotection: The case of software. *International Studies Quarterly* 49(1): 45–71.

Shah, Sonali. 2006. Open beyond software. In *Open Sources 2.0*, eds. C. DiBona, D. Cooper and M. Stone. California: O'Reilly Media Inc.

Singel, Ryan. 2008. *DIY robotics: The rise of open source hardware*. http://www.wired.com/gadgets/miscellaneous/news/2008/03/etech_hardware. Accessed 16 Feb 2009.

Thompson, Clive. 2008. Build it. share it. profit. Can open source hardware work? *Wired* 16(11): 166–176.

UNCTAD. 2004. Free and open source software: Policy and development implications. New York and Geneva: United Nations Conference on Trade and Development.

Vallance, Ryan R. 2000. Bazaar design of nano and micro manufacturing equipment. Paper presented at Nanotechnology Workshop, July 14, University of Kentucky.

van Reijswoud, Victor, and Arjan de Jager. 2008. *Free and open source software for development*. Monza Mi, Italy: Polimetrica.

Wong, Kenneth. 2004. Free/open source software : Government policy. *UNDP Asia-Pacific Development Information Programme—ePrimers on Free/Open Source Software*.

Chapter 21
Southern Roles in Global Nanotechnology Innovation: Perspectives from Thailand and Australia

Donald C. Maclurcan

As in Thakur's account of the conditions under which open access could help developing countries in nanotechnology, the importance of local innovative capacity arises in Donald MacLurcan's report on interviews with Thai and Australian scientists on the "nano-divide." Two different divides can be distinguished in the interview transcripts: an innovation divide that "relates to where nanotechnology knowledge is generated and retained"; and a nano-orientation divide in the topics that get onto the research agenda. The increasing concentration of nanotechnology knowledge in the North appears to create new possibilities for Southern dependency. But nanotechnology also brings the potential for Southern innovation, according to those he interviewed. Participating in applications is not beyond the capacity of developing countries, and they may be particularly active at the smaller technological scales that are important in their economies.—eds.

21.1 Introduction

The term 'nano-divide' has become a catch-phrase for describing various kinds of global nanotechnology inequities (Moore 2002; Mehta 2002; Van Amerom & Ruivenkamp 2006; Hassan 2005). Whilst common understandings appear to broadly relate to who will benefit from the fruits of nanotechnology innovation, there has been little in-depth exploration as to what the term 'nano-divide' really means. Moreover, despite early signs of surprisingly high levels of active Southern country engagement in nanotechnology (Court et al. 2004; Maclurcan 2005), little detail has been garnered about how the nano-divide might play out. Questions remain, for example, about whether nanotechnology offers truly new challenges and opportunities for the South or whether the issues faced are generic and re-hashed. Or is the divide such a bad thing after all? Compounding these matters has been a significant lack of Southern input in international debates about the technology's trajectory (Maclurcan 2005). In this paper I aim to provide clarity about what the 'nano-divide' actually means, and explore two propositions relating to the role of Southern countries in global nanotechnology innovation: that the South will be 'left behind'; and that nanotechnology provides opportunities for the South to play an active role in global innovation.

21.2 Methods

This paper reports on a 2004 qualitative study undertaken in Thailand and Australia[1]. A group of key informants[2], 16 from Thailand and 15 from Australia, were interviewed about their understandings and perspectives relating to nanotechnology, as part of a wider study on nanotechnology and the South.

Whilst the study sought exploratory, rather than representative, perspectives on how nanotechnology might be understood in the South, a key informant process was used to ensure a range of perspectives were considered (Mee et al. 2004). Given the argument that studies assessing nanotechnology's impacts relating to the South must go beyond consultations based purely on scientific perspectives (Invernizzi & Foladori 2005), this study included interviewees with expertise in ethics, law, social science, science policy and development. Effort was made to ensure the involvement of people with experience across the 'development process', from grassroots activism through to government policymaking and industry leadership, with interviewees coming from academia, as well as private, government and non-government (NGO) sectors. Nineteen of the key informants (61%) were engaged in work that involved nanotechnology. All key informants from Thailand were Thai citizens. Tables 21.1 and 21.2 outline each key informant's relevant position and affiliation, as of the time of their interview.

Key informants were identified through web and literature searches as well as a simplified process of co-nomination (Loveridge 2002).

Table 21.1 Thai Interviewees

Name	Position, Affiliation	Sector
Gothom Arya	Chairman, Appropriate Technology Association, Thailand	NGO
Tanit Changthavorn	Intellectual Property Specialist, National Centre for Genetic Engineering and Biotechnology, National Science and Technology Development Agency	Government
Tawatchai Charinpanitkul	Associate Dean for Research Affairs, Faculty of Engineering, Chulalongkorn University	Academic
Suwabun Chirachanchai	Associate Professor, Petroleum and Petrochemical College, Chulalongkorn University	Academic
Nares Damrongchai	Policy Researcher, Asia Pacific Economic Cooperation Centre for Technology Foresight	Government
Joydeep Dutta	Associate Professor, Microelectronics, Asian Institute of Technology	Academic
Worsak Kanok-Nukulchai	Professor, Structural Engineering, Asian Institute of Technology	Academic
Promboon Panitchpakdi	Director, Raks Thai Foundation	NGO
Pakdee Pothisiri	Senior Deputy Permanent Secretary, Ministry of Public Health	Government
Pinit Ratanakul	Executive Director, College of Religious Studies, Mahidol University	Academic
Pathom Sawanpanyalert	National Professional Officer (Health Systems Development), World Health Organisation, Thailand	NGO
Sirirurg Songsivilai	Chairman & Co-Founder, Innova Biotechnology Co. Ltd	Private
Nadda Sriyabhaya	President, Stop-Tuberculosis Association, Thailand	NGO
Wiwut Tanthapanichakoon	Director, National Nanotechnology Centre, National Science and Technology Development Agency	Government
Pairash Thajchayapong	Advisor to the Prime Minister on Science and Technology	Government
Yongyuth Yuthavong	Senior Researcher, National Centre for Genetic Engineering and Biotechnology, National Science and Technology Development	Government

Table 21.2 Australian interviewees

Name	Position, Affiliation	Sector
Leigh Berwick	Investment Manager (Nanotechnology, Invest Australia	Government
Vijoleta Braach Maksvytis	General Manger, Global Aid, Commonwealth Scientific Industrial Development Organisation	Government
Paul Bryce	Director, APACE-VFEG (Appropriate Technology for Community and Environment Inc—Village First Electrification Group)	NGO
Melinda Cooper	Research Fellow, Department of Sociology, Macquarie University	Academic
Bruce Cornell	Senior Vice President and Chief Scientist, AMBRI Pty Ltd	Private
Patricia Coyle	Medical Doctor, Department of Anaesthesia, Royal Prince Alfred Hopsital	Government
Peter Deutschmann	Director, Australian International Health Institute, University of Melbourne	NGO
Mike Ford	Associate Director, Institute for Nanoscale Technology, University of Technology, Sydney	Academic
Mike Lynskey	Chief Executive Officer, The Fred Hollows Foundation	NGO
Benno Radt	Research Fellow, Department of Chemical and Biomolecular Engineering, University of Melbourne	Academic
Michael Selgelid	Sesqui Lecturer in Bioethics, Faculty of Medicine, University of Sydney	Academic
Greg Tegart	Executive Advisor, Asia Pacific Economic Cooperation Centre for Technology Foresight	Government
Chris Warris	Researcher, Australian Academy of Science	NGO
Terry Turney	Director, Nanotechnology Centre, Commonwealth Scientific	Government
John Weckert	Professor, Centre for Applied Philosophy and Public Ethics, Charles Sturt University Wagga	Academic

Linguistic, financial and temporal limitations, as well as nanotechnology's nascent stage at the time of the study, restricted the ability for wider public engagement, particularly outside of Bangkok, Thailand. Despite every effort to ensure diversity, the majority of Thai key informants spoke fluent English and had, at some stage, received educational training abroad. The results of this study must be interpreted with these limitations in mind.

The study was supplemented by a two-page, mailed survey to members of the Thai nanotechnology research community.[3] Members of this community were identified through available literature, particularly via leads arising from the 2004 document: Final Report: Survey for Current Situation of Nanotechnology Researchers and R&D in Thailand", published by researchers at Chulalongkorn University (Uniserch 2004). The survey received a high return rate, with participants constituting approximately 10% of the Thai nanotechnology research community at the time.[4]

A study of a small number of key informants and nanotechnology practitioners in Thailand can in no way be seen as indicative of attitudes across the nonhomogenous South, particularly given Thailand's lack of a colonial history. However, Thai perspectives can be useful for exploring and considering nanotechnology and the South, given the situation Thailand faces in terms of both development and nanotechnology.

Thailand is classified by the United Nations Development Program as a "middle income country" (United Nations Development Program 2007) and is ranked 74th out of 175 countries on the Human Development Index (United Nations Development Program 2003).[5] In recent decades it has experienced remarkable progress in human development (United Nations Development Program 2007) However, Thailand's greater population continues to face significant challenges. As of 2004, 21% of the Thai population earned less than $2 a day (World Bank Independent Evaluation Group 2007), whilst financial inequality had increased over the past 40 years, particularly between urban and rural areas (United Nations Development Program 2007). Stark inequities are also evident in the distributed burden of the HIV/AIDS epidemic and general access to health services (United Nations Development Program 2007). Various populations still suffer from very high levels of child malnutrition and maternal mortality, whilst overuse of pesticides is a threat to many in rural areas (United Nations Development Program (2007). Despite the fact that the vast majority of Thais live in rural locations, the country is experiencing rapid urbanisation as well as an ageing population (United Nations Development Program 2003).

In terms of it engagement with emerging technology, Thailand has supportive infrastructure and strong hopes for biotechnology R&D (Sahia 1999). In an early study of Southern nanotechnology capabilities, Thailand was identified as a "middle ground" Southern country (Court et al. 2004). This analysis is supported by early evidence of nanotechnology R&D (Unisearch 2004, Lin-Liu 2007, Thajchayapong & Tanthapanichakoon 2003; Tanthapanichakoon 2005; Panyakeow & Aungkavattana 2002), including the establishment of a national centre (Liu 2003) and development of a national nanotechnology strategy (Sutharoj 2005). In these early stages, the Thai government's nanotechnology budget was approximately US$2 million per year (Changsorn 2004) and based on the following motivation:

> ...*the government of Thailand is determined to promote and accelerate nano science and technology as a crucial instrument of sustainable economic growth and international competitiveness* (Tanthapanichakoon 2004).

Consequently, the initial focus has been on developing waterproof, more durable silks; 'smart packaging' to monitor and maintain the state of food; more productive wine fermentation; 'self-sterilising' rubber gloves; and new car body materials (Changsorn 2004).

Just as Thailand faces significant challenges with biotechnology innovation (Thajchayapong & Tanthapanichakoon 2003), so too do people claim Thailand faces significant challenges for nanotechnology innovation (Tanthapanichakoon 2005; Sandhu 2008). From the perspective of ELSI, Thailand has a history of controversy in biotechnology, ranging from issues of morality (Changthavorn 2003) and environmental concerns (Kachonpadungkitti & Macer 2004), through to issues of intellectual property such as 'biopiracy' (Kerr et al 1991; Meléndez-Ortiz & Sánchez 2005), and compulsory licensing (Knowledge and Ecology International 2008). Yet already, the ELSI of Thai nanotechnology has created controversy around the issue of 'atomically modified organisms' (ETC Group 2004).

The justification for engaging the perspectives of key informants from Australia is grounded in Pieterse's notion of 'reflexive development' (Pieterse 2000; 2001; 1998). Building on Beck's ideas around 'reflexive modernisation' (Beck 1992; 1996), Pieterse argues that the goals and methods of development are, both historically and ever-presently, shaped in a reflexive process, increasingly driven by dispersed decision-making and control, and maximised by shared, often transnational, reflexivity or mechanisms of 'feedback' (Pieterse 1998). With the critique of science viewed within a reflexive approach as part of the developmental politics (Pieterse 1998). Pieterse argues that there is value in perspectives about science and development from both the North and South, particularly given "the perplexities of progress are shared..." (Pieterse 2001).

In this light, and given the role of the North in shaping and driving debates about nanotechnology and the South (Court et al. 2004), the value of considering perspectives from the North alongside some from the South, must not be underestimated. Therefore, rather than use the views of Australian key informants to compare the situation in Thailand to that of Australia, in this paper I use the perspectives of Australian key informants to provide views on nanotechnology innovation in the South, from a country in the North.

Whilst Australia has been firmly entrenched in international nanotechnology debates, having developed the world's first 'nanomachine' in 1997 (Cornell at al. 1997), as of 2004, Australia also lacked a formal national nanotechnology initiative and its global output was below expected levels (Warris 2004). Furthermore, Australia has also faced a slow uptake of engagement with ELSI debates in areas such as health and safety (Priestly et al. 2007), and low levels of public understanding and knowledge about nanotechnology (Bowman and Hodge 2007). The similarities at the time of this research, in terms of nanotechnology's nascent nature in both Thailand and Australia, lend weight to the validity of considering Thai and Australian key informant perspectives concurrently.

Considering the importance of the interviewee's own framework of meanings, the 31 interviews were semi-structured, which allows for a broad framing but individual divergence (Britten 1995). Each interview lasted between 20 and 80 min, was face-to-face,[6] and interviewees were offered professional translation services. All data was analysed using NVivo^{TM} software, noted for its ability to assist in developing an emergent analysis (Reid et al. 2005). In Section 1 I explore how the nano-divide is understood and the implication of the divide's constructs in terms of the roles to be played by various countries in global nanotechnology innovation.

The literature often presents Southern countries as 'passive' agents in global nanotechnology innovation— with an inability to develop endogenous nanotechnology capabilities. In Section 2 I explore the nature of that passivity and barriers and challenges facing Southern endogenous innovation.

Others in the literature point to nanotechnology providing opportunities for the South to play new roles in the global R&D process. In Section 3 I entertain the proposition of Southern countries as 'active' agents in the nanotechnology process.

21.3 Understanding the Nano-Divide and its Constructs

Interviewees regularly referred to the term 'nanodivide' but assumed its meaning and knowledge of how it is constructed are commonly understood. Given the apparent difference in understandings that emerged, in this section I seek to piece together interviewee comments to establish some relevant clarity. The first part considers how the term 'nanodivide' is interpreted and understood. The second explores some early characteristics of the nanotechnology innovation divide in terms of its leaders and those 'left behind'. Addressing these two points establishes a context and framework for my assessment, in this paper, of Southern roles in the global nanotechnology innovation.

Generally speaking, the literature does not clarify what is meant by a 'nano-divide'. Yet it is clear that, for interviewees, the term 'nano-divide' can have two different meanings. The first, that I term the 'nanoinnovation divide', refers to inequity based on where knowledge is generated and retained and a country's capacity to engage in these two processes. Tegart presented this divide as one between the "information rich and information poor..." (Tegart). Those on the 'leading' side of this divide are seen as able to actively contribute to and direct nanotechnology's trajectory, whilst those who are 'left behind' are seen as playing passive roles, unable to exert influence over any sphere of nanotechnology's global trajectory. The second meaning, that I term the 'nano-orientation divide', refers to inequity based on the areas in which nanotechnology research is targeted, as compared to the areas in which it would address basic human needs. In this sense, Arya spoke of a differentiation between nanotechnology addressing 'real' and 'felt' needs (Arya). For many, this translated into a belief that nanotechnology would be governed more by market push- rather than social pull-factors.

Whilst the divides differ in their nature, where research is targeted is often initially dependent on where knowledge is being generated and retained. Given that most of the world's research into emerging technology occurs in the North, comments on the orientation divide generally related to global inequities in terms of limited Northern research focused on Southern problems. However, the prospect of the South as active agents in global nanotechnology innovation prompts additional consideration for inequities in the orientation of Southern generated knowledge—a matter I will explore in a future paper.

According to both Thai and Australian interviewees, there is an increasing concentration of nanotechnology R&D generation and ownership in the hands of "limited leading countries" (Damrongchai). These countries were classified as 'leaders' largely because of the high levels of early nanotechnology investment (Tegart; Damrongchai; Charinpanitkul), and also happened to be some of the more wealthy countries in the world (Kanok-Nukulchai). Leaders in nanotechnology innovation were said to include[7] the U.S., Japan, Taiwan, Germany, Australia, Sweden, the U.K., France, Switzerland and Hong Kong.[8]

Given certain countries are leaders of nanotechnology innovation it stands to reason that certain countries be 'left behind' (Sawanpanyalert; Kano-Nukulchai; Warris; Berwick; Tanthapanichakoon). The previously outlined pressure to be at the forefront of nanotechnology innovation is often driven by a belief that if a country neglects nanotechnology it will be in an unenviable position later on (Tanthapanichakoon, Turney), having to try to "catch-up" (Berwick). According to Tanthapanichakoon, even a country with endogenous nanotechnology capabilities could fall behind if it did not seek to constantly develop its research position (Tanthapanichakoon). The insinuation here is that, rather than all countries gaining from nanotechnology, no matter what the nature of their engagement, those that do not develop and maintain competitive innovative capabilities will actually lose out. Furthermore, as shall be explored in Section 2, there is a perceived potential, particularly from Thai interviewees, for nanotechnology to actually reinforce the underdevelopment of some countries by creating greater technological dependency (Arya; Charinpanitkul, Yuthavong).

The proposition that countries will play different 'roles' in global nanotechnology innovation prompts a greater exploration of exactly what kind of roles are envisaged for the South. Will the South be left behind by nanotechnology or will the situation present new opportunities allowing Southern countries to become agents in global nanotechnology innovation? As shall be seen in Sections 2 and 3, this question leads to an exploration of the barriers and possibilities for Southern nanotechnology innovation.

Four key issues relating to barriers and possibilities were presented as largely determining Southern roles in global nanotechnology innovation, these being "their understanding, their commitment, the resources and infrastructure" (Sawanpanyalert). Whilst many perspectives, particularly Australian, went beyond considering the development 'problem' as one solely influenced by issues of domestic Southern capacity by also looking at global externalities and contexts, these four issues form the crux of my discussion in this paper.

21.4 The South Left Behind

Interviewees presented a number of different explanations for why Southern countries might be left behind in nanotechnology's global development. In the first part of this section I look at some of the envisaged scenarios, exploring varying levels of engagement and different kinds of roles. I then progress to addressing the underlying assumption—that the South cannot play an active role in global nanotechnology innovation—by exploring the perceived challenges to developing innovative capabilities.

For some Australian interviewees, the possibility for nanotechnology's development in the South was either non-existent, a "contradiction in terms" (Coyle) or "[not] a direct link, by any means" (Cornell). Those who struggled to see any link suggested that nanotechnology was irrelevant to the South and that not only was endogenous R&D unlikely, but that they may not even play the role of 'recipient', given that "…existing, basic, often very cheap, sometimes even free, technologies or medicines, are not available in developing countries…" (Selgelid).

However, many of the interviewees, led by the Australians, saw the link between nanotechnology and the South via some form of passive diffusion, where the role of the South was as 'recipient', rather than innovator, particularly in the "very, very poor countries in Africa" (Tegart). The common implication was that nanotechnology will most likely reach the South as a result of Northern influence. Selgelid's response highlighted this mindset when he commented, "nanotechnology would be great if someone really made it and provided it to developing countries" (Selgelid). Bryce's reasoning was along similar lines as he saw diffusion coming via a "serendipitous process" (Bryce).

Northern-dictated aid was seen as likely mechanism for Southern engagement with nanotechnology. Berwick, a policy officer with Invest Australia, saw potential for nanotechnology to be incorporated in "world aid programs and assistance programs to help developing nations just help themselves develop further" (Berwick). Similarly, Cornell referred to potential areas of application in U.N. aid packages or U.S. or European initiatives operating in the South (Cornell). Australian interviewees also saw the potential for international aid organisations to invest in Southern nanotechnology, commonly citing organisations such as the Bill and Melinda Gates Foundation. Braach-Myksvitis noted that this kind of activity has already commenced, with the Global Research Alliance[9] having been approached by a number of Northern foundations wishing to ensure the benefits of nanotechnology reach the South in this way (Braach-Maksvytis). Damrongchai, a Thai technology policy officer, agreed that there was potential for nanotechnology to enter the South via aid, citing potential applications such as single-life diagnostic kits, and methods by which to increase food preservation (Damrongchai). Damrongchai's comments were distinct, with most Thai interviewees ambivalent about nanotechnology's potential delivery through aid and development assistance, having given little consideration to the idea. Panitchpakdi was one who spoke about the potential for greater Southern dependency as a result of becoming recipients of nanotechnology-based aid whereby it would "…all depend on the countries that are advanced [if they] are willing to share" (Panitchpakdi). For Damrongchai, the idea of a Northern-controlled situation raised considerable concerns about donors exerting political influence and impressing conditionality upon Southern recipient countries (Damrongchai).

As an extension of aid, technology transfer was also discussed as a mechanism by which the South's role would remain passive.[10] Some interviewees suggested that, with the right education and training, nanotechnology could be transposed from the North to the South (Tegart, Bryce). Importing products and technologies was viewed as a possible means of engagement for those countries without nanotechnology R&D capabilities (Arya). Kanok-Nuckulchai believed that less developed countries, although unable to "build technology themselves" even if they commence nanotechnology activities now, can still start accumulating the knowledge and, once they have sufficient human resources and infrastructure, "...can absorb and transfer some of technology..." (Kano-Nukulchai).

Others saw this form of technology as entrenching the passivity of the South in global nanotechnology processes via a continuation of Southern technological dependency (Cooper). Some interviewees envisaged this dependency in terms of a 'trickle-down' of nanotechnology from the North to countries without endogenous innovative capabilities (Ratanakul). Cornell, for example, imagined the South would only benefit "as a consequence" of 'spin-offs' from Northern nanotechnology advances in areas such as water desalination, cheap nutritional foods and low cost fuels (Cornell). Weckert was similarly cynical about the way in which many Southern countries would engage with nanotechnology, suggesting that until developing countries "...get a bit more economically advanced", their engagement may be limited to Northern companies who "see some big economic advantage" of distributing nanotechnology in the South (Weckert). Tegart added that a number of U.S. companies might already be viewing the South as a "potential market" for Northern products (Tegart). In this respect, a number of Thai interviewees confirmed that 'nano-products' have already entered the Thai market (Damrongchai, Chirachanchai). What emerged in the interviews was a strong fear of 'import dependency', with some saying that Thailand could end up "...buying a lot of things" (Yuthavong) and 'losing a lot of currency' by buying-in high-cost technology through both products and services (Kano-Nukulchai; Yuthavong). Furthermore, Thai interviewees constantly referred to an "import threat" (Charinpanitkul), with one practitioner worried about a flood of "cheap products from China" and Charinpanitkul adding that the danger lies in imported nanotechnology products entering the Thai market and the population ignoring domestically-produced products once they "...get[s] used to those [international] products" (Charinpanitkul). For Damrongchai, it is this continual buying of products from the North that perpetuates underdevelopment (Damrongchai).

Another kind of passive role presented the South as 'nano-manufacturers'. Thai interviewees saw a role for the South to "partner in the manufacturing stage" with Northern counterparts (Sawanpanyalert). Weckert suggested this was merely the outsourcing of work by companies in the North (Weckert). Dutta saw the move to outsourced nano-manufacturing that would occur in Southern countries as a natural progression, explaining:

> for niche products, [where the] investments are lower, they [developed countries] will have to transfer the technology where you need more labour, where you need larger space to manufacture. You cannot keep it in developed nations, it is too expensive... (Dutta)

Others agreed it likely that countries with strong nanotechnology programs would exploit countries playing passive roles in its development (Tegart; Ford). There was recognition that Northern nanotechnology R&D partnerships with the South would seek to benefit from reduced costs (Bryce) and lower levels of regulation in the South (Ford). Ratanakul was sceptical that such partnerships would allow Southern countries to play an active role in global nanotechnology innovation, saying:

> The problem is these Western scientists are doing research for their own benefit... When they finish the[ir] research they go back and then they create new technology, based on the research, and they sell it [to] us. (Ratanakul)

The suggestion here is that the North will value-add to nanotechnology products (possibly manufactured in the South) and that those not absorbed by Northern markets or considered 'too risky', will be off-loaded to the South, as often happens with pharmaceuticals. In this light, Deutchmann spoke of his concern that "....junk products would be dumped at a cheap price on the developing world..." (Deutchmann).With Cornell believing that many developing countries will take "...whatever is available at a reasonable price" (Cornell).

At the heart of nearly all of these scenarios is the assumption that nanotechnology R&D, and therefore a potentially active role in global nanotechnology innovation, is limited to the developed countries and beyond the realm of developing countries. Turney and Berwick respectively spoke of a "cultural perception" and "natural tendency" to expect that nanotechnology will be a 'developed world technology' (Berwick; Turney). Interviewees such as Chirachanchai provided support for this hypothesis by speaking of the "advanced countries" assuming leadership roles in global nanotechnology innovation (Chirachanchai). Additionally, there was a belief that "developing countries are going to miss the boat... [and not take] advantage of nanotechnology to exploit their local, competitive advantages..." (Turney). Dutta said that if Southern nanotechnology is to face barriers they will be barriers of Northern perception, commenting: "...some of the people may think 'what the hell can you do in nanotechnology in Thailand'" (Dutta). But there was also evidence of domestic challenges relating to Southern perceptions. Kanok-Nukulchai, amongst others, noted that his initial impressions about developing nanotechnology in a country such as Thailand were "distant", with ventures into research feeling "contradictory" (Kanok-Nukulchai). Others spoke of the barrier of internal cynicism. They suggested that even in a country making efforts to become active in global nanotechnology innovation, there is potential for people to think "it is just too difficult..." (Turney), or that the research is "too late" to catch the North, leading to the cessation of activity (Chirachanchai). Driving these perceptions was a belief that "inhouse" development of nanotechnology is too difficult for many Southern countries, given the weak capacity for innovation (Yuthavong). A range of factors were presented as challenges to developing capacity, commencing with issues of Southern nanotechnology awareness and commitment, moving to issues of basic capacity through to challenges in growing and maintaining competitive R&D capabilities. In explaining the challenges to Southern countries playing active roles in global nanotechnology innovation, Thai interviewees often used examples from the Thai experience.

Preceding the issue of basic capacity was a belief that a lack of awareness, understanding and commitment could inhibit the ability for Southern countries to enter global nanotechnology innovation. In this light, a number of Australian interviewees said that the initial barriers to Southern innovation include awareness of what nanotechnology is actually about (Ford), and recognition of its opportunities and future importance (Tegart; Warris). This lack of awareness was compounded by poor understanding, particularly amongst Southern leaders, that Tegart saw as a major challenge for the South (Tegart). In contrast, Thajchayapong suggested that there is a great deal of public awareness around nanotechnology and a general acceptance of nanotechnology's merits amongst policy makers in the E.U., Japan and the U.S. (Thajchayapong).

According to many interviewees, lack of awareness and understanding about nanotechnology fits within a bigger picture in the South where nanotechnology innovation is 'prioritised out' by more immediate needs (Turney) to do with the "basic requirements of life" (Cornell). In this respect, Kanok-Nukulchai explained that nanotechnology is perceived as a "luxurious" investment, particularly in light of its embryonic state of development and the long-term nature of 'returns' (Kanok-Nukulchai).

Following on from this, some interviewees saw a challenge for the South in gaining political commitment for nanotechnology (Tegart; Panitchpakdi). This could translate to inadequate resource allocation (Tegart; Sriyabhaya), with a particular concern that Southern nanotechnology R&D would be under-funded (Tegart; Warris; Panitchpakdi; Dutta; Sriyabhaya; Radt). As Tanthapanichakoon noted about the Thai situation, "...we still do not have a very strong budget or input into nanotechnology" (Tanthapanichakoon). This perspective was supported by a number of Thai nanotechnology practitioners who claimed low budgets were inhibiting the progress of their work.

In addition to awareness, understanding and commitment, "...so many other challenges" were seen as reducing the ability to develop nanotechnology capabilities (Lynskey). As Warris noted, "...[developing countries] have got to get certain things in order before they can get into more high-tech applications, such as nanotechnology" (Warris). In this light, interviewees pointed to challenges with respect to both basic knowledge and capacity.

In terms of basic knowledge, there was a belief that Thailand lacks the fundamental knowledge to engage in nanotechnology (Sawanpanyalert). Chirachanchai elaborated by saying that, in the past, "understanding at the molecular level has been neglected" (Chirachanchai). According to others, the fact that "the science is not quite there" (Thajchayapong) was believed to place the country "a bit far behind from the very beginning" (Chirachanchai). More generally, Cornell was sceptical about the ability for developing countries to lead innovation, saying: "the idea of them being at the frontier of any of these areas is somewhat difficult to perceive, given the fact that it is born, really, at the very cutting-edge of developed country science..." (Cornell).

With respect to basic capacity, interviewees specifically referred to 'human resources' as the "biggest concern" (Damrongchai) and the "greatest obstacle" (Charinpanitkul) to Southern nanotechnology innovation. In this respect, Thailand was already said to be experiencing a shortfall of researchers (Kano-Nukulchai, Yuthavong, Dutta), compacted by the belief that Thailand will face difficulties in finding people with an 'interest' in nanotechnology (Arya, Sriyabhaya). But Lynskey suggested that this lack of a "critical mass" of relevant human resources actually means that the idea of a country such as Eritrea developing endogenous nanotechnology capabilities is "completely unrealistic" (Lynskey). These challenges were placed in the broader context of a general shortage of science and technology researchers in the South[11] (Tanthapanichakoon) and critical weaknesses in terms of educational capacity (Tegart). Cornell said that this will be a particular barrier for Southern nanotechnology, given the lack of "long-standing commitment to education" in areas such as the molecular sciences that form the basis for developing nanotechnology capabilities (Cornell). On the other hand, strong levels of human resources were seen as the backbone to nanotechnology innovation in a number of Northern countries such as Japan and Taiwan (Kanok-Nukulchai; Thajchayapong).

Yet there was a belief that shortfalls in the specific kind of human resources required to drive nanotechnology innovation will mean even greater challenges for the South (Radt). Dutta spoke of particular Thai shortfalls with respect to those with basic knowledge in quantum physics and chemistry, or those with an ability to shift into nanotechnology from other fields (Dutta). For Turney, this was part of a general problem in terms of the "level of education" of current or potential practitioners within the South (Turney). In addition to researchers, technicians were seen to be "critical people... in the exploitation of much of this technology" and a further area in which the South was viewed as being in a much weaker position than the North (Tegart). Kanok-Nukulchai noted that it may take some time to cultivate the expertise nanotechnology demands (Kanok-Nukulchai).

Futhermore, interviewees spoke of a strong potential for Southern nanotechnology researchers to be drawn to the North via the commonly expressed 'brain-drain' phenomena (Tegart; Ford). Brain drain was seen as a threat to retaining workers in a country such as Thailand if it did not have enough nanotechnology infrastructure and facilities, with a belief that the "best brains" would start to look to Singapore, the U.S. or Europe (Damrongchai). Others, including several Thai nanotechnology practitioners, saw this as part of the bigger challenge of creating a 'research culture' and general support for Thailand's scientific community. These challenges were seen, across Thai and Australian interviewee perspectives, as presenting a significant barrier to developing and retaining a critical mass of researchers in the South (Tegart, Damrongchai; Ford). Ratanakul believed this was already a major challenge for Thailand, stating that ...the government has not been thinking of the measures to prevent the well trained Thai scientists from being lured away by affluent nations... (Ratanakul).

The other threat posed by brain drain relates to the challenges of re-integrating returned researchers into Southern contexts. Charipanikul believed that it would be quite difficult for Thais with nanotechnology expertise to find employment upon return from overseas work or training (Charinpanitkul). Dutta explained that whilst "...the Thai government has spent a lot of funds to train people abroad..." their adaptation time upon return is too long "...because there are no active groups working here" (Dutta).

In addition to human resources, many saw infrastructure as a basic requirement, and thereby major challenge, for the Southern development of nanotechnology capabilities. Interviewees first discussed problems with the amount of infrastructure available, referring only to 'hard' infrastructure in terms of equipment and instrumentation. Like a number of others, Dutta believed that "...nanotechnology needs quite a bit of infrastructure", given the scale on which the research occurs (Dutta). But Cornell noted more strongly that if a country wants to seriously engage in nanotechnology R&D then "you could fill telephone books with the kind of infrastructure that you need" (Cornell). In elaborating, Cornell outlined the need for:

> an ability to work with ceramics, with plastics, organic chemistry development, with fine metals, thin-film deposition, you must have electronic foundries... you need everything that currently supports a modern industrial economy and that goes from screen-printing, paints, chemistries, lubricants, polymers, waxes, solvents, all of the moulding industries, the etching industries, electrochemical industries, (Cornell).

In this respect, interviewees referred to the "limited capabilities" within the South (Tegart, Tanthapanichakoon). According to Cornell, developing countries have "...not yet advanced to the point whereby this kind of equipment, this kind of capability, is naturally part of their world" (Cornell). For Cornell, this was partially explained by the common, developing country absence of a military industrial complex that he saw as the foundation for driving innovation; pointing to the case of the U.S. where he saw nanotechnology driven in this manner (Cornell).

The second aspect of infrastructure requirements relates to the quality and cost of instrumentation. Practitioners specifically mentioned the essential need for observation and characterisation instrumentation such as Tunnelling Electron Microscopes and thinfilm coaters. In this respect, Thailand was seen to lack some of the required equipment (Dutta), with Kanok- Nukulchai noting a perception that the technology is too advanced:

> ...when we talk about nanotechnology most people think... it is something we cannot see, something that need[s] a lot of high-tech equipment and when we look back at Thailand we are not that advanced in terms of technology (Kanok-Nukulchai).

Prohibitive costs were presented as the main barrier to the acquisition of such instrumentation (Berwick), with Cornell believing that nanotechnology requires "a fairly large investment in fairly expensive equipment" (Cornell).

In addition to challenges with respect to basic capabilities, interviewees spoke of challenges relating to the development and maintenance of competitive, nanotechnology R&D capabilities. There was a strong belief, for example, that access to appropriate instrumentation was a key barrier to the development of Southern nanotechnology strength. Radt believed many of the developing countries actually have the instruments required to undertake nanotechnology R&D but saw the barrier more as a matter of access to, and maintenance of, these instruments (Radt). Dutta partly agreed, saying of the existing instrumentation, much of it is "underused" (Dutta).

The challenges around developing basic nanotechnology capabilities suggest coordination and strategic planning is required. However, coordination is another area in which interviewees saw challenges for Thailand, with a genuine concern that research will be "unfocused and resource[s] will be scattered" (Kanok-Nukulchai). Chirachanchai, for example, saw problems in ensuring that each research effort was part of an overall strategy for Thai nanotechnology (Chirachanchai). In this light, Tanthapanichakoon highlighted that Thailand "...does not have a national strategy and all the labs, or centres, are working on their own interests or on their own subjects, without coordination..." (Tanthapanichakoon). Furthermore, Tanthapanichakoon said that Thailand suffers from a significant breakdown in communication between many of the government agencies that would need to be working together when it comes to nanotechnology innovation (Tanthapanichakoon).

Accompanying coordination of research is the Southern challenge of strategic planning. Tegart believed the initial planning difficulty is in assessing capabilities and then selecting focused areas for research (Tegart). Dutta's concern, that people talk generally about nanotechnology without a concentrated focus in any particular direction (Dutta), was seen as part of a bigger fear that Thai nanotechnology lacks a clear and comprehensive vision for the future (Chirachanchai).

But, developing focused nanotechnology research can be made more difficult if the ability to develop knowledge is restricted by nanotechnology innovation's global leaders. Interviewees strongly argued that some of the greatest barriers to Southern innovation relate to "who's involved and actually creating the technologies" (Cooper) and, stemming from this, the "big issue" (Damrongchai) of control over intellectual property rights.

Concerns about the inhibitive impact of Northern nanotechnology patenting upon Southern attempts to develop innovative capabilities were seen to be uniquely enhanced given nanotechnology relates to the fundamental building blocks of all material things (Damrongchai). Added to this, the potential disappearance of the 'cost-barrier' for nanotechnology R&D (as shall be discussed in Section 3), makes the issue of patent control "extremely important" because "...the powerbroker will be the knowledge" (Braach-Maksvytis).

Furthermore, a major concern held by interviewees was that a great deal of nanotechnology patenting would be speculative in order to claim future applications. Pothsiri was worried that "a Western country, particularly in the private sector... may try to play around with this kind of thing without making any attempt to find a new innovation" (Pothsiri). Additionally, a number of interviewees said their concerns lay with the increasing move towards, or ambiguous nature of, "process" patents[12] in light of a greater research focus on atomic self-assembly (Selgelid, Changthavorn). In these respects, nanotechnology was seen as leading to corporate monopolies (Ratanakul), locking up research in the North and thereby providing "...another key barrier to developing country uptake" (Braach-Maksvytis) by blocking potential avenues for Southern R&D (Cooper).

One Thai interviewee spoke very strongly about how nanotechnology patenting will maintain and promote the technological divide through ongoing oppression of the South. Arya presented intellectual property rights as the "new economic power... [and] new instrument of domination", with patent holders often over-exploiting their position of strength (Arya). He went on to say that the control of proprietary knowledge is driving greater oppression through a divide that, in addition to being technological, includes an:

> ...economic, social and also political divide, because those who have the new technologies will also invest, not only for the products to serve mankind but the products which can be used for domination, for hegemony, weapons of new kinds and so on and so forth. (Ayra)

General concerns were also held for the ability for Southern countries to translate nanotechnology research into patented knowledge. Charinpanitkul said that nanotechnology patenting in Thailand "...will be a big obstacle" given patent understanding, even in the university, is insufficient (Charinpanitkul). Added to this, Changthavorn pointed specifically to a lack of nanotechnology understanding amongst Thai lawyers (Changthavorn). Charinpanitkul saw the lack of knowledge as severe and inhibitive, highlighting, with respect to nanotechnology patent applications, "...we do not know even what style or what wording we should add..." (Charinpanitkul).

A contributing factor to weak Southern patenting and another major challenge to the South playing active roles in global nanotechnology innovation is the potentially poor level of private sector engagement with nanotechnology R&D (Turney). Thajchayapong said that the science of nanotechnology actually demands greater participation from industry (Thajchayapong), with private sector participation suggested to be a crucial driver of early nanotechnology success in Japan and Taiwan (Charinpanitkul).

Aside from foreseeable financial 'return', available financing and other financial incentives were seen as the initial drivers of private sector engagement in nanotechnology. However, building on earlier concerns about a general lack of funding, there was a belief that risks, particularly those associated with intellectual property, could make access to nanotechnology finance and capital a serious problem in some of the Southern countries (Turney). In this light, many saw the countries of the North in comparatively strong positions. Tanthapanichakoon, for example, suggested that a country such as the U.S. is in "the best position" when it comes to nanotechnology R&D largely because of the "good system of venture capital" in place (Tanthapanichakoon).

A second challenge facing the development of Southern-owned proprietary knowledge is the difficulty of technology transfer from academia to industry. The initial problem is that in a country such as Thailand a great deal of the nanotechnology research is 'fundamental' rather than 'applied'. As Tegart noted, whilst there are some Thai researchers completing PhD's in nanotechnology "they are rather pure science topics... and not very focused on practical and applied areas...." (Tegart). Secondly, Thai nanotechnology practitioner responses demonstrated weak professional links between academia and industry. Thajyapong said these weak links were most visible in the poor levels of communication about nanotechnology between industry, government policymakers, researchers and academia (Thajchayapong). Finally, there was a belief that Southern firms are limited in their ability to absorb nanotechnology R&D from academia (Pothsiri).

Even leveraging from international partnerships to overcome capacity issues encountered skepticism. Interviewees highlighted barriers in terms of the lack of Southern infrastructure (Cornell), and massive amounts of competition to be "at the forefront" in nanotechnology (Kanok-Nukulchai). To elaborate, Tanthapanichakoon used the example of the New Energy and Industrial Technology Development Organisation of Japan whose policy, in terms of collaborative research, is "...no grants at all in nanotechnology field[s]" (Tanthapanichakoon).

It is important to note that most of the challenges raised for the South to play an active role in global nanotechnology innovation were not considered specific to nanotechnology but, rather, generic to all high-tech fields (Radt). Some of the examples presented include: the low levels of investment in emerging technologies (Panitchpakdi); the Northern concentration of proprietary knowledge (Tanthapanichakoon)—with nanotechnology said to raise issues similar to those at the forefront of the biotechnology debate in this respect (Damrongchai); there were also shared problems in the hesitation of the private sector to engage in nanotechnology R&D and difficulties associated with technology transfer (Tegart).

If many of the challenges are generic, is there anything to suggest that global roles in emerging innovation will change at the hands of nanotechnology? Will this divide be any different to preceding technological divides? Selgelid thought not, saying he saw no reason why the general situation relating to inequality would be any different for nanotechnology:

> The North-South divide is really complex and I do not see why there should be anything special or unique about the North-South divide or rich-poor divide as far as nanotechnology [is concerned]. I would imagine the same kind of dynamics that are driving inequality in all kinds of other domains would just apply to this domain, as well... (Selgelid).

Others were concerned that nanotechnology's innovation divide could be "exaggerated" (Deutchmann) and worse than the divide currently witnessed with ICT (Weckert). This view was partially justified by a belief that nanotechnology enters a platform of existing and widening divides (Tegart; Cornell; Weckert), particularly those for biotechnology and ICT (Tanthapanichakoon). In this light, Yuthavong[13] saw potential for an extreme shifting of R&D concentration away from developing countries, saying that nanotechnology R&D is "moving too fast" for many developing countries to "...really capture the benefits fully" (Yuthavong), with Sawanpanyalert adding that Northern countries are in a much more favourable position to respond and adapt their capabilities (Tegart, Sawanpanyalert, Tanthapanichakoon). However, as shall be explored in Section 3, others saw nanotechnology presenting new opportunities for the South to play an active role in global innovation.

21.5 New Opportunities

Whilst previous arguments suggest that an active role in nanotechnology innovation is beyond the South, others suggest that the barriers are more matters of perception (Turney, Dutta) and that nanotechnology can also be viewed as an opportunity for the South (Yuthavong). As noted in previous research, despite seemingly universal understandings, nanotechnology conjures a range of perceptions, some of which fail to consider nanotechnology in some of its more simple forms (Maclurcan 2009). 'Further consideration' is suggested as leading to more 'circumspect' perspectives (Tegart), with Selgelid's responses highlighting this point—after considering the issues in greater detail, Selgelid stated "...there is the possibility that nanotechnology is not out of the reach, or should not be out of the reach of developing countries" (Selgelid). In this respect, a number of interviewees, particularly those from Thailand, saw alternative paths that involved the South as 'nano-innovators', actively contributing to nanotechnology's global trajectory. Nearly one-third of interviewees from both Australia and Thailand specifically referred to nanotechnology providing 'opportunities' or holding 'potential' for Southern innovation (Tegart, Damrongchai; Berwick, Yuthavong, Chirachanchai; Dutta; Ford). In fact, some even suggested that Southern countries actually "...have the advantage" in terms of nanotechnology innovation (Turney) and that, on the back of various technologies, the South would "be at the same level" as the developed countries at some future stage (Dutta). In this section I provide supporting arguments for many of these claims.

Two elements, each of which I will explore in this section, contribute to the argument that the South can play an active role in global nanotechnology innovation. The first element is that early signs of Southern nanotechnology commitment could set a platform for more active engagement in global innovation, including hope for engagement that includes some of the Least Developed Countries (LDCs). In this respect, interviewees also outlined constraints as to the kind of nanotechnology activity that might be possible. The second element is the suggestion that Southern countries might be able to overcome a number of the previously raised barriers and challenges to developing innovative capabilities.

Interviewees introduced the issue of Southern nanotechnology commitment by citing nine developing countries active in nanotechnology R&D and commenting on the strength of each country's commitment. China was the one Southern country presented as a playing a most active role in nanotechnology and was often grouped with the North, given it is "...moving so fast and putting so many resources into nanotechnology..." (Tanthapanichakoon). Warris noted that "...China had the highest ratio [of nanoscience compared to their total science] in the world", highlighting that it had identified nanotechnology as an area of increasing importance

(Warris). Following China, India, South Korea and South Africa were all seen as playing highly active roles ahead of Thailand, Malaysia, Vietnam, the Philippines and Indonesia who were viewed as playing moderately active roles. Although Tegart saw the Philippines, Vietnam and Indonesia as "much further behind" there was chance for strong presence from these three (Tegart). As Tanthapanitchakoon noted, there is a lot of enthusiasm in a country such as Vietnam where the government are "very keen to promote their nanotechnology program", in order to "catch up" (Tanthapanichakoon).

Thailand provides an interesting case given it is a 'middle-range' developing country seeking to make its mark in global nanotechnology innovation. Both Tegart and Turney spoke about the willingness and drive from Thailand to harness nanotechnology through the development of innovative capabilities (Tegart; Turney). Thai interviewees confirmed the strong desire and ambitions, with Pothsiri noting that the Thai government's policy is "...to promote this kind of innovation to be... something that we would be able to do by ourselves...." (Pothsiri). Arya said that the Thai government's hopes are actually targeted at ensuring nanotechnology contributes up to one per cent of the Thai Gross Domestic Product in the coming ten years (Arya). Strategically, given its central location amongst South East Asian nations, Thailand is seeking to be a "hub" for nanotechnology (Charinpanitkul), with Tegart positive that Thailand could be among the leaders in South East Asia if it receives strong government support (Tegart).

Both interviewees and practitioners saw government support and 'endorsement' for nanotechnology as real strengths for Thailand. Pothsiri, speaking about the Thai nanotechnology climate said that "right now the chance is quite good [to build capacity] because there is a policy commitment from the government" (Pothsiri). Interviewees said that the policy commitment had already resulted in initial funding, with Damrongchai adding that there is enough money available to make a substantial investment in nanotechnology R&D (Damrongchai). In addition to early funding, interviewees mentioned the establishment of a national nanotechnology centre and the earmarking of specific agencies to drive nanotechnology forward (Pothsiri). Interviewees also noted that this policy commitment was translating into support for nanotechnology across a number of sectors. Thajchayapong highlighted examples from his own experiences with the Ministries of Commerce and Defence:

> ...two weeks ago, I was delighted that I was invited by [the] Ministry of Commerce Permanent Secretary and I was explaining to them about nanotechnology, in front of 150 or 200 people... [similarly, at] the military school, they had about 150 student[s] listening to nanotechnology... (Thajchayapong)

However, interviewees were quick to highlight that these developments all stemmed from the Thai Prime Minister who was pinpointed as the main driving force for nanotechnology in Thailand (Damrongchai, Kanok-Nukulchai, Sriyabhaya, Changthavorn). Turney more explicitly noted that "the Prime Minister is actually driving this, personally, as something he wants to see happen" (Turney), with Tanthapanichakoon adding that the Prime Minister individually realized the importance of emerging technology, such as nanotechnology, for Thailand's future

(Tanthapanichakoon). Thajya-pong, highlighting the importance of political leadership in a country such as Thailand, outlined the circumstances in which the Prime Minister initiated Thailand's first serious foray into nanotechnology:

> ...about two years ago he [the Prime Minister] went to the science park. He visited us and he was surprised. He used the words to the effect, 'I did not realise that you have done so much' and then he mentioned about nanotechnology. And that is how we say, 'o.k., if Prime Minister use the word nanotechnology we have to respond' and we set up the centre (Thajchayapong).

Thailand's high level of commitment, combined with the previously mentioned nanotechnology activity in other countries, suggests the existence of foundations upon which Southern countries could play active roles in global nanotechnology innovation. A few of the Australian interviewees admitted surprise at the early nanotechnology capabilities in some of the developing and transitional countries (Tegart, Braach-Maksvytis). Braach-Myksvitis saw the "early start" from developing countries as something new in the science and technology arena (Braach-Maksvytis). In this respect, Cooper said that nano-innovation does not have to follow the same distributive pattern as biotechnology innovation, believing that early widespread engagement "... totally changes the picture" (Cooper). In presenting hopeful visions of Southern countries playing active roles in global nanotechnology innovation, some referred to other Southern successes such as the development of the Indian pharmaceutical industry in the 1970s that produced cheap drugs geared at local needs (Cooper) and South Africa's recent ability to build a critical mass of scientific researchers (Lynskey).

Although the examples presented thus far mainly deal with the more advanced developing countries, there was a belief that the nanotechnology revolution could extend to the LDCs or 'very poor countries' (Ratanakul; Dutta)]. Charinpanitkul, who assessed Thailand's nanotechnology capabilities in 2004, said that countries like Laos could be conducting nanotechnology research within 5 years (Charinpanitkul).

However, interviewees also placed limitations on the scope and nature of nanotechnology activity that might be possible in some Southern countries.

In terms of scope, there was a belief that Southern countries might only be able to engage with nanotechnology innovation at certain stages of the R&D cycle. Although Sawanpanyalert did not see Southern countries as necessarily able to work on 'early-stage' nanotechnology research, he said that there are opportunities for 'later research' (Sawanpanyalert). In this respect, Southern innovation was seen as dependent on partnerships with the North (Sawanpanyalert; Turney; Yuthavong), particularly in an area such as drug development where developing countries "cannot do it alone" and will require "assistance from developed countries" (Yuthavong). Dutta pointed to countries such as Switzerland and Sweden as partners for components of the R&D phases that cannot be completed in Thailand (Dutta).

However, Braach-Maksvytis, harking back to the way nanotechnology is perceived, noted that the scope of a country's contribution to global nanotechnology innovation "...depends on what end of the scale you are talking about" (Braach-Maksvytis). Low-tech scale production of nanopowders for cosmetics, plastics and the polishing of silicon chips could be globally widespread and developed entirely within the South (Braach-Maksvytis), whilst "niche products for which you need very high investments" might remain outputs from the North (Dutta).

In terms of the nature of nanotechnology research, the ability to conduct both fundamental and applied scientific research across all fields, was seen as beyond many of the Southern countries (Arya). Arya explained this situation in greater detail, in the Thai context:

> ...Thailand is not in a good position to compete at the fundamental research level. We do not have the capacity, we do not have the potential. So, whereas we can do some fundamental research for nanomaterials or nanotechnology, we have to look, more, at the applied research. (Arya)

As shall be explored in a future paper, the need to focus on applied research was closely followed by a need for Southern countries to find their own niche areas of application, building on niche knowledge (Arya).

As alluded to in Section 2, a range of capacity issues contribute to determining a countries' level of engagement with nanotechnology (Charinpanitkul, Sawanpanyalert). Nonetheless, arguments were made in six areas for why Southern countries might be able to overcome some of the previously raised barriers and challenges to developing innovative capabilities.

21.5.1 Availability and Demand of Human Resources

The first argument related to the level of available human resources to drive nanotechnology R&D. With respect o the Thai situation, Tanthapanichakoon noted that, from the outset, the Prime Minister demanded particular effort in the area of "human resources development in nanotechnology" (Tanthapanichakoon). The general belief was that Thailand possesses an adequate workforce to commence nanotechnology R&D initiatives (Damrongchai), with an estimated number of "not less than 100 researchers[14]... [and] nearly twenty laboratories that have been working on nanotechnology, scattered around [the] universities" (Thajchayapong).

These views are coupled with a belief that nanotechnology might not require a big labour force (Kano-Nukulchai) - that a small number of scientists, with differing backgrounds but unified in their focus on nanotechnology "can be quite a big force" (Yuthavong). The implication is that countries could consider nanotechnology innovation even if they only had a few researchers in nanotechnology-related fields (Warris). Lynksey, although previously sceptical about the hopes for endogenous innovation, said that getting a 'critical mass' is possible for countries that have a credible, political approach—even some of the LDCs, although this would be "....a couple of generations away" (Lynskey). Yuthavong agreed, stating that Thailand can develop a critical mass of scientists "if it gets the right policy and right directives" (Yuthavong).

Furthermore, discussing nanotechnology with some of the Thai interviewees unearthed a surprising number of people who claimed that they, or their colleagues, had been working in nanotechnology for many years. This was coupled with a belief that, in Thailand, a number of things have already been done on the nanoscale (Songsivilai). Charinpanitkul, who produced Thailand's first national assessment of nanotechnology capabilities, cited Thai quantum dot research before the 1990s and, more recently, carbon nanotube research from the late 1990s (Charinpanitkul). Surveyed practitioners claimed to have been working in nanotechnology for an average of 6.2 years, with some claiming as much as 14 years. To explain this, some interviewees referred to support for 'nanotechnology' as a recent Thai phenomenon:

> ...back 10 years or more... I was studying in that field [nanotechnology] and researching the molecular assembly of molecules... but at that time the environment in Thailand was not very supportive of doing research into nanotechnology... (Damrongchai).

In this light, Damrongchai believed that Thailand has enough workers with the appropriate skills and experience to seriously engage in nanotechnology innovation (Damrongchai). Songsivilai added that the diverse backgrounds of the Thai scientific workforce are an advantage in terms of the skill-sets nanotechnology R&D requires (Songsivilai). In further support, my assessment of Thai practitioners who claimed to be working in nanotechnology showed a highly qualified workforce, with nearly all having completed doctorates. Kanok-Nukulchai said that within 3 to 5 years someone in Thailand will receive a Masters or PhD degree in nanotechnology (Kanok-Nukulchai). In the broader context of Southern countries, Berwick suggested that "...[a] mastery of the basic principles of nanotechnology can be applied across the board..." (Berwick).

21.5.2 *Infrastructure*

In addition to the hopes of addressing human resource barriers, interviewees suggested nanotechnology could bypass some of the infrastructural challenges commonly associated with many emerging technologies. Dutta suggested that developing bionanotechnology applications, for example, does not necessarily need an extensive biotechnology centre (Dutta). Similarly, Radt claimed that nanotechnology "could transform any developing country to catch up with the scientific communities [sic]" considering that, in an area such as drug development, it does not necessarily rely on "very expensive and long-lasting studies" (Radt).

Additionally, some interviewees challenged the belief that nanotechnology R&D requires special instrumentation or world class facilities, claiming, rather, that many developing countries have much of the standard instrumentation that is needed (Dutta, Radt). With these points in mind, Radt distinguished nanotechnology from the capital intensive nature of the previously emergent computer chip industry that needed "big clean rooms, big plant[s]... [and] a lot of extremely expensive instruments to start" (Radt).

Furthermore, interviewees believed that nanotechnology R&D can utilise existing infrastructure and previous approaches "...because nanotechnology is based on the basic sciences: chemistry, biology and engineering..." (Radt). With similar thoughts in mind, interviewees presented nanotechnology as a "natural progression" for Thailand, given its background in biotechnology research [Sawanpanyalert, Tanthapanichakoon). Thus, interviewees often spoke of nanotechnology R&D as merely a modification or "upgrading" of existing activities and infrastructure (Pothsiri). As Chirachanchai noted, "...what we have now, we can apply it but just change the point of view from 'trial and error' to understanding at the molecular level" (Chirachanchai). Coyle suggested that, although "the terms may be new...", it is, in fact, putting into effect much existing knowledge, with a new emphasis on the way to do things (Coyle). In this respect, Chiranchanchai said it is actually more a matter of identifying people's roles and that the infrastructure will naturally develop when people come together (Chirachanchai).

21.5.3 International Knowledge

Building on the argument that nanotechnology can utilise existing infrastructure, interviewees presented a strong belief that developing countries can "leapfrog" their R&D capabilities and position within global innovation (Warris; Tanthapanichakoon, Yuthavong; Dutta) When interviewees referred to 'leapfrogging', they spoke of utilising existing knowledge and research efforts by scientists in other countries, learning "...from what has already been done..." (Berwick), in order to adapt and extend the lessons learnt (Chirachanchai). Thai interviewees, in particular, saw opportunities for countries to "jumpstart" their R&D activities (Ratanakul), avoiding the need to start from "square one" (Tanthapanichakoon; Ratanakul). But Braach- Maksvytis said that the potential for Southern countries to leapfrog in nanotechnology innovation also paves the way for a new trajectory, whereby countries could "steer a new, emerging science area into a very real and very practical outcome" (Braach-Maksvytis). Braach-Maksvytis elaborated;

> ...rather than going down the track of implementing technologies and devices that developed countries have now and then just repeating that, to actually look beyond already and use the new worlds that new technologies open up (Braach-Maksvytis).

The ability to leverage global nanotechnology developments was seen as a major positive for a country such as Thailand (Tegart, Arya, Warris, Weckert). Tanthapanichakoon expressed the Thai thinking that there is no need to "...reinvent the wheel... we want to know what the world has and what kinds of existing technologies we might make use of in order not to start from scratch" (Tanthapanichakoon). This is made all the more easy given, "...a lot of papers can get translated, [and] put on the Internet a lot quicker now..." (Berwick). Furthermore, there was a belief from Thai nanotechnology practitioners that, because nanotechnology is 'new' and not yet well defined, it is not too difficult to catch up, with Dutta adding that Thailand's 'late' start "does not necessarily mean that we have lost anything..." (Dutta).

On the contrary, Kanok-Nuckulchai suggested that the very nature of backwardness means that 'falling behind' in nanotechnology may not be such a big issue for some countries "because they can wait until the leading country develop[s] these technologies and they can follow" (Kano-Nukulchai). In fact, in looking to leap ahead, the South's 'backwardness' in nanotechnology R&D was even viewed as potentially advantageous in terms of saving "time and money" (Dutta).

There was a general recognition that support and cooperation from the North can be useful in Southern efforts to 'catch up'. In this respect, Turney noted early signs of Northern support for Southern nanotechnology programs, highlighting the case of Japan that is currently "...funding educational programs within countries like Malaysia and Thailand" (Turney). In this instance, Turney saw the net benefit as a flexible, highly-skilled workforce from which all countries can benefit (Turney).

21.5.4 Research and Development Costs

Whilst entry costs were previously raised as a barrier for Southern innovation, many saw nanotechnology offering "a door for countries into research even if they do not have as large funding bodies as perhaps highly developed countries" (Radt). In addition to the benefits outlined in 5.3.1, 5.3.2 and 5.3.3, Ford said that there is considerable potential for Southern innovation "...for the simple reason that a lot of nanotechnology is extremely cheap" (Ford). Some interviewees agreed that big budgets are not necessary to commence working in nanotechnology (Chirachanchai), as highlighted by the Thai situation where the initial budget of 6 million baht[15] was "...enough, at that stage" (Thajchayapong).

However, it is important to note that low budgets do not necessarily equate to 'low-end' nanotechnology in terms of the sophistication of R&D output. Rather, many nanotechnology approaches allow "... very useful and high-impact experiments with relatively small budgets..." (Radt). For example, given it utilises natural chemical reactions, self-assembly was seen as a potentially inexpensive means by which the South could play an active role in cutting-edge nanotechnology innovation (Chirachanchai). As Radt noted with nanotechnology-based drug delivery systems:

> ...most of the nanotechnological systems are self-assembling systems, therefore, often it is possible to simply put reactants together in the right order... it is possible to produce these systems in parallel in a large scale... self assembling systems are easy to scale-up and that is why they are potentially useful and cheaper (Radt).

In addition to cost reductions for drug development arising from chemical self-assembly, Chirachanchai saw nanotherapeutics as potentially inexpensive because increased molecular understanding, combined with advances in simulation, would minimise the need for 'trial-and-error' research (Chirachanchai).

A number of these cost-cutting factors contributed to Dutta referring to what he termed "poor man's nanotechnology" (Dutta), whereby nanotechnology could offer entry points to cater for all sorts of residual Southern capacity:

> If I think about a country like Laos... it is lacking resources... I would suggest that, 'look I am doing nanotechnology the poor man's way'... I am using colloidal nanoparticles and trying to find out applications of each ones. Laos could possibly also concentrate on things like this (Dutta).

21.5.5 Comparative Advantage

Low costs are also assisted by traditional and new means of comparative advantage. Advantages were said to revolve around traditionally proposed areas, such as the low costs of labour, easy access to clinical trial participants and natural resource abundance. However, a number of advantages were suggested as specifically relating to nanotechnology. Thai nanotechnology practitioners directly referred to their nanotechnology research utilising local biological products and raw materials, such as herbal products with natural resource bases (Charinpanitkul, Tanthapanichakoon). According to one practitioner, the link to resources also included the ability to produce nanomaterials from local materials. This was seen as a potential cost-saving advantage for a country like Thailand, with Chiranchanchai suggesting that if Thailand produces a nanoscale product, such as a base for drug delivery, "...the products, compared to other countries, may be cheap if we use our natural abundance for that" (Chirachanchai).

21.5.6 Private Sector Support

In terms of some of the capacity matters raised in Section 2, Damrongchai challenged the belief that researcher retention would be an issue for a country such as Thailand, saying that there are enough researchers who would want to remain in Thailand (Damrongchai). There was also a belief that the early signs in Thailand suggest adequate private sector support for nanotechnology. According to some, industry is showing early interest in nanotechnology (Charinpanitkul, Thajchayapong), signalling a positive change in perceptions amongst Thai businesses that now see it as a "promising field" (Charinpanitkul) that is resulting in the emergence of nanotechnology entrepreneurs (Turney).

In this light, although Tanthapanichakoon agreed that Thailand faced technology transfer challenges and did not have any industrial products,[16] he noted that some product samples and prototypes are being produced in labs (Tanthapanichakoon). Tanthapanichakoon added that the commercialisation stage of mass production was probably another 2 to 3 years away (Tanthapanichakoon).

21.6 Conclusion

My initial research shows the existence of two, slightly different 'nano-divides'. The 'nano-innovation divide' relates to where nanotechnology knowledge is generated and retained, whilst the 'nano-orientation divide' relates to where nanotechnology research is targeted. The nano-innovation divide is the platform on which the nano-orientation divide is determined. Early levels of nanotechnology funding in the North appear to be driving the innovation divide, which is framed in the absolute terms of: countries that lead R&D and those that are 'left behind'. In this sense, the nano-innovation divide was negatively viewed, with no one raising the possibility that the divide might be a good thing, either for the world in terms of maintaining an effective international division of labor or, perhaps more importantly, for a country in terms of it entirely averting engagement with nanoinnovation. This suggests an immediate mainstreaming of nanotechnology debates into a competitive, nationalistic 'growth' narrative that leaves little space for post-development critiques or notions of alternative trajectories.

In terms of new possibilities for Southern dependency, fears were held for an increasing concentration of proprietary knowledge in the North, reinforcing an inequitable IP regimen and accelerating the global division of labour. Compounding these issues are the proposed capacity challenges facing a country's efforts to develop endogenous nanotechnology capabilities. Although most of these challenges do not appear exclusive to nanotechnology, some key informants perceive them as inhibitive for Southern engagement in global innovation.

However, it would appear likely that nanotechnology also creates new potential for Southern innovation. There was a strong belief that nanotechnology innovation is misperceived as being beyond the capacity of developing countries. This was supported by reference to the early levels of Southern nanotechnology awareness and commitment, and subsequently raised the possibility that Least Developed Countries could become active in global nanotechnology innovation. However, the results show that opportunities for Southern nanotechnology innovation are seen as strongly moulded by the global constraints placed on Southern innovation, with interviewees noting constrained possibilities in terms of the scope and nature of nanotechnology activity that might be possible in some Southern countries. In particular, there are significant differences in potential 'scale'; from those countries where nanotechnology innovation might occur on a relatively small scale and be targeted towards a certain phase in the R&D process, to those for whom the scale of nanotechnology innovation is more broadly encompassing and comparable to R&D efforts in the North. Although some are generic, such as the reverse argument that IP is actually a key instrument in promoting nanotechnology innovation in the South, most of the factors raised in arguments supporting the ability for Southern countries to engage in nanotechnology innovation relate specifically to nanotechnology.

Overall, Australian interviewees were more likely to consider Southern nanotechnology within a global context of external challenges to building endogenous capacity, identifying market-dictated challenges, such as the orientation of nanotechnology research and development, its cost and its availability, as well as other structural barriers relating to the global patent system. From this perspective, the development 'problem' is seen to include global institutions in the North, implying that any positive, sustainable alteration of the divides' trajectory must address international imbalances in systems of trade as well as other structural inequalities.

Thai interviewees were more likely to consider nanotechnology within a domestic context of internal challenges to building endogenous capacity. They identified capacity issues such as developing adequate human resources and technical infrastructure, as well as ensuring political support and appropriate financing. From this perspective, the development 'problem' is seen more as a result of Southern inabilities, with Southern countries viewed as 'backward' or 'lagging' and needing to 'catch-up'. My research has reinforced the critical influence of how nanotechnology is perceived in terms of debates on how it will unfold. Building on the findings of previous research (Maclurcan 2009), interviewees commonly made arguments based on fluid definitions for nanotechnology and the possibilities it raises. This point explains a number of apparent contradictions, particularly in areas such as nanotechnology's infrastructural and cost requirements, where interviewees presented an unusually wide range of views. The conflicting views also raise the interesting question as to the role of hype in nanotechnology discourse.

Equity is a central ethical issue in the international debates around nanotechnology. However, current thinking, as supported by this research, suggests that greater equity can only occur on a platform of more widespread nano-innovation, where nations become more globally competitive. Ultimately, this thinking limits subsequent debates to a narrative underpinned by the contested philosophies of 'sustainable economic development' and denies the many, creative alternatives emerging worldwide.

Acknowledgments This paper was presented at the Third Nano Ethics Workshop held at the University of Aarhus, October, 2008. The author is grateful to the Nano Ethics Network for their support in making this presentation possible.

Notes

1. This paper shares its methodology with previous research I have published (Maclurcan 2009).
2. Key informants, or experts, are defined as "...those who can provide relevant input to the process, have the highest authority possible and are committed and interested"
3. Hereafter referred to as 'nanotechnology practitioners'.
4. This figure is estimated, based on a previous report's claim that the overall number of nanotechnology practitioners in Thailand was around 100 (Unisearch 2004).
5. An index combining normalized measures of life expectancy, literacy, educational attainment, and GDP per capita for countries worldwide.
6. Noted as an advantageous method in future-oriented research (Garrrett 1999).
7. In descending order of interviewee level of citation.
8. China was not considered a nanotechnology 'leader', although its grouping with the North, in terms of nanotechnology prowess, will be discussed later.

9. An alliance of nine knowledge-intensive technology organizations from around the world (www.research-alliance.net).
10. The role of licensing will be explored in a future paper as a strategy for developing endogenous innovation capabilities.
11. 2.7 per 10,000 population in Thailand (according to Tanthapanichakoon).
12. "A claim or claims to a process for the manufacture of a product, which may itself be the subject of a patent though it does not necessarily have to be" (UNCTAD 2004).
13. Who, as of 2007, is Thailand's Minister of Science and Technology.
14. Thajchayapong draws this statistic from a 2003 study titled: "Final Report: Survey for Current Situation of Nanotechnology Researchers and R&D in Thailand", that was conducted by the Unisearch group at Chulalongkorn University in Bangkok (Unisearch 2004).
15. approximately equivalent to $165,000.
16. Although Panitchpakdi claimed Thailand has already produced a nanoscale diagnostic kit (Panitchpakdi).

References

Beck, U. 1992. *Risk society: Towards a new modernity*. Sage, London.
Beck, U. 1996. Risk society and the provident state. In: Lash S, Szerszynski B, Wynne B (eds.) *Risk, environment and modernity: Towards a new ecology*. Sage, London: 27-43.
Bowman, D. and G. Hodge. 2007. Nanotechnology and public interest dialogue: Some international observations. *Bulletin of Science, Technology and Society* 27(2):118-132
Britten, N. 1995. Qualitative research: Qualitative interviews in medical research. *British Medical Journal* 311(6999): 251-253.
Changsorn, P. 2004. Firms see lower costs, more profit in nanotech. *The Nation*. November 22.
Changthavorn, T. 2003. Bioethics of IPRs: What does a Thai Buddhist think? Paper presented at Roundtable discussion on Bioethical Issues of IPRs, Selwyn College, University of Cambridge.
Cornell, B.A. et al. 1997. A biosensor that uses ion-channel switches. *Nature* 387(6633):580-583.
Court, E, A.S. Daar, E. Martin, T. Acharya, and P.A. Singer. 2004. Will Prince Charles et al. diminish the opportunities of developing countries in nanotechnology? February. from http://www.nanotechweb.org/articles/society/3/1/1/1
ETC Group. 2004. Scientists prepare to use nanotechnology to poison us all?—Jazzing up jasmine: Atomically modified rice in Asia? News Release, 3. Retrieved April 4, 2004, from www.etcgroup.org/en/materials/publications. html?pub_id=117
Garrett, M.J. 1999. *Health futures: A handbook for health professionals*. World Health Organisation, Geneva.
Gutierrez, O. 1989. Experimental techniques for information requirements analysis. *International Journal of Information Management* 16(1):31-43.
Hassan, M.H.A. 2005. Nanotechnology: Small things and big changes in the developing world. *Science* 309(5731):65-66.
Invernizzi, N. and G. Foladori. 2005. Nanotechnology and the developing world: Will nanotechnology overcome poverty or widen disparities? *Nanotechnology Law and Business* 2(3):101-110
Kachonpadungkitti, C. and D.R.J. Macer. 2004. Attitudes to bioethics and biotechnology in Thailand (1993-2000), and impacts on employment. *Eubios Journal of Asian and International Bioethics* 14(2004):118-134.
Kerr, W.A., J.E. Hobbs, R. Yampoin. 1991. Intellectual property protection, biotechnology, and developing countries: Will the TRIPs be effective? *AgBioForum* 2 (3&4):203-211.
Knowledge Ecology International. 2008. Thailand's compulsory licensing controversy. Retrieved September 19, 2008, from http://www.keionline.org/index.php?option=com_content&task=view&id=90

Lin-Liu, J. 2003. Thailand's leader plants the seeds for a future in nanobiotech. *Small Times*. Retrieved June 11, 2004, from www.smalltimes.com/document_display.cfm?document_id=5588

Liu, L. 2003. Current status of nanotech in Thailand. *Asia Pacific Nanotech Weekly* 1(19):1–4.

Loveridge, D. 2002. *Experts and foresight: Review and experience*. The University of Manchester, Manchester.

Maclurcan, D.C. 2005. Nanotechnology and developing countries: Part 2—what realities. *AzoNano online journal of nanotechnology*. Retrieved October 30, 2005, from http://www.azonano.com/Details.asp?ArticleID=1429

Maclurcan, D.C.2009. Nanotechnology and the global south: Exploratory views on characteristics, perceptions and paradigms. In: Arnaldi S, Lorenzet A, Russo F (eds) Technoscience in progress: managing the uncertainty of nanotechnology. IOS, Amsterdam, pp 97–112

Mee, W., R. Lovel, F. Solomon, A. Kearnes, and F. Cameron. 2004. Nanotechnology: The bendigo workshop report. CSIRO Minerals, Melbourne

Mehta, M.D. 2002. Nanoscience and nanotechnology assessing the nature of innovation in these fields. *Bulletin of Science, Technology and Society* 22(4):269–273.

Meléndez-Ortiz, R. and V. Sánchez (eds.). 2005. *Trading in genes: Development perspectives on biotechnology, trade, and sustainability*. Earthscan, London.

Moore, F.N. 2002. Implications of nanotechnology applications: Using genetics as a lesson. Health Law Rev 10(3): 9–15.

Panyakeow, S. and P. Aungkavattana. 2002. Nanotechnology status in Thailand. In: Tegart G (ed) *Nanotechnology the technology for the 21st century: vol. II the full report*. APEC Center for Technology Foresight, Bangkok, pp 163– 168 Nanoethics

Pieterse, J.N. 1998. My paradigm or yours? Alternative development, post-development, reflexive development. *Development and Change* 29(2):343–373.

Pieterse, J.N. 2000. After post-development. *Third World Quarterly* 21(2):175–191.

Pieterse, J.N. 2001. *Development theory: Deconstructions/ reconstructions*. Sage, London.

Priestly, B.G., A.J. Harford, M.R. Sim. 2007. Nanotechnology: A promising new technology—but how safe? *The Medical Journal of Australia* 186(4):187–188.

Reid, A. L.N. Wood, G.H. Smith, and P. Petocz. 2005. Intention, approach and outcome: University mathematics student's conceptions of learning mathematics. *International Journal of Science and Math Education* 3 (4):567–584.

Sahai, S. 1999. Biotechnology capacity of LDCs in the Asian Pacific Rim. *AgBioForum* 2(3&4):189–197.

Sandhu, A. 2008. Thailand resorts to nanotech. *Nature Nanotechnology* 3(8):450–451.

Sutharoj, P. 2005. Nanotechnology: Ten-year plan for Asean leadership. *The Nation*. Retrieved June 30, 2005, from http://www.nationmultimedia.com/2005/06/27/byteline/index.php?news=byteline_17840267.html

Tanthapanichakoon, W. 2005. An overview of nanotechnology in Thailand. *KONA: Powder and Particle* 23: 64-68.

Thajchayapong, P. and W. Tanthapanichakoon. 2003. Current status of nanotechnology research in Thailand. In nano tech 2003 + Future (International Congress and Exhibition on Nanotechnology), vol. S3–2 (ed) N. E. a. I. T. D. Organization, New Energy and Industrial Technology Development Organization, Tokyo, Japan, p 2.

UNCTAD. 2004. Process Patents: Burden of Proof. CY564-Unctad, 1.November 29: 496.Unisearch. 2004. Final report: Survey for current situation of nanotechnology researchers and R&D in Thailand. Chulalongkorn University, Bangkok.

United Nations Development Program. 2003. Human development report 2003: Millennium development goals: A compact among nations to end human poverty. Oxford University Press, New York.

United Nations Development Program. 2007. Thailand human development report 2007: Sufficiency economy and human development. United Nations Development Program, Bangkok.

van Amerom, M. and M. Ruivenkamp. 2006. Image dynamics in nanotechnology's risk debate. In second international Seville seminar on future-oriented technology analysis: impact of FTA approaches on policy and decision-making, Seville, Spain, p 6.

Warris, C. 2004. Nanotechnology benchmarking project. Australian Academy of Science, Canberra.

World Bank Independent Evaluation Group. 2007. Development results in middle-income countries: An evaluation of the world bank's support. The World Bank Group, Washington, D.C.

Chapter 22
How Can Nanotechnologies Fulfill the Needs of Developing Countries?

David J. Grimshaw, Lawrence D. Gudza, and Jack Stilgoe

If the nanotechnology research community wants to address the nano-orientation divide that Maclurcan identifies, it must find a way to re-orient itself to the needs of low-income communities. To illustrate the process of matching local needs in a developing country with nanotechnology capabilities, we reprint a chapter by David Grimshaw, Lawrence Guzda, and Jack Stilgoe, reporting on their efforts to find such a match in one community in Zimbabwe. Starting from the community's articulation of its own needs, they did not quite reach an example or list of nanotechnologies that could help. But they did discover the importance of local expertise in monitoring that possibility, and the distancing of Northern expertise through the pull of commercial opportunities there. As with the gap that Slade reported in Chapter Four between public values and practical actions, using nanotechnologies to address the Millennium Development Goals, as some have advocated, will not be a simple task.—eds.

Originally published in Savage et al. (eds.), *Nanotechnology Applications for Clean Water*, 2008: 535–549. Reprinted with permission from Elsevier.

22.1 Nanotechnologies and Developing Countries

We live in a rapidly changing world. Technological advances are increasing productivity and income, quality of life, and life expectancy in the developed world, that is. The truth is that technological development is focused on meeting the wants of rich consumers. Scant attention is paid to the vital needs of people in the developing world. Each new technology that comes along tends to result in a wider gap between the rich and the poor in the world. Yet some innovations fail to be applied in developing countries where there is the need. The founder of the Intermediate Technology Development Group (now known as Practical Action) observed that: "new technologies are developed only when people of power and wealth back the development" (Schumacher 1979). The challenge is to ensure that nanotechnologies are applied to areas of need in developing countries.

Porritt (2006) has argued that to enable sustainable development we need to work with the market system and not against it. This means understanding the market mechanisms, understanding the innovation processes, and then working with the key stakeholders to enable business models that will deliver on human need rather than on consumer wants. With existing technologies this becomes a challenge because the business models, including the supply chain logistics, are already well established. In the case of new technologies there is a window of opportunity before products are released into the market to negotiate new business models.

In a global economy, many topical issues—for example, sustainable development, climate change, and democracy—are all influenced by the role of science and technology in society. A major challenge is to release public value from science and technology and to channel that public value into developing countries to help reduce poverty (Wilsdon et al. 2005). The concept of public value used here refers to value generated by science and technology that is not solely reaped by the market. The central topic of releasing public value from science in a global context is one of the most significant and challenging issues facing societies throughout the world today.

Low-income countries are not only poor in terms of measures of human well-being but are also poor in terms of indicators of technology. They spend a small proportion of GDP on research and development: less than 1 percent compared to high-income countries that spend around 2.5 percent. The number of scientists in low-income countries is less than 50 per 100,000 people compared to over 3,000 in high-income countries. Technology has failed to meet the needs of the poor, with 1.2 billion people living on less than US$1 per day. At the centre of these deliberations is the essence pointed out by Sachs (2005) that "the single most important reason why prosperity spreads, and why it continues to spread is the transmission of technologies and the ideas underlying them."

The challenge faced might be reframed as being one of "how do we enable nanotechnologies to deliver products which fulfil human needs rather than consumer wants?"

22.2 How Can Nanotechnologies Deliver Public Value?

The role of technology in development is perhaps even more important in the new century than it was in the last. In the era of globalisation, new technologies are rapidly reshaping the livelihoods and lifestyles of people throughout the world. The pace of technological change is increasing, and is beyond the capacity of society to understand and regulate its impacts—even when the implications are profound and far reaching, as is the case with nanotechnologies.

Most scientific and technological research is now in the private sector, producing research for Northern wants rather than Southern needs. Small-scale farmers and the informal sector give little attention to small-scale technological innovation.

Knowledge and communication-based industries are rapidly reshaping the global economy. Many believe that these trends are contributing to a new "knowledge divide" between the information-rich and the information-poor. There is an increasing sense of urgency—in the North and in the South—over the need to regain control over the ways nanotechnologies are developed and used. It is not recognised widely enough that the poor are able to innovate themselves, and innovations arising from developing countries need to be increasingly recognised and supported.

Traditional views of technology that rely on a linear model of innovation and diffusion are not appropriate to programmes that aim to respond to new technologies (Dantas 2005). The predominant traditional view has been based on technological determinism. As Winner (1986) suggested, "the adoption of a particular technical system requires the creation of a particular set of social conditions as the operating environment of that system." Such thinking leads to a technological push philosophy as embodied in the motto of the 1933 Chicago World Fair, "Science explores, technology executes, man conforms" (Fox 2002). The worldview on which this philosophy is based is predominantly "Northern," where the power is vested in global enterprises with large research and development budgets and where markets have developed to approximate to monopoly conditions. An example of this is the domination of Microsoft in the market for software. Practical Action views technology as not only meaning the hardware or technical infrastructure, but also the information, knowledge, and skills that surround it, and the capacity to organise and use these.

Thinking from science and technology and the international development traditions were brought together in a recent book edited by Leach, Scoones, and Wynne (2005). The traditional deficit model of public engagement was criticised and a number of themes were articulated, including the issue of how risk is framed and communicated. Wilsdon et al. (2005) call for the direction of science to be built on notions of public value. The notion of public value raises issues about equity, efficiency, and the very purpose of science. Those concerned with the ethics of development (Gasper 2004) and the philosophy of science are also making valuable contributions to this debate.

An alternative view of technology is required. Grove-White et al. (2000) have suggested that technologies need to be seen as social processes. This alternative view must recognise the role of the user (Southern poor) and the context provided by the cultural and political environment in which the user is based. The distinction being made here has been labelled "technology in use" by Edgerton (2006) who argues that the historical emphasis on technology innovation is misleading. Much technology that is in use in the world is adapted or imitated rather than innovated.

The quest to ensure that all people have access to clean drinking water is now enshrined in the Millennium Development Goals. Often approaches to providing water for poor communities have been driven either by economics or by technology. The economics route might typically centre on the importance of regulations, institutions, and open markets whereas the technology approach might focus on designing a water pump, filter system, or novel application of nanotechnology. Yet we know that the technology for providing clean water has been known about and in use for thousands of years (e.g., the Romans around 300 BC). Failure to solve the issue might also be seen as a cultural or indeed political or managerial problem.

22.3 Nanodialogues in Zimbabwe

In 2006, researchers from Demos, Practical Action, and the University of Lancaster collaborated on a process designed to engage Zimbabwean community groups and scientists from both the North and South in debates about new (nano) technologies (Grimshaw et al. 2006).

The dialogue was one of four experiments, collectively referred to as the nanodialogues, in public engagement with nanotechnologies, funded by the Office of Science and Technology's "Sciencewise" programme. Sciencewise was created to foster interaction between scientists, government, and the public on impacts of science and technology.

Governments, companies, and NGOs are all talking about nanotechnology as "The Next Big Thing." Alongside the promise of new worthwhile opportunities comes uncertainty about risks, ethics, and the benefits to those people who are too often left out of conversations about the ends of technology—the poor. The potential benefits of the applications of nanotechnologies in developing countries are exciting. But the conversation linking the needs of people in developing countries to the resources and scientific knowledge of researchers around the world needs to be nurtured.

Epworth is a suburb of Harare, but it feels rural. It is just outside the Harare city limits, which means it is cut loose from the support of the city. In 2005 it was the scene of some of the harshest of the slum clearances that formed Robert Mugabe's "Operation Murambatsvina" ("Drive Out Trash"), which left thousands homeless. It is framed by outcrops of rock that have been worn away to resemble meticulously stacked balls. The balancing rocks are famous—they appear on the 10,000 dollar banknote. In the distance, you can see the electricity pylons of Harare's suburbs.

22 How Can Nanotechnologies Fulfill the Needs of Developing Countries?

Fig. 22.1 Spring Water, Epworth, Zimbabwe

But the telegraph poles around Epworth carry no cables. Plans for electricity and telephone lines were abandoned before completion.

Epworth gets its water from a combination of shallow wells and springs (Figure 22.1). The water brought up from the well looks clean enough, but with the pollution from the city, it's impossible to tell what it contains. "We're supposed to check," shrugs our guide, who acts as one of the community leaders. Nearby, a new well is being created. At the bottom of a six-metre pit, a man is filling a bucket with wet sand. His colleagues pull up the bucket and pile the sand around the pit's edge. It has taken two days so far, and will take another three. Then they need to seal it and put a lid on it. The well is next door to a pit latrine. It is far from ideal, which is why new sanitation methods are so important. Though Epworth is cut off, it is near enough to the city to be cramped. There is little space, and the well needs to be dug where there is water.

Any conversation about technology in Epworth has to start from here. In Zimbabwe, there is a headline context—a failing state and an economy that is both shrinking and sliding out of control—and there is an everyday context. In this everyday context, the idea of nanotechnology is not on its own likely to generate excitement. Ask what technologies people would like to see to help them get clean water and they mention rope-and-washer pumps, which replace disease-ridden open wells, and can be made and fixed using old tyres.

People in the developing world don't have much of a voice in science and technology. They are less likely to enjoy the benefits of new technologies and more likely to suffer from their downsides. The Royal Society and Royal Academy of Engineering (2004) took issue with the sweeteners often offered to the developing world by nano-marketeers:

> Much of the "visionary" literature... contains repeated claims about the major long-term impacts of nanotechnologies upon global society: for example, that it will provide cheap sustainable energy, environmental remediation, radical advances in medical diagnosis and treatment, more powerful IT capabilities, and improved consumer products... However, it is equally legitimate to ask who will benefit and, more crucially, who might lose out?... Concerns have been raised over the potential for nanotechnologies to intensify the gap between rich and poor countries because of their different capacities to develop and exploit nanotechnologies, leading to a so-called "nano-divide."

Other contributions, such as the Meridian Institute's "Global dialogue on nanotechnology and the poor," have stimulated wider discussion about possible benefits. One academic study, collecting the insights of people thinking about nano and development, concluded that the top three applications are energy, agriculture, and water. For our second experiment, we chose to explore the relevance of nanotechnology in the provision of clean water. Demos worked with Practical Action, the development NGO, which for the past 40 years (under its former name of the Intermediate Technology Development Group) has been making technology work for people in poor countries. Its vision is of appropriate, usable, sustainable technologies, driven by human needs rather than markets.

In Harare, we put together a three-day workshop with local mushroom farmers, brick makers, and water scientists. The nonscientists were representatives of communities that work with Practical Action. Three were from Epworth and three from Chakohwa, a rural community near Chimanimani, in the mountains of eastern Zimbabwe. The scientists were from government agencies, universities, and charities. The participants named our workshop Nanokutaurirana, a Shona neologism meaning "Nanodialogue." But for the first day and a half, the word nanotechnology was not mentioned. We wanted people first to define what the problem was.

Their description of the problem had multiple roots. Water is a market commodity, it is unaffordable, it is scarce, it is a long way away, and the responsibility for collecting it normally falls to women and girls. Where wells exist, they are crammed next to latrines and difficult to seal off from contamination. Near Harare, in addition to a recent cholera outbreak, there is chemical pollution from factories.

Away from the city, the rural community reported that water was contaminated by natural salt deposits. By the end of day one, we had a rough map of the issues and the connections between their social, technical, and political dimensions. The more the problem came into view, the further removed nanotechnology seemed as a solution. The community representatives had been let down in the past by well-intentioned technologies. Water pumps had arrived with instructions in English or German. When handles had broken or filters had clogged, they had been unable to find the parts or the expertise to fix them. As one of the community representatives asked, "When the NGO goes away, who has the knowledge to run and maintain their technology?"

In recognition of these characteristics of the problem domain we took a systemic approach. Many complex problems in science, engineering, or indeed other fields have some characteristics in common. Hard systems approaches have sometimes failed, for example, in the case of the Challenger disaster in 1986 when the space shuttle exploded moments after take-off killing all seven crew. Was this an engineering failure or one of managerial or political failure? McConnell (1988) says the emphasis at NASA had shifted from technological considerations to managerial, commercial, and political ones. This is a good illustration of how the way we frame problems affects the outcome in terms of the activities that take place to solve the problem situation. Two lessons are taken from this story: first that in complex problem situations a systemic approach has proved worthwhile; and second that "what in fact made the situations ill-defined was that objectives were unclear and that both what to do and how to do it were problematical" (Checkland and Scholes 1990). The dialogue took a soft systems approach.... The essence of the soft systems approach is that it allows a natural dialogue to take place with the facilitators using the methodology to capture and keep in a systematic way the outputs of each session.

The problem situation was captured during the workshop held in Zimbabwe. Before the workshops a root definition (The provision of clean drinking water for a poor rural community by Practical Action using a low cost product of nanotechnology that has been developed by Northern Scientists.) and CATWOE (Customers, Actors, Transformation, Worldview, Owner, Environmental Constraints) were conceptualised, ready to be tested with the real dialogue during the fist two days of the workshops....

Our approach was to build on Practical Action's experience of engaging people in developing countries in debates about new technologies.

Figure 22.2 depicts the problem situation in the form of a rich picture. During the first day of the workshop this rich picture was drawn by the organisers as a reflection of the problem presentation. The idea of the rich picture very simply is that it can convey relationships and connections much more clearly than prose.

In the problem situation identified there were several subsystems. The model in Figure 22.3 illustrates the three subsystems. Figure 22.4 shows some possible

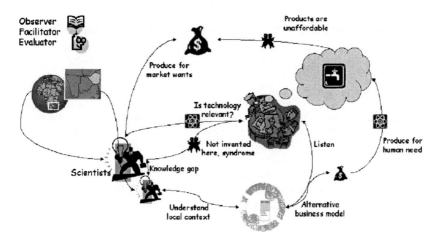

Fig. 22.2 Rich Picture of the Problem Situation

Fig. 22.3 Sub-systems

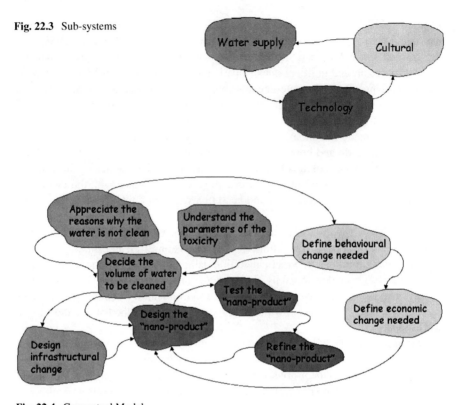

Fig. 22.4 Conceptual Model

interactions between these three subsystems, with each subsystem being shown in a different colour. The conceptual model shows a set of activities that would realise the root definition.

Academics might ponder such questions as: Is the methodology any good? Does it work? But in the "real world" most people recognise the need to find a methodology that works for them and get on with it.

> If a reader tells the author, "I have used your methodology and it works," the author would have to reply "How do you know that better results might have been obtained by an ad hoc approach?" If the assertion is: "The methodology does not work," the author can reply, ungraciously but with logic, "How do you know the poor results were not due simply to your incompetence in using the methodology?" (Checkland 1972)

For these communities, local technology was not a matter of pride; it was a matter of what worked. The system that shapes the problem needs to map onto the system that provides the solution. So the rope-and-washer pump makes sense. It is not so much a thing as a system. It is not owned or sold by any one company and it is flexible enough to fit different societies. The participants were well aware that, as one put it, "All these new technologies are old in other countries."

As the historian David Edgerton (2006) describes, whereas the West obsesses about the increasing "pace of innovation... most change is taking place by the transfer of techniques from place to place." Technological systems—the way things are used, abused, and controlled—are political. There are reasons why they end up the way they do, and there are ways in which we can talk about better or worse technologies. We can judge new technologies according to the extent to which they lock people into certain systems (as, e.g., GM crops and centralised nuclear power do) or provide an open platform for new sorts of use (e.g., Linux or micro-renewable energy).

In our first experiment, with the Environment Agency, our participants exposed the politics of technology by talking about issues of detectability (whether we will be able to find nanoparticles once we release them) and reversibility (whether we will be able to backtrack). They realised that, even after we understand the effects of nanoparticles in a lab, when we release them, we will know much less about their impacts. So innovation becomes experimentation as technological systems become bigger and more complicated.

Technologies carry with them some definition of social need and some promise of a technical fix. They define both a problem and a solution. And the systems of research, innovation, and regulation of which they are a part can harden this definition. So whereas in the United Kingdom we may take the system—transport, maintenance, markets, and a stable economy—for granted, in Epworth, this needs close scrutiny. Rather than starting from the technology, we need to start from the local context and think about alternatives.

Edgerton (2006) argues that the politics of new technologies have tended to narrow down consideration of alternatives. "Alternatives are everywhere, though they are often invisible." Public discussion reveals these alternatives. Technologies do not force people to do things, but as they open new doors, there is a danger that old ones can close. Whereas good intentions are focused on nano and development, they may lose sight of what else can more easily benefit poor people. In the course of our public engagement work, as we have reflected the public context of nanotechnology back to institutions, we are often asked whether public concerns are specific to nanotechnology or more general. Our response has usually been that nanotechnology is currently a good place to start the conversation, but that the sentiments speak way beyond this. In Zimbabwe, the issue of whether we should be talking about nano at all never seemed more pressing.

Halfway through our workshop's second day, we introduced nanotechnology. Luckily, a few weeks before, Hille et al. (2006) had provided some examples of nano-products working in developing countries. Their report was careful to point out that the diffusion of these technologies was some way off, but it provided some examples of nanoscale water filters working in South Africa.

Our participants understandably shared little of the West's excitement for nanotechnology. Even for the scientists there was little prospect of riding nanotechnology's funding wave. So the group asked about applicability, alternatives, environmental impact, cost, maintenance, and the capacity to manufacture and maintain the technology locally. They asked whether these technologies were fixed or adaptable for local needs, whether they would mean an increase in employment for Zimbabwean scientists, or greater reliance on the West. And they asked at what scale these could be used. Were they the sorts of filters that would be used centrally, at government treatment plants, or could they be put in schools and controlled by communities? The experiment revealed the huge gulf between research and diffusion. We began to see the steps that need to be taken to connect innovation to human need in a place like Epworth.

The temptation is to see the problem as intractable, to say that science has nothing to offer and that Zimbabwe needs to provide solutions for its problems. But this would deny the huge potential that exists for constructive collaboration. Around the world, there are efforts under way to direct emerging technologies towards pressing human needs. A more positive approach might ask how these efforts might yield greater benefits.

As a step towards this, we asked our Zimbabwean participants to produce a set of recommendations for U.K. scientists. They concluded that innovation does need to point in a different direction, but collaboration can be hugely positive "when there is a story to tell"—that is, when it starts with some concrete benefit in mind. The Zimbabwean scientists recognised that in many cases, given the asymmetry of resources, Western scientists would have to lead research, but this research should recognise the value of local knowledge and work to empower Southern scientists and build their capacity. They also recommended immediate steps that could be taken, such as opening up access to all scientific journals.

Back in the United Kingdom, we went to visit Mark Welland in Cambridge. We were keen to see what lines could be drawn between the needs of people and science, to stretch the connection back to the research base. Welland runs Cambridge University's Nanoscience Centre, but he is also co-director of the Yousef Jameel Science and Technology Research Center at the American University in Cairo. His research team is driven by scientific curiosity, but he encourages his colleagues to reflect on public value as part of their work. At one of his centre's nanoscience seminars, we told the scientists about our time in Zimbabwe and asked for their thoughts and questions.

In Zimbabwe, scientists saw community participation as a vital, if hugely complicated, part of what it means to do good science and engineering. In the United Kingdom, systems work against community or public engagement. Talking to the young researchers at the Interdisciplinary Research Collaboration (IRC) in Cambridge about our experiment, it became clear that many of them have an appetite to use their skills to contribute to human needs. But advancing in their scientific career often feels like a routine progression through certain stages in which they have "no control over their own research projects, social impacts or otherwise."

At the moment, the gravitational pull for these scientists is towards certain sorts of innovation—marketable technologies or a narrow definition of world-beating basic research. We need a broader understanding of innovation, which places greater value on the needs of people in the developing world. The young scientists in Cambridge recognised the scale of the challenge that this poses to established systems, but were unsure how to continue the conversation and change things from inside.

22.4 Balancing Risk and Opportunity

Bürgi and Pradeep (2006) observed that "the convergence of the newly emerging technologies of the twenty-first century has the potential to revolutionize social and economic development and may offer innovative and viable solutions for the most pressing problems of the world community and its habitat. However, a better understanding of the potential benefits and hazards of nano-scale science and technology is essential because it will provide policymakers with better tools to take responsible choices."

Research on the ethical, legal, and social implications is, according to Mnyusiwalla et al. (2003), lagging behind the science. As evidence for this view they quote the low number of citations in the literature and the fact that in the United States, there are research funds available that are not being used. For example, the National Nanotechnology Initiative allocated US$16-28 million to social implications but spent less than half that amount. One of the main reasons quoted for the lack of awards was the paucity of good quality proposals.

In our society, where there is a risk, the insurance industry will be key players in identifying and analysing that risk. The Benfield Group (Benfield is the world's leading independent reinsurance intermediary and risk advisory business) concluded their assessment by stating that "the industry's current focus is on risk management and containment, for example in the manufacture and transportation of nanotechnology" (French 2004). A study by Munich Re (the largest re-insurance company) echoing the emphasis on risk management states, "up to now, losses involving dangerous products were on a relatively manageable scale whereas, taken to extremes, nanotechnology products can even cause ecological damage which is permanent and difficult to contain" (Schmid 2002).

An increased risk potential of nanotechnology, according to Schmid (2002), can arise from:

- New types of loss arising from new material properties such as magnetic fluids.
- Increase in major claims.
- Liability cases arising out of changing legislation designed to protect the wider environment.
- Adverse sociopolitical effects and irreversible ecological damage.

Materials fabricated on the nanoscale have properties that are different from those that are manufactured at a normal scale. For example, the precise way in which the atoms are arranged often leads to unusual optical and electrical properties. Carbon at the nanoscale can conduct electricity better than copper. In other cases the small size may have the effect of being more toxic than normal. A distinction can be made, in terms of risk assessment, between active and passive nanoparticles. Passive particles, such as a coating, are likely to present no more or less a risk than other manufacturing processes according to French (2004). However, she goes on to assert that in the case of active nanoparticles, their ability to move around the environment leads to risks associated with control and containment.

In the United Kingdom, nanotechnology is being seen as an opportunity to have an earlier and more open debate about emerging technologies, to avoid the antagonism and distrust generated with genetically modified (GM) foods. The government is supporting the Royal Society and Royal Academy of Engineering's (2004) call for "a constructive and proactive debate about the future of nanotechnologies... at a stage when it can inform key decisions about their development and before deeply entrenched or polarized positions appear." The nanodialogues are a set of opportunities for early public debate. One of these aims to engage communities in Zimbabwe in discussions about emerging technology.

Views about the relevance of application areas for poor people converge on two sectors, namely, water and energy. These were the sectors, according to an international group of experts convened by the Meridian Institute to advise a Rockefeller project, thought to be where applications of nanotechnology are likely to bring potentially beneficial products that could offer solutions for poor people. According to one recent study, the top three applications that would help developing countries are energy storage, production, and conversion; agricultural productivity enhancement; and water treatment (Salamanca-Buentello et al. 2005).

We chose water treatment as a focus for our dialogue. First, in development terms it is a well-established priority. Second, technology is at a stage where it may be able to make a significant contribution to filtration and decontamination. The Millennium Development Goal is to halve the proportion of people without sustainable access to safe drinking water and basic sanitation by 2015. Our dialogue sought to introduce the views and values of people for whom clean water is an everyday problem into debates about possible technical solutions. By involving scientists who are engaging in leading research, we can move the debate upstream. We hope one of the outcomes will be a sustained dialogue between scientists and end users that enables new technology to deliver on human needs rather than be driven by market wants.

22.5 Future Directions

The emergence of nanotechnology has coincided with greater openness in science and innovation policy. For government, public engagement has become a way of avoiding a repeat of past mistakes. Depending on who you ask, nanotechnology might be the Next Big Thing, the Next Asbestos, or the Next GM. But before its impacts have been felt, nanotechnology has become a test case for a new sort of governance. It is an opportunity to reimagine the relationship between science and democracy. For public engagement to matter it must go beyond risk management. New conversations with the public do not provide easy answers. They ask difficult but important questions, and take us into a vital discussion of the politics of science.

The concept of new technology presents many challenges to those concerned with how it can be used to reduce poverty in the world. The promise of many new technologies has been high yet their ability to deliver sustainable change in the lives of poor people has been limited. At the same time the very models and assumptions underpinning much of international development have been economic growth. The essay presented a case for using a new paradigm based on enabling choices to be made that fulfil the needs of people. This requires a move away from the old paradigm, which is supply driven, delivering products to a market at a price that will maximise profits for the owners of the intellectual capital.

The dialogues held in Zimbabwe are one small step towards this new paradigm. They connect the needs of poor people with scientists who are in the process of developing new applications of nanotechnologies. The next step will be to move beyond dialogue towards the engagement of the scientist with relevant stakeholders in developing countries.

References

Bürgi, B.R. and T. Pradeep. 2006. Societal implications of nanoscience and nanotechnology in developing countries. *Current Science* 90 (5): 645-658.
Checkland, P.B. 1972. Towards a systems based methodology for real world problem solving. *Journal of Systems Engineering* 3 (2): 87-116.
Checkland, P.B. and J. Scholes. 1990. *Soft systems methodology in action.* Chichester: John Wiley.
Dantas, E. 2005. The system of innovation approach and its relevance to developing countries, SciDevNet Policy Briefs, April. available at http://www.scidev.net/dossiersjindex.cfm?fuseaction=printarticle&dossier=13&policy=61. (accessed April 26, 2005).
Edgerton, D. 2006. *The shock of the old: Technology and global history since 1900.* London: Profile Books.
Fox, N. 2002. *Against the machine: The hidden Luddite tradition in literature, art and individual lives.* Washington D.C.: Shearwater Books.
French, A. 2004. *Nanotechnology: New opportunities, new risks.* London: Benford Group. B6.
Gasper, D. 2004. *The ethics of development.* Edinburgh: Edinburgh University Press.
Grimshaw, D.J., J. Stilgoe, and L. Gudza. 2006. The role of new technologies in potable water provision: a stakeholder workshop approach, Rugby: Practical Action. http://practicalaction.org/docs/ia4/nano-dialogues2006report.pdf. (accessed August 15, 2007.
Grove-White, R., P. Macnaghten, and B. Wynne. 2000. Wising up: The public and new technologies, Research Report. Centre for the Study of Environmental Change, Lancaster University.
Hille, T., M. Munasinghe, M. Hlope, and Y. Deraniyagala, 2006. Nanotechnology, water and development. Meridian Institute. http://www.merid.org/nano/waterpaper (accessed July 5, 2006).
Leach, M., I. Scoones, and B. Wynne. 2005. *Science and citizens: Globalisation and the challenge of engagement.* London: Zed Books.
McConnell, M. 1988. *Challenger: A major malfunction.* London: Unwin Hyman.
Mnyusiwalla, A., A.S. Daar, and P.A. Singer. 2003. Mind the gap: science and ethics in Nanotechnology. *Nanotechnology* 14: 9-13.
Porritt, J. 2006. *Capitalism as if the world matters.* London: Earthscan.
Royal Society and Royal Academy of Engineering. 2004. *Nanoscience and nanotechnologies: Opportunities and uncertainties.* London: Royal Society.
Sachs, D. 2005. *The end of poverty.* Penguin Books: London.
Salamanca-Buentello, F., D.L. Persad, E.B. Court, D.K. Martin, A.S. Daar, and P.A. Singer. 2005. Nanotechnology and the developing world. *PLoS Medicine* 2 (4): 300-303.
Schmid, G. 2002. *Nanotechnology: What is in store for us?* Munich: Munich Re Group.
Schumacher, E.F. 1979. *Good work.* London: Jonathan Cape.
Wilsdon, J., B. Wynne, and J. Stilgoe. 2005. The public value of science: Or how to ensure that science really matters. London: Demos.
Winner, L. 1986. *The whale and the reactor.* Chicago: University of Chicago Press.

Chapter 23
Technical Education and Indian Society: The Role of Values

Raghubir Sharan, Yashowanta N. Mohapatra, and Jameson M. Wetmore

While much of this volume focuses on the role of equity in technologies, political structures, the scientific community, and economic systems, one of the most lasting ways to create a more equitable world may come through the redesign of education. This section closes with an essay by Raghubir Sharan, Yashowanta Mohapatra, and Jameson Wetmore in which they argue that engineers should receive an education in values. Values are, of course, always built into technologies and are omnipresent in systems of education. Thus the authors suggest that a more conscious attempt to consider them—by both faculty and students—is in order. Including such a component, however, begs the question: Whose values? The Indian educational system has been struggling with this issue. There are economic benefits to emulating the American system of engineering education, but such an education may not translate into technologies that directly benefit local people. Engineering communities in each country must decide who they want to prepare their students to help. The authors here contend that education should help budding engineers understand the plight of the poor and disenfranchised so that they are better able to use their talents to help remedy such situations.—eds.

J.M. Wetmore (✉)
Center for Nanotechnology in Society, Arizona State University, Tempe, AZ, USA
e-mail: Jameson.Wetmore@asu.edu

23.1 Introduction

The relationship of any society with education, technology, and values manifests in different ways. For example, the relationship between society and education is so natural that determining the time when the two could be differentiated would be difficult. Similarly, the realization that values are important for survival of a society must have come very early. It is only the technology (in the modern sense of the last 300 years or so) that is the newcomer on this scene. It has reshaped the relationship between education and value. This reshaping is still under formation, and the explosive growth of technology has not yet permitted this formation to settle down.

To illustrate the issues that are of importance in this formation this paper presents a case study of possible assimilation of nanotechnology in India. The discussions have led in a direction that even before we take up this case of assimilation of nanotechnology, several other issues would have to be sorted out. This has necessitated a detour to consider the assimilation of different technologies in a society which is not a generator of these technologies. This is done through some examples.

For instance, the steam engine helped to consolidate wealth and power in the West. When introduced in India through railways, it empowered a wide variety of people by piercing through class and caste barriers (almost as a side effect of its main purpose of making the colonial government more powerful). Even within India, as in any other societies with gross inequalities, the promise of technology speaks relatively different languages depending on the social ladder and one's economic condition. For the virtually dispossessed, as in the remote tribal villages of India, technology means the possibility of government largesse in the form of basic amenities. For the aspiring middle class technology may increase the possibility of joining the elite. And for the elite technology offers the promise of closing the gap in global standards. At first it seems that it can be no different for nanotechnology, and yet there are differences and different possibilities.

India has experienced the huge transformative power of technology (including its perils) in the last six decades since Independence. Take the example of the Green Revolution which filled her granaries and produced pockets of prosperous farmers, or the more recent communications revolution in which self-employed tradesmen (carpenters, tailors, and vendors) bypass the need for access to elite infrastructure using new "democratizing" technologies to carry on their businesses. Along similar lines there is an expectation that any significant technology, nanotechnology included, would unleash through its internal logic of expression in a societal context the empowerment of people. The two important questions are whether there is anything intrinsically special in nanotechnology, which can help reduce inequalities; or what can be done consciously to shape it that way. We will argue that one needs to pay more careful attention to both explicit and implicit values that we lend to our technology education in this regard.

23.2 The Case of Nanotechnology

For our purposes here, nanotechnology is different from other technologies in three important senses as evident in its practice. These differences are significant not so much in their distinctiveness but in their extent and degree. First, the adjective "nano" doesn't really limit the scope or description. It somehow promises to engulf, in future, all technologies by the virtue of the claim that it has struck the "right" scale for intervention to be able to change or influence them all. This belief, or "mythic," has taken the proportions of a paradigm, if technology has ever had a paradigm, and acts as the source of ideas about the future that are both imaginable and unimaginable (for example, the idea of "Singularity" (Kurzweil 2005) in growth curve). Second, the intrinsic inter-disciplinarity of nanotechnology opens up a seemingly limitless innovation space necessitating explicit and frequent decisions about the direction and growth of nanotechnology, not just in terms of how and where it will be used, but what the technology itself will be. Third, even in its nascent state of emergence, one can predict that its products are likely to revolutionize the three most important and inter-related material conditions of any civilization—energy, environment, and health-care—with the possibility of overcoming traditional limitations of technologies in these areas. In other words, the three significant characteristics of nanotechnology are its rapidly expanding ambition and scope, boundary-less space for innovation, and a certain focus on areas that are essential for civilization. All these characteristics point to the crucial role that values would/should play in the development of the field.

There is, therefore, a great need to understand the mechanisms by which nanotechnology develops to be able to incorporate desired values during its creation. We are all too familiar with the limitations of large scale technological intervention in the past laying bare the skewed set of implicit values or cultural specificity of their origin. Most powerful technologies lay their claim to "success" on the basis of some combination of technical virtuosity and the logic of the market or commerce. In contrast, there is a heightened expectation from nanotechnology that it will be able to transcend these limitations through the "right" kind of intervention. A number of groups and governments have argued that the expanding scope of nanotechnology should in some essential way factor in sensitivity to societal values and moral issues—inequalities included (See Chapters 4, 5, 7, 11, 12, and 18 in this volume). One of the primary ways it is argued this should happen is by educating a young army of practitioners in not only the technical details of nanotechnology, but in values and the importance of technology to social contexts as well. This goal has led, for instance, to not only the creation of stand alone centers dedicated to studying the social aspects of nanotechnology—like the Centers for Nanotechnology in Society at Arizona State University and the University of California, Santa Barbara—but also to the inclusion of such topics in the training of students at all levels. Fortunately, much of the cutting edge work being done in nanotechnology and the nanosciences today is either located at, or organically linked to, groups in educational institutes of technology.

23.3 Values and Direction in Indian Technical Education

To make this approach more clear, it may be helpful to describe in brief the origin of this article. In 2008 the National Nanotechnology Infrastructure Network (NNIN)[1] and the Indian Institute of Technology Kanpur[2] (IITK) set out to jointly organize the first in a series of international winter schools in nanoscale science and engineering for graduate students. Twenty students from top universities from the U.S. and India attended the course. Besides being on nanotechnology, the winter school had the goals of fostering global awareness, sharing knowledge, and promoting cooperation among academics and future leaders. The instructors were drawn from engineering, philosophy, and social sciences. Interestingly, the course consisted of two teaching components. Half of it was conducted in the classroom and technical labs. The other half was a field trip to a remote village called Paralakhemundi in the Indian state of Orissa where students were challenged to think about the context, application, and social impact of technology. This combination of cutting edge technological and societal concerns provided the perfect setting for value-related questions to emerge naturally in the minds of the participants and a chance for instructors, students, and local villagers to discuss them.

The winter school turned out to be a fertile ground for dormant seeds of possible tension between the values implicit in content and context. The tension was captured with a simple question: Is there anything in the technology education at a premier Institute (such as IIT Kanpur, an elite institute modeled along the lines of MIT (Leslie and Kargon 2006)) which would make a young participant decide to make a "journey" from *IITK to Paralakhemundi* forgoing the much trodden path of *IITK to the U.S.A.* Should graduates from India's best technical institutes use their knowledge and training to benefit those living in Indian villages or the western world? Clearly, the training and educational experience at IIT Kanpur prepares one for the latter. There is, however, a need for more people willing to take the former.

The contrast between the two symbolic journeys sum up the value-laden tensions in the development of new technologies and the role that education plays in them. We ask, therefore, what can be done to technology education programs so that they present the multiple journeys available to their participants in clearer terms? Or, as E.F. Schumacher put the title of his famous book, can we better provide *A Guide for the Perplexed* (1977)? We are not looking for a recipe for building educational institutions to do this. But certain directions or options can emerge if we consider the means by which our values implicitly or explicitly determine the way we construct our educational programs and embed them in our society.

Values and value-education acquire a central place in any discussion on reforms in education. This happens because people are generally unhappy with the values prevalent in their society. They want the situation corrected through proper value education, particularly, to the younger generation. This desire to educate the young in values existed in the past, is present today, and will remain strong even tomorrow.

Two questions arise at this stage. The first question is: Is the value education of the younger generation, by itself, capable of correcting lack of values in a society? The second question is: What will be a good way to impart value-education within the precincts of the formal educational system? Both these questions have been raised and answered by great minds over centuries. Hence it will be difficult to say something new. One can only systemize the existing thoughts in the context of present day concerns. Hence, let us summarize some of the Indian context that underlies this essay.

The last 100 years, overall, has been a great century for Indian Society. There is no doubt that during this period several depressing events have taken place. But two beneficial events overshadow all the depressing events. The first such event was based on a unique successful nonviolent political struggle by an oppressed society against its oppressors. This was pursued under the leadership of Mahatma Gandhi in the first half of the twentieth century and culminated in political independence for India in 1947. The second remarkable event is the success of the democratic method of governance, in spite of handicaps like mass illiteracy, in the second half of the twentieth century.

These events have deeply influenced the development of the educational system, including technical education, in India. An understanding of these influences will bring a remarkable fact to focus: In spite of several similarities with the Americo-European systems of technical education on which Indian technical educational system has been modeled, there are some substantial basic differences. These differences affect the effectiveness of including value education in technical education in India.

23.4 Embedding Values in Technologies and Education

To see how this works one must first understand how values are embedded in both technologies and education. Technologies and systems of education don't just "arrive." They are created by people, institutions, and relationships. It is through this social process of creation that the values of the creators get intertwined into the design of new artifacts, systems, and institutions.

For example, in a given instance a group of people build a new technology based on: 1. their experience of building other technologies; and 2. the material and intellectual infrastructure available. Both of these facets of the production of technology help to ensure that the values of the people (as well as the society that shapes them) are built into it. Their past experiences shape what they see as desirable and not desirable. And the material and intellectual resources that they are exposed to shape what they believe is possible as well as what they think the potential users might be like. These two facets have a large impact on how they envision the technology fitting into the social fabric of society and it is this fit that will determine which values are privileged and which ones are not.

For those who share the same values as the creators of a technology, this process can work fine. But when people have different goals, tensions can be raised. This often happens when a technology is not created in a given society, but rather reaches it through transfer, market forces, and/or diffusion. The new society need not accept the system of values associated with the technology, but the differences will have some effect. The society that receives the technology may decide to shift its values to incorporate the technology, rework it to accommodate their values, live in tension with the conflicts that the technology raises, or simply reject the technology altogether.

A similar process happens in the development of education. Education systems are imbued with the values of their creators. These values are embedded in the variety of majors, courses, and specific instruction offered. Ultimately the education system will shape the goals, beliefs, and aspirations of its students. One does not often think of systems of education being transferred like technologies, but they frequently are. This happens on a small scale when instructors share course materials and syllabi. But it can also happen on a large scale. In the 1950s, when the Indian government sought to increase the technical abilities of its country, it worked to transplant the technical educational system which was prevalent in America (epitomized by MIT, Cornell, Caltech, Purdue, etc.) to India. This resulted in educational institutes like IIT Kanpur, BITS Pilani, and a large number of others. As with the transfer of technologies, the transfer of systems of education can create tensions. The U.S. system of science and engineering education that was transferred to India promoted a set of values that were somewhat foreign to the average Indian. Of course this is part of the reason why the education techniques were imported, but that doesn't mean that it didn't raise issues.

In the end, let us consider the process by which values and society construct each other. For this the first thing that should be noted is that unlike modern technologies, whose sources are sparsely localized, the sources of quality thinking about values are widely spread over almost all the cultures. In other words, any culture worth the name must possess a set of values that it has not just developed, but which are intimately interwoven into it. Thus it is quite possible that a society might be severely handicapped in constructing cutting edge technologies but very proficient in constructing values. Hence, it is reasonable to think that the residents of Paralakhemundi might be weak in nano-technology but very strong in knowledge of ways of living a flourishing life free of depression, alienation, and frustration.

Thus while technology and systems of education both carry values with them, because they reflect the values of their creators, these may not always be the values that are appropriate for a given context (or some might argue any context in some cases). If we want to increase the likelihood that appropriate values are promoted throughout a given society, we need to make sure that we construct systems of education that train the people making technologies to understand, respect, and promote these values. With these thoughts in mind, let us construct various alternate possibilities that exist for guiding us in the design of a value component for technical education.

23.5 Difficulties in Imparting Value Education

23.5.1 What Should an Ideal Life Be?

If we could agree on what an ideal life should be, then the task of designing a methodology for imparting value-education would be easy. We would have to locate gaps between the ideal life and existing pattern of life and devise ways to fill these gaps. This approach runs into difficulties however, because usually we have different, very often conflicting, views of the ideal life. For example, let us consider the lives of Alexander and the Buddha as ideal lives. Can one simultaneously live both these lives?

According to the people of his community, Alexander was a brave soldier, a great strategist, and a brilliant leader.[3] He was a man of strong desires and possessed the will to get the desires fulfilled. Acquirement of absolute power was his aim. But, according to the people of the communities that he invaded and overran, Alexander was a plunderer who was ruthless and needlessly cruel. Nevertheless, Alexander represents a way of life. He exists in the thoughts and actions of those individuals, communities, corporations, and nations that want to go on generating wealth and acquiring power till they become the wealthiest and most powerful.

The idea of the Buddha stands for values and compassion for all beings. It demands shedding of desires. It emphasizes meditation and importance of right thoughts, right action, right speech, right livelihood, and so on. It leads towards a noble life which is based on concern for others (*par hit chinta*).[4] It points to all the advantages of a peaceful, serene life. But one difficulty with this view is that it believes that all the rest in the world will also be noble. When confronted with the likes of Alexander it easily gets enslaved. Based on our experience of centuries, we already know that slavery is a highly undesirable state and must be avoided at all cost.

This tension gives rise to our dilemma. If we become very powerful then there is danger that we may enslave others; if we become very passive then there is a danger that we may get enslaved. A very difficult act of balancing on the strength of well engrained values is eternally required to handle this dilemma.

23.5.2 Balancing the External and Internal Demands of an Individual

To do the balancing just mentioned one has to note that each individual has to interface with an external world and has to live with internal self. At the external world level he or she has to balance two extreme requirements. First, wealth must be generated (*artha*). And second, it must be done ethically by establishing harmonious relationship in society (*dharma*). At the internal level, one is torn between two extremes: the natural desire to enjoy sense pleasure (*kama*) and the need to control and win over all the desires. A philosophy to suitably harmonize and balance the external (*artha* and *dharma*) and internal (*kama*) behaviors of individuals has existed in India for millennia. This way of looking at life permits us to analyze the dominant culture of the present day.

Imagine a policy that considers only generation of wealth (*artha*) and satisfaction of sense pleasure (*kama*) as worthy of emphasis and underplays the importance of policies for concern for others (*dharma*) and deliverance from desires. In practice this would translate to the life of consumerism and indulgence in sense pleasure, which we are lamenting today. Let us enquire how this policy of overemphasis on generation of wealth and enjoyment of sense pleasure has come about. What is the basis of the thought process which has naturally led to this policy? For this let us change the focus from "external and internal concerns of the individual" to a "spirit of exploration."

23.5.3 Spirit of Exploration

Consider a society where exploration is considered of immense value. Like Alexander, one wants to visit other lands. It was by this process, for instance, that Europeans discovered a sea route to India, and then America, Australia, New Zealand, and other parts of the world. As everybody will agree, exploration is a very exhilarating activity. Survival in exploration requires knowledge of geography, astronomy (for navigation in sea), anthropology (to understand the culture of new people one meets on exploration), technology (to build boats and guns), rules for team behavior (for organization and management), and commerce (to create extra benefits from exploration).

One also finds that exploration, by its nature, brings us in contact with other persons and sharpens the concept of boundaries between "we" and "they." Particularly in the process of generation of wealth it becomes convenient to have two dharma (harmonious social relationships): one for "between us" and the other for "between us and them."

This division between the rule of conduct for "us" and "them" becomes visible even in the writings of great thinkers like John Stuart Mill. He felt that other societies have to cross a threshold of development before these can be included in the "consideration of the greatest good to the greatest number of people." Thus utilitarianism (a noble ethical principle), instead of weakening the divide between "us" and "them" of which it was capable, reinforced this divide by providing "ethical" support to it.

It is important to note that "them" here consists of two components. The first component consists of that large mass of humanity which was considered to be below "the threshold of civilization." For convenience, let us call this component "others." The other component is "nature."

The educational system that developed under this culture of divides cared for the greatest good for the greatest number (amongst a very small selected population) sometimes at the cost of "others" and "nature." With time, the condition of "others" has changed. Now they want to be counted in the small selected population for which the greatest good is given by adopting the educational system developed by "us." This leaves poor "nature" alone to bear the brunt because the balance between "satiation of desires" and "control of desires" is tilted towards more and more consumption and greater exploitation of nature. The way out would be to design a new educational system which will endeavor to restore this balance, thereby providing relief to nature. Additionally, this new educational system should attempt to weaken the divides so that all mankind is included in any consideration of progress.

23.5.4 The Tasks

With the above observations in view let us consider the tasks for value education.

(i) The first task is to blur the boundary between "us" and "them" while talking about equality, justice, and fraternity. It is extremely important to note that boundaries between "us" and "them" appear at various scales. These exist between cultures, between nations, between communities, between families, and even inside the same individual. Go to any society or even an intellectual group and there will be "we" and "they" partitions. Effective value education has the difficult task of removing all these boundaries.

(ii) In addition to balancing between Alexander-like and Buddha-like attributes and blurring of boundaries of "us" and "them," there is another task for value-education. This is to note that in any society there will always be some privileged and some deprived. Ideally, the value-education for the two has to be different. The privileged have to be taught to develop "altruism" and "concern for others" and the deprived have to learn "self-respect" and "enterprise." This complicates the task of value-education. In a way value-education has to help the privileged to develop an ability to say: That's enough, I don't want any more. Suppose a nation decides that it has too much power, it should not accumulate any more. Suppose a corporation says that it has too much wealth, it should not acquire any more. Suppose an individual says that he or she has too much material goods and should not acquire any more. Such thoughts have power to transform the world into a better place. Unlike for the privileged, a value education for the deprived has to inculcate enterprise so that they develop self-assurance and self respect. These simple looking things are difficult to do. These give rise to difficulties in imparting effective value education through curricular means in an existing distorted educational system. But one should not give up Instead, one should become an optimist.

The optimism would provide strength in the same way that it has provided strength to some proponents of values in the past. With this in mind we consider, in the next section, some viewpoints from two very different thinkers: Norbert Wiener (1894–1964) and Mahatma Gandhi (1869–1948). The former was a renowned and influential American scientist and the latter was a very influential human being from India. They came from very different backgrounds but both cherished universal values. Wiener was a major contributor to science and technology in America in the first half of the twentieth century. Mahatma Gandhi, on the other hand, promoted the simultaneous development of new value and a new society in India in the first half of the twentieth century. The voices of both these thinkers have been somewhat forgotten and have lost some of their impact in the second half of the twentieth century. Our task is to think of ways of "social amplification" to extract these feeble (weak) messages of value from the pervading "noise" of consumerism and excessive reliance on technological solutions.

23.6 The Views of Mahatma Gandhi

The collected works of Gandhi run ninety volumes of over 250 pages each (available at www.gandhiserve.org/cwmg/cwmg.html). Our interest, from this vast collection, is primarily in one book: *Hind Swaraj* or "Indian Home Rule." This book was written by Gandhi in 10 days in 1908 on a ship which was sailing from England to South Africa. The contemporary relevance of this book has been emphasized by Parel (1997) who has republished *Hind Swaraj* along with some other relevant papers. Parel writes: "The key to an understanding of *Hind Swaraj* lies in the idea that worldly pursuits should give way to ethical living. This way of life has no room for violence in any form against any human being, black or white (xvi)." This requires that divides be blurred. But there are difficulties in blurring boundaries. Let us continue with Parel who uses writings from *Hind Swaraj* to make this point:

> By modern or Western civilization ([Gandhi] often used these terms interchangeably) he meant that "mode of conduct" which emerged from the Enlightenment, and more exactly, from the Industrial Revolution. "Let it be remembered," he wrote in 1908, "that western civilization is only a 100 years old, or to be more precise fifty." The Industrial revolution for him was much more than a mere change in the mode of production. As he interprets it, it brought into being a new mode of life, embracing a people's outlook on nature and human nature, religion, ethics, science, knowledge, technology, politics and economics. According to this outlook, nature was taken to be an autonomous entity operating according to its own laws, something to be mastered and possessed at will for the satisfaction of human needs, desires and political ambitions. This outlook brought about an epistemological revolution which in turn paved the way for secularization of political theory. The satisfaction of the desire for economic prosperity came to be identified as the main object of politics...
>
> Modern political theory provided the general ethical framework within which the changes occurring in the scientific, technological and economic fields are to be integrated. Two types of political theory emerged, one for industrial societies and the other for the rest of the world. Liberalism and liberal institutions were thought appropriate for industrialized societies; imperialism and colonialism for the non-industrialized societies such as India... Even the saint of liberalism, J.S. Mill, accepted this civilisational partition of the world. He would in all sincerity use the very doctrine of liberty to justify imperial rule over India. (Parel 1997, xviii–xix)

Thus a divide, aided by arguments of modern civilization, was created between human beings at that time. Gandhi was non-violently opposing this divide. A century has passed since *Hind Swaraj* was written and since then much progress has taken place in the world. It is our view that, in spite of this "progress," violence has not vanished, new divides have appeared, and Gandhi has become more relevant. But following Gandhi will require, in an essential way, drastic reduction in consumption of material goods, control of sensual desires, and work towards ensuring justice to even those who differ with us. According to Gandhi, these are prerequisites for practicing non-violence.

We are not ready to make these sacrifices. We want to continue with consumption, satiation of desires, and "justice (only) for people on our side of divide." But we do want a non-violent world. We are thus confronted with a difficult choice. The way out will be to learn: How to balance between "Alexander-like" and "Buddha-like" tendencies? The learning from this would lead to meaningful contents for education in values for all persons including scientists and technologists.

23.7 The Views of Norbert Wiener

The excerpt from Norbert Wiener below is from his book: *The Human use of Human Beings: Cybernetics and Society* (1954). Norbert Wiener had obtained a PhD in philosophy from Harvard University in 1912, which was rather remarkable since he was only eighteen at the time. Most of his working life was spent with the electrical engineering department of MIT in the United States. He became a celebrity in the scientific community because his contributions were at the core of several advancements in control engineering, stochastic processes, cybernetics, artificial intelligence, computers, communication technology, and many others. He is also considered to one of the founding fathers of the Information Age. After the Second World War, in which he made critical technological contributions, he was tormented in the witch-hunt of the scientific community in America. This book was written twice during this period; first in 1950 and then again in 1954. The agony and experience of this period comes out clearly in the excerpt being given below (our emphasis):

> ...However, there is nothing in the industrial tradition which forbids an industrialist to make a sure and quick profit, and to get out before the crash touches him personally.
> Thus the new industrial revolution is a two–edged sword. *It may be used for the benefit of humanity*, but only if humanity survives long enough to enter a period in which such a benefit is possible. *It may also be used to destroy humanity, and if it is not used intelligently it can go very far in that direction.* There are, however, hopeful signs on the horizon. Since the publication of the first edition of this book, I have participated in two big meetings with representatives of business management, and I have been delighted to see that awareness on the part of a great many of those present of the social dangers of our new technology and the social obligations of those responsible for management to see that the new modalities are used for the benefit of man, for increasing his leisure and enriching his spiritual life, rather than merely for profits and the worship of the machines as a new brazen calf. There are many dangers still ahead, but the routes of good will are there, and I don't feel as thoroughly pessimistic as I did at the time of publication of the first edition of this book. (Wiener 1954, 162).

This was written in 1954. Can we say with confidence that the hopes of Wiener have come true in corporate America of 2010? In case the answer is no, then perhaps time has come to listen to Mahatma Gandhi, Norbert Wiener, Martin Luther King, Henry David Thoreau, Ralph Waldo Emerson, Leo Tolstoy, Fyodor Dostoevsky, Kabir, and the like. Such thoughts would help us in the design of courses on value education and guide us to develop meaningful technology.

23.8 From IITK to Paralakhemundi

In the classroom sessions of the 2008 NNIN/IITK winter school students got a brief introduction to Gandhi's teachings about technology, society, and values. They then took a day and a half journey via a series of buses, trains, and SUVs to get from IITK to Paralakhemundi, a small town in Eastern India surrounded by even smaller villages. The students that participated in the program came from some of the most elite technical institutes in the U.S. and India and had extensive experience working on cutting edge issues in nanoscale science and technology. When they arrived at the Association for India's Development (AID) Rural Technology Resource Center and began to explore the villages around Paralakhemundi they found people who had very different concerns. They were interested in finding ways to obtain clean water, ensure safe and nutritious food, and make a decent wage.

The students, considered some of the best up and coming problem solvers in the U.S. and India, didn't have immediate answers to these concerns. Instead they marveled at the techniques the local people had developed. The local people had two resources that are uncommon in the Western world: lots of land and lots of people. The forms of work and the technologies used in Paralakhemundi capitalized on these resources. Brick making was done by hand with clay found in people's backyards, all aspects of farming were done with animal labor, and clothes were made with thread spun on spinning wheels. The values instilled in Western technical education, including the replacement of humans with machines, the use of scarce resources, and (perhaps the most important of them all) efficiency, did not have a place in the rural villages. The solutions the local villagers found worked better in their context than solutions that might have been developed at a technical university in the United States

This does not mean the students didn't see room for improvement. But they didn't see those improvements coming directly from the training they had received. On the last night with the villagers, all of the students and a number of locals sat around a campfire. When asked what could be done to help the area, the students responded that the villagers and local aid organizations should tap into the expertise, local knowledge, and enthusiasm of the students enrolled at JITM, the local technical school.

A number of locals thought this improbable. Just as many at IITK hope to gain skills so they can move to the U.S.A., the goal of many students at JITM is to learn enough to make the journey to the big city of Bangalore where they can make decent wages in the IT industry. But just as one local announced that no student at JITM would ever want to stay to help the local villages there came a voice of dissent: "That's not true. Some of us do want to help." A handful of students had made the mile walk in darkness from JITM to meet with the students of the winter school. They announced that they had come because they were interested in applying their skills to the problems of those that lived around them. They stayed for a couple more hours discussing the local issues and what can be done with their colleagues from the U.S. and the visiting Indian technical students.

Whether this small exchange will lead to lasting change for the area is far from clear. But it is a sign that value education is possible and that there are students interested in learning. All the students involved had come to understand the profound impact that technology can have on people and were humbled to find out that the values they had been taught might not be appropriate for some situations. They became sensitive to the idea that local people, cultures, practices, and values need to be considered with the development of any new technology. Creating education programs and institutions that facilitate this kind of learning are not only needed, there are students hungry for them.

23.9 Conclusion

We desire that our wealth, technology, education, and value systems should help us in living a meaningful life. This requires a balance between the relentless chase for wealth and power through technology, on one hand, and the blurring of divides between "us" and "them" through compassion on the other. This balancing has to be done with a feeling of justice for even those whom we don't like. Attempting this balance is a worthy aim for value education. However, it is also possible to limit value education by considering only codes of conduct and ethical laboratory practices as has been argued in a recent paper (Harris 2008). Let us make a wise choice under the full knowledge that true value education may slow down material progress. This choice will also help us to decide whether we should design engineering curriculum to facilitate intellectual journeys from "IITK to Paralakhemundi" or from "IITK to the United States."

Notes

1. The NNIN is a U.S. National Science Foundation funded network that promotes cutting edge research and education in nanoscale science and engineering through its fourteen laboratory facilities across the United States.
2. IIT Kanpur is one of the first seven premier Institutes of Technology in India. Each one of them was set up with some foreign collaboration. In the case of IIT Kanpur, it was collaboration with the United States through the Kanpur Indo-American Program (KIAP). The program was funded by the U.S. Agency for International Development (USAID), which subsequently invited California Institute of Technology (Caltech); Carnegie-Mellon; Case Institute (now part of Case-Western Reserve); the University of California, Berkeley; Purdue University; Ohio State University; the University of Michigan; and Princeton University to join in advising and assisting IIT Kanpur. The collaboration operated for about a decade from 1960 to 1970.
3. The person we are referring to is of course commonly known in the West as "Alexander the Great." Since we are opening the question as to whether his approach to life was indeed "great," we've shortened his name accordingly.
4. "Par Hit Chinta" is a Sankrit phrase that is also used in Hindi. It roughly translates to: 'concern for the welfare of others.'

References

Harris, Charles E, Jr. 2008. The good engineer: Giving virtue its due in engineering ethics. *Science and Engineering Ethics* 14: 153–164.

Kurzweil, Raymond. 2005. *The singularity is near*. New York: Viking.

Leslie, Stuart W., and Robert Kargon. 2006. Exporting MIT: Science, technology, and nation-building in India and Iran. *Osiris* 21: 110–130.

Parel, Anthony J. 1997. *Gandhi: Hind Swaraj and other writings*. New Delhi: Cambridge University Press.

Schumacher, E.F. 1977. *A guide for the perplexed*. New York: Harper Collins.

Wiener, Norbert. 1954. *The human uses of human being: Cybernetics and society*. Boston, MA: DaCapo Press.

Part V
Lessons for Action

Chapter 24
Keeping the Dream Alive: What ELSI-Research Might Learn from Parliamentary Technology Assessment

Rinie van Est

The last section of the book offers lessons for action. Rinie van Est, a staff member at the Dutch parliamentary technology assessment organization, the Rathenau Institute, proposes that the notion of moving towards equity and equality through nanotechnology is as big a vision as the visions of technological transformation put forth by many nano-advocates. In order to move in the direction of the vision, van Est argues for three important steps. First, we must reflect on emerging equity and quality issues early in the development of new technologies rather than after they are produced. Van Est challenges researchers to develop constructive suggestions to strengthen equity and equality. Second, we need to mobilize the public rather than wait for people to get interested in nanotechnology. Van Est envisions people of all walks of life weighing in on what they want our nano-enabled future to be. To make this possible we need to stimulate public participation on many levels, including the active engagement of civil society organizations and experts like social scientists, policy makers, and politicians. Third, and perhaps most important for this volume, van Est encourages ELSI-researchers to play a more active role in the public and political debate so their insights may have a larger impact in society. Publishing in academic journals. . . or even Yearbooks. . . may not reach a sufficient audience to make change possible. To further the cause of equity in a meaningful way, academe needs to connect with the larger world.—eds.

R. van Est (✉)
Rathenau Institute, The Hague, The Netherlands; Eindhoven University of Technology, Eindhoven, The Netherlands
e-mail: q.vanest@rathenau.nl

This chapter was peer reviewed. It was originally presented at the Workshop on Nanotechnology, Equity, and Equality at Arizona State University on November 21, 2008.

24.1 Introduction

Nanotechnology—the control and modification of matter at the atomic and molecular scale—is now an *important science and a growing business*. Nanotechnology is widely expected to enable new advances in diverse areas of human endeavour: from new materials and cosmetics, to solar panels and drug delivery systems. Besides marketable products, the enabling character of nanotechnology inspires new ways of looking at the future in various key areas. Or to paraphrase the Belgian scientist De Man (2005): promises at the nanoscale lead to gigascale dreams, (implicit or explicit) social visions. These visions shape current research agendas and, thus, are instrumental in shaping future society. The changes and promise that these innovative technologies bring need to be assessed from a public interest point of view. The wish to develop nanotechnology for the common good is the biggest dream of all.

The rise of nanotechnology has been accompanied by increasingly active debate about the governance of science and technology. Both in the United States and Europe this has led to initiating public dialogue and research on the ethical, legal, and social implications of nanotechnology (so-called ELSI-research). ELSI-research is designed to identify, at an early stage, the impact of science and technology and thus avoid future social, environmental, and health problems and stimulate the development of socially desirable solutions, for example, embedded in the design of the technology itself or in regulatory practices. This promise has been embraced by politicians, policy makers, and scientists in both the United States and Europe. The first ELSI-research program was established in the United States in 1990 as part of the Human Genome project. Over the last decades many European countries have set up similar programs. For example, in 2004 the Centre for Society and Genomics was established as part of the Netherlands Genomics Initiative (cf. Zwart and Nelis 2009).[1] In the field of nanotechnology ELSI-research programs have been flourishing. Work on ELSI-research has been funded by the National Nanotechnology Initiative, but also as part of the European Framework Programs.[2] Given those expectations, and the enormous potential impact of this technology, we need to reflect on whether ELSI-research, in particular on equity and equality issues related to nanotechnologies, in its current form is able to live up to those expectations. The comments offered in this paper are guided by this author's experience as a practitioner and scholar in the field of (parliamentary) technology assessment.

24.2 New Big Socio-Technical Visions

The wide significance of nanotechnology is captured by the notion of NBIC-convergence. NBIC entails the idea that nanotechnology, biotechnology, information technology, and the cognitive sciences complement and stimulate each other. It is instructive to provide three examples of how nanotechnology promotes convergence, and the way it has inspired various grand socio-technical visions of the future. Moreover, we will also point at various concerns that these visions have led to.

First, the expectations about nano-electronics have inspired big electronics companies all over the world to develop at the end of the 1990s a new vision for the information society. This industry vision has been promoted under different headings, like "ambient intelligence" in Europe, "ubiquitous network society" in Japan, and "pervasive computing" in the United States. Common to these concepts is the prospect of "smart" environments that deliver all kinds of services to people, like reminding you of an appointment or detecting driver's emotions to prevent car accidents. Here (bio)sensors, RFIDs, and tiny computers are incorporated practically anywhere: in walls, in clothing, or even in the human body itself. Over the last decade research guided by this new vision has received a lot of government funding. The social implications are enormous. In fact, since this vision signifies a new phase in the development of IT, it also implies a new phase in the debate on IT and privacy. The major concern, of course, is that these developments in the field of nano-electronics will mean the end of privacy.

Second, in a similar vein, the emergence of the new field of synthetic biology reveals the influence of nanotechnology on biotechnology. Synthetic biologists regard a cell as a collection of "nanomachines" which can be copied, redesigned, and improved. This approach marks the dawn of a new era in biotechnology, and one which demands a reconsideration of issues like bio-safety and intellectual property rights.

A third debate that has emerged with nanotechnology is the debate on human enhancement. The *Converging Technologies for Improving Human Performance* workshop organized by the U.S. National Science Foundation (NSF) in 2001 actually coined the term *NBIC*. According to the NSF, NBIC-convergence creates various opportunities for improving human performance. For example, it predicted fast broadband interfaces between human brains and machines and complete control of the genetics of humans (Roco and Bainbridge 2002). This sparked a heated debate among a growing circle of scientists worldwide regarding the degree to which "human enhancement" can be seen as ethically acceptable (cf. Van Est et al. 2006). The ETC Group (2003) from Canada was one of the first international non-governmental organizations (NGOs) that called for attention to the social impact of technological convergence. This organisation used threatening terms like "super-colliding technologies" and basically repeated the worries expressed by Bill Joy in April 2000 in *Wired*. His article, "Why the future doesn't need us," sketches three doom-scenarios: robots producing robots with superhuman intelligence; designer viruses that unleash a worldwide plague epidemic; and nanobots, artificial plants, or bacteria that can reproduce rapidly and cover entire regions of the world.

These three examples show that nanotechnology cannot just simply be defined as studying and manipulating matter on a scale smaller than 100 nm. A broad interpretation of nanotechnology is needed that acknowledges that the technical challenge of going to the nanoscale has led to new research venues and paradigms that together are expected to form a powerful wave of emerging NBIC-technologies. Making nanotechnology public thus entails making the social visions and meanings of this new wave of technologies public. The next section argues that decades of experience in the field of parliamentary technology assessment in Europe shows there are three types of connected activities that are vital for achieving this (Bütschi et al. 2004).

24.3 Technology Assessment: Investigation, Interaction, Intervention

The modern practice of parliamentary technology assessment (TA) includes more than doing science and writing reports. Since the early 1990s, an increasing number of citizens and stakeholders have become involved in TA processes (Joss and Bellucci 2002). At the beginning of this century the TA community woke up to the fact that good communication skills and (personal) links with politicians and journalists are crucial for improving the impact of TA. A quite recent definition of technology assessment sums up these lessons learned: "Technology assessment is a scientific, interactive and communicative process which aims to contribute to the formation of public and political opinion on societal aspects of science and technology" (Bütschi et al. 2004, 14).

In the 1960s, TA developed as a special branch of policy analysis that dealt with the social impact of science and technology. In 1972 the practice of parliamentary TA became established when the American Office of Technology Assessment (OTA) was created by law to strengthen the position of the Congress in dealing with science and technology. In Europe the first parliamentary TA offices were established in France, Denmark, and the Netherlands in 1986. Britain and Germany, who followed in 1989, saw TA as a method of expert policy analysis (Vig and Paschen 2000). Denmark and the Netherlands, however, saw TA as a more general and "open" process for involving the public in policy dialogues and building societal consensus on issues of technological change. Besides scientific study, the Danish Board of Technology and the Dutch Rathenau Institute saw TA as a process for involving experts, stakeholders, and citizens in policy dialogues on issues of technological change. This involvement has taken various forms, including citizens' panels, scenario workshops, consensus conferences, and technology festivals (cf. Joss and Bellucci 2002; Slocum 2003). Setting up public participatory processes required new skills from TA practitioners, in particular process management skills. As a consequence, developing quality criteria for setting up interactive processes became important (Van Eijndhoven 1997). Examples are social fairness (fair rules for selecting participants), process fairness (fair rules for allowing participants to be heard), and transparency about the methods that are used and how the results are achieved (Bütschi et al. 2004, 36–40).

24.4 Focus on Intervention

Over the last decade, parliamentary TA practitioners have sought ways to improve the impact of their work (cf. Decker and Ladikas 2004). Within the parliamentary TA community it was realized that it is not sufficient to just communicate the results of the TA project at the end of the project (Bütschi et al. 2004). Frequently the results of the project came out too early or too late to be relevant for the policy making process. This was, for example, because parliamentarians were not yet aware of the issue or because the political debate had already taken place. Involving experts, stakeholders, and citizens in the project, and expecting them to spread the word in the outside world does not help much to create an impact either. To have a significant impact those actors that can create change should be gently pushed to become engaged and intervene. Parliamentarians (legislators), policy makers, researchers, business managers, journalists, et cetera, need to become aware of both the issues and the project, understand the social implications, and take actions in line with their position in the network. This would mean, for example, that in the case of suspected health impacts from nano-particles, a journalist would write about it in a newspaper or a legislator would raise this issue at a governmental or regulatory level.

Linking the outcomes of a TA project to the relevant actor networks again requires extra skills from TA-practitioners. First, they need to constantly keep track of the social, political, and scientific issues. For example, in the case of nanotoxicity this means keeping up to date with the (lack of) knowledge on the environmental and health threats of nanoparticles. It also means keeping up to date with the way in which different governmental agencies, firms, politicians, and stakeholders are dealing with these issues. This doesn't require one study, but a whole series of TA activities—studies, workshops, interviews with the media, building up personal contacts with relevant network actors, et cetera—over a considerable period of time. For example, the Rathenau Institute[3] has actively dealt with the safety-issues around nano-particles since 2003. Moreover, TA-practitioners should build up personal contacts in these various worlds and partially develop, or at least understand, the vocabulary and ideas that reflect these worlds. To get heard by external audiences, the TA practitioner needs to go public. It is not sufficient to write a good report; it should be communicated through the media to a larger audience. This requires communication skills, like being able to write an opinion article that really gets accepted by a newspaper or defending in a convincing way your views during public debates. Interestingly, this public role demands from the TA practitioner that he or she is regarded in the public domain as a scientific expert. In other words, the TA practitioner needs to assume the role of a public intellectual, a scientist that plays an active role in the public and political debate (cf. Bijker 2003).

In summary, to stimulate a public and political discussion, there is a need for information about the social meaning of nanotechnology. Moreover, there is also a need for the involvement of experts, stakeholders, and citizens. But scientific investigation and discussion within TA projects is not sufficient. Real public engagement starts when governmental agencies, industry groups, scientists, and citizens take up their responsibility for dealing with a public issue. This means that seducing these actors to tackle public issues has become an important goal for TA. The remainder of this paper reflects on the meaning of these three vital TA activities—investigation, interaction, intervention—for the ELSI-research on equality and equity issues in nanotechnology, and the role of researchers interested in these issues.

24.5 Need for Scientific Investigation and Reflection

Moral values like equity, fairness, and social justice play an important role in many public disputes about science and technology. There are important questions like: Who will benefit and who will be harmed from these developments? In what cases is it right that some actors benefit more from the technology than others? Nanotechnology is by no means an exception to this rule. Distributive justice concerns appear regularly (cf. Van Est et al. 2004; Sandler 2009). Will people all over the world benefit from the advances in nanotechnology, or will it strengthen the gap between North and South and create a so-called nano-divide? Will nano-medicines improve the access to healthcare? Will the use of nanotechnology to enhance human performance lead to a loss of solidarity with handicapped people? Will, in the very long-term, humans even be genetically divided?[4]

This yearbook justifies the need for reflection on equity and equality issues in the field of nanotechnology. It clearly puts these issues on the academic agenda. However, we should be aware of two challenges in the way this agenda is being set. One challenge is to pay serious attention to newly emerging equity and equality issues. A second challenge is to signal and stimulate developments that have the potential to increase social equity.

24.6 Need for Upstream Reflection and Construction

With respect to the first challenge, it was interesting to notice that most of the presentations given at the Workshop on Nanotechnology, Equity, and Equality concerned and reflected on technologies that have already reached the market, such as genetically modified food crops and mobile phones. It is important to study these existing technologies because they provide insights into the social and economic consequences of modern technologies. These studies illustrate how equity and equality issues work out in practice. But besides studying existing technologies it is also important to pay attention to emerging technologies and the related emerging equity and equality issues. There is thus the challenge of upstream reflection. The rise of ELSI research over the last two decades is actually based on a strong faith in upstream reflection. For example, in the early 1990s, concern with the impact of new gene technologies led the United States to set up the ELSI program of the Human Genome project, as:

> A new approach to scientific research by identifying, analyzing and addressing the ethical, legal and social implications of human genetics research at the same time that the basic science is being studied. In this way, problem areas [would] be identified and solutions developed before scientific information is integrated into health care practice (National Human Genome Research Institute 2008).

Since nobody can predict the future this is easier said than done (cf. Collingridge 1980). Upstream reflection on emerging equity and equality issue is not an easy challenge. Social scientists have to find out the added value and limits of upstream reflection by trial and error.

As the above quote shows, the promise of ELSI research is not only to identify and thus avoid problems, but also to signal and stimulate the development of desired solutions. However, scholarship in general focuses on reflection and not construction. There is an emphasis on showing and explaining how things went wrong. Reflection, thus, is mostly critical. There, however, is also a clear need for constructive suggestions in order to strengthen equity and equality (cf. Schot and Rip 1996). History provides ample examples. The safety bike played a key role in the emancipation of women at the end of the nineteenth century (Bijker 1995). This did not happen overnight, however, nor without a struggle. During the 1880s and 90s, female cyclists were often accosted. But at the end of the century, women had gained through cycling increased mobility and personal freedom. Moreover, they had become freed from their restrictive Victorian clothing style. In a similar way modern navigation systems in cars make it easier for men and women to travel over long distances alone. Large public water systems, as well as individual water wells in poor countries, have increased access to water worldwide. Sewage systems and public health technologies, like vaccines, have increased the health and longevity of many people. The internet has strongly increased the ability of lay-people to access expert information. Many patients have become experts on their disease. It is important, therefore, to study the complex and often unpredictable ways in which technologies have played a role in increasing social equality. Thus, besides upstream reflection, scholars should also look for solutions.

24.7 Need for Upstream Public Participation

Besides reflection, there is a need to actively engage relevant actors in discussing the social meaning of emerging technologies. In particular in Europe, the need for public engagement to move "upstream" is recognized (cf. Wilsdon and Willis 2004). For example, the Royal Society and the Royal Academy of Engineering (2004) published a highly influential report on nanoscience and nanotechnologies which stated that in order to have an impact on technology development, public engagement should take place early in the development process, before stakeholders adopt entrenched positions and opinions become polarized.

Upstream public engagement is clearly inspired by the American ELSI programs (Van Est 2010). Besides upstream reflection, the Europeans also wanted participation processes to move upstream, so that not only expert-knowledge, but also voices from stakeholders, NGOs, and citizens could enter the debate on science. MacNaghten et al. (2005) signaled some other limitations of ELSI research. They feared that ELSI-research was "framed as being able to scrutinize only the impacts or effects of the technology rather than deeper social and political considerations" (MacNaghten et al. 2005, 6). Moreover, they argued that besides the risk issue, more fundamental social issues around ownership, control, and social ends should be part of the debate.

Earlier public debates, like the one on genetically modified food, show that civil society organizations (CSOs)[5] play a key role in bringing considerations of social and environmental justice to the fore, providing them with a social context and putting them on the public and political agenda (cf. Jasanoff 1997). For example, in the field of synthetic biology the ETC group (2007, 2008) stresses that the debate should not be limited to issues of bio-security and bio-safety. Other important socio-economic issues are the creation of new intellectual monopolies, the implications of commodification in synthetic biology for the conservation of genetic resources, the politics of biodiversity, and international trade. In other words, CSOs regularly put forward the key question of how innovation might be governed in a way that conforms to the aim of a just and sustainable global socio-economic development. The early engagement of CSOs and citizens in discussing emerging technologies is important in addressing issues related to equity and equality.

24.8 Emerging Publics

However, the early involvement of relevant actors is not self-evident. A first bottleneck for public participation is that politicians, policy makers, or CSOs might not yet have heard of the development of nanotechnology. And if they did so, they may not have given it high priority. In fact, the agenda of most actors is already overbooked and dominated by downstream issues. Most organizations lack the resources and institutional mechanisms to keep up to date with upstream issues. Accordingly there is a need for these institutes to get away from their daily routine in order to develop future agendas and raise their own awareness and knowledge about upstream issues. Below two specific examples are described of how awareness and knowledge about nanotechnology can be raised. One example relates to the Dutch parliament, the other concerns a European project to empower CSOs.

Parliamentarians (legislators) in particular are absorbed with topical matters. In the Netherlands, the Theme Committee on Technology Policy presented a special institutional setting within the Dutch parliament that enabled a somewhat less political and more reflexive and long-term perspective on developments within science and technology. One of the long-term issues the Theme Commission picked up was nanotechnology. Together with the Rathenau Institute, this parliamentary committee organized a public meeting *Small technology—Big consequences*, on October 13, 2004, which was effective in putting nanotechnology on the Dutch political agenda.

With respect to empowering CSOs, the NANOCAP project, sponsored by the European Commission, provides a nice example. NANOCAP stands for Nanotechnology Capacity Building NGOs, and involves labour unions and environmental organisations. This project gave these organizations the opportunity to deepen their knowledge about nanotechnology and enabled them to develop their strategy and start playing an active role in the political and public debate.

One major problem is that the idea that "publics" are out there waiting and ready to become involved is fiction. We might better speak of new emerging publics (cf. Dijstelbloem 2008). Since nanotechnology and its related social debate are still in an early phase, the social issues involved are to a large extent unknown. Accordingly, it is still rather unclear which social groups will be afflicted by nanotechnology, and thus, which groups should become engaged in the debate. The concept of emerging publics thus can be applied to existing CSOs, which may not yet be aware of nanotechnology. But it also counts for groups that have not yet identified themselves nor organized themselves. Upstream engagement thus requires as a first action the identification of relevant emerging publics. With respect to nanotoxicity, small-sized businesses that are entering the nanotech-market might be regarded as a new emerging public. In the Netherlands, mothers with hereditary breast cancer lobbying for legalizing embryo selection are another example. In this respect, Nicolas Rose (2006) speaks of "biological citizenship": groups organizing themselves around certain disorders. With regards to nanotechnology it is important to keep a close watch on the arrival of new groups of engaged citizens, because they literally embody new emerging social issues, and are thus able to represent these issues in the public debate in an authentic way.

24.9 Engage Engaged Publics

Moving participation upstream thus forces us to reflect on the meaning of the words "public" and "publics." There is a tendency among policy makers in Europe to equate public participation with citizen participation. In this model, the individual citizens selected are expected not to have a stake in the respective issue. These "pure" citizens, then, as a group are expected to mirror the general public (cf. PAGANINI 2007). The consensus conference method—a public enquiry centered around some fifteen citizens who are charged with the assessment of a socially controversial topic—fits this picture. These types of methods to involve citizens are very valuable. And it is good to see that such participatory methods, which have been developed within the TA community, have become part of the ELSI research program on nanotechnology in the United States.

It should be stressed, however, that public participation includes more than engaging disinterested citizens. It is at least as important to look for (emerging) engaged citizens and CSOs. The challenge of upstream public engagement requires a new perspective on the word "public." With respect to downstream issues, we have grown accustomed to the idea that public engagement refers to CSOs and citizens. This is because it could be assumed that experts were already involved in the issue. This assumption no longer counts for upstream issues where the (public) engagement of experts is not self-evident. In that case, stimulating public participation should also (maybe even first of all) be focused on involving experts, like social scientists, policy makers, and politicians.

24.10 Need for Connecting to the Real World

The science community's core activity is research and publishing their results in academic journals. It is not self-evident that these voices will be heard outside the narrow peer communities or reach a lay public. In particular, for ELSI-research this is problematic because of its promise to identify and avoid problems, and stimulate the development of desired solutions. MacNaghten et al. (2005) criticized the lack of impact of ELSI-research, and stressed the need to link back the results to the decision-making of scientists, industry, and policy makers. To fulfill its promise, ELSI-research needs to take up the challenge of making an impact on the real world. This requires a more active and civically engaged role for academic researchers.

Social scientists have expanded their research methods. For example, setting up public panels has become an accepted qualitative method within ELSI-research. Experience in the field of TA shows that just publishing the results of these quantitative and qualitative studies within academic journals will not lead to a big impact in the real world. To achieve a big impact, the results need to be purposefully translated and communicated towards, for example, the policy arena or a wider public. Moreover, researchers need to build personal connections with these different worlds to have an impact on these worlds. In other words, building up a network is a vital aspect of research that tries to stimulate actors to get involved in social justice issues around nanotechnology. Various ELSI-research programs, like the Center for Nanotechnology in Society at Arizona State University (CNS-ASU) and the Technology Assessment program (TA NanoNed) of the Dutch national nanotechnology initiative, have put a lot of effort over the last years to establish contacts and collaboration with the science and engineering communities. As a result some scientists and engineers have become engaged in public issues around nanotechnology. The time has come to broaden the scope of engagement to include relevant actors outside academia. A good example is the briefing of the U.S. Congressional Nanotechnology Caucus on the societal implications of nanotechnology that CNS-ASU organized on March 9, 2009, in collaboration with the Woodrow Wilson Center.

Achieving an impact outside academia also requires ELSI-researchers to play a more active role in the public and political debate. At the 2001 annual meeting of the Society for Social Studies of Science, Bijker (2003, 446) argued that "STSers should be the public intellectuals of the next decade." There indeed is a need for public intellectuals who stimulate the public and political debate on emerging nanotechnologies by clarifying the role these may play in the future and the social challenges such technologies pose for the future. ELSI-researchers are in the perfect position to play a role at the forefront of this emerging public debate and inspire other public actors to get involved.

One barrier to this involvement, however, is the current "publish or perish" culture, which forces researchers to specialize and write for narrow audiences. Writing opinion articles for newspapers or popular magazines does not count for academic promotion. Even monographs are often not rated. ELSI-researchers, therefore, need free space and support to be able to assume their public responsibility. In the Netherlands, for example, researchers working at the Center for Society and Genomics—an ELSI-research program on genomics in the Netherlands— are allowed to spend 1 day a week on public-oriented activities like giving public lectures or writing opinion articles for newspapers.

24.11 Conclusions

The political promise of ELSI-research is to identify and avoid problems, and stimulate the development of desired solutions in the early phase of the development of nanoscience and nanotechnology. Based on my experiences in the field of (parliamentarian) technology assessment, I believe that ELSI-research in its current form will not live up to those big expectations. One option is to lower political expectations and communicate to policy makers that this is simply academic research. To keep the dream of governing nanotechnology from a public perspective alive, however, the scope of ELSI-research has to be broadened and sharpened.

With respect to scientific investigation there is a need to address upstream developments and analyze cases of technologies that contributed to increased social justice. With regards to interaction, setting up public panels has already become an accepted method within the ELSI-research program. The challenge of upstream public engagement also relates to involving experts, CSOs, and newly emerging publics. Finally, it is important that the valuable insights of ELSI-researchers have a wider impact in society. To achieve that, new linkages with relevant networks and players outside academia need to be built up. This requires that academic researchers leave their ivory towers and become public intellectuals.

Acknowledgements I would like to thank the editors for their constructive comments on this paper.

Notes

1. See http://www.society-genomics.nl/
2. See http://www.nano.gov/html/society/ethical_legal_society.html; and http://cordis.europe.eu/science-society.

3. The Rathenau Institute is an independent organization financed by the Dutch Ministry of Education, Culture and Science and is a part of the Royal Netherlands Academy of Arts and Sciences. The Rathenau Institute concerns itself with issues on the interface between science, technology and society, and provides the Dutch Parliament with timely and well-considered information. The Institute was founded in 1986, and currently has some 45 staff. The Rathenau Institute has two core tasks. Its traditional role is studying and stimulating social and political debate about the impact of science and technology on society from the point of view of the public (Technology Assessment, or TA). Since 2004 the Institute has also been investigating how the science system performs and how it responds to scientific and social developments. This task is called Science System Assessment, or SciSA. For further information see http://www.rathenauinstituut.com/
4. Silver (1998) predicts a division of the world into a genetically enhanced elite ("GenRich") and genetically deprived proletarians (so-called "Naturals") in three hundred years time.
5. Civil society organizations (CSOs) are organizations whose membership represents a variety of public interests and responsibilities and which may include trade unions and employers' organizations ("social partners"), non-governmental organizations, professional associations, charities, grass-roots organizations, organizations that involve citizens in local and municipal life, churches, and religious communities (European Commission 2006).

References

Bijker, Wiebe, E. 1995. *Of Bicycles, bakelites, and bulbs: Toward a theory of sociotechnical change.* Cambridge, MA: The MIT Press.

Bijker, Wiebe, E. 2003. The need for public intellectuals: A space for STS. Pre-presidential address, Annual Meeting 2001, Cambridge, MA. *Science, Technology & Human Values* 28: 443–450.

Collingridge, David. 1980. *The social control of technology.* London: Pinter.

Danielle Bütschi, Rainer Carius, Michael Decker, Søren Gram, Armin Grunwald, Petr Machleidt, Stef Steyaert, Rinie van Est. 2004. The practice of TA; Science, interaction and communication. In *Bridges between science, society and policy: Technology assessment – Methods and impacts.* eds. Michael Decker and Miltos Ladikas. Berlin: Springer.

Decker, Michael and Miltos Ladikas, eds. 2004. *Bridges between science, society and policy: Technology Assessment – Methods and impacts.* Berlin: Springer.

De Man, Hugo. 2005. Ambient intelligence: Gigascale dreams and nanoscale realities. IEEE International Solid State Circuits Conference, February 6–10, San Francisco, USA.

Dijstelbloem, Huub. 2008. *Politiek vernieuwen: Op zoek naar publiek in de technologische samenleving.* Amsterdam: Van Gennep.

ETC Group. 2003. *The big down: From genomes to atoms. Atomtech: Technologies converging at the nano-scale.* Winnipeg, Canada: ETC Group.

ETC Group. 2007. *Extreme genetic engineering. An introduction to synthetic biology.* Winnipeg, Canada: ETC Group.

ETC Group. 2008. *Commodifying nature's last straw? Extreme genetic engineering and the post-petroleum sugar economy.* Winnipeg, Canada: ETC Group.

European Commission. 2006. *Science and society action plan. European Research Area.* Brussels: European Commission, DG Research.

Jasanoff, Sheila. 1997. NGOs and the environment: From knowledge to action. *Third World Quarterly*, 18 (3): 579–594.

Joss, Simon, and Sergio Bellucci, eds. 2002. *Participatory technology assessment: European perspectives.* London: Centre for the Study of Democracy.

Joy, Bill. 2000. "Why the future doesn't need us." *Wired* 8 (4) April: 238–262.

Macnaghten, Phil, Matthew Kearnes, and Brian Wynne. 2005. Nanotechnology, governance, and public deliberation: What role for the social sciences? *Science Communication* 27 (2) December: 1–24.

National Human Genome Research Institute. 2008. About ELSI: About the ethical, legal, and social implications (ELSI) program. http://www.genome.gov/10001754. (accessed September 23, 2008).

Rinie van Est, Christien Enzing, Mark van Lieshout, Anouschka Versleijen. 2006. Welcome to the 21st century: Heaven, hell or down to earth? A historical, public debate and technological perspective on the convergence of nanotechnology, biotechnology, information technology and the cognitive sciences. In Robby Berloznik, Raf Casert, Robby Deboelpaep, Rinie van Est, Christien Enzing, and Anouschka Versleijen, eds. *Technology Assessment on converging technologies*. European Parliament Report.

Roco, Mihail C., and William S. Bainbridge, eds. 2002. *Converging technologies for improving human performance: Nanotechnology, biotechnology, information technology and cognitive science*. Arlington, VA: National Science Foundation (NSF)/Department of Commerce (DOC). June.

Rose, Nicolas. 2006. *The politics of life itself: Biomedicine, power, and subjectivity in the twenty-first century*. Princeton, NJ: Princeton University Press.

Royal Society and Royal Academy of Engineering. 2004. *Nanoscience and nanotechnologies: Opportunities and uncertainties*. London: The Royal Society.

Sandler, Ronald. 2009. *Nanotechnology: The social and ethical issues*. Washington: NJ: Woodrow Wilson International Center for Scholars, Project on Emerging Nanotechnologies.

Schot, Johan, and Arie Rip. 1996. The past and future of constructive technology assessment. *Technological Forecasting and Social Change* 54: 251–268.

Silver, Lee. 1998. Remaking Eden: Cloning and beyond in a Brave New World. New York: Avon.

Slocum, Nikki. 2003. *Participatory methods toolkit: A practitioner's manual*. Brussels: King Baudouin Foundation and Flemish Institute for Science and Technology Assessment.

The PAGANINI project. 2007. *Participatory governance and institutional innovation: The new governance of life. A summary report of the PAGANINI project*. Austria: Department of Political Science, University of Vienna.

van Eijndhoven, Josee. 1997. Technology assessment: Product or process? *Technological Forecasting and Social Change*. 54: 269–286.

van Est, Rinie, Ineke Malsch, and Arie Rip. 2004. *Om het Kleine te Waarderen. Een schets van nanotechnologie: publiek debat, toepassingsgebieden en maatschappelijke aandachtspunten*. The Hague: Rathenau Institute.

van Est, Rinie. 2010. The broad challenge of public engagement in science. In *Science and technology policy in the making: Observation and engagement*, ed. Erik Fisher, Special Issue of *Science and Engineering Ethics*.

Vig, N. and H. Paschen. eds. 2000. *Parliaments and technology: The development of technology assessment in Europe*. Albany, NY: State University of New York.

Wilsdon, J. and R. Willis. 2004. *See-through Science: Why public engagement needs to move upstream*. London: Demos.

Zwart, H. and A. Nelis. 2009. What is ELSA genomics? *EMBO reports: Science & Society series on convergence research* 10: 540–544.

Chapter 25
Nanotech Ethics and the Policymaking Process: Lessons Learned for Advancing Equity and Equality in Emerging Nanotechnologies

Evan S. Michelson

Drawing on his experience in the Woodrow Wilson Center Project on Emerging Nanotechnologies, as well as his experience in strategic planning and international development, Evan Michelson urges those interested in equity to engage with decision-makers. Michelson offers a number of lessons to help those with a desire to make the development of nanotechnology more equitable have a significant effect. He recommends that advocates aim to influence nano agenda setting as early as possible, develop proof of concept examples for pro-poor applications, build public and policymaking constituencies, communicate explicitly the anticipated equity and equality impacts, and use diverse empirical research methods. Now is the time to get these issues on the nanotechnology agenda, Michelson suggests, while relationships and institutions are still emerging.—eds.

E.S. Michelson (✉)
Robert F. Wagner Graduate School of Public Service, New York University, New York, NY, USA
e-mail: Evan.michelson@nyu.edu

Originally presented at the Workshop on Nanotechnology, Equity, and Equality at Arizona State University on November 21, 2008.

25.1 Introduction

One of the most interesting aspects associated with the appearance of nanotechnology in society is that the advancement of this technology is occurring alongside the development of a broader awareness as to the potential equity and equality impacts of new technologies. While analogies comparing the societal implications of nanotechnology with other emerging technologies—such as genetically modified organisms—may inevitably fall short, the articles presented in this collection are aimed at ensuring that the diffusion of nanotechnology in society avoids the pitfalls of earlier processes of technology adoption and do not exacerbate existing social inequalities (Sandler and Kay 2006). Furthermore, concerns about the disproportionate lack of access to, and use of, new technologies by certain segments of the population (minorities, the poor, and the disabled) emphasize the need to bring these ethical issues to the forefront at the early stages of nanotechnology's development.

Substantial social science research is already being conducted as to how nanotechnology is impacting society through various sectors of the economy—from consumer products to medical applications to food and food packaging—and through various points of intersection with the oversight system—from international standard-setting bodies to national regulatory responses to state and local guidelines. Considering the articles in this book, other forward-looking assessments of how nanotechnology will impact various economic, regulatory and policy considerations (Davies 2008; Luoma 2008; Schultz and Barclay 2009), and previous assessments of the equity and equality concerns raised by other emerging technologies, I will develop a set of "lessons learned" that can serve as a guide in analyzing nanotechnology's potential societal implications.

The very diversity of the situations in which nanotechnology is likely to be applied—from creating new water filtration systems for the developing world to innovating new medical treatments for the developed world—guarantees that questions surrounding equity and equality will likely touch more people's lives over a longer period of time. In a recent overview of the bioethical consequences of nanotechnology, Michelson et al. (2008) identify a broad range of these equity and equality concerns that have the potential to "promulgate, exacerbate, or provide new variations on familiar issues," such as environmental justice, informed consent, and human enhancement. Additionally, there are other equity and equality issues associated with disability rights, animal rights, and access to new technologies in the developing world that must also be illuminated if the full range of nanotechnology's potential influence is to be understood. As the chapters in this book indicate, the importance of equity and equality in nanotechnology needs to become more apparent for policymakers.

While equity and equality commitments may not always be made explicit or directly referenced, they inevitably underlie much of the policy debates that have recently become the topics of vigorous debates. This chapter attempts to lay out a series of strategies for action and to emphasize the ways in which policy-related debates about nanotechnology can more effectively elucidate the equity and equality components of nanotechnology.

25.2 Strategies for Action

There are a variety of approaches that researchers, decision-makers, and analysts can adopt to more fully integrate notions of equity and equality in nanotechnology. The steps outlined below—mostly drawn from recent examples of work supported and conducted by the Project on Emerging Nanotechnologies at the Woodrow Wilson International Center for Scholars—should be taken as an initial set of actions that can help illuminate the importance of equity issues to a diverse community of scholars and practitioners. They cover several key domains—including focusing on issue framing and agenda setting; employing diverse social science research methodologies; developing pro-poor, proof of concept technical applications; building public and policymaking constituencies; and explicitly communicating anticipated equity and equality impacts—that are central to making progress as to how equity and equality considerations can be addressed.

25.2.1 Focus on Issue Framing and Agenda Setting

The manner in which the topics of equity and equality are introduced into debates about nanotechnology is likely to have a significant effect on the degree of importance afforded to these issues. For instance, policymakers typically frame discussions about nanotechnology's societal implications first through a technological perspective—raising questions about technical characteristics, feasibility, and cost—and then by turning to equity and equality questions—such as distributive impacts and access—as subsequent considerations. Under this approach, it is more likely that equity and equality issues will fade into the background and be overtaken by a more immediate focus centering on the novelty of a nanotechnology breakthrough, its potential economic implications, or even its environmental, health, and safety (EHS) implications. There are many reasons for EHS issues overtaking equity issues: EHS considerations can be more easily quantified than ethical considerations; existing institutions and mechanisms are better equipped to address EHS considerations; and conversations about EHS issues, while still contentious, are likely to lack deeper assessments about the more foundational moral questions related to equity and equality. Such constraints, however, reduce the likelihood that equity and equality issues will be adequately addressed in the policymaking process.

Reformulating this approach is possible by, first, making equity and equality concerns paramount and then introducing other science and technology issues. Such a transition has the potential to make a broader set of constituencies interested in guiding and responding to nanotechnology's societal implications, in part because such reframing would make evident the connections between nanotechnology and a wider range of social, political, and economic concerns. This approach would likely require the application of theoretical frameworks and paradigms from social science and humanities disciplines that sometimes fall outside the scope of science and technology policy—for example, philosophy, international development, anthropology, history, and psychology—in order to provide a more diverse and interdisciplinary sense of how equity and equality issues have been addressed in other settings and with respect to other topics of inquiry. Determining how equity and equality issues are framed—either as primary or secondary considerations—and how the agenda for answering these questions is established—including which theoretical paradigms are used to structure discussions in these areas—is an important step in ensuring that the topics of equity and equality are adequately represented in policy deliberations. For example, Sandler (2009) points out that instead of giving primacy to questions related to technical norms, funding streams, and scientific practicality, new frames are needed that elevate issues like ethical capacity, social justice, sustainability, and human flourishing. Similarly, Parens et al. (2009) note that policy making around these issues will remain difficult in the absence of an overarching theoretical framework that integrates ethical discussions across a range of emerging technologies, from nanotechnology to synthetic biology to neuroscience. Such a broad "ethics of emerging technologies" would help to illuminate the multiple and far-reaching equity and equality effects of policy decisions related to these areas of technological development.

25.2.2 Employ Diverse Empirical Research Methodologies

In researching nanotechnology's equity and equality impacts on society, it is necessary to employ a diverse set of empirical methodologies that can answer different research questions at different levels of aggregation over different periods of time. This need for methodological pluralism is underpinned by the notion that equity and equality issues can affect society at varying rates of speed and in rather diverse cultural, geographic, and social contexts. Research methodologies are needed that range from large scale, quantitative randomized controlled experiments to qualitative, in-depth case studies. Additionally, research methods from a diverse range of fields in the social sciences and humanities can be applied to gain insight into questions about nanotechnology's ethical and social implications. These diverse methodologies are needed exactly because considerations of equity and equality are themselves heterogeneous, touching upon such questions as: Who benefits and loses from the advancement of a particular application of nanotechnology? What societal needs and desires are being fulfilled by the technological application? What plans, if any, exist to address emerging environmental or distributional issues once the technology is developed and marketed? Which institutions—government, industry, non-governmental organizations, or others—are monitoring these concerns across timescales and geographies? Studies that track ethical beliefs over time, investigate a particular ethical controversy in great depth, or analyze the process by which differing ideas of equity are reflected in legislation all require different types of methodological designs from different areas of social science.

The ideas presented in this volume succeed in demonstrating the effective use of many different kinds of research methodologies. These methods include case studies of nanotechnology-based applications in unique country contexts, the use of databases to provide quantitative analyses of nanotechnology's regional impacts, focus groups to offer insights about public responses to questions of fairness and equity, fieldwork to illuminate perceptions of nanotechnology's diffusion in society, and text-based content analysis to determine the public values espoused in government strategy and planning documents. Theoretical models are being developed that look to build a framework within which these empirical findings, employing a diversity of methodologies, can be understood, interpreted, and integrated (Hill 2007; Office of Science and Technology Policy 2008). Expanding and adding to these empirical methods is imperative to enriching future debates focused on issues of equity and equality in nanotechnology.

Of particular interest moving forward is the expanded use of ethnographies and case studies that detail how decisions related to equity and equality are implemented, communicated, and influence local, national, and international decision-making bodies. Along these lines, a casebook of practice that illustrates commonalities and differences in how different individuals and institutions respond to equity and equality considerations would be an invaluable resource to the field. Finally, detailing the strengths and drawbacks of these different methodologies—and addressing issues such as how research samples and cases were chosen, what issues were included and excluded in the study, and why these choices were made—would go a long way in advancing the procedures for how research in this field is conducted in the future.

25.2.3 Develop Pro-Poor, Proof of Concept Technical Applications

In addition to the social sciences, technical researchers have a role to play in ensuring the nanotechnology's societal impacts are adequately addressed (Fisher et al. 2006; Fisher and Mahajan 2006). A significant technical barrier that exists with respect to an important global equity issue—nanotechnology's potential to address problems in the developing world—is the lack of operational, field tested products. While nanotechnology offers great promise to improve drinking water quality, expand the use of solar energy, and revolutionize food packaging to improve shelf-stability, there is a lack of high profile proof of concept applications in these areas remain few and far between. While pro-poor, proof-of-concept applications are beginning to emerge with respect to next-generation water filtration systems, this is not always the case for other applications of nanotechnology. Corporations, government funders, investors, and researchers need to be sensitive to this general lack of operational prototypes. Without these early- and intermediate-stage demonstrations of the technology's effectiveness, there is the real risk that the hype surrounding nanotechnology's promise will morph into skepticism about its ability to deliver transformational, socially relevant products and services and lead nanotechnology to follow in a lineage of great technological promises that fail to substantially improve society. This delay in commercializing high-impact applications that address critical needs in the developing world has the potential to reduce the scope of the technology's value and create the impression that nanotechnology is predominantly a technology for improving consumer products—such as cosmetics, clothing, and automobiles—that are manufactured mostly for consumers in the developed world (Damrongchai and Michelson 2009). Without such a "killer app" in the developing world, it will become more difficult to "convince the public that the wait, the investment, and the uncertainties about risks are worth it" (Rejeski 2007).

In order to encourage the development and long-term commercial viability of these pro-poor applications, government science agencies in the developed world can make a concerted effort to ensure that their funding of nanotechnology research includes resources to bring promising technologies—including shelf-stable foods and drugs, affordable solar energy systems, and smart sensors for improving transportation effectiveness—from the laboratory to market. These efforts could be the focus of government venture capital programs (Osama 2008a) aimed at boosting university spin-offs, supporting small- and medium-sized enterprises, or breaking down intellectual property barriers. South-south collaborations (Osama 2008b) can also be used as a mechanism to foster this kind of intermediate product development by way of support from international institutions like the United Nations, academies of science, and professional societies.

25.2.4 Build Public and Policymaking Constituencies

To broaden the framework in which nanotechnology's equity and equality implications are discussed, it is also necessary to build public and policymaking constituencies and coalitions that center on different applications, regulations, and topical areas. A good example of this approach is the push for expanded financial and human resources to investigate the EHS implications of nanotechnology. In the United States, coalitions of non-profit organizations, companies, industry associations, think tanks, academic institutions, and federal agency representatives have called for more attention and funding to address potential EHS risks. Although these groups have often acted alone, there has been a surprising alignment of stakeholder interests and positions, fostered by organizations like The National Academies (National Research Council 2009), the International Council on Nanotechnology (2008), and the Centers for Nanotechnology in Society (Barben et al. 2008). Specific, high-profile instances of such policy relevant constituency-building also include the collaboration between DuPont and the Environmental Defense Fund in creating the widely used *Nano Risk Framework* (Medley and Walsh 2007; Walsh and Medley 2008) and the collaboration between the Project on Emerging Nanotechnologies at the Woodrow Wilson Center and the Grocery Manufacturers Association in creating a dialogue around the regulatory process for nanomaterials in food packaging (Taylor 2008).

The scope and extent of these public and policymaking constituencies need to be expanded to include a specific focus on the equity and equality implications of nanotechnology. The challenge is that while stating this point is easy, operationalizing it in practice is hard. There are multiple barriers to building these constituencies. One is that this process of building constituencies requires a forward-looking, long-term mindset by various parties—a mindset that can be easily overrun by short-term political or economic calculations. In addition, many groups fail to see that their self-interests can be advanced by aligning and partnering with organizations of different ideological bents. Consider, for instance, how the environmental movement has begun to work closely with religious organizations as partners in favor of stricter environmental protection and stewardship (Tucker and Grim 2001; Murphy 2006; Sierra Club 2008). When applied to nanotechnology, the expansion of these types of coalitions should continue to involve representatives from regulatory bodies, corporations, and public interest groups, and the findings generated from these collaborations need to occur in an open forum and be distributed widely. As noted earlier, some efforts modeling this constituency-building approach, particularly in the areas of EHS risk management and food packaging, have already emerged. One approach might be to build on these existing aspects of cooperation and expand them to include more explicit discussions of equity and equality concerns.

25.2.5 Explicitly Communicate Anticipated Equity and Equality Impacts

Finally, different communication and outreach strategies will need to be adopted by stakeholders looking to disseminate information about the equity and equality impacts of nanotechnology. Attempts to reach the public through various media outlets, by way of "old" media like newspapers, television, and radio, and "new" media like podcasts, blogs, Twitter, and RSS feeds, will require the use of messages about nanotechnology's equity and equality issues in a way that is understandable to people who may be encountering these topics for the first time. Similarly, researchers and policymakers who may have substantive technical or legal expertise may lack sufficient training and experience in considering equity and equality issues. These stakeholders will also need to be able to discuss information about how developments in nanotechnology could challenge society.

One communication device that has worked effectively is the use of regularly updated inventories and maps that track how nanotechnology is being commercialized and where such commercialization is occurring. For example, the Nanotechnology Consumer Products inventory maintained by the Project on Emerging Nanotechnologies tracks the number of consumer products on the global market by way of a versatile interface that fills a critical knowledge gap for a range of interested parties, from the general public to decision-makers to scientists. As others have demonstrated (Hansen et al. 2008), this inventory can be analyzed along multiple dimensions, including the degree to which the commercialization of these products touch upon equity and equality issues. An analysis of this sort—which could focus on the degree to which producers are designing products for the poor and for the wealthy—would be an important step forward.

Similarly, on-line maps that indicate regional innovation clusters are also valuable in tracking which areas of the country serve as leaders in commercializing nanotechnology and which regions are lagging behind. The use of these kinds of diverse communication tools can be adapted for monitoring equity and equality issues for a variety of audiences. One option is to track the geographic distribution of nano-based health interventions to determine if their use is limited to wealthier areas or whether these applications have a more broad-based distribution. If certain kinds of nanomaterials are deemed to be pollutants, such inventory and mapping tools could also be employed to follow whether manufacturing and disposal sites are disproportionately located in poorer areas.

25.3 Conclusion

Integrating these findings and lessons learned into the nanotechnology in society and public policy research agenda going forward is an important step to ensuring that the applications of nanotechnology are integrated into society in a more sustainable manner. Working to ensure that issues of equity and equality receive attention from decision-makers at this early stage is also important given the context within which nanotechnology is emerging: through a networked, globalized system of research and development, in which the benefits and risks of new technologies are more quickly experienced across a variety of geographic and cultural settings (Wagner 2008). Strategies are needed that focus on issue framing and agenda setting; develop pro-poor, proof of concept applications; build public and policymaking constituencies; explicitly communicate anticipated equity and equality impacts; and employ diverse empirical research methodologies. The hope is that these lessons can help spur action and improve research associated with nanotechnology's equity and equality implications.

References

Barben, Daniel, Erik Fisher, Cynthia Selin, and David Guston. 2008. Anticipatory governance of nanotechnology: Foresight, engagement, and integration. In *The handbook of science and technology studies*, 3rd ed., eds. Edward J. Hackett, Olga Amsterdamska, Michael E. Lynch, and Judy Wajcman. Cambridge, MA: The MIT Press.

Damrongchai, Nares, and Evan S. Michelson. 2009. The future of science and technology and pro-poor applications. *Foresight* 11 (4): 51–65.

Davies, J. Clarence. 2008. *Nanotechnology oversight: An agenda for the next administration*. Washington, DC: Project on Emerging Nanotechnologies, Woodrow Wilson International Center for Scholars. http://www.nanotechproject.org/process/assets/files/6709/pen13.pdf. Accessed 25 Aug 2009.

Fisher, Erik, and Roop L. Mahajan. 2006. Contradictory intent? US federal legislation on integrating societal concerns into nanotechnology research and development. *Science and Public Policy* 33: 5–16.

Fisher, Erik, Roop L. Mahajan, and Carl Mitcham. 2006. Midstream modulation of technology: Governance from within. *Bulletin of Science, Technology & Society* 26: 485–496.

Hansen, Steffen Foss, Evan S. Michelson, Anja Kamper, Pernille Borling, Frank Stuer-Lauridsen, and Anders Baun. 2008. Categorization framework to aid exposure assessment of nanomaterials in consumer products. *Ecotoxicology* 17: 438–447.

Hill, Christopher T. 2007. The post-scientific society. *Issues in Science and Technology* Fall 2007. http://www.issues.org/24.1/c_hill.html. Accessed 25 Aug 2009.

International Council on Nanotechnology. 2008. *Towards predicting nano-biointeractions: An international assessment of nanotechnology environment, health and safety research needs.* Houston, TX: International Council on Nanotechnology. http://cohesion.rice.edu/CentersAndInst/ICON/emplibrary/ICON_RNA_Report_Full2.pdf. Accessed 25 Aug 2009.

Luoma, Samuel N. 2008. *Silver nanotechnologies and the environment: Old problems or new challenges?* Washington, DC: Project on Emerging Nanotechnologies, Woodrow Wilson International Center for Scholars. http://www.nanotechproject.org/process/assets/files/7036/nano_pen_15_final.pdf. Accessed 25 Aug 2009.

Medley, Terry, and Scott Walsh. 2007. *Nano risk framework.* Washington, DC: Environmental Defense Fund and DuPont Corporation. http://www.edf.org/documents/6496_Nano%20Risk%20Framework.pdf. Accessed 25 Aug 2009.

Michelson, Evan S., David Rejeski, and Ronald Sandler. 2008. Nanotechnology. In *From birth to death and bench to clinic: The Hastings Center bioethics briefing book for journalists, policymakers, and campaigns,* ed. Mary Crowley Garrison, 111–116. NY, The Hastings Center. http://www.thehastingscenter.org/Publications/BriefingBook/Detail.aspx?id=2192. Accessed 25 Aug 2009.

Murphy, Brian. 2006. Green gospels: Environmental movement aims for religious mainstream. *USA Today.* July 6. http://www.usatoday.com/news/religion/2006-07-06-greengospels_x.htm. Accessed 25 Aug 2009.

National Research Council. 2009. *Review of federal strategy for nanotechnology-related environmental, health, and safety research.* Washington, DC: The National Academies Press.

Office of Science and Technology Policy. 2008. *The science of science policy: A federal research roadmap.* Washington, DC: Office of Science and Technology Policy. http://www.ostp.gov/galleries/NSTC%20Reports/39924_PDF%20Proof.pdf. Accessed 25 Aug 2009.

Osama, Athar. 2008a. *Washington goes to Sand Hill Road: The federal government's forays into the venture capital industry.* Washington, DC: Foresight and Governance Project, Woodrow Wilson International Center for Scholars. http://www.wilsoncenter.org/topics/docs/ResearchBrief_Osama_final.pdf. Accessed 25 Aug 2009.

Osama, Athar. 2008b. *Fostering South-South research collaborations.* Boston, MA: The Frederick S. Pardee Center for the Study of the Longer-Range Future, Boston University. http://www.bu.edu/pardee/files/documents/BU-Pardee-Policy-Paper-002-Research.pdf. Accessed 25 Aug 2009.

Parens, Erik, Josephine Johnston, and Jacob Moses. 2009. *Ethical issues in synthetic biology: An overview of the debates.* Washington, DC: Synthetic Biology Project, Woodrow Wilson International Center for Scholars. http://www.synbioproject.org/library/publications/archive/synbio3/. Accessed 25 Aug 2009.

Rejeski, David. 2007. Nanotechnology: Waiting for the killer app. *Nanotechnology Now* September 27. http://www.nanotech-now.com/columns/?article=117. Accessed 25 Aug 2009.

Sandler, Ronald. 2009. *Nanotechnology: The social and ethical issues.* Washington: Project on Emerging Nanotechnologies, Woodrow Wilson International Center for Scholars. http://www.nanotechproject.org/process/assets/files/7060/nano_pen16_final.pdf. Accessed 25 Aug 2009.

Sandler, Ronald, and William D. Kay. 2006. The GMO-nanotech (dis)analogy? *Bulletin of Science, Technology, and Society* 26: 57–62.

Schultz, William B., and Lisa Barclay. 2009. *A hard pill to swallow: Barriers to effective FDA regulation of nanotechnology-based dietary supplements.* Washington, DC:

Project on Emerging Nanotechnologies, Woodrow Wilson International Center for Scholars. http://www.nanotechproject.org/process/assets/files/7056/pen17_final.pdf. Accessed 25 Aug 2009.

Sierra Club. 2008. *Faith in action: Communities of faith bring hope to the planet*. Washington, DC: Sierra Club. http://www.sierraclub.org/ej/downloads/faithinactionreport2008.pdf. Accessed 25 Aug 2009.

Taylor, Michael R. 2008. *Assuring the safety of nanomaterials in food packaging: The regulatory process and key issues*. Washington, DC: Project on Emerging Nanotechnologies, Woodrow Wilson International Center for Scholars. http://www.nanotechproject.org/process/assets/files/6704/taylor_gma_pen_packaging1.pdf. Accessed 25 Aug 2009.

Tucker, Mary Evelyn, and John A. Grim. 2001. Introduction: The emerging alliance of world religions and ecology. *Daedalus* 130 (4): 1–22.

Wagner, Caroline. 2008. *The new invisible college: Science for development*. Washington, DC: The Brookings Press.

Walsh, Scott, and Terry Medley. 2008. A framework for responsible nanotechnology. In *The Yearbook of Nanotechnology in Society: Vol. 1: Presenting Futures*. eds. Erik Fisher, Cynthia Selin, and Jameson M. Wetmore. Berlin, Germany: Springer.

Chapter 26
Building Equity and Equality into Nanotechnology

Susan E. Cozzens

This book has addressed a number of incredibly complex issues which can appear too daunting to tackle. But while equity and equality are significant challenges, this does not mean that important progress cannot be made. This final chapter outlines a practical program for building equity and equality into nanotechnology using actions based on the results of previous studies. It is directed at nanotechnology policymakers at a national level, but offers lessons for anyone who makes decisions about technologies, especially the "pro-poor," "fairness," and "equalizing" approaches it advocates. It is written to encourage readers not just to consider the issues of equity and equality as they promote, research, design, regulate, and disseminate nanotechnology, but to recognize that there are concrete steps that can be taken to make the world more equitable and that nanotechnology can play an important role in doing that. Working to promote equity, equality, and development through nanotechnology is a formidable challenge and no efforts will ever be perfect, but keeping in mind some reasonably simple lessons can go a long way towards making the world a more equitable place.—eds.

S.E. Cozzens (✉)
School of Public Policy, Georgia Institute of Technology, Atlanta, GA 30332-0345 USA
e-mail: susan.cozzens@iac.gatech.edu

This paper was originally presented at a Workshop on Real Time Technology Assessment for Nanotechnology, at the S.NET first annual meeting, Seattle, Washington, September 2009.

Imagine that you are a staff member serving a policymaker who is drawing up plans for a national nanotechnology initiative. You want to make sure the initiative provides benefits for everyone in your country, not just certain groups, and that it helps to build what the Europeans call "social cohesion"—the sense in the country that people are in it together, not pulling against each other. You will never reach perfection on either of these goals, but you want to make sure that your initiative makes things better, not worse. You may want to take on this task because it is the right thing to do—a moral responsibility. Or you may want to take it on because it is politically important—because otherwise many of the people who vote in your country will be left out. What steps do you take? This essay lays out what we know based on current research.

I call the process described here an "equity and equality assessment" (EEA) The "equity" part refers to fairness according to one common theory of distributive justice, which requires that the most disadvantaged members of society get some benefit when benefits are distributed.[1] The "equality" part refers to the widely shared idea that no one should benefit or lose because of an "ascriptive" characteristic—something about them that they cannot control, like gender, race, ethnicity, or ability status.[2]

For your equity and equality assessment, it makes a huge difference what kind of country you work in. Is it an affluent country with high levels of education, a highly skilled workforce and a high standard of living? Or is it a middle income country with high inequality, that is, some rich people and many poor ones? Is your country sharply divided in terms of ethnicity or race? Does it have a one-ethnicity elite group? Does your country have a big science and engineering research community, or a small one? Is the country itself big or small? All these elements greatly affect what you plan to do to increase equity and equality, and I will try to point out some of those differences in what follows.

I assume that you are already engaged in a process of setting technological priorities in your nanotechnology program, using common criteria such as building on local capacities and achieving commercial success. The checklist here gives several additional factors to consider in deciding what your country will concentrate on in nano development.

26.1 Inventory Your Vertical and Horizontal Inequalities

Before you even get started on nanotechnology, you should spend some time raising your awareness of your national inequalities. In making an inventory of them, you should think of these in two dimensions: vertical and horizontal.

Vertical inequalities fall along the rich-poor dimension. If you are working in a rich country, you probably have a big and politically powerful middle class, and you may tend to forget the 10–20% of your population that lives below the poverty line. If you are in a poor country, you may face huge gaps between rich and poor, with a few rich families enjoying the standard of living of the affluent world and many

Fig. 26.1 Gini diagram, indicating the households that are the target in pro-poor policies
Source: http://en.wikipedia.org/wiki/File:Economics_Gini_coefficient2.svg, accessed August 5, 2010.

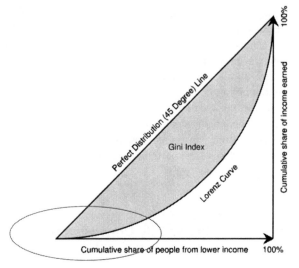

poor ones struggling for survival in urban slums or poor villages. In either case, if you want the benefits of the technology you are shaping to reach everyone, you cannot aim just for the rich or the middle class. You have to keep the poor in mind as well. This is an approach I will call "pro-poor" (Cozzens 2008). Pro-poor policies aim to provide benefits to households at the bottom of the income distribution (see Fig. 26.1).

But as you know, social cohesion in your country does not depend only on the distribution of income. You also have inequalities between culturally defined groups. These are called "horizontal inequalities," and they carry their own kinds of tensions. For example, there are gaps between women and men in income and power—again, smaller in some countries and larger in others. Will the way you are planning your initiative make these worse or better?

In some cases, your country has one or more ethnic groups that have been historically advantaged politically and economically, and others that have been disadvantaged. You may have experienced violence or wars drawn along these lines. Will your program favor one group or another? What about people with different physical abilities? Will the benefits of your programs reach them?

Policies and programs that try to equalize benefits and opportunities across these horizontal divisions can be called "fairness" approaches. Pro-poor and fairness approaches are two of the three major types of policies you will want to build into your nanotechnology initiative.

Inequalities are uncomfortable topics, especially for people who are on the up side of them—which you most likely are if you have the power to set priorities for major new technology initiatives. They tend to be very active political issues in countries with big gaps and less active political issues in countries with strong, egalitarian social welfare systems that are reducing the gaps. You will probably think more about these inequalities in doing your EEA than you have for a while. Because you are in a position of power, it may be difficult for you to see them at first, and some of them may startle you. So you could need some professional help, and you may therefore want to talk to some local social scientists, development organizations, and civil rights advocates about how the differences affect daily lives in your country.

These vertical and horizontal inequalities are concepts that you probably do not regularly consider when thinking about technology programs. When you are committed to doing an equity and equality assessment, you need to keep them on a 3 × 5 card next to your computer, or incorporate them into your screen saver, to refer to them constantly.

26.2 Add a Better Life Dimension to the Criteria for National Nanotechnology Priorities

Since you are working hard on your initiative, you are already putting together a lot of priorities diagrams, with current local capabilities on one axis and commercial opportunities on the other. Unless you are in a country with really big nanotechnology capabilities that can afford to study almost every topic within nanotechnology, you are using these diagrams to try to set priorities. When you find the spots where the capabilities and commercial opportunities match up best, you intend to invest resources in expanding the efforts.

Your equity and equality assessment will lead you to add a couple of diagrams you might not otherwise draw. One is a "better life" diagram, which plots local capabilities against the opportunity to make life better for poor people in your country.[3] South Africa, for instance, is working on using nanotechnology for community water filtration systems, lighter and stronger concrete for simple houses, and more powerful tuberculosis treatments. Another example is the Monad Nanotech Company in India, which is working on a nanotechnology-based technology to eliminate pathogens in water, using any visible light source for energy (Meridian 2006).

Your better life diagram can still have a dimension pointing to commercial opportunities. Which of these technologies might also have a significant market? Since the needs of the poor are generally neglected in research programs, North and South, and since there are plenty of poor people out there, commercial opportunities are often wide open in areas like tropical diseases and water and sanitation. For example, inexpensive mobile phones have reached a very wide market because they offer something valuable at a price that even poor people can afford.

Within the better life analysis, you will want to think about the horizontal groups. Are you thinking about better lives for everyone in poor communities or only for some? For example, community water filtration may not provide much benefit for women, who walk long distances and bring water to their homes in head-jugs— literally an unequal burden. Are there opportunities for nano-enabled materials to allow affordable piping directly to village homes? Adding this element to your priorities would help with gender balance.

You can extend the same logic to middle class members of horizontal groups. Will making materials lighter remove one barrier to opening up construction jobs to women workers, for example? Can better wheelchair access ramps be designed or lighter wheelchairs? Again, taking these kinds of uses into account is not incompatible with business opportunities, but instead can extend the range of opportunities you consider, giving you more chances for matches with your current local research and technology capabilities.

26.3 Listen to Disadvantaged Groups

This kind of analysis of priorities is typically done through a process. Europeans sometimes call that process "foresight" and have honed methods for doing it (European Commission 2005); at CNS-ASU we call it Real Time Technology Assessment or anticipatory governance (Barben et al. 2008; Guston and Sarewitz 2002). One of the keys to those methods is involving a broad group of stakeholders. In Europe, this usually means not just scientists, engineers, and relevant industrial firms, but also civil society groups with a special interest in the area. A foresight exercise in Europe on alternative fuels, for example, would include environmental groups.

Equity and equality assessment of necessity pushes this envelope. None of the groups who typically sit at the foresight tables know what people living in poverty want or need, and often they have strong incentives not to worry about it. Likewise, those tables are too often full of men from dominant ethnic or racial groups. Your interest in the topic will provide the incentive, but the knowledge is still somewhere else.

The most effective way to re-orient your country's nano priorities toward a better life for all is to listen to people in the groups you are trying to reach. Go to poor communities and talk to both men and women, young and old, powerful and disempowered, those who usually have a voice and those who are often silenced. Meet with organized groups from the disability community. Work with spokespeople for ethnic minorities.

Be eclectic and creative in the ways you open up this communication. Putting one person from a disadvantaged group on a high-level advisory panel—even though it is a good thing to do—is not likely to produce the flow of information you need. That person's voice will likely be drowned out by the majority, and he or she may not have the comfort level necessary to speak frankly. You may need to work with professional intermediaries, such as medical professionals or social workers who are immersed in the lives of the communities you want to reach, but confident enough in the culture you operate in at the top of government to speak up. You also need to establish some discussions that are held under the cultural ground rules of the people you are trying to reach. This is an exercise in listening, not "public education." The voices you hear will not be united, but the range of views will help you to orient the nano program—and hopefully make it truly national.

26.4 Pay Attention to Production, Sales, and Service Jobs

Nano initiatives typically claim that they will produce jobs in the local economy. Since this is a goal that politicians always want, it is a good selling point. It is your job as a staff member to take this factor realistically into account in setting nano priorities. If you are concerned about reducing economic inequality in your country, you will want to focus on generating large numbers of mid-wage jobs, the ones that are widely accessible in terms of skills and bump up the middle of the income distribution. This is an approach that I call "equalizing" (Cozzens 2008).[5] It is the third key type of instrument in re-distributional technology policy, along with pro-poor and fairness approaches.

Some technologies that have emerged in the past have produced huge numbers of mid-wage jobs, for example, electronics. Others have produced very few, for example, health biotechnology. Electronics production was taken up in emerging Asia, using medium-skill labor. Pharmaceutical production has remained largely in the United States, Europe, and Japan, and uses high-skilled labor. When you are considering priorities in your program, think about how the technologies you are thinking of developing will be produced. Do you have the right skills available in the labor force to do that production? Should you think about skills development alongside technology development to keep the employment benefits of a new technology at home? Or are you actually supporting the development of job-less production processes—perhaps not quite what the country is expecting, nor what your political leaders have promised.

Whether or not your country produces a new product, if it reaches a mass market and is used in your country, it will generate jobs in sales and service. You will want to start thinking early in your program about having the right skills in your economy to have those jobs filled by locals. Because of global inequalities in nano capabilities, nano production and sales will happen through foreign direct investment (FDI) in many countries. If you are in one of those countries, you have a policy opportunity to set conditions for that investment that include local employment. There is considerable international experience in doing this (see Santos, and Wan 2009; Kinyondo and Mabugu 2009; Lee et al. 2009). What that experience shows is that if your country does not negotiate hard for lasting local benefits, including skills-building, the benefits of the new technology can be siphoned off.

The employment issue is not just about the number of jobs, however, but also about who gets them—the equality issue. Jobs still tend to be highly segregated by gender and ethnicity, and wages still tend to be paid differentially along these lines in most countries of the world. If you are stimulating industry that is going to be creating new jobs, what can you do now to make sure that they are open to everyone and pay equal wages for equal work and skills? Can you put regulations in place as conditions of your support? Can you set a positive tone and provide leadership for this new industry to set equity and equality in work as its own goals? Experience shows that these goals will not be reached automatically—they have to be articulated and implemented.

26.5 Set the Conditions for Widespread Commercialization

As you invest in nanoscience, companies from all over the world will be watching and looking for opportunities to turn the new knowledge your country produces into businesses. As a national policymaker, you want companies from your country to have the best chance of doing that. Innovation policy in your country probably encourages intellectual property expertise, and you have patent laws to give firms a temporary monopoly on profits from the new technologies they invent.

However, that goal may be in tension with the goal that led to your equity and equality assessment—making sure the benefits of new technologies get to everyone. In particular, some of the companies with the temporary monopoly may use the patent system to try to undermine competing businesses and gather profits from the technology well beyond the length of time the system allows, for instance, by filing interlocking patents on related processes that keep other businesses from producing the product even after its patent has expired.

What previous studies show is that the widest set of opportunities for new businesses will come when the basic technological knowledge is "open source"—that is, when anyone with the right skills can use the knowledge freely to set up a business. (The skills part of this is very important and keeps lots of people from being able to take advantage of the opportunities, and sometimes special, expensive infrastructure is needed.)

So if you are really interested in creating new companies and opening up opportunity broadly rather than just strengthening a few big companies, you will want to set the conditions for as much of your country's new nanotechnology as possible to be protected from monopoly use. There are a number of ways to do this under current forms of intellectual property protection, for example, "copy-left"-type agreements (which prevent knowledge hoarding); non-exclusive licensing; and humanitarian patent pools.[6] In addition, you need to work on establishing policies and practices that hold firms to the time limits on technology protection. This is a time of great inventiveness in intellectual property provisions, and you will want to be inventive within your nano initiative as well. You need to ask yourself: What parts of the knowledge your program will produce will be available to multiple users? How can you make sure it continues to be widely accessible?

This is not an area that is completely under the control of the science and technology ministry in all probability, and certainly not entirely under the control of your one national government as the patent rules are negotiated internationally now. Therefore, you will want to stay in dialogue with those who lead your national efforts in this international dialogue, to make them aware of the equity and equality implications of the positions they negotiate.

26.6 Spread Risks Evenly

You are very aware of the debate over the health and environmental risks of nanotechnology, because it is so prominent in nanotechnology discussions. While you work to limit these risks in general, in your equity and equality analysis, you will also need to think about whether those risks are evenly or unevenly distributed. Privileged communities tend to protect themselves fairly vigorously in debates over risk, making disadvantaged communities and groups even more vulnerable to risks shifting in their direction.[7]

Here is how a recent analysis (Cozzens and Wetmore, 2010) has summarized the issues of uneven risk:

- The issue of unequal distribution of health or environment risks for nanotechnologies is shrouded at present by the general uncertainty surrounding this topic. It is clear, however, that those engaged in the production of nano-materials will have more concentrated exposure.

In short, workers are at greater risk. Foladori and Zayago Lau (Chapter 11, this volume) point out that labor unions are seldom at the table in technology assessment processes; this observation extends to the foresight and priority-setting processes discussed earlier in this essay. The main antidote for this particular uneven risk then is awareness and participation. You may need to work with your labor ministry on this issue. Your goal is to meet international standards in your workplaces, even as those standards are emerging.

- Countries with strong regulatory regimes may also provide greater protection for workers than for consumers; but if production takes place in countries with weak regulatory regimes, those protections are less likely.

Countries of the South beware: you do not want to let yourselves be the guinea pigs in nanotechnology experiments. Get engaged in setting the conditions for FDI, including environmental and health standards. If expertise is short in your country for enforcement, band together regionally to get people with the right skills onto the job.

- Nano-materials used in other production processes, such as agriculture, may be less visible and therefore stimulate less regulatory scrutiny. And again, the protection against risk may be unequally distributed across countries.

Again, Foladori and Zayago Lau (Chapter 11, this volume) point to these dangers. Most of the regulatory discussion is happening in the North, and a lot of it is rooted in the chemical industry, where nanotechnology business is currently concentrated. Even within countries of the North, there are huge disparities in expertise and power in this arena between the corporate actors and workers. When one speaks of agricultural workers who are exposed to the products produced in one industry but being used in another, the disparities are even greater. New agricultural production products may look like the old ones, even if their effects are drastically different and the risks are much higher. And when we extend the arena from farm workers in affluent countries to those in less affluent countries, the situation is even more drastic. The opportunities for risk-shifting are very prominent, and the mechanisms for reporting and addressing emerging problems are weak. You need to be aware of these issues for your context, and work on addressing them in the design of your program.

- Finally, the risks of nano-materials in consumer products may be experienced first by affluent buyers who can afford the new products. But ultimately they will spread through public water, landfills, etc., to the whole population, whether or not they have received the benefits provided by the new products.

As expensive nano-socks are washed, their silver nano-particle treatment ends up in the general water supply (Benn and Westerhoff 2008). Toys treated with the silver nanoparticles are thrown into the local landfill, along with nano-enhanced tennis balls. The nanoparticles that made these luxury products special are much, much smaller than anything a local water system in an affluent country was designed to filter out; and poor countries may not even have filtration in their city water systems,

Fig. 26.2 Dharavi Slum Mumbai (image from www.travelblog.org)

let alone in the countryside. The result in the affluent country is that everyone pays the environmental and health price for the luxury goods. This is unfair enough, but in a poor country the poorest of the poor often either live on riverbanks or take their water supplies directly from rivers, which will increasingly become sources of undetectable, but probably dangerous, nanoparticles. So we had better get the general environmental hazard problem addressed before any of this happens.

The solution for you, here, is simple, straightforward, and probably something you were doing anyway: engaging in the international discussion on environmental and health hazards and regulation. Just be sure to give voice in those discussions to the general populations and specific groups who are particularly vulnerable.

26.7 Consider Who Is Paying the Costs of Regulation

Since regulations around nanotechnology risks are certainly coming, your EEA needs to pay attention to the fact that the costs of regulation itself can be unevenly distributed. Let me illustrate this point with a story from our recent studies. The story is set in Europe, where consumers and governments are very cautious about the risks of new technologies. For a long time, so many consumers were adamantly opposed to genetically-modified (GM) crops appearing in their food supply that European governments did not allow any to be grown in Europe. A few years ago, however, the European Commission approved growing GM corn, under a very special set of conditions.

As part of a larger study (Cozzens et al. 2008), we looked at the distributional consequences of GM corn in the Czech Republic, a small place where only 60 farmers had taken up the opportunity to plant the crop (Gatchair et al. 2009). There is a version of GM corn that is specially prepared to resist infestation by the European corn borer, so if the farm was in an area where that pest is prevalent, GM corn was an attractive option. However, the European regulations said that the farmer cannot grow it quite the same way he or she grows non-GM corn. In particular, the farmer has to leave large boundaries around GM fields and plant buffer zone crops to keep the GM corn from spreading.

All this costs money and productivity for the Czech farmers who want to prevent the corn-borer. Here is the equity question: Why should individual Czech farmers have to bear that cost, when the benefits of the environmental protection are widespread throughout the European population? This is another unevenness question. It really has to do with the design of the regulatory policy option: Europe subsidizes the growth of lots of crops, so why not help out this Czech farmer, taking the cost of environmental protection as a public cost rather than passing it on into his private farm budget?

Here is the lesson for you. You are clearly going to be thinking about regulation around the nanotechnologies that you will be encouraging with your initiative. And your government will be designing general regulations that will cover imported as well as domestically produced nanotechnologies. What costs are generated by those regulations? Are those costs disproportionately paid by groups that you are trying to help, not hinder? Small businesses come to mind. If yes, then how can you design the law or the resources that come with it in ways that even out the costs or impose them on the parties that get the benefit?

These are complicated questions, but there is a literature in economics that deals with them in part (see Littrell 1997; Faunce 2007). Precisely because the regulatory issues around nanotechnology loom so large, these questions are a necessary element of the equity and equality assessment.

26.8 Consider Global Relationships

Your EEA also needs to consider relationships your initiative establishes between research institutions, companies, and consumers in other parts of the world. Wait, is it unfair at this stage to ask about global relationships here? This essay started with the assumption that you were working towards certain goals for *your* country, wanting to make sure that everyone benefited and that the initiative enhanced social cohesion *within your borders*. Where does this global stuff come in?

The fact is, however, that in what the essay has already discussed, your country has appeared on one side or the other of several global relationships. In your better life analysis, you are considering marketing opportunities beyond the borders of your own country. In your planning for production, sales, and service jobs, you have considered your relationship with multi-national firms and the conditions you can set for foreign direct investment. In your plans for commercialization, you have recognized that your country is part of a set of international agreements and that you need to work on those at the same time that you are working with national provisions for open source protection. And finally, in the risk and regulation area, your national workers, households, firms, and regulators are clearly part of an international arena that both constrains and enables what you do.

If you are a staff member from a developing country, then, Point 8 merely highlights the consolidation of all these observations: global nanotechnology development offers both opportunities and dangers for your country, both economically and environmentally. The most shocking stories about distributional consequences of emerging technologies that appear in our research feature multi-national companies based in the North creating problems in the South. One route is vertical integration that sets up a flow of profits to the North by buying up firms in the South (Gatchair et al. 2009).[4] Another is by preventing competition in the South in ways that also restrict the distribution of benefits from a technologically advanced product. Eli Lilly's battles with Argentinian firms overproduction of recombinant insulin serve as an example (Bortagaray 2008). Policies in the South that build up local capability are constantly endangered by appropriation of those capabilities through vertical integration with the North, as well as through brain drain. In this set of problems, there are no easy solutions, but awareness is the first step in prevention.

If you are a staff member in an affluent country, of course, firms based in your country are the bad guys in these stories. Have your policies encouraged them in this? There is a good chance that the answer is yes.

What I recommend for countries of the North is including global economic growth in your equity and equality assessment. Most of the space that exists in the world economy for growth is in the South, bringing those large populations living in poverty up to a higher standard of living. Without that growth, economies of the North will stagnate. Otherwise Northern countries will just be competing with each other for the same pool of economic resources. We all have a stake in higher growth rates in poorer countries. This is hard economic reality, as well as an expression of commitment to basic human rights.

The policy makers you work for probably know this intellectually but have trouble selling it politically. Your equity and equality assessment can help them with the political problem by identifying specific global opportunities for your national firms that build economic capacity in Southern partners while they maintain economic strength at home. Pure competition thinking will not get you there, but abstract arguments are not an effective antidote. Look for the concrete opportunities and report them.

26.9 Experiment with Re-Distributional Policy and Program Designs

Along the way in the early points of this essay, some options appeared for designing policies and programs that increase equity and equality. As a group, I call these "re-distributional." They start from an existing unequal distribution of something people value—such as income, health, or economic opportunity—and change the distribution by doing something new that is more fair and creates more equal outcomes than what came before it. This concept works very well in social democracies that already do a lot of income re-distribution, but it also works in more strongly market economies that do less of that traditional "social policy." Some scholars call these "social innovation" approaches.

What I have proposed is that there are three main types of social innovation policies and programs (Cozzens 2008):

- *Pro-poor* approaches focus on people at the low end of the income distribution. They either make life better for poor households or focus on lifting them out of poverty. The technologies you identified in your "better life" analysis fall into this category. Pro-poor approaches in innovation policy are often visible in developing countries, and less so in affluent ones.
- *Fairness* approaches focus on reducing and eventually eliminating inequalities across horizontal groups in how much they have of what they value. In a fair society, for example, high-technology jobs would be taken up by people from different ethnic groups in proportion to their numbers and educational levels—something that does not happen in the United States now (see Chapter 3 this volume). Fairness approaches in science and innovation policy are very visible in affluent countries, but also appear in middle-income countries.

- *Equalizing* approaches focus on creating middle-income jobs and changing the structure of the economy in ways that reduce income inequality. When you work on local jobs as conditions for foreign direct investment, you are using this approach

These three approaches, used throughout the nine points above, form a useful starting point for you as you work to incorporate the results of your equity and equality assessment into the design of your national nanotechnology initiative. But they surely do not exhaust the possibilities. I urge you, then, to be creative in your approaches, add to this list, and share your results widely. The world has a lot still to learn in this area of science, technology, and innovation policy

As I noted at the outset, building equity and equality into your nanotechnology initiative will not be easy, and it will not create dramatic changes in the short run. But the small steps you take now serve to point the arrow in a positive direction. In the long run, then, this will be a centrally important part of your work on nanotechnology development.

Acknowledgment This essay reflects lessons learned in research under several grants from the National Science Foundation, in particular 0354362 and 0726919. The latter was collaborative with Work Package Four of a larger project supported by the European Commission, ResIST (http://www.resist-research.net/home.aspx). Cozzens gratefully acknowledges the contributions of a large set of collaborators to these projects. Any opinions, findings, conclusions, or recommendations expressed here are those of the author and do not necessarily reflect the views of any agency that funded the work.

Notes

1. We work here with the contractarian approach outlined by John Rawls in Justice as Fairness (Rawls 1999).
2. For a discussion of distributive justice in science and technology policy, see Cozzens (2007).
3. See Cozzens et al. (2005) for a description of this technology-based development strategy.
4. See http://www.csir.co.za/enews/2009_feb/msm_05.html, accessed December 13, 2009; http://www.scidev.net/en/news/tb-drug-delivery-gets-nanotech-boost.html, accessed December 13, 2009; http://www.aggregateresearch.com/article.aspx?ID=6279&archive=1, accessed December 13, 2009.
5. The 2008 chapter uses the term *egalitarian* for this policy option, but I have switched to *equalizing* to be more descriptive.
6. For example, see http://www.doctorswithoutborders.org/press/release.cfm?id=3970&cat=press-release , accessed December 13, 2009; http://lists.essential.org/pipermail/ip-health/2005-May/007932.html, accessed December 13, 2009.
7. See the considerable literature on environmental racism, at both national scale within the U.S. and global scale, for example, Bullard (2005).

References

Barben, Daniel et al. 2008. Anticipatory governance of nanotechnology: Foresight, engagement, and integration. In *The handbook of science and technology studies, 3rd ed.,* eds. Edward J. Hackett, Olga Amsterdamska, Michael Lynch, and Judy Wajcman. Cambridge, MA: MIT Press.

Benn, Troy, and Paul Westerhoff. 2008. Nanoparticle silver released into water from commercially available sock fabrics. *Environmental Science and Technology* 42: 4133–4139.
Bortagaray, Isabel. 2008. Recombinant insulin case study in Argentina. Working paper for Project Resultar.
Bullard, Robert D. 2005. *The quest for environmental justice: Human rights and the politics of pollution*. San Francisco, CA: Sierra Club Books.
Cozzens, Susan E. 2007. Distributive justice in science and technology policy. *Science and Public Policy* 34: 85–94.
Cozzens, Susan E. 2008. Equality as an issue in designing science, technology, and innovation policies and programs. In *Confluence. interdisciplinary communications 2007/2008*, ed. Willy Østreng. Oslo: Centre for Advanced Study at the Norwegian Academy of Science and Letters. http://smartech.gatech.edu/handle/1853/24604. (accessed August 4, 2010).
Cozzens, Susan E., Kamau Bobb, Kendall Deas, Sonia Gatchair, Albert George, and Gonzalo Ordóñez. 2005. Distributional effects of science and technology-based economic development strategies at state level in the United States. *Science and Public Policy* 32: 29–38.
Cozzens, Susan E., Isabel Bortagaray, Sonia Gatchair, and Dhanaraj Thakur. 2008. Emerging technologies and social cohesion: Policy options from a comparative study. Paper presented at the PRIME Latin America Conference, September 24–26, 2008. http://prime_mexico2008.xoc.uam.mx/papers/Susan_Cozzens_Emerging_Technologies_a_social_Cohesion.pdf.
Cozzens, Susan E. and Jameson M. Wetmore. 2010. Equity. In *Encyclopedia of nanotechnology in society*. ed. David Guston. Thousand Oaks, CA: Sage.
European Commission. 2005. *Deliberating foresight: Knowledge for policy and foresight knowledge assessment*. Brussels: Author. EUR21957.
Faunce, Thomas A. 2007. Nanotherapeutics: New challenges for safety and cost-effectiveness regulation in Australia. *Medical Journal of Australia* 186: 189–191.
Foladori, Guillermo, and Edgar Zayago Lau. 2010. The role of organized workers in the regulation of nanotechnologies. In *Yearbook of nanotechnololgy in society vol.2: The challenges of equity, equality, and development*. ed. Susan E. Cozzens and Jameson Wetmore. Dordrecht: Springer.
Gatchair, Sonia. 2010. Potential implications for equity in the nanotechnology workforce in the U.S. In *Yearbook of nanotechnololgy in society vol.2: The challenges of equity, equality, and development*. ed. Susan E. Cozzens and Jameson Wetmore. Dordrecht: Springer.
Gatchair, Sonia, Isabel Bortagaray, and Lisa Pace. 2009. Genetically modified corn. Working paper for Project Resultar.
Guston, David, and Daniel Sarewitz. 2002. Real-time technology assessment. *Technology in Society* 24: 93–109.
Kinyondo, G., and M. Mabugu. 2009. The general equilibrium effects of a productivity increase on the economy and gender in South Africa. *South African Journal of Economic and Management Sciences* 12: 307–326.
Lee H.Y., K.S. Lin, and H.C. Tsui. 2009. Home country effects of foreign direct investment: From a small economy to a large economy. *Economic Modeling* 26: 1121–1128.
Littrell, Earl, and Thompson, Fred. 1997. The cost of regulation. *Interfaces* 27: 22–28.
Meridian Institute. 2006. *Workshop on nanotechnology, water, and development: Workshop summary*. Denver, CO: Meridian Institute.
Rawls, John. 1999. *A theory of justice*, revised ed. Cambridge, MA: Harvard University Press.
Santos-Paulino, Amelia U., and Guanghua H. Wan. 2009. Special section: FDI, employment, and growth in China and India. *Review Of Development Economics* 13: 737–739.

Index

A
Ability status, 434
Ableism, 89–103, 216, 219, 227
Academics, 8, 112, 188, 386, 396
Access
　easy, 373
　lacked, 279
　limited, 83, 286
　minority, 71, 75, 82
　open, 331–345
　security, 312
Accountability, 11, 157, 323
Advanced materials, 213
Advantages and disadvantage, 53, 170, 185, 187–188, 211, 224, 235, 294, 320–321, 323, 325, 336, 338, 343, 359, 373, 399
Aerogel, 340, 342
Affected non-users, 222
Affluent nations, 362
African Agricultural Technology Foundation (AATF), 262–263
African Americans, 4
Ageism, 93
Agenda setting, 315–316, 323, 325, 425, 430
Agricultural production, 440
Agricultural workers, 114, 440
Agriculture/agricultural applications, 6, 36, 279
Alexander (the Great), 399–401, 403, 405
Ambient intelligence, 411
Anticipatory governance, 146–147, 156–159, 161, 265, 437
Anti-environmentalism, 93
APEC Centre for Technology Foresight, 116
Applied research, 139, 213, 293, 318, 326, 369
Appropriate technology, 212–215, 226–227, 351–352
Argentina, 184, 296, 317

Arizona State University (ASU), 236, 395, 418, 437
Arnstein's ladder of citizen participation, 118
Asia/Asian/Asians, 42, 49, 52–53, 56, 58–59, 62–65, 167, 226, 260, 280, 351–352, 367, 438
Association for India's Development (AID), 404
Atomic Force Microscope (AFM), 138, 177
Atomic self-assembly, 364
Australia, 110, 115–118, 190, 263, 349–375, 400

B
Bangalore, 404
Benefits/beneficial, 6, 13, 15, 66, 73, 78, 81–83, 85, 98, 101, 112, 118–120, 122, 141, 166, 178, 183–185, 187, 192–193, 201–202, 210, 212–216, 219–220, 222, 224–226, 235, 240, 244–246, 252–253, 261, 265, 280–281, 283, 285, 287, 295, 299–300, 304, 315, 322, 324–325, 332–333, 336, 339–342, 345, 357, 365, 372, 382–384, 389, 390, 397, 400, 426, 430, 434–435, 438–439, 441–443
Berkeley Software Distribution (BSD), 337–338
Biology, 17, 24, 30, 36–37, 42, 90–91, 320–321, 326–327, 343, 371, 411, 416, 426
Biosafety, 256–258, 261, 264, 266–267
Biotechnology, 3–19, 27, 90–91, 137–139, 141, 168–169, 232, 235, 251–267, 294, 321, 326, 336, 351, 353–354, 365, 368, 370–371, 410–411, 438
BITS Pilani, 398
Blacks, 51–54, 62–65, 219
Blue-collar jobs, 148, 160
Boston, 166, 168, 172, 174–176

Brain-drain, 299, 361
Brazil, 117, 184, 278, 292–295, 309–327, 332, 334, 341
Brazilian Center for Physics Research (CBPF), 295
Brazilian Ministry of Science and Technology (MCT), 292–296, 299–303, 305–306, 318, 322, 324–325
Brazilian Nano Network, 301
Brazilian Synchrotron Light Laboratory (LNLS), 295, 297
Brazilian Technical Regulations Association, 301, 303
Brussels, Belgium, 200
The Buddha, 399
Burden, 112, 210, 213–214, 217, 220, 222, 224, 225–226, 300, 342, 353, 436
Business opportunities, 129, 166, 437

C

Caltech, 137, 398, 405
Cancer, 4, 14, 70–79, 81–85, 182, 219, 417
Capitalist class, 185–187
Carbon nanotubes, 370
Career development, 25, 28
Caste-ism, 93
Centre for Society and Genomics (the Netherlands), 410
21st Century Nanotechnology Research and Development Act, 232
Cesar Lattes Center of Nanoscience and Nanotechnology, 295
Chemistry, 27, 30, 36–37, 42, 49, 130, 133, 280, 321, 361–362, 371
Chemotherapy, 74
China, 114, 140, 167, 178, 188, 311, 322, 334, 358, 366–367, 375
Chumby HDK License, 338
Citizen participation, 118, 233–235, 417
Civil society, 73, 116, 191, 198, 286, 416, 420, 437
Civil Society Organizations (CSO), 416–420
Classification, 31–34, 42–43, 91, 155, 162, 168, 183–186, 188, 190, 322
Classism, 242
Clean water, 97, 279–280, 382–384, 390, 404
Climate change, 71, 93, 222, 273, 380
Co-authorship, 25, 174–175, 178
Cognitive science, 90–91, 139, 232, 235, 410
Collaboration, 24, 27, 29, 31, 35–36, 42, 53, 71, 76, 78, 141, 225, 262, 284–285, 296, 336, 342, 345, 388, 405, 418, 428–429

Colonialism, 10, 402
Colorado School of Mines, 236
Commercialization, 13, 15, 48, 51, 84, 119, 121, 166–168, 170, 172, 176, 178, 183, 187, 189–190, 192–193, 210, 255, 284, 313, 322–324, 429, 439, 443
Commodities, 80, 114, 279, 286, 320
Community organizations, 257
Competitiveness, 53, 57, 77, 90, 94, 96, 101–102, 112, 121, 146, 167, 184–185, 286, 293–294, 299–300, 303–304, 311, 354
Computer science, 4, 18, 24, 52
Concentration, 65–66, 110, 120–121, 165–178, 292, 296, 304, 312, 320, 325, 334–336, 356, 365, 374
Conflict, 6, 12, 15, 32, 122, 221, 284, 375, 398–399
Consensus conference, 232, 236, 245–246, 412, 417
Consortium for Science, Policy and Outcomes (CSPO), 70
Constructive Technology Assessment (CTA), 233
Consultative Group for International Agricultural Research (CGIAR), 261–263
Consumerism, 93, 96, 399, 402
Consumer(s), 78, 301, 429
Context of development, 216
Context of use, 214–216
Contingency planning, 157–158
Converging technologies, 91, 101, 235, 411
"Copy-left"-type agreements, 439
Cornell University, 175
Corporate monopolies, 364
Creative Commons Attribution License, 340, 346
Creative destruction, 129, 131, 139–140, 151
Critical mass, 7, 78, 361, 368–369
Current Population Survey, 55
Customers, 112, 185, 217, 338, 385

D

Danish Board of Technology, 412
Da Silva, Lula, 293, 304
Data-mining techniques, 33, 39
Deaf culture, 93, 95
Decision making, 111–112, 115–119, 121, 159, 221–222, 224–225, 233, 235, 241–243, 245, 247, 252–254, 256, 259, 265, 267, 288, 312–313, 354, 418, 427
Democratic deliberation, 210

Democratization, 183, 302
Demos, 110, 382, 384
Denmark, 119, 412
Department of Health and Human Services Office of Minority Health (OMH), 76
Desalination, 280, 358
Design/designers, 10–11, 19, 52, 82, 134, 147, 158, 177, 182, 188–189, 204, 209–227
Deunionization, 147–148
Developing countries, 16, 148, 167, 184, 214, 227, 252–253, 266, 277–289, 300, 303, 310–312, 315–316, 325, 331–346, 357–368, 370–371, 374, 379–391, 443–444
Development
  career, 25, 28
  context of, 216
  drug, 368, 370, 372
  economic, 77, 167, 170, 178, 200, 309–327, 375, 389, 416
  organizations, 136
  sustainable, 189, 202, 281, 380
Diagnostic kits, 120, 357, 376
Diffusion, 121, 130–132, 140–141, 148, 150–151, 156, 176–178, 215, 282, 324, 332, 334, 341, 357, 381, 387–388, 397, 424, 427
Digital Millennium Copyright Act (DMCA), 343
Disability/disabled, 10, 16, 92–97, 99–102, 110, 115, 212, 216–217, 227, 234, 287, 424, 437
Disruption/disruptive, 116, 152, 158, 183, 215, 278, 282
Distributive justice, 90, 254, 263, 265, 267, 414, 434, 445
Diversity, 19, 29, 31, 48–49, 52–53, 62–66, 80, 174, 176–177, 216, 220, 235, 242–243, 245, 247, 344–345, 353, 416, 424, 427
Domination, 114, 364, 381
Donor(s), 256, 260–261, 266–267, 358
Dostoevsky, Fyodor, 403
Downs Syndrome, 241
Drug(s), 74, 81, 96, 154–155, 177, 182, 282, 368, 428
  delivery, 74, 162, 278, 282, 333, 372–373, 410, 445
  development, 368, 370, 372
DuPont, 169, 187, 189, 428

E
Ecological damage, 389
Economic crisis, 146, 184, 187
Economic development, 77, 167, 170, 178, 200, 309–327, 375, 389, 416
Economic policy, 146
Economics, 146, 156, 161–162, 211, 382, 402, 435, 443
Education
  engineering, 51–53, 398
  graduate, 49, 55, 57, 62, 64–65
  public, 284, 437
  STEM, 4–5, 7–8, 16–19, 52–53
  technical, 393–405
  value, 396–403, 405
Efficiency, 85, 94, 102, 120, 153, 155, 160, 166, 341, 381, 404
Electronics, 27–28, 36, 48, 113, 128, 137, 153, 155, 160, 169, 175, 194, 223, 294–295, 305, 333, 351, 411, 438
Embrapa, 295
Emerging technology, 11–17, 72, 76, 85, 140, 142, 166, 255, 299, 311, 322, 353, 356, 367, 390
Emerson, Ralph Waldo, 403
Endogenous growth models, 146, 149, 161
Energy, 8, 27, 113–114, 116, 120, 128, 130, 134–137, 153, 156, 160, 187, 200, 252, 278–279, 281–282, 289, 294, 304, 311, 313, 321, 323, 326, 333, 365, 384, 387, 390, 395, 427–428, 436
Engagement, 26–27, 76, 110, 112, 117–118, 158, 284–285, 287–288, 313, 315, 324, 350, 353–354, 356–358, 364, 366, 368–369, 374, 381–382, 387–388, 390–391, 414–419
Engineering, 4–5, 9, 16, 18–19, 24–30, 36–37, 41, 48–49, 51–52, 56–61, 63, 66, 72, 90, 113, 133–134, 153, 162, 189–190, 232, 301, 310, 321, 342–343, 351–352, 371, 383–384, 388, 396, 398, 403, 405, 415, 418, 434
Engineering education, 51–53, 398
Engineers, 4–5, 8, 17, 24–25, 48–49, 51–52, 57, 62, 73, 177, 243, 418, 437
Enhancement, 73, 94, 97, 110, 115, 154, 166, 183, 232, 235–237, 242, 244–246, 284, 287, 294, 299, 323, 390, 411, 424
Enlightenment, 402
Entrepreneurship, 7, 134
Entry costs, 372
Environmental Defense Fund (EDF), 187, 189, 428

Environmental health and safety (EHS), 28, 425
Environmental protection, 77, 119, 191, 221, 429, 442
Environmental Protection Agency (EPA), 77, 119
Equalizing, 209–228, 231–248, 251–267, 296, 438, 445
Equinet, 273
ETC Group, 110, 113–115, 117, 187, 189–190, 232, 283, 354, 411, 416
Ethical Legal and Social Implications (ELSI), 232, 243, 354, 409–420
Ethics, 72, 89–103, 243, 284–287, 350, 352, 381–382, 402, 423–430
Ethnicity, 11, 32, 48–49, 54, 56, 63, 66, 78, 85, 434, 438
Europe, 116, 134, 167, 178, 200, 257–258, 319, 341, 361, 410–412, 415, 417, 437–438, 442
European Chemicals Agency, 203
European Commission, 200–202, 205, 417, 420, 437, 442, 445
European Framework Program, 410
European Union, 110, 184, 186, 200, 311, 335
Evolutionary economics/economists, 63, 65, 129, 139, 146, 159, 161
Exclusionary mechanisms, 217
Expert(s)/expertise, 19, 27, 41–42, 53, 117, 191, 198, 214, 224–225, 233, 236, 243, 246, 258, 261–262, 272, 279, 293, 301, 311, 338, 342–343, 350, 361–362, 375, 384, 390, 404, 412–413, 415, 418–419, 429, 439–440
Exposure to risks, 110, 115, 204, 226, 244, 284, 324
Extended-release vaccines, 282

F
Faculty, 7–9, 27, 78, 351–352
Fairness, 3–19, 23–43, 47–67, 69–85, 89–103, 105–106, 231–248, 312, 412, 414, 427, 434–435, 438, 444–445
Family context/family status, 8
Federal University of Pernambuco, 296
Female, 8, 15, 24–27, 29, 32–43, 52, 56, 98, 148, 238, 415
Feminism, 5, 10–11, 18
Feminist design, 216–218
Feynman, Richard, 17, 137–138
Financial capital, 132, 134, 136, 185–186
Financial system, 132, 134
Financiers, 136, 140
Food, 16, 95, 97, 112–117, 155, 190–192, 194, 198, 201, 226, 255–256, 260, 272–273, 279, 281–282, 289, 294, 304, 321, 326, 333, 354, 357, 404, 414, 416, 424, 427–429, 442
Forecasting, 119, 157
Foresight, 90, 100–102, 116, 158, 161, 351–352, 437, 440
France, 114, 140, 178, 192, 296, 356, 412
Friends of the Earth Australia (FoEA), 112–113, 115, 118
Fuels, 321, 326, 358, 437
Fundação Osvaldo Cruz, 295
Fundamental research, 369
Funding
 organizations, 7, 15, 27, 76, 121, 167, 200, 318, 372
 priorities, 71, 223, 241
Futures, 158, 210, 223, 225

G
Gandhi, Mahatma, 397, 401–403
90/10 Gap, 300
Gates Foundation, 261, 357
GDP, 122, 140, 148, 310, 317, 319, 375, 380
GDP-ism, 93
Gender, 3–19, 24–26, 29–43, 49, 64, 98, 215–220, 233, 236–239, 246, 254, 258, 434, 437–438
General Public License (GPL), 337–338, 345–346
General Purpose Technology (GPT), 130, 136, 141, 147, 150, 152–153, 169, 311
Genetically Modified Organism (GMO), 273, 424
Genetic modification (GM), 253
Geo-engineering, 90
Georgia Institute of Technology, 236, 326
Germany, 114, 140, 188, 322, 356, 412
Gini coefficient, 172, 174, 316, 318, 435
Globalization, 129, 170, 220
Global North, 19, 121, 214
Global South, 16
Golf, 105
Governance, 15, 89–103, 110–112, 118–119, 132, 146–147, 156–161, 254, 256, 259, 265–267, 278, 284–285, 288–289, 313, 323, 325–326, 390, 397, 410, 437
Governance systems, 132
Government laboratories, 26, 168, 174, 176–177
GPT adoption cycle, 150–151, 161

Greenpeace, 116
Green Revolution, 254, 260–264, 267, 394
Grocery Manufacturers Association, 428
Growth theory, 146–147, 159, 161

H

Harvard University, 403
"Haves" and "have nots", 240, 246
Health
    care, 80–84, 110, 116, 137, 241, 243, 279, 313, 395, 415
    impacts, 193, 413
Hegemony, 364
Hierarchical organizations, 6, 10, 30
High-skilled jobs, 139–140, 438
High technology industries, 48, 54–55, 57, 62–64
Hind Swaraj, 402
Hindu, the, 93
Hispanic, 49–54, 62–65
HIV/AIDS, 333, 353
Homo sapiens, 93
Hong Kong, 356
Horizontal inequality, 54
Human enhancement, 110, 115, 232, 235–237, 242, 284, 287, 411, 424
Human Genome Project, 17, 410, 415
Humanitarian needs, 288
Humanitarian patent pools, 439
Human nature, 246, 402
Human resources, 284, 292, 295–296, 302, 332–334, 342, 345, 358, 361–362, 369–370, 375, 428
Human rights, 16, 100, 444
Human subjects research, 84
Human Use of Human Beings: Cybernetics and Society, The, 403

I

Income, 56, 80, 110, 139, 146–148, 153, 177, 187, 233, 236–239, 246, 280, 292, 300, 304, 312, 316, 318, 334, 341, 345, 353, 380, 434–435, 438, 444–445
Income growth, 147
India, 93, 140, 278–279, 296, 332, 334, 341, 367, 394, 396–402, 405, 436
Indian Institute of Technology Kanpur (IITK), 396, 404–405
Indicators, 4, 26, 70, 171, 176, 316, 380
Indigenous people, 96
Indonesia, 367
Industrial design, 212
Industrial revolution, 128–129, 133–134, 183, 200, 402–403
Industry, 5, 7, 11–12, 14–15, 24, 26, 31–32, 39–40, 42, 49–50, 52–53, 55–57, 62–66, 110, 111–113, 116–119, 121–122, 128, 134–135, 137, 142, 146, 148, 152, 169, 182–183, 185, 193–194, 198, 200, 202–204, 217, 224, 240, 293–294, 299, 311, 313, 318–319, 322–324, 332–333, 337–338, 340–342, 344–345, 350, 364–365, 368, 370, 373, 389, 404, 411, 414, 418, 426, 428, 438–440
Industry voluntary initiatives, 204
Inflation, 146, 318, 343
Information and communication technologies (ICT)/information technology, 52, 91, 128, 139, 150, 169, 171, 200, 220, 232, 235, 242, 295, 311, 341, 410
Informed consent, 424
Infrastructure, 78, 116, 128, 130, 133–135, 214, 264, 287, 292–293, 295–296, 305, 310, 333, 342–343, 345, 353, 356, 358, 361–362, 365, 370–371, 375, 381, 394, 396–397, 439
Inmetro Nanometrology Center, 295
Innovation
    management, 24
    policy, 70–71, 73, 110, 118, 146, 210, 390, 439, 444–445
Institutional arrangements, 132, 136, 141, 232, 312–313
Instrumentation, 138, 168, 175, 319, 362–363, 370
Insurance, 82–85, 184, 186–187, 190, 237–238, 240, 242, 244, 389
Intellectual property (IP), 26, 31, 72, 114, 118, 121–122, 169, 211, 256, 261–263, 284–285, 287, 304, 332–336, 338–339, 342–344, 351, 354, 363–364, 374, 411, 428, 439
Intellectual property rights (IPR), 256, 334–336, 341, 343–344
Interdisciplinarity/interdisciplinary, 11–12, 14–15, 24, 27, 29–30, 35, 41–43, 51, 63, 72–73, 138, 141, 388, 426
International Council on Nanotechnology (ICON), 190, 428
International Organization for Standardization (ISO), 91, 185, 190
International Service for the Acquisition of Agri-biotech Applications (ISAAA), 255–258, 262–263, 266–267

Inventors
  female, 24, 30, 33–35, 37, 40–42
  male, 33, 35–37, 42

J
Japan, 31, 114, 140, 167, 178, 188, 296, 311, 319, 322, 335, 356, 360–361, 364–365, 372, 411, 438
JITM, 404
Jobs
  blue-collar, 148, 160
  high-skilled, 139–140, 438
  low-skilled, 140
  white-collar, 148, 160
Journals, 29, 388, 418
Joy, Bill, 411
Justice, 12, 49, 90, 234–235, 240, 254, 259, 263, 265, 267, 401–403, 405, 414, 416, 418–419, 424, 426, 434, 445

K
Kabir, 403
Kenya, 251–267
Keynesian economics, 146, 161
Killer app, 428
King, Martin Luther, 403

L
Labeling standards, 183
Labor
  skilled labor, 148–152, 160–161, 438
  unskilled labor, 54, 148, 151
Laboratories, 8, 26, 36, 53, 70, 128, 132, 135, 137–139, 168, 174, 176–177, 183, 186, 194, 201, 210, 260–261, 281, 285, 293, 295, 297, 324, 326, 336, 369, 405, 428
Latin America/Latin American, 167, 178, 191, 280, 293–295, 311, 316–318, 325
Latino/a Americans, 4, 219
Leapfrog, 371
Legal control/legal issues, 183
Legitimacy, 156
Libertarian, 263
Life sciences, 4, 7, 17–18, 24–27, 37, 42
Lifestyles, 9, 381
Linux, 337, 339, 341, 387
Lisbon Strategy, 202
Livelihoods, 112, 257, 279, 381, 399
Lobbyist, 243
Local economies, 214
Loka Institute, 115–116, 118
Low-skilled jobs, 140
Luddites, 220

M
Malaysia, 367, 372
Male, 7–9, 14, 24–26, 30, 32–42, 55–56, 58, 98, 217
Managers, 135, 140, 217, 221, 255, 263, 267, 284, 324, 336, 352, 382, 385, 413
Manufacturers/manufacturing, 55, 57, 65, 90, 113, 115–116, 120, 134, 138, 154, 169, 182–186, 188–189, 191–192, 200–204, 213, 221, 252, 278, 289, 313, 324, 335–336, 338, 342, 358–359, 376, 387, 389, 428, 430
Manufacturing Engineering Center at Cardiff University (UK), 342
Market
  commodity, 286
  labor markets, 48, 50, 54–65, 152, 177, 300
Marxist, 5, 9–10, 14–15
Massachusetts Institute of Technology (MIT), 4, 403
Material embodiment, 210
Materiality, 154, 210
Materialized inequity, 210–211, 216, 219, 222–225
Materials science, 27, 168, 321
Medical care technology, 219
Medical devices, 27, 36, 138, 169, 226
Medicine, 27, 30, 36, 77–79, 82, 84, 113, 154, 200, 224, 272–273, 282, 287, 311, 321, 326–327, 333, 352
Membrane technology, 280
Men, 4, 8–11, 14–18, 24–25, 27–31, 33–34, 36, 42, 98, 217, 237, 415, 435, 437
Metropolitan/metropolitan areas, 166, 168, 170–171, 176, 178
Mexico, 184, 296, 317
Microelectronics, 113, 223, 295, 351
Military, 10, 112, 116, 121, 130, 136, 153, 224, 241, 244, 284, 304, 362, 367
Millennium Development Goals (MDGs), 278
Mill, John Stuart, 400
Minority, 10, 27, 29, 63, 71, 75–84, 146, 176, 242, 245
Modernization/modernization, 286
Molecular assembly, 370
Molecular manufacturing, 90, 213
Mozambique, 342
Multicultural, 5, 9–10
Multidisciplinarity, 168
Multinational firms, 169, 287

Index

N
Nanodistricts, 170–177
Nano-divide, 350, 355–356, 374, 384, 414
Nanofiltration, 280
Nanomachines, 411
Nano-materials, 50, 138, 141–142, 154, 162, 334, 440–441
Nanomedicine, 69–85
Nano-particles, 154, 413, 441
Nanotechnology, Biotechnology, Information Technology and Cognitive Science (NBIC):Converging Technologies for Improving Human Performance workshop, 91
Nanotechnology Capacity Building NGOs (NANOCAP), 417
Nanotechnology Consumer Products Inventory, 429
Nanotechnology initiative, 5, 50, 66, 76, 121, 141, 167, 182, 184, 213, 296, 354, 389, 410, 418, 434–435, 445
National Academies (US), 428
National Cancer Institute (NCI) (US), 70–71, 77, 84
National Citizens Technology Forum (NCTF), 232–233, 235–236, 239, 241, 245–247
National Council for Scientific and Technological Development (CNPq) (Brazil), 293
National Institutes of Health (NIH) (US), 7, 17, 79–80, 82–84
National Institutes of Health Revitalization Act (USA), 84
National Institute for Space Research (INPE) (Brazil), 295
National Nanotechnology Infrastructure Network (NNIN), 342, 396, 404–405
National Nanotechnology Initiative (NNI) (USA), 5, 50, 66, 76, 121, 141, 167, 182, 184, 213, 354, 389, 410, 418, 434, 445
National Nanotechnology Program (Brazil), 232, 278, 293–294, 300–301, 303
National Science Foundation (NSF) (US), 24–25, 37, 48, 50, 70, 91, 264, 405, 411
National Science and Technology Council (NSTC), USA, 48–50, 72, 76, 116, 232
National systems of innovation, 310
Native Americans, 241
Neoclassical paradigm, 146
Network, 6, 10, 13, 15, 27, 30, 50, 53, 63, 65, 78, 112, 117, 128, 134–135, 152, 158, 175, 186, 191, 257, 263–264, 267, 293–298, 301–303, 305–306, 318–319, 322, 324, 326, 332, 336, 339, 342, 346, 396, 405, 411, 413, 418–419, 430
   organizations, 6, 15
Neuro-engineering, 90
Neuroscience, 426
New growth theory, 146–147, 159, 161
New York, 168–169, 172, 174–176
New Zealand, 400
NGOs, 110, 115–118, 122, 184, 187–188, 190–194, 244, 256–258, 262–264, 267, 278, 382, 411, 416–417
Non-exclusive licensing, 439
Non-governmental organizations, 110, 182, 187, 284, 325–326, 411, 420, 426
Normative, 5, 92–93, 98–100, 156, 265, 312, 316
Novartis, 263

O
Observer principle, 115
OECD (Organization for Economic Cooperation and Development), 116–117, 137, 140, 148, 184
Office of Technology Assessment (OTA) (US), 412
Open Invention Network (OIN), 339
Open source, 337, 344, 439, 443
Open Source Software (OSS), 331–346
Opinion poll, 76
Organizations/organizational characteristics, 43
Orissa, India, 396

P
Paint(s), 36–37, 169, 182, 201, 223, 282, 362
Paralakhemundi, India, 396, 398, 404–405
Parliamentary technology assessment, 409–420
Participation
   citizen participation, 118, 233–235, 417
   public participation, 111, 191, 232–235, 246, 284–285, 287–288, 301, 415–418
Participatory design, 212, 220–222
Participatory policy analysis, 233
Partido dos Trabalhadores, 292, 304
Patents/patenting
   rules, 439
   thicket, 339
People of color, 10
Performance enhancement, 94, 166
Persons with disabilities, 10, 95, 100
Pervasive computing, 411
Petrobrás, 295

Pharmaceutical, 6, 120, 162, 213, 223, 305, 359, 368, 438
Philippines, 367
Photovoltaic(s), 281–282
Physical capital, 154
Physicians, 80, 82–83, 204
Physics, 8, 17, 24, 27, 30, 36, 41, 49, 115, 137, 295, 321, 361
Pluri Annual Plan, 292, 302
Policy(ies)
  economic, 146
  innovation, 70–71, 73, 110, 118, 146, 210, 390, 439, 444–445
  participatory, analysis, 233
  science and innovation, 70–71, 73, 390, 444
  Science and Innovation Policy (SIP), 70–71, 73, 390, 444
  sciences, 233
  science and technology (S&T), 233, 310
  science, technology, and innovation, 445
Political action, 341
Poor countries, 10, 216, 226–227, 357, 368, 384, 415, 441
Post doctoral scholars/postdocs, 7–8
Poverty, 183, 213, 215, 223, 234, 257, 278–279, 281, 292, 304, 312, 380, 390, 434, 437, 444
Practical action, 281, 380–382, 384–385
Prediction, 96, 113, 118, 147, 157–159, 161–162
Privacy, 110, 115, 284, 287, 411
Private sector engagement, 364
Productivity, 6, 8–9, 11, 25, 27, 29, 49, 51, 53, 63, 65, 72, 90, 94, 102, 120, 130, 134, 141, 146–153, 160, 162, 184, 235, 292, 311, 323–324, 380, 390, 442
Proletariat, 134
Promotion, 6, 8–9, 25, 98, 118, 120, 148, 343, 419
Pro-poor, 425, 427–428, 430, 435, 438, 444
Proprietary knowledge, 364–365, 374
Prostheses, 333
Publication(s), scientific, 25–26, 28, 310, 317, 324, 326
Public awareness, 360
Public health, 77, 113, 119, 122, 192, 219, 221, 272, 351, 415
Public interest group, 429
Public participation, 111, 191, 232–235, 246, 284–285, 287–288, 301, 415–418
Public-Private Partnerships (PPPs), 261–263, 267
Public value, 69–85, 380–381, 388, 427
Public Value Mapping (PVM), 69–85
"Publish or perish", 419
Purdue University, 405

Q

Quality of life, 70, 234–235, 240, 244, 247, 323, 380
Quantum dot, 162, 370

R

R&D investment, 139, 149, 161, 315–317, 325
R&D (research and development), 6, 26–28, 48, 50–51, 55, 72–73, 90, 94, 111, 114, 121, 133, 139–141, 149–151, 161, 167–169, 172, 175–178, 182–184, 188–189, 194, 202, 214–216, 232, 234–235, 241, 254–255, 260, 266, 278, 284, 287, 293–294, 315–317, 319, 322–325, 332, 353, 355–360, 362–373, 374, 376, 380–381
Race, 11, 48–49, 52, 54, 56–57, 62–64, 66, 78, 85, 93, 112, 122, 215–220, 233, 236–239, 241–243, 254, 258, 434
Racism, 93, 216, 218–219, 445
Rathenau institute, 412–413, 416, 420
REACH, 77, 203
Real Time Technology Assessment (RTTA), 213, 233, 433, 437
Redistribution/re-distributional, 129, 216, 444
Regional/regional distribution, 177
Regulation of risks, 118–119, 442–443
Religion, 49, 254, 258, 402
Renato Archer Research Center (CENPRA), 295
Research agenda(s), 5, 18, 85, 112, 223, 243, 299, 303–304, 410, 430
Researchers, 4, 7–9, 14–15, 17–18, 26–28, 30–31, 36–37, 42, 67, 70, 78–79, 84–85, 111–112, 158, 175, 178, 210, 223–224, 227, 235, 245, 263, 278, 287, 293–296, 301, 303, 305–306, 310, 339–340, 343–344, 351–353, 361–362, 365, 368–369, 373, 376, 382, 388, 413–414, 418–419, 425, 427, 429
Research institutions, 111–112, 170, 294–295, 310, 314–315, 318–319, 325–326, 443
Research Triangle, North Carolina, 48, 174–175
Responsible codes of practices, 204
RFID, 411
Rio de Janeiro, 297, 319
Risk assessment, 119, 190, 195, 201, 203–204, 227, 278, 326, 389

Index

Risk awareness, 315–316, 323, 325
Robot, 334
Rockefeller foundation, 70, 261
Route 128, 165
Royal Academy of Engineering, 113, 189–190, 383, 415
Royal society, 113, 189, 383, 390, 415
RTTA (Real Time Technology Assessment), 213, 233, 265, 437
Rural, 56, 172, 178, 257, 263, 281, 312, 353, 382, 384–385, 404
Russia, 178, 188, 341

S

S&E employment, 24, 49
Safety, 28, 90, 102, 110, 117, 119, 141, 166, 183, 188, 191–193, 201–204, 219–221, 256, 272, 283–285, 289, 301, 303, 325, 336, 354, 411, 415–416, 425
São Paulo, 296–298, 319, 324
Scanning Tunneling Microscope (STM), 138, 182
Scenarios, 101, 145–162, 240, 357, 359, 411
Schumacher, E.F., 380, 396
Science capabilities, 169
Science and engineering research, 232, 434
Science and Innovation Policy (SIP), 70–71, 73, 390, 444
Science policy, 70–73, 79, 84, 233, 245, 291–306, 315, 350
Science, technology, and innovation policy, 445
Science and technology (S&T) policy, 233, 310
Science and Technology Studies (STS), 233, 259
Secretariat of Science and Technology for Social Inclusion (Brazil), 292, 300
Security, 32, 70, 90, 95–96, 117, 187, 289, 312, 333, 343, 416
Self assembling systems, 372
Self-regulatory systems, 189
Sensors, 182, 281, 411, 428
Sewage systems, 415
Sexism, 93, 216, 219
Silicon Valley, 48, 166, 193–194
Silicon Valley Toxics Coalition (SVTC), 193–194
Singapore, 178, 361
Singularity, 395
Skill biased technological change, 54, 146
Skilled workforce, 372, 434
Skill premium, 148–151, 160

Skin tones, 218
Small and medium-sized enterprises (SMEs), 169–170, 428
Snowboarding, 336
Social cohesion, 80, 90, 227, 434–435, 443
Social innovation, 444
Social need, 387
Social programs, 318
Social science/social scientists, 5, 7, 19, 71, 350, 424–426
Social stratification, 28, 246
Sociological research, 8
Solar energy, 156, 427–428
South Africa, 189, 255, 266, 278, 296, 367, 387, 402, 436
South Korea, 188, 322, 367
South-South collaboration, 114, 262, 334, 428
Speciesism, 93
Species-typical, 90, 92–97, 110
Spillovers, 167, 322, 324
Stakeholders, 71–72, 75–77, 97, 116, 182, 222, 224–225, 232, 234, 244, 256, 278, 315, 323, 325–326, 344, 380, 391, 412–413, 415–416, 429, 437
Standard of living, 434–435, 444
Standards, 72, 74, 91, 101, 118–119, 183, 188, 220, 273, 285, 333, 343–345, 394, 440
States, U.S., 5, 8–9, 24, 27, 31–43, 47–67, 70, 73, 91–92, 119, 121, 135–136, 141, 167–170, 172–173, 175, 178, 189–190, 232, 234, 242, 287, 357–358, 360–362, 364, 396, 398, 404–405, 411, 418, 445
STEM education, 4–5, 7–8, 16–19, 52–53
STS, 5, 10–11, 18
Sun Microsystems, 338
Super soldiers, 241
Supply/demand, 7, 9, 17, 49, 51, 54, 148, 151–152, 160
Surveillance, 115, 120, 203–204
Survey, 9, 16, 33, 43, 55, 96, 118, 146, 232–233, 236–239, 244, 247, 341, 353, 376
Sustainable development, 189, 202, 281, 380
Sustainable economic development, 375
Sweden, 7, 356, 368
Sweet potato, 255–256, 261–262, 266
Switzerland, 296, 356, 368
Synthetic biology, 90–91, 411, 416, 426

T

Taiwan, 189, 356, 361, 364
TAPR Open Hardware License, 338, 346

Team, teams/mixed teams/inventor teams, 6, 12, 14–15, 19, 24, 29, 31–38, 40–43, 158, 236, 239, 248, 303, 388, 394, 400
Technical fix, 387
Technicians, 51, 198, 361
Techno-economic paradigm, 129–132, 134, 136–137, 141–142
Technological breadth, 169
Technological change, 53–54, 139, 146–152, 157, 161, 262, 302–303, 381, 412
Technological dependency, 356, 358
Technological innovation, 24, 29, 112–114, 121, 130, 254, 295, 299, 310, 381
Technological performance, 310
Technological revolution(s), 127–142, 182
Technology assessment, 111, 119, 156, 188, 191, 213, 233–234, 254, 265, 409–420, 437, 440
Technology development trajectories, 211, 216–217, 226–227
Technology licenses, 287
Technology steering mechanisms, 215
Technology system, 132, 136–137, 139
Technology transfer, 71, 85, 227, 295, 326, 358, 364–365, 373
Techno-poor impaired, 94–95
Temporary monopoly, 439
Thailand, 184, 349–376
Thoreau, Henry David, 403
Tolstoy, Leo, 403
Toxic effects, 188
Toxic properties, 115, 183
Trade, 26, 112–114, 116, 129, 146, 148, 183–184, 189–192, 194, 198–205, 256, 282, 286, 294, 304–305, 319–320, 334, 343, 375, 420
Trade-Related Aspects of Intellectual Property Rights (TRIPS Agreement), 114
Trade union, 184, 189, 191–192, 194, 199–205, 304, 420
Training, 49, 51–53, 57, 62–63, 78, 140, 156, 160–161, 203–204, 262, 296, 334, 337, 342, 353, 358, 362, 395–396, 404, 429
Transhuman, 93
Transparency, 6, 191–192, 245, 412
Transport/transportation, 116, 128, 130, 134, 153, 155, 201, 203, 221–222, 283, 387, 389, 428
Tunneling Electron Microscope(s) (TEM), 27, 177
Typewriter, 96

U

Ubiquitous network society, 411
Uganda, 271–273
Uganda National Bio-safety Committee, 272
Uganda National Bureau of Standards (UNBS), 273
Uganda National Council for Science and Technology (UNCST), 272
U.K. (United Kingdom), 342, 388
Underdevelopment, 356, 358
Underrepresented minorities, 4, 7, 14, 16
Unemployment, 56, 61, 136, 139–140, 146, 305
Union, 56, 60, 110, 147–148, 152, 184, 186, 188–195, 199–205, 220–221, 301, 304, 311, 335, 417, 420, 440
United Kingdom (U.K.), 110, 114, 140, 189, 342, 387–388, 390
United Nations, 95, 100, 116–117, 198, 256, 278, 286, 353, 428
United Nations Convention on the Rights of Persons with Disabilities, 95, 100, 116–117, 198, 256
United Nations Food and Agriculture Organization (FAO), 198
Universities, 6, 15, 19, 24–27, 50, 53, 111–112, 139, 169–170, 174, 176–178, 260, 278, 293, 295, 310, 319, 322–323, 345, 369, 384, 396
University of California at Berkeley, 236
University of California, Santa Barbara, 395
University of New Hampshire, 236
University of Wisconsin-Madison, 236
Unskilled labor, 54, 148, 151
Urban, 56–57, 60, 172, 222, 312, 353, 435
User groups, 211, 216, 222, 224
Utilitarian, 263

V

Vaccine, 71, 262, 282, 286, 415
Values, 38, 70–80, 82, 84–85, 101–102, 111, 162, 215–216, 219, 223–226, 232–235, 301, 304, 390, 393–405, 427
Venture capital, 26, 28, 136, 139, 364, 428
Vertical inequality, 54
Vietnam, 367
Vision, 90, 363, 384, 411
Voluntary certification, 185, 190

W

Wage inequality, 53–54, 146–148, 151–152, 154, 156, 160–161
War on Cancer, 4, 83
Washington, D.C., 168, 172, 174–175

Index 457

Water, 97, 120, 128, 133–134, 153, 182, 187, 193, 200, 226–227, 252, 263, 272, 278–282, 285, 289, 295, 304, 321, 323, 327, 333, 339, 358, 382–385, 387, 390, 404, 415, 424, 427, 436, 441–442
Water filtration, 120, 252, 424, 427, 436
Water treatment, 279–281, 289, 323, 390
White-collar jobs, 148, 160
Whites, 49, 51–54, 62–65
Wiener, Norbert, 401, 403
Windsurfing, 336
Women, 4–19, 23–43, 55, 93, 96, 98, 217–218, 227, 237, 258, 384, 415, 435–437
Woodrow Wilson Center's Project on Emerging Nanotechnologies, 119
Workers, 5, 10, 48, 50–52, 54–58, 64–66, 96, 114–115, 120, 148–150, 152, 154, 181–195, 200–201, 203–204, 220–222, 292, 301, 303–304, 323, 361, 370, 437, 440, 443
Workforce, 4, 8, 24, 47–67, 186, 243, 247, 311, 320, 369–370, 372, 434
Workplace, 49, 66, 110, 115–116, 157, 204, 220–222, 301, 303, 440
World Health Organization, 117, 198, 280
World Intellectual Property Organization (WIPO), 43, 169, 343
World Trade Organization (WTO), 198, 343

Z

Zimbabwe, 382–388, 390–391